D0768942

# Chemistry of Natural Protein Fibers

# Chemistry of Natural Protein Fibers

Edited by

**R. S. Asquith**

*The Queen's University of Belfast*
*Belfast, Northern Ireland*

PLENUM PRESS • NEW YORK AND LONDON

Library of Congress Cataloging in Publication Data

Main entry under title:

Chemistry of natural protein fibers.

Includes bibliographical references and index.
1. Proteins. 2. Fibers, I. Asquith, Raymond Smith, 1925-
QD431.C452 547'.75 76-52636
ISBN 0-306-30898-3

A Division of Plenum Publishing Corporation
227 West 17th Street, New York, N.Y. 10011



Printed in the United States of America

To

**Professor Helmut Zahn**

on the occasion of his sixtieth birthday

# Contributors

**R. S. Asquith,** Department of Industrial Chemistry, The Queen's University, Belfast, Northern Ireland

**C. L. Bird,** Department of Colour Chemistry, The University, Leeds, England

**J. H. Buchanan,** National Institute for Medical Research, Mill Hill, London, England

**J. C. Fletcher,** Wool Industries Research Association, Headingley, Leeds, England

**Fred Kidd,** Scottish College of Textiles, Galashiels, Scotland

**N. H. Leon,** Unilever Research, Isleworth Laboratory, Isleworth, Middlesex, England

**H. Lindley,** Division of Protein Chemistry, CSIRO, Parkville (Melbourne), Victoria 3052, Australia

**M. S. Otterburn,** Department of Industrial Chemistry, The Queen's University, Belfast, Northern Ireland

**J. A. Swift,** Unilever Research, Isleworth Laboratory, Isleworth, Middlesex, England

**C. S. Whewell,** Professor and Head of Department of Textile Industries, University of Leeds, Yorkshire, England

**K. Ziegler,** Deutsches Wollforschungsinstitut, Aachen, West Germany

# Preface

This volume arose originally from the complaints of the editor's students, both undergraduate and postgraduate, that there was no modern book on protein fibers which told enough about protein science and chemical technologies related to fibers. By and large this is probably a reasonable *cri de coeur*.

The undergraduate on a technological course, lacking information on the basic scientific techniques used to carry out the research on which his fiber technology is based, can find it difficult to obtain this information. The pure science undergraduate often lacks knowledge of the application of these techniques in protein fiber technology. The young graduates, commencing research related to some aspect of protein fibers, are drawn from a wide range of scientific disciplines, having been trained as biochemists, chemists, physicists, technologists, and histologists, to name but a few. Generally these new research workers pass through a preliminary "lost" period in which they have to evaluate their background in relation to the wide and differing fields of research in protein fiber science to which they are now exposed. As time goes on they then either develop a wide knowledge covering science and technology or remain in a specific part of their original discipline, with a narrow knowledge of its application in the field of the research degree they are taking.

I hope this book will reduce the length of the "lost" period of a few younger scientists commencing research on protein fibers. Possibly it could attract others away from a narrow attitude of one scientific technique for solving all problems and expand their interests. Should the book be successful in these attempts, it will have achieved much of value. If, at the same time, it assists research workers in obtaining information, both within their own fields and also in the work peripheral to their interests, it will extend its value.

The authors who have collaborated in the production of this book are not only men of achievement within the research field, but also are enthusiasts who have taught, guided, and inspired many younger workers. Each brings to the book a different point of view and different deep interests in specific parts of this wide field of science and technology. As editor, I would like to express my gratitude to them for their willing cooperation, even when I had to delete some of their writings. I hope they will feel that the finished volume comes up to the high expectations they all had of it.

In such a work as this, the problem becomes not what should go in, but what should be left out. I have confined the overall scheme, rightly or wrongly, to general techniques, fiber structure, and chemical processes applied to the protein fibers. I hope, as a result, that the reader will be stimulated by it, and will feel that my colleagues and I have done justice to the work of those natural scientists and chemical technologists, past and present, who have devoted their working lives to these studies.

R. S. Asquith

*Belfast*

# Contents

*Chapter 1*

**The Basis of Protein Chemistry**

*J. C. Fletcher and J. H. Buchanan*

*Chapter 2*

## The Chemistry and Reactivity of Silk

*M. S. Otterburn*

*Chapter 3*

## The Histology of Keratin Fibers

*J. A. Swift*

*Chapter 4*

## The Chemical Composition and Structure of Wool

*H. Lindley*



*Chapter 5*

## Chemical Reactions of Keratin Fibers

*R. S. Asquith and N. H. Leon*

*Chapter 6*

## Crosslinking and Self-Crosslinking in Keratin Fibers

*K. Ziegler*

*Chapter 7*

**The Dyeing of Wool**

*C. L. Bird*



*Chapter 8*

**The Chemistry of Wool Finishing**

*C. S. Whewell*

*Chapter 9*

**Other Animal Fibers**

*Fred Kidd*

# 1

# The Basis of Protein Chemistry

**J. C. Fletcher**
**and J. H. Buchanan**

## 1.1. Introduction

The purpose of this introductory chapter is to present a survey of basic protein chemistry which will enable those readers who are beginning work on fibrous proteins to consolidate the knowledge they already possess and to become acquainted with the specialized literature. At the same time it is hoped that an account of the methods available for the investigation of the structure and reactivity of proteins may usefully complement the description in later chapters of their application to fibrous proteins.

Proteins are natural polymers built up from amino acid residues linked through peptide bonds.

$$\begin{array}{c} \qquad\qquad\qquad R_2 \\ \qquad\qquad\qquad | \\ \text{—NHCHCONHCHCO—} \\ | \qquad\qquad\qquad \\ R_1 \qquad\qquad\qquad \end{array}$$

The amino acids commonly present in proteins are listed in Table 1-1. The successive levels of structure in a protein are classified according to a convenient, but possibly too rigid, terminology introduced by Linderstrøm-Lang.[1,1a] The linear sequence of residues in a peptide chain constitutes the

---

**J. C. Fletcher** • Wool Industries Research Association, Headingley, Leeds, England
**J. H. Buchanan** • National Institute for Medical Research, Mill Hill, London, England

**Table 1-1. Common Amino Acids Found in Proteins, $R-CH(\overset{\oplus}{N}H_3)COO^{\ominus}$**

| | R | Mol. wt. | Abbreviation: Three-letter[a] | Abbreviation: Single-letter[b] |
|---|---|---|---|---|
| Glycine | $-H$ | 75.07 | Gly | G |
| Alanine | $-CH_3$ | 89.10 | Ala | A |
| Valine | $-CH(CH_3)_2$ | 117.15 | Val | V |
| Leucine | $-CH_2CH(CH_3)_2$ | 131.18 | Leu | L |
| Isoleucine | $-CH(CH_3)CH_2CH_3$ | 131.18 | Ile | I |
| Serine | $-CH_2OH$ | 105.09 | Ser | S |
| Threonine | $-CH(CH_3)OH$ | 119.12 | Thr | T |
| Cysteine | $-CH_2SH$ | 121.16 | Cys | C |
| Cystine | [c] | 240.30 | Cys Cys (linked) | |
| Methionine | $-CH_2CH_2SCH_3$ | 149.21 | Met | M |
| Proline | [d] | 115.13 | Pro | P |
| Aspartic acid | $-CH_2CO_2H$ | 133.11 | Asp } Asx[e] | D } B[e] |
| Asparagine | $-CH_2CONH_2$ | 132.12 | Asn } Asx[e] | N } B[e] |
| Glutamic acid | $-CH_2CH_2CO_2H$ | 147.13 | Glu } Glx[e] | E } Z[e] |
| Glutamine | $-CH_2CH_2CONH_2$ | 146.15 | Gln } Glx[e] | Q } Z[e] |
| Phenylalanine | 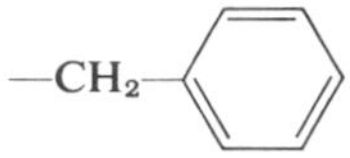 $-CH_2-C_6H_5$ (benzene ring) | 165.19 | Phe | F |
| Tyrosine | 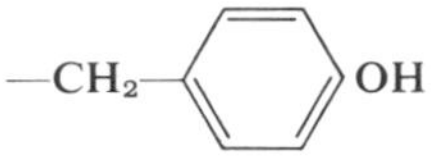 $-CH_2-C_6H_4-OH$ | 181.19 | Tyr | Y |
| Tryptophan | 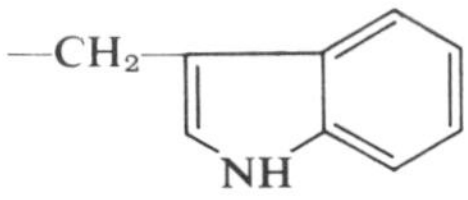 $-CH_2-$ (indole ring, NH) | 204.23 | Trp | W |
| Arginine | $-CH_2CH_2CH_2NHC(=NH)NH_2$ | 174.20 | Arg | R |
| Histidine | 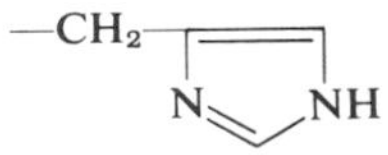 $-CH_2-$ (imidazole ring, N, NH) | 155.16 | His | H |
| Lysine | $-CH_2CH_2CH_2CH_2NH_2$ | 146.19 | Lys | K |

[a] Data from IUPAC-IUB Commission on Biochemical Nomenclature, Abbreviated Designation of Amino Acid Derivatives and Peptides, Tentative Rules, *J. Biol. Chem.* **241**, 2491 (1966); *Biochem. J.* **102**, 23 (1967).

[b] Data from IUPAC-IUB Commission on Biochemical Nomenclature, A One-Letter Notation for Amino Acid Sequences, Tentative Rules, *J. Biol. Chem.* **243**, 3557 (1968); *Biochem. J.* **113**, 1 (1969).

[c] $-CH_2SSCH_2CH(\overset{\oplus}{N}H_3)COO^{\ominus}$.

[d]
```
CH2—CH2
|      |
CH2   CHCO2⊖
  \   /
   NH2.
   ⊕
```

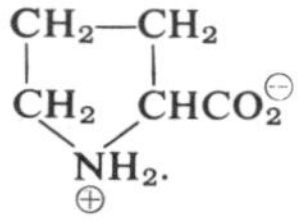

[e] Used when acid and amide are not distinguished.

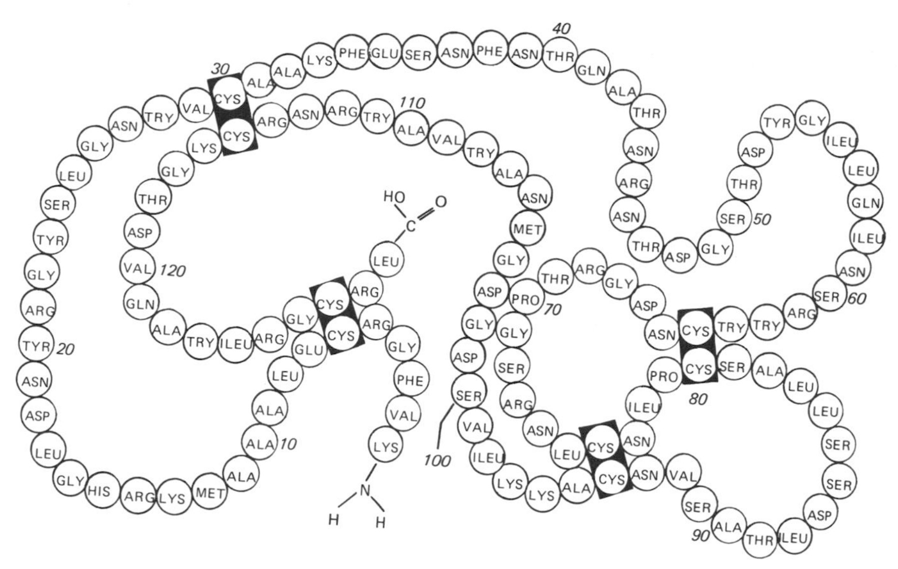


Fig. 1-1. The primary structure of egg-white lysozyme, from R. E. Canfield and A. K. Liu, *J. Biol. Chem.* **240**, 1997 (1965). (Courtesy American Association of Biological Chemists, Inc.)

*primary structure* of the chain (Fig. 1-1). Regular local folding of certain sections of the chain occurs to form *secondary structures* stabilized by hydrogen bonds between —NH— and —CO— groups, while other sections exist in a randomly coiled state (random in the sense of the absence of a definite structural regularity). The properties of synthetic homo- and copolymers of amino acids and their derivatives, and information from known three-dimensional structures of proteins, suggest that amino acids can be divided into two groups[2,2a]: those which, when they predominate in a given section of the chain, promote folding into an $\alpha$ helix ("helix-promoting residues"), and those which cause the chain to adopt a form other than $\alpha$-helical ("helix-breaking residues"). Local interactions between side-chain and main-chain atoms thus appear to constitute an important conformation-determining factor, but a detailed interpretation of the dependence of higher structure on the primary structure of the chain requires consideration of many other factors. (For the $\alpha$-helix structure in keratin, see Chapter 4.)

Secondary structures of two types, $\alpha$ helix and $\beta$ sheet, occur in the molecules of globular proteins, although to a limited extent. This is compatible with the general requirement for these proteins to assume a relatively compact form with polar side chains orientated outward toward the polar environment and nonpolar groups buried in the interior. Such interactions involving amino acid side chains produce a global folding of the chain into a

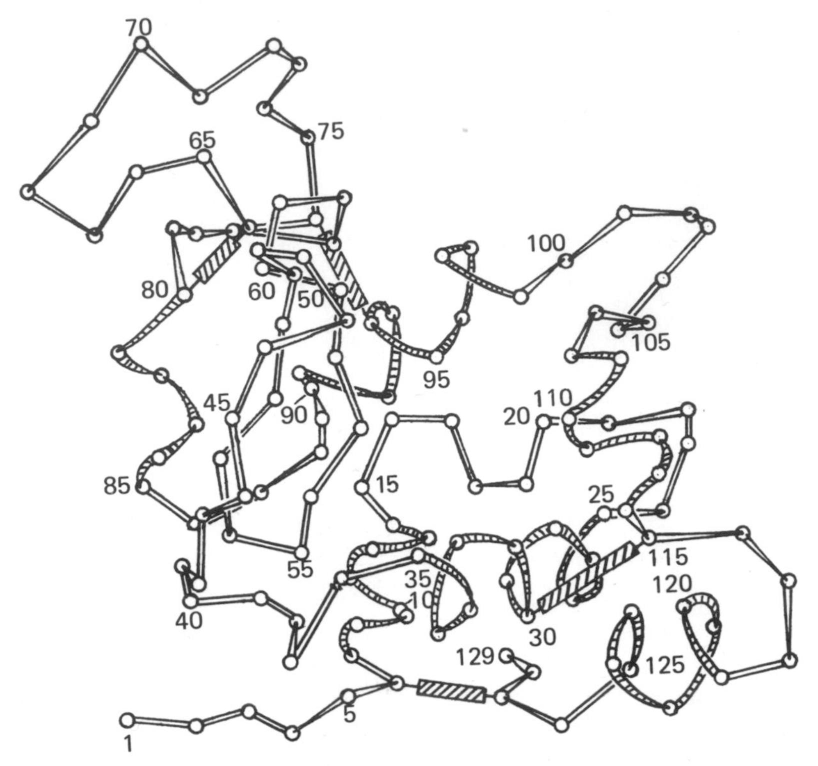


Fig. 1-2. Schematic drawing (by W. L. Bragg) of the main chain conformation of lysozyme, from C. C. F. Blake *et al.*, *Nature*, **206**, 757 (1965). [Courtesy Macmillan (Journals) Ltd.]

*tertiary structure* of minimum free energy (Fig. 1-2) in which the pairing of half-cystine residues to form disulfide bonds, when these occur, is probably a consequence of the equilibrium folding rather than a controlling influence on it.[3] Finally, structural units of the same or different types may associate to form *quaternary structures*,[4–5a] as in the hemoglobins.

Changes in the environment of the protein molecule, which can influence all the noncovalent interactions involved in the structure, may bring about dissociation of quaternary structures and greater or lesser perturbations of secondary and tertiary structures. These phenomena underline the complex process of *denaturation*.[6,7]

The foregoing remarks relate more specifically to globular proteins, which exist in solution in an aqueous environment. Fibrous proteins belong to the class of structure proteins and occur essentially in the solid state. They give characteristic X-ray diffraction patterns that indicate considerable axial structural order. This feature still provides the firmest basis for their classification[8] into the keratin–myosin–epidermin–fibrinogen group and the collagen group, and differentiates them from the nonfibrous structure proteins such as resilin and elastin. Secondary structures occur in the ordered (crystalline) regions of fibrous proteins to a greater extent and in a greater variety of forms than in globular proteins, and associate to form higher structures such as multiple helixes and stacked sheets.

Work on fibrous proteins has been hampered by the difficulty of obtaining homogeneous, soluble preparations from many of them, and the recognition and characterization of their fundamental structural units has become possible only relatively recently. A general review of structure proteins is given by Seifter and Gallop.[9] Other recent reviews deal with keratins,[10,10a] the fibrin–fibrinogen system,[11] and collagen.[12,12a]

The presence of prototropic groups in the side chains of certain amino acids is responsible for the acid–base characteristics of proteins, which determine the dependence of structure and behavior on the concentration of hydrogen and other ions in the aqueous environment. Examples of the manifestation of these electrochemical properties are the separation of proteins by differential migration in an electric field, the variation in the solubility of globular proteins with pH, and the interaction between fibrous proteins and dyes. The reviews by Steinhardt and Beychok[13] and by Crewther et al.[10] may be recommended.

### 1.1.1. Outline

Methods for separating proteins not only serve the purpose of isolation, but also provide criteria of homogeneity. At the same time, they may give information about the parameter on which separation depends (e.g., molecular mass, in the case of sedimentation methods). Some structural features of the molecule can be probed by spectroscopic methods, and the results of X-ray diffraction measurements may permit an almost complete description of the three-dimensional structure of a crystalline protein.

From a chemical standpoint, a protein is initially characterized in terms of its amino acid composition. The tactical plan for the determination of the primary structure provides a framework for the exhibition of many characteristic reactions: the fission of disulfide bonds, the cleavage of peptide chains by chemical and enzymatic methods, and the reactions of terminal amino and carboxyl groups by which the amino acid sequences of peptide fragments are determined. Protein chemists are continuing to seek ever more specific reactions of amino acid residues in intact proteins to aid them in the solution of their various problems.

## 1.2. Isolation of Proteins and Their Characterization by Physical Methods

### 1.2.1. Separation Procedures

Proteins occur as components of complex systems in which they interact to a greater or less extent with one another and with other substances, such as nucleic acids, carbohydrates, and lipids. Physical disruption of the ordered structures of cells and tissues is frequently necessary before protein components can be extracted with water, dilute acids, aqueous buffer solutions, or water–solvent mixtures. The conditions of extraction must be such that the proteins of interest are not denatured and usually must be found empirically. The extraction of proteins from natural sources is reviewed by Keller and Block.[14] Eastoe and Courts[15] deal with the isolation of connective tissue proteins. The special problems posed by keratins, in which disulfide bonds must be broken before proteins can be extracted, are considered elsewhere in this volume.

In the initial stages of the isolation of proteins from crude extracts, fractional precipitation by the adjustment of ionic strength or solvent concentration is used whenever possible, since large quantities of material can be handled. Subsequently, differential migration methods of higher resolution but lower capacity can be exploited. Many of these methods are applicable on both an analytical and a preparative scale. A general account of protein fractionation procedures is given by Sober et al.[16]

#### *1.2.1.1. Chromatographic Methods*

Because of their relatively small capacities for large molecules and the tendency of proteins to be strongly adsorbed, ion-exchange resins have limited utility for the fractionation of proteins. Ion-exchange celluloses have proved much more suitable: they are hydrophilic in nature, have a high capacity for proteins, and the several types available cover a wide range of properties.[16a]

Separation of proteins on a molecular-size basis (gel chromatography, gel filtration) is effected by chromatography on gels of dextran, agarose, or polyacrylamide, cross-linked to varying extents.[17,18] The incorporation of dipolar ionic sites in cross-linked dextrans[19] has been shown to give materials possessing considerable potential for the chromatographic separation of proteins.

In affinity chromatography, a substance with an affinity for a particular component of a mixture of proteins is immobilized on a solid support, which then adsorbs that component selectively.[19a]

### *1.2.1.2. Electrophoretic Methods*

In moving-boundary electrophoresis[20] the sample itself stabilizes the solution against thermal convection. This is a very precise and reproducible method for the determination of the number and amounts of proteins present in a mixture and has been used extensively in work on keratins,[10] but to a large extent it has been replaced by zone-electrophoretic techniques that are more readily adapted to preparative work. The technique of zone stabilization by sucrose gradients[21] was later modified by the superposition of a pH gradient, in which each protein migrates to a zone corresponding to its isoelectric point; this is known as "isoelectric focusing." A heterogeneous mixture of low-molecular-weight aliphatic polyaminopolycarboxylic acids is used to generate the pH gradient.[22] If a dilute polyacrylamide gel is employed for zone stabilization in place of sucrose, the capacity of the method can be increased by a factor of about ten.[23]

A different principle underlies the technique of isotachophoresis.[24] A mixture of proteins sandwiched between a "leading ion" of highest mobility and a "terminating ion" of lowest mobility is caused to move through a supporting gel by the application of an electric field. A common counterion species is present. Migration of ions takes place according to principles first formulated by Kohlrausch, and at equilibrium all ion species are moving with the same velocity, separated into contiguous zones in order of their mobilities. Suitable chosen "spacer ions" permit separation of the zones. Ionic strengths are higher than in isoelectric focusing, and there is less risk of precipitation. The proceedings of a symposium on isoelectric focusing and isotachophoresis are available.[24a]

Various electrophoretic procedures have been described that utilize columns or beds of inert supporting material (e.g., ethanolized cellulose powder).[25] An important new development was the introduction of supporting media of such a structure that they exert a sieving effect and thus contribute an additional separatory effect based on molecular size. Gels prepared from hydrolyzed starch were first to be used.[26] In "disc" electrophoresis[27] the isotachophoretic principle is used to concentrate the protein components as thin discs in a large-pore polyacrylamide gel before subjecting them to electrophoresis in a smaller-pore "sieving" gel of the same material. This technique possesses remarkable resolving power, and several preparative versions of it have been described.[28,29] Electrophoresis in polyacrylamide gels containing 0.1% sodium dodecyl sulfate offers a quick and reliable method for the determination of molecular weights over the range 10,000 to 200,000 with an accuracy of 5–10%.[30,30a,31] The application of disc electrophoresis in an 8 *M* urea gel to the determination of the molecular weights of keratin derivatives has been described by Jeffrey.[32]

A unified theory for molecular-sieve chromatography and electrophoresis in gels has been proposed by Morris and Morris.[33a]

### 1.2.2. Size and Shape of Protein Molecules

#### *1.2.2.1. Gel Filtration*

The use of gel filtration for the separation of proteins according to their size has been referred to above. The behavior of a protein molecule on a gel column is determined by hydrodynamical considerations; a smooth relationship between elution volume and Stokes' radius is expected from theoretical models. For carbohydrate-free globular proteins, which are generally similar in shape and density, the Stokes' radius is proportional to the molecular weight, and estimates of this quantity with an uncertainty of about 10% can be obtained from gel columns calibrated with a series of similar proteins. Proteins that are highly asymmetric or have unusually low or high densities behave anomalously. This difficulty can be overcome by reducing the disulfide bonds and working in a denaturing and disaggregating solvent such as 6 $M$ guanidine hydrochloride, thus ensuring that all proteins adopt qualitatively similar randomly coiled conformations.[34,34a] Similar results, although on an empirical basis, have been obtained in phenol–acetic acid–water mixtures.[35] The analytical aspects of gel chromatography are reviewed in detail by Ackers[18] and Andrews.[36]

#### *1.2.2.2. Sedimentation Analysis*

Information about the number and sizes of the protein species present in a solution, and the interactions between them, can be obtained from measurements of the concentration changes that occur when the solution is subjected to high force fields in the ultracentrifuge. Sophisticated optical techniques enable these concentration changes to be monitored while the experiment is in progress.

In high-speed *sedimentation velocity* experiments, a solute sediments toward the bottom of the cell, leaving a layer of pure solvent behind. The *sedimentation coefficient* (velocity per unit centrifugal field) is obtained from the rate of movement of the boundary between the solvent and the solution. A relation between the sedimentation coefficient and the molecular weight of the solute can be obtained by equating the centrifugal force and the frictional force when the solute molecules are moving with constant velocity, and then expressing the frictional coefficient (a measure of the shape and degree of hydration of the molecule) in terms of the translational diffusion coefficient, which can be measured in a separate experiment with an artificially formed boundary in a special cell. Alternatively, the frictional coefficient may be estimated from viscosity measurements. For analytical purposes, the method of sedimentation velocity has tended to be replaced by gel filtration, which possesses greater resolving power and is easily operated under preparative conditions.

At lower rotor speeds, complete depletion of the upper layers of the solution does not occur. Eventually *sedimentation equilibrium* obtains throughout the cell; sedimentation is balanced by diffusion and no net transport of material occurs. If the chemical potential of the sedimenting solute is equated to the potential of the centrifugal force field, relations can be obtained that give the molecular weight directly. The extended times formerly required to obtain equilibrium have been shortened by the introduction of synthetic boundary cells, which allow the run to be started with the distribution of solute already nonuniform, and by the development of better optical systems, which facilitate precise measurements with shallower cells.[37] "Transient-state" techniques (e.g., the method due to Archibald, which makes use of the fact that the material flux across either end of the column of liquid in the cell is zero at all times during the run) may be applied to obtain information before equilibrium is established.

The literature on ultracentrifugation to 1963 is summarized in the volume edited by Williams.[38] A recent introductory book is that by Bowen.[38a] The theory underlying ultracentrifugation and other physical measurements is given by Tanford[39] and Cann.[39a]

If sedimentation is carried out in gradients of salt or sucrose, the utility of the technique is extended considerably. Velocity-zonal sedimentation in pre-formed gradients permits the recovery and examination of the separated zones.[40] If sedimentation is continued until the sedimenting species reach positions where the gradient density is equal to their buoyant density, the resulting equilibrium (isopycnic) zoning can give information about the molecular weights of the solvated species, as well as their buoyant densities.[41] The run may be started with a homogeneous solution, the gradient being generated as the run proceeds, or the sample may be layered over a pre-formed gradient. Equilibrium zoning is particularly applicable to conjugated proteins (glycoproteins, lipoproteins) when the inhomogeneity resides in the prosthetic group rather than in the protein itself. The volume edited by Hirs and Timasheff[41a] gives a practical account of many of these methods.

Molecular masses of proteins are frequently expressed in *daltons*, a name proposed for the unified atomic mass unit (one twelfth of the mass of an atom of the nuclide $^{12}C$).

#### *1.2.2.3. Other Physical Methods*

Information about the sizes and shapes of protein molecules, some of it of a rather indirect sort, can be obtained from measurements of the viscosity, osmotic pressure, or light-scattering behavior of their solutions. For well-characterized proteins, the more exact information furnished by X-ray diffraction and chemical sequence studies has tended to make these methods of lesser importance. The relevant theory and technical details can be found in the compendia edited by Hirs and Timasheff,[41a] Alexander and Block,[42,43] and Leach.[44] Edsall's review,[45] written in 1953, still provides a useful introduction to the various physical methods and their interrelations.

Electron microscopy has provided important information about the molecular dimensions of fibrous proteins.

### 1.2.3. Determination of Secondary Structure

#### *1.2.3.1. Optical Rotatory Dispersion*[46,41a]

The optical rotatory power of polypeptides and proteins arises from two sources of asymmetry—that of individual L-amino acid residues, and that due to any asymmetry of the molecule as a whole, such as the presence of helical structures. The latter type of asymmetry, when it occurs, is the dominant form. The dependence of optical rotation on wavelength for an α-helical structure was calculated theoretically by Moffitt, and predictions from this relation were found to be in accord with the behavior in solution of those amino acid homopolymers that can exist in both helical and random-coil forms. Although the theoretical foundation of the Moffitt equation has been shown to be inadequate, it is useful on an empirical basis for estimating the extent to which helical structures are present in soluble proteins.[47,48] The parameter of the Moffitt equation that characterizes the amount of α helix in a protein is, in fact, highly correlated with the content of "helix-promoting" amino acids.[49] Recent work on the optical rotatory dispersion and circular dichroism of polypeptides and proteins has been reviewed by Bayley.[50]

#### *1.2.3.2. Nuclear Magnetic Resonance Spectroscopy*

Despite much work on the proton-nmr spectra of free amino acids and their homopolymers, progress in the interpretation of the complex, broad-line spectra of proteins has been slow. Most success has been achieved in the investigation of the changes in overall conformation resulting from reactions of particular types of amino acid residues, and the mechanism of binding of small molecules to proteins.[51,52] Technical advances, which include repetitive scanning in association with a computer of average transients (CAT) and operation in the pulsed Fourier transform (PFT) mode, promise rapid developments in the protein field, particularly in their application to magnetic resonance spectra arising from the nuclide $^{13}C$, which has a natural abundance of 1.1%.[52a]

### *1.2.3.3. Determination of Protein Structure by X-Ray Diffraction*

In the analysis of crystal structures by X-ray diffraction, a photographic plate or a direct recording device is used to measure the intensities of the maxima in the scattered radiation. The most important part of the information carried by the radiation, the phase differences at the maxima, is lost, and it is not possible to deduce the details of the structure without ambiguity from the intensity data alone. For small molecules, a plausible approximation to the structure can generally be assumed and then successively refined until the calculated scattering pattern fits that actually observed. For molecules as complex as proteins, this approach is not feasible.

The phase problem can be solved by the method of isomorphous replacement—the preparation and structure analysis of at least two, and preferably more, derivatives of the protein which contain a heavy atom or group of atoms and which crystallize in the same form as the parent protein. It is this method which has enabled X-ray crystallographers to contribute detailed and precise information about the three-dimensional structures of proteins during the past decade. Systematic methods for the preparation of heavy-metal derivatives are being actively sought, but to date most success has been achieved by trial and error.[53]

By the method of isomorphous replacement the structure of a crystalline protein can be obtained in detail down to a resolution of approximately 2.0 Å, at which the peptide chain can be followed through helical and non-helical regions. (Resolution here refers to the smallest wavelength used in the Fourier synthesis.) Resolution beyond this stage is possible using essentially conventional refinement techniques, bearing in mind that the amount of data required for a given resolution increases as the inverse cube of the resolution. The ultimate limit is set by natural imperfections in the protein crystal; it varies for different proteins but is in the region of 1.5 Å.

Complete determination of the amino acid sequence of a protein by X-ray diffraction alone is not feasible, because of the effectively identical geometry of certain amino acids (e.g., threonine and valine). Thus the complementarity of chemical sequence studies and X-ray diffraction, emphasized by Crick and Kendrew[54] in a classic expository article, still holds true.

Fibrous proteins behave toward X rays as collections of small crystallites. Relatively diffuse photographs are obtained, and it is necessary to try and deduce the symmetry properties of the structure and then to construct models embodying this symmetry so that their calculated scattering properties may be compared with those of the protein.

An account of X-ray diffraction for the nonspecialist and its use in the determination of protein structure is given by Holmes and Blow.[55] A detailed account of progress up to 1964 is given by Dickerson.[56] Later work is summarized by Blow and Steitz[57] and Blundell.[57a] The conformation of fibrous proteins is the subject of a book by Fraser and MacRae.[58]

## 1.3. Amino Acid Analysis

### 1.3.1. Preparation of the Hydrolysate

Hydrolysis with constant-boiling hydrochloric acid *in vacuo* or under nitrogen[58a] produces a mixture of amino acids whose composition approximates that of the protein or peptide. Certain amino acids are progressively destroyed. These include serine and threonine[59] and cystine.[60,61] Valine and isoleucine are liberated from peptide linkage relatively slowly. Adequate corrections can usually be determined by hydrolyzing the protein for different periods of time and extrapolating the results of analyses to zero or infinite time. Tyrosine can be protected from partial destruction by the inclusion of phenol or thioglycollic acid in the hydrolysis mixture.[62] Roach and Gehrke[63] have summarized previous work on acid hydrolysis and have shown that if the temperature is increased to 145°C, the time of hydrolysis can be reduced to 4 h without loss of accuracy. For large protein molecules, the exact amino acid composition may only become known when the complete primary sequence is worked out.

The amounts of tryptophan, asparagine, and glutamine present in a protein cannot be determined by acid hydrolysis, as these amino acids are completely converted to other products. A procedure for the total enzymatic hydrolysis of soluble proteins by successive digestion with papain, leucine aminopeptidase, and prolidase which makes a useful complement to acid hydrolysis has been described by Hill and Schmidt.[64] Bennett et al. describe a similar procedure using enzymes bound to Sepharose.[64a] Procedures applicable to keratin have been described by Milligan et al.[65] and Cole et al.[66] The difficulties peculiar to the complete enzymatic digestion of collagen are discussed by Bensusan et al.[67,67a]

Alkaline hydrolysis of proteins finds limited application in the determination of tryptophan and methionine sulfoxide.[68,68a] Times up to 50 h may be required for maximum yields of tryptophan.

### 1.3.2. Analysis of the Hydrolysate

Automatic analyzers are now employed almost exclusively for the amino acid analysis of proteins and peptides, although in amino-acid-sequence work where a large number of smaller peptides must be analyzed, more rapid but less precise methods, such as high-voltage paper electrophoresis, are invaluable. The ease and rapidity with which a full column analysis can be completed owes much to the work of Stanford Moore and W. H. Stein, who developed early investigations by Elsden and Synge into a comprehensive scheme of analysis first on columns of starch, then on columns of ion-exchange resin, and finally into an automatic procedure. Their meticulous descriptions of successively improved methods of separation encouraged the rapid dissemination and development of amino acid analysis.

A key step was the development of conditions for reproducible reaction between the amino acids and triketohydrindene hydrate (ninhydrin, I). This reaction, the mechanism of which is as shown,[69,70] yields the colored com-

(I) + R(H)C(NH$_2$)—COO$^\ominus$ ⟶

→ ($-H_2O$, $-CO_2$) → NH$^\oplus$=CHR → (hydrolysis) →

—NH$_2$, O$^\ominus$ + R—CHO → ($+$(I), $-H_2O$) → (II)

pound Ruhemann's purple (II) in substoichiometric yields that are slightly different for the different amino acids. Reproducible color yields are obtained provided that oxidative side reactions are eliminated by the inclusion of a reducing agent in the system.[71,72]

The automatic amino acid analyzer described by Spackman et al.[73] utilizes the two-column system of the manual procedure[74] and was the prototype for several commercial amino acid analyzers. Single-column systems which use a continuously varying buffer gradient,[75] a discontinuous buffer change,[76] or a combination of the two methods[76a] have also been described. The basic techniques are described in the reviews by Hamilton[77] and Spackman.[78] Recent developments have concentrated on increasing the speed and sensitivity of the ninhydrin analyzer[78a,78b] and on the use of 4-phenylspiro[furan-2(3H),1′-phthalan]-3,3′-dione (fluorescamine) for the fluorometric analysis of amino acids in the picomole range.[79,80] Imino acids such as proline, hydroxyproline, and sarcosine must be treated with *N*-chlorosuccinimide or another halogen source before they will react with fluorescamine.[81] Modern high-speed automatic analyzers have necessitated an increased emphasis on data processing.[82] Compilation of currently published amino acid analyses is being undertaken by Kirschenbaum.[82a]

### 1.3.3. Separation of Amino Acids on Paper and on Thin Layers

High-voltage paper electrophoresis is extremely useful for the rapid quantitative analysis of small peptides on a nanomolar scale. The cooled-tank apparatus of Michl[83] and the cooled-plate apparatus of Gross[84] are both used, and the techniques are reviewed by Dreyer and Bynum[85] and Blackburn,[86] respectively. Quantitation is effected by staining the paper with cadmium–ninhydrin reagent, cutting out the spots, and eluting the color with methanol for measurement. A combination of high-voltage electrophoresis and chromatography suitable for multiple separations is described by Corfield and Simpson.[87]

The separation of amino acids by paper chromatography[88] has been a technique of immense value to protein chemists ever since its invention, but the elaborations necessary for quantitative work have seldom been found worthwhile. The enormous literature of the subject is admirably documented.[89–93] A reliable two-dimensional separation is that described by Levy and Chung.[94] Thin-layer chromatography of amino acids on cellulose or silica gel is rapid and effective for qualitative purposes.[95–98] The difficult problem of direct quantitative evaluation of spots on paper and thin layers is receiving serious attention.[98–99a] Double isotopic labeling of the dansyl derivatives of amino acids is the basis of a method of amino acid analysis in the picomole range.[100,100a]

### 1.3.4. Amino Acid Analysis by Gas Chromatography

The speed and sensitivity offered by gas chromatography make it an inevitable competitor to even the most sophisticated system for ion-exchange resin chromatography of amino acids. The early work has been reviewed by Weinstein[101] and Blau.[102] The involatility of the amino acids makes mandatory the preparation of derivatives in which the ionizable groups are substituted. Derivatization must be quantitative, and it seemed originally as if a compromise would have to be made between excessive volatility and ease of chromatographic separation. *N*-Trifluoroacetyl derivatives of the esters of amino acids with *n*-amyl alcohol,[103] *n*-butyl alcohol,[104–106] or the *N*-heptafluorobutyl propyl esters[106a] have proved suitable for the separation and estimation of all the common amino acids. A direct esterification procedure which reduces the total overall analysis time to less than 1 h has been described by Roach and Gehrke.[107] Islam and Darbre[108] have found it possible to eliminate losses in the preparation and chromatography of the methyl esters: the increased volatility of these derivatives offers advantages for amino acids such as arginine and cystine, which give derivatives of comparatively low volatility.

Gehrke et al.[109,110] have described the quantitative analysis of all the protein amino acids after conversion to their *N*-trimethylsilyl trimethylsilyl esters, a technique also applicable to the analysis of sulfur-containing amino acids.[111] Smith and Shewbart[112] have made a quantitative comparison of the effectiveness of various trimethylsilylating reagents for amino acid analysis.

### 1.3.5. Miscellaneous Amino Acid Determinations

Special problems are presented by (1) amino acids that cannot be accurately estimated in acid hydrolysates by standard procedures (e.g., tryptophan, cystine), (2) amino acids specific to particular proteins (e.g., hydroxyproline in collagen),[15,113–115] and (3) amino acids not normally present in proteins which are produced by chemical modification of the protein. In the case of wool keratin, lanthionine, lysinoalanine, and cysteic acid are particularly important, and the amounts present serve to some extent as indicators of alkaline and oxidative damage.

#### *1.3.5.1. Tryptophan*

Tryptophan is destroyed by acid hydrolysis (see, however, Noltmann et al.[68]) but can be determined in alkaline or enzymatic hydrolysates by chromatography or by a specific color reaction such as that with *p*-dimethylaminobenzaldehyde.[116] This last reagent has been extensively applied to the determination of tryptophan in keratins: results are summarized by Cole et al.[66] Spande and Witkop[117] describe a spectrophotometric method depending on the shift in absorption maximum when tryptophan in the intact protein is oxidized by *N*-bromosuccinimide. Other methods for the determination of tryptophan are summarized by Scoffone and Fontana.[118] The use of *p*-toluenesulfonic acid for the hydrolysis of proteins, followed by determination of tryptophan on the amino acid analyzer, has been reviewed by Liu.[118a]

### *1.3.5.2. Cystine*

Cystine plus cysteine may be determined as cysteic acid, with an average recovery of 92%, by oxidizing the protein with performic acid before hydrolysis.[119] The recovery of methionine sulfone is quantitative. Spencer and Wold,[120] by hydrolyzing the protein in 6 *N* hydrochloric acid made 0.2–0.3 *M* in dimethyl sulfoxide, obtained quantitative recovery of cystine plus cysteine as cysteic acid. Fletcher et al.[121] determined the cystine plus cysteine content of wool by a modified isotope dilution method. Methods that are based on the reduction of disulfide to thiol, and which can thus be used for the determination of cystine and cysteine separately, are considered in Section 1.4.7.

### *1.3.5.3. Lanthionine and Lysinoalanine*

Lanthionine[122] and lysinoalanine[123–125] are found in alkali-treated proteins. Lanthionine gives a color with ninhydrin under acid conditions[126] and this reaction can be used for its determination in column effluents, provided that it is separated from proline. Robson et al.[127] describe such a procedure and also the separation of lysinoalanine from lysine at pH 5 on a 50-cm resin column. Tasdhomme[128] has separated lanthionine, lysinoalanine, and other modified amino acids in the presence of the normal protein amino acids in a single-column run.

The desirability of making multiple analyses for these amino acids in processed wools has stimulated the development of paper electrophoretic and paper chromatographic methods. Dowling and Crewther[129] estimate as little as 5 μmol of lanthionine per gram of wool by paper chromatography with good precision and accuracy. Most other methods are based on the oxidation of lanthionine to its sulfoxide, which facilitates separation. As lysinoalanine is stable under the conditions of oxidation, it can be determined in the same analysis.[130] A summary is given by Derminot et al.[131]

### *1.3.5.4. Cysteic Acid*

The cysteic acid content of wool is an index of oxidative treatment undergone by the fiber, and its determination is of some technological importance. Ziegler[132] and Bauters et al.[133] discuss the problem and describe methods based on paper electrophoresis.

#### *1.3.5.5. Total Amide Residues*[134]

Although the asparagine and glutamine contents of a protein can be determined by the analysis of enzymatic hydrolysates (Section 1.3.1), it is frequently desirable to obtain the total amide value by less elaborate means. Leach and Parkhill[135] showed that liberation of ammonia from amide groups was complete after hydrolysis with 12 *N* hydrochloric acid at 37°C for 10–12 days. An equivalent result was obtained by determining, at intervals up to 8–10 h, the ammonia present in hydrolysis mixtures of the protein with 2 *N* hydrochloric acid at 100°C and extrapolating to zero time to correct for the slow liberation of ammonia from labile amino acids.

## 1.4. Disulfide Bond Cleavage

An essential preliminary to the determination of the primary structure of a protein is the cleavage of inter- or intrachain disulfide bonds and, for multichain proteins, the separation of the constituent chains. In the case of the keratins, soluble protein fractions cannot be obtained until almost all the disulfide bonds have been broken. All three common methods of cleavage—oxidation, reduction, and sulfitolysis—have their particular advantages and disadvantages. When thiol groups are formed, it is convenient to block them by direct alkylation or by reaction with a compound containing an activated double bond.

Methods for the determination of disulfides and thiols are frequently required in connection with disulfide bond cleavage and are discussed briefly in this section. Comprehensive reviews of the sulfur chemistry of proteins have been given by Cecil and McPhee[136] and Cecil,[137] and the reactivity of thiol groups in proteins has been reviewed by Friedman.[137a]

### 1.4.1. Oxidation

Organic peracids are used to oxidize cystine residues to cysteic acid. Performic acid is preferred to peracetic acid because, using performic acid, oxidation is virtually complete.[138] About 85–90% of cystine is oxidized to cysteic acid, the remainder of the cystine being converted to unidentified oxidation products. The oxidation of methionine to methionine sulfone is quantitative. Upon oxidation, tryptophan gives a number of products, which precludes its estimation in oxidized proteins. The high oxidation potential of performic acid causes it to liberate free halogen from halogen acids, and if care is not taken to remove these before oxidation, halogenation of tyrosine will occur.[139]

### 1.4.2. Reduction

Goddard and Michaelis[140,141] originally used mercaptoacetic (thioglycollic) acid for the reduction of the disulfide bonds in keratin. Since then a number of thiols have been used to reduce protein disulfide bonds [e.g., 2-mercaptoethanol,[142,143] 2-mercaptoethylamine,[144] and 1,4-dimercaptobutan-2,3-diol (dithiothreitol, DTT)[145,146)].

The reaction of thiols with disulfides is represented by the following equations:

$$R'S - SR'' + R'''S^- \rightleftharpoons R'S - SR''' + R''S^- \quad (1\text{-}1)$$

$$R'S - SR''' + R'''S^- \rightleftharpoons R'''S - SR''' + R'S^- \quad (1\text{-}2)$$

The equilibrium constant of reactions (1-1) and (1-2) is near unity and a considerable molar excess ($\times$ 100–400) of thiol reductant is required to drive the reaction to near-completion.[142] Because of the large reduction potential of DTT, a smaller molar excess ($\times$ 13) is required.[147]

Reduction is usually carried out at pH 8–10 in 6–10 *M* urea and since at equilibrium 8 *M* urea is 0.02 *M* with respect to cyanate,[254a] it is necessary to remove the latter ion (cf. Section 1.7.1.) by recrystallization from ethanol–water, treatment with acid, or by anion exchange.[148]

By contrast with thiol reductants, tributylphosphine (TBP)[149] has the advantage that only a small excess is required for complete fission of disulfide bonds:

$$R'_3P + H_2O + RS - SR = 2RSH + R_3P \rightarrow O \quad (1\text{-}3)$$

and the product of the reaction does not interfere with the subsequent alkylation of the reduced protein.[150] If enzyme digestion is to be carried out after disulfide bond fission with TBP, traces of this substance must be removed.[66] It is noteworthy that urea is not required when using TBP.

Leach et al.[151] described an electrolytic method for the reduction of the disulfide bonds in wool keratin. Small amounts of thiol are used as carriers and 100% reduction is possible at pH 9.

### 1.4.3. Sulfitolysis

The reaction of sulfite with protein disulfide bonds has been the subject of a number of studies.[152–155] The equilibrium between sulfite or bisulfite and disulfides[156,157] is represented by the equation

$$R'S - SR'' + HSO_3^- \rightleftharpoons R'S - SO_3^- + R''SH \quad (1\text{-}4)$$

and by alkylating the thiol produced, the reaction can be made to go to completion.[158] *S*-Sulfo derivatives are stable in neutral solution but are unstable in acid media.[159]

### 1.4.4. Oxidative Sulfitolysis

Sulfitolysis of disulfides results in asymmetric fission (Eq. 1-4), which adds an undesirable degree of complexity if the products have to be fractionated. Oxidative sulfitolysis[159] converts both fission products to their *S*-sulfo derivatives. The reaction is represented by (Eq. 1-5) and various oxidizing agents (i.e., air,[160] $Cu^{2+}$,[161] sodium dithionate, or iodosobenzoate[162]) can be used in the reoxidation step:

$$RS-SR + SO_3^{2-} \rightleftharpoons RS-SO_3^- + RS^- \qquad (1\text{-}5)$$

Reaction (1-4) is reversible and excess thiol converts an *S*-sulfo group to the corresponding thiol:

$$R'S-SO_3^- \xrightarrow[R''SH]{\text{excess}} R''S-SR'' + R'SH + HSO_3^- \qquad (1\text{-}6)$$

Nucleophilic displacement by thiols can be used to reform disulfide bonds (e.g., in aldolase[160] and in the synthesis of insulin[163]).

### 1.4.5. Cyanolysis

Cyanide reacts with disulfides to form *S*-cyano derivatives:

$$RS-SR + CN^- \rightleftharpoons RSCN + RS^- \qquad (1\text{-}7)$$

This reaction is not of practical value for the cleavage of the disulfide bonds in proteins, but Catsimpoolas and Wood[164] have shown that it is potentially useful for the specific cleavage of peptide chains at the amino groups of cystine: the *S*-cyanocysteine residue formed in the unsymmetrical reaction slowly cyclizes to an acyliminothiazolidine that undergoes spontaneous hydrolysis.[211]

Jacobson et al.[165] have subsequently shown that virtually complete cleavage at the amino groups of half-cystine residues can be obtained by reduction of the protein with dithiothreitol and conversion of cysteine residues to *S*-cyanocysteine residues with 2-nitro-5-thiocyanobenzoic acid. Cleavage is effected by exposure to 6 *M* guanidinium chloride–0.1 *M* sodium borate at 37°C for 12 h.

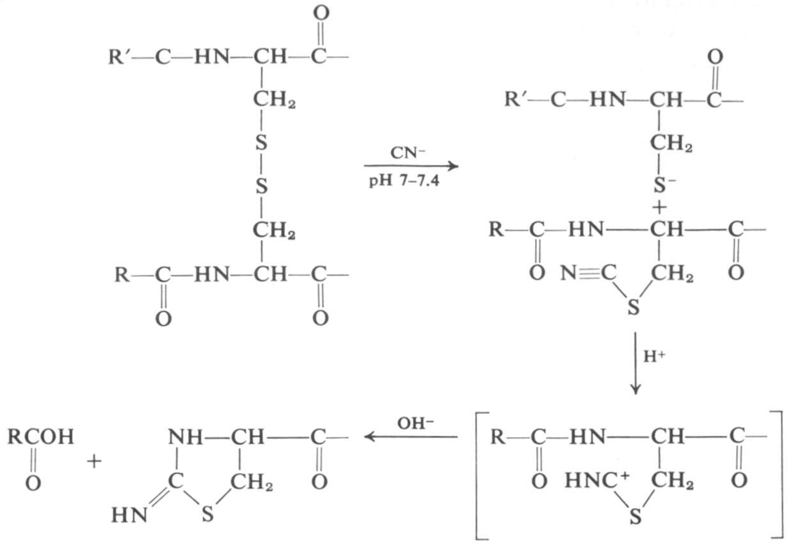

$R'-C-HN-CH(CH_2-S-S-CH_2)-C(=O)-$ / $R-C(=O)-HN-CH-C(=O)-$ $\xrightarrow[\text{pH 7–7.4}]{CN^-}$ $R'-C-HN-CH(CH_2S^-)-C(=O)-$ + $R-C(=O)-HN-CH(CH_2-S-C\equiv N)-C(=O)-$

$\downarrow H^+$

$\left[R-C(=O)-HN-CH(CH_2-S-C^+NH)-C(=O)-\right] \xrightarrow{OH^-} RCOH(=O) + NH-CH(CH_2-S-C(=NH))-C(=O)-$

Iminothiazolidine

### 1.4.6. *S*-Alkylation

Proteins with their disulfide bonds reduced to the thiol form can only be handled at low pH values and in an inert atmosphere, and not much work has been done on them.[166] Of the numerous reactions available for blocking thiol groups, carboxymethylation with iodoacetic acid is used most frequently. Alkyl iodides have the disadvantage that iodide ion is produced, and, unless all light is excluded, iodine is liberated and this will iodinate tyrosine, histidine, and tryptophan and oxidize methionine to the sulfoxide.[143] Too much excess reagent should be avoided because methionine and *S*-alkylcysteine residues readily form sulfonium salts, and alkylation of tyrosine, lysine, and histidine side chains can occur.[167,168] The solubility characteristics desired in the modified protein may influence the choice of a blocking reagent: ionizable groups such as *S*-carboxymethyl and *S*-aminoethyl promote solubility to a greater extent than does *S*-carbamidomethyl. *S*-Aminoethylation is effected with ethyleneimine[169] in preference to 2-bromoethylamine and the modified cysteine residues are susceptible to digestion with trypsin (Section 1.5.2). Bradbury and Smyth[169a] have used 3-bromopropionic acid to obviate losses due to cyclization of *S*-carboxymethylcysteine.

Thiol blocking with compounds containing an activated double bond, such as acrylonitrile,[170] has the advantage that iodide ion is not generated; by careful control of reaction time and pH the reaction with amino groups can be minimized.[171]

Other reagents for blocking thiol groups may be chosen for their suitability from an analytical viewpoint, or for tracing the blocked half-cystine residues through separation procedures. *N*-Ethylmaleimide and its derivatives[172] are examples. For the characterization of thiol-containing sequences in myosin, Weeds and Hartley[173] converted the thiol groups to mixed disulfides by treatment with excess $^{35}S$-cystine.

### 1.4.7. Estimation of Cystine and Cysteine in Proteins

A comprehensive review of this topic has been given by Leach.[174] Methods for the combined cystine plus cysteine content of proteins are mentioned in Section 1.3.5.2.

Methods that permit the separate determination of thiol and disulfide groups in proteins depend on the reactivity of thiol groups—disulfides are determined via the thiols produced, after cleavage of the S—S bond. Because of the instability of cystine and cysteine, methods that do not involve hydrolysis of the protein are generally to be preferred. Such a method is that in which the reaction between thiol groups and organic mercury compounds is monitored by a polarographic technique[174]:

$$P - SH + R{-}Hg{-}X = P - S{-}Hg{-}R + H^+ + X^- \qquad (1\text{-}8)$$

Partial oxidation products of cystine interfere. A radiometric modification of this method can be used on samples as small as single wool fibers.[175]

Several spectrophotometric methods are available which are applicable to intact, soluble proteins. Ellman's reagent, 5,5′-dithiobis (2-nitrobenzoic acid) reacts with protein thiol groups in weakly alkaline solution to generate an equivalent amount of highly absorbing thiol anion.[176–177a]

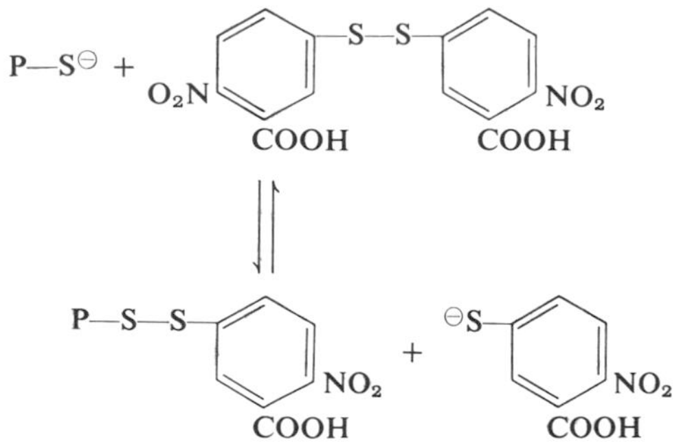

P—S⊖ +
S—S
$O_2N$
$NO_2$
COOH COOH

⇌ (1-9)

P—S—S
$NO_2$
COOH
+
⊖S
$NO_2$
COOH

The thiols formed in a similar way from 2,2′- and 4,4′-dithiodipyridine[178] tautomerize to thiopyridones possessing characteristic ultraviolet absorption. Protein thiol groups add to 4-vinylquinoline to give derivatives that absorb at 318 nm.[179]

Complete hydrolysis of the protein after reaction with the thiol reagent, followed by estimation of the modified half-cystine residues by amino acid analysis, is the basis of the methods in which thioglycollic acid,[68] 4-vinylpyridine,[180] or 2-vinylpyridine[180a] are used. Close control of the conditions of hydrolysis is required, as the cysteine derivatives tend to be labile.

Hydrolysis of the protein before the application of a reaction characteristic of thiols is exemplified by the phosphotungstic acid reduction method[174] associated with the names of Folin and Shinohara. Although this method is not very accurate, it is rapid and has frequently been used to give comparative results on a large number of samples, particularly in work on keratins.

Liu and Inglis have estimated cystine in wool as the *S*-sulfocysteine derivative.[180b]

### 1.4.8. Differential Labeling

The varying reactivity of cystine residues in proteins is a subject of continuing interest. Lindley[154] reviewed the work on the sulfitolysis of wool keratin and concluded that the reaction of disulfide bonds in wool with sulfites is complicated by the histological structure of wool. As a consequence of this complexity, differential labeling has found greater use in better characterized proteins and has clearly shown how the reactivity of cystine residues can vary under different conditions.[144,155,181]

The technique of differential labeling with iodoacetate and $^{14}$C-iodoacetate has, however, been used to study the distribution of reactive cystine residues in the low- and high-sulfur regions of wool keratin.[182,183,315] The same pair of labels were used to locate the thiol and disulfide bonds of papain.[147] Chiancone et al.[184] have studied the reaction of *p*-mercuribenzoate with masked and free thiol groups of hemoglobin.

## 1.5. Peptide Chain Cleavage

### 1.5.1. Cleavage by Enzymes

Native proteins are relatively resistant to proteolytic enzymes. Useful structural information can be gained from attack limited to certain susceptible bonds,[185] and among fibrous proteins the resistance of ordered structures to enzymatic attack has been utilized for the isolation of protein fractions of high helix content from wool keratin[186,187] and for the fission of myosin into light and heavy meromyosins.[188] If extensive digestion of a protein is required, it must first be denatured.[189] Preliminary cleavage of disulfide bonds is also usual, unless the object is the isolation of fragments in which these bonds are intact.

An enzyme catalyzes the hydrolysis of those peptide bonds adjacent to amino acid side chains that satisfy its particular specificity requirements, and digestion with enzymes of well-defined specificity to produce a limited number of peptide fragments is extensively used in the determination of amino acid sequences. The enzymes most frequently used are the following:

1. Trypsin, which splits at bonds involving the carboxyl groups of the basic amino acids lysine and arginine. The optimum pH range is 7–9. Trypsin is the most specific proteolytic enzyme and is, consequently, usually the first choice for peptide chain degradation.
2. Chymotrypsin, which splits chiefly at bonds involving the carboxyl groups of the aromatic amino acids tyrosine, phenylalanine, tryptophan, and, more slowly, of leucine. The optimum pH range is 7–9. The course of hydrolysis in the presence of chymotrypsin or trypsin can be followed by the uptake of alkali needed to maintain the pH of digestion.[190]
3. Pepsin, which splits chiefly at bonds involving an amino or a carboxyl group of an aromatic amino acid residue, and more slowly at a wide range of other bonds. Pepsin is particularly valuable for the isolation of peptide fragments containing intact disulfide bonds, since disulfide interchange is minimized at the optimum pH of digestion (1.5–2.0), and also for the digestion of insoluble cores from previous enzyme digests which can be dissolved in formic acid and the solution diluted when the enzyme is added.[191]

Enzymes of lower specificity which are available for the secondary cleavage of larger peptides are:

4. Thermolysin,[192] which splits bonds involving the amino groups of residues with hydrophobic side chains (pH 8.5).
5. Elastase,[193] which splits bonds involving either the amino or carboxyl groups of residues with neutral aliphatic side chains (pH 8.5).
6. Papain (pH 5–7.5).
7. Subtilisin (pH 7.5–8.5).
8. Pronase (pH 6.5).

The last three enzymes possess a low degree of specificity (for details, see Hill[185] and Smyth[194]). The different rates of cleavage at different bonds often result in the formation of overlapping fragments.

### 1.5.2. Cleavage by Enzymes Acting on Modified Proteins

A useful extension of the technique of digestion with enzymes of high specificity is afforded by chemical modification of amino acid side chains. Thus trypsin, which acts at peptide bonds involving arginine or lysine residues, can be restricted to act at only one of these types of residue if the other is chemically modified so that its side chain no longer fulfils the specificity requirements of the enzyme. Further, if the chemical modification can be reversed, the high specificity of trypsin can be utilized again to split the peptides separated from the first digest, after the modifying group has been removed. Besides deleting sites of enzyme action in this way, one may modify inert side chains so that they become new sites of attack (e.g., the conversion of cystine residues to trypsin-sensitive residues of *S*-aminoethylcysteine by reduction and reaction with ethyleneimine).

#### *1.5.2.1. Modification of Lysine*

Acetylation with $^{14}$C-acetic anhydride[195] and succinylation[196] have been used for the isolation of peptides containing internal lysine residues, but in neither case can the blocking group subsequently be removed. The most useful procedures available at present are trifluoroacetylation with *S*-ethyl trifluorothiolacetate,[197] readily reversible by treatment with alkali in the cold, or modification by one of a series of dicarboxylic acid anhydrides which introduces groups whose ease of removal in acid solution (with anchimeric assistance from the second carboxyl group) increases in the order[198] maleyl $<$ *exo-cis*-3,6-endoxo-$\Delta^4$-tetrahydrophthaloyl $<$ citraconyl $<$ dimethylmaleyl.

Other reversible modifications proposed are acetoacetylation with diketen,[199] reversed by treatment with hydroxylamine and tetrafluorosuccinylation,[200] reversed by treatment with cold alkali.

#### *1.5.2.2. Modification of Arginine*

Modification of the guanidinium group of arginine to prevent the attack of trypsin at the peptide bond involving the carboxyl group can be effected by reaction with 1,3-diketones, although lysine can also react to a lesser extent and deamination of $\alpha$-amino groups can occur. Benzil,[201] cyclohexan-1,2-dione,[202,202a] and malonaldehyde[203] require extremes of pH for reaction. Reagents effective under milder conditions have been described: biacetyl,[204,205] which reacts as the trimer[206]; glyoxal[207]; and phenylglyoxal.[208] The maximum extent of reaction in all cases appears to be about 80%. The modified arginine residues do not regenerate arginine under the normal conditions of acid hydrolysis. Patthy and Smith[208a] have recently described the highly specific modification of arginine residues with cyclohexan-1,2-dione in borate buffer at pH 8–9. The blocking group can be quantitatively displaced with hydroxylamine buffer at pH 7.

#### *1.5.2.3. S-Aminoethylation of Half-Cystine Residues*

Thiol groups react with ethyleneimine to produce *S*-aminoethylcysteine residues.[169] The peptide bonds involving the carboxyl groups of these residues are attacked by trypsin but at a rate considerably less than that for peptide bonds involving lysine.

### 1.5.3. Cleavage by Chemical Methods

Witkop[209,210] classifies chemical methods of chain fission as *preferential* (competitive or hydrolytic) and *selective* (noncompetitive, nonhydrolytic) and gives an extensive discussion of the mechanistic principles involved. A selective and quantitative method of chemical fission, especially one restricted to an amino acid that occurs relatively infrequently in proteins, provides a valuable complement to enzymatic methods, and this is evidenced by the extensive use made of the cyanogen bromide method of fission at methionine residues. Intensive work has produced *some* method for almost every amino acid with a functional side chain,[211] but only a few of these have proved satisfactory in application.

#### *1.5.3.1. Cleavage at Methionine*

The reaction with cyanogen bromide takes place under mild conditions (0.1 *N* hydrochloric acid or 70% formic acid at 20°C) to produce cleavage on

```
            NH—CHRCO—                          NH—CHRCO—
           /                                  /
—NHCH—C                            —NHCH—C
      |    \\                             |    \\
      |     O                             |     O
     CH2—CH2          ——→                CH2—CH2          ——→
          |                                   |
          S + C≡N                           ⊕S—C≡N
          |   |                               |
         CH3 Br                              CH3    Br⊖

          ⊕
          NH—CHRCO—                                 O
         //                                        //
 —NHCH—C                           —NHCH—C
      |    \                             |    \
      |     O          ——→               |     O + H2N—CHRCO—
      |    /                             |    /
     CH2—CH2     Br⊖                    CH2—CH2
        + S—C≡N
          |                             Homoserine lactone residue
         CH3
```

the carboxyl side of the methionine residue, which is converted to a residue of homoserine lactone.[212,213] Reduction of any methionine sulfoxide to methionine is essential for complete cleavage.[214]

After separation of the cleavage products, their acid hydrolysates are treated with alkali to convert homoserine lactone to homoserine before amino acid analysis is carried out.[215]

#### *1.5.3.2. Cleavage at Tryptophan and Tyrosine*

*N*-Bromosuccinimide can be used to cleave peptide bonds on the carboxyl side of tryptophan, tyrosine, and histidine residues, the ease of cleavage decreasing in that order. The conditions under which the different types of bond are cleaved are set out by Ramachandran and Witkop.[216] An extent of reaction exceeding 50% has seldom been found, but almost complete cleavage at the single tyrosyl residue in a lysine-rich histone has been reported.[217]

#### *1.5.3.3. N → O Acyl Shift: Cleavage at Serine and Threonine*[218]

Treatment of proteins with concentrated sulfuric acid or hydrogen fluoride results in a conversion of peptide linkages that involve the amino groups of serine and threonine residues to ester linkages through the $\beta$-hydroxyl group:

$$\text{—NHCHRCO—}\underset{\underset{\text{HOCH}_2}{|}}{\text{NHCHCO—}} \longrightarrow \text{—NHCHRCO—}\underset{\underset{\text{OCH}_2}{|}}{\text{NH}_2\text{CHCO—}} \quad (1\text{-}10)$$

This N → O shift is rapidly reversed under alkaline conditions, but if the amino groups liberated are first blocked by acetylation, formylation, or carbamylation, the peptide chain may be cleaved at the ester bond by acid or alkaline hydrolysis. The extent of rearrangement can be as high as 90% of the serine residues, but is much lower for threonine.

#### *1.5.3.4. Splitting Out Aspartic Acid with Dilute Acid*

Hydrolysis with boiling dilute acid of such a concentration that the ionization of the participating $\beta$-carboxyl group is just suppressed (0.03 *N* hydrochloric acid or 0.25 *N* acetic acid) leads to the preferential splitting out of aspartic acid from peptide chains.[219] The side chain of asparagine must undergo deamination before splitting can take place, and this may be utilized to split successively at aspartyl and asparaginyl residues.[220]

This method has been used for the isolation of protein fractions from wool.[221]

#### 1.5.3.5. *Hydrolysis with Concentrated Hydrochloric Acid*

Bonds involving the amino groups of serine and threonine are preferentially cleaved in concentrated hydrochloric acid, but general hydrolysis of peptide bonds is rapid under these conditions, and prolonged treatment leads to the accumulation of dipeptides. An acid concentration of 10 *N* at 37°C has frequently been used, but Naughton et al.[222] have shown that conversion of aspartyl residues into the $\beta$ form is minimized in 5.7 *N* acid. There is also less risk of disulfide interchange at this concentration of acid. As a rough rule, these authors state that, in 5.7 *N* acid, hydrolysis for 10 min at 105°C is equivalent to hydrolysis for 24 h at 37°C. To produce a greater range of overlapping fragments for sequence determination, Alexyev et al.[223] added portions of peptide at intervals of time to a common portion of 6 *N* hydrochloric acid at 100°C.

The various aspects of the partial hydrolysis of proteins with acids are discussed extensively by Hill.[185]

## 1.6. Separation of Mixtures of Peptides Produced by Chain Cleavage

In the fractionation of complex mixtures, an initial subfractionation on as large a scale as possible, even if optimum resolution is not attained, is often a desirable preliminary to the use of techniques of higher resolving power. Column methods, which can be scaled up to deal with substantial loads, are an obvious first choice.

### 1.6.1. Column Chromatography

Gel filtration on cross-linked dextrans or polyacrylamide gels is effective when the number of peptides is not too great (e.g., when cyanogen bromide has been used to cleave chains at methionyl residues[215]). High concentrations of urea or formic acid may be employed when the components present have a tendency to aggregate.

Frequent use is made of separations in volatile buffers on columns of sulfonic acid cation-exchange resins[224] and quaternary ammonium anion-exchange resins.[225] When the latter type of resin is used, care is necessary to avoid sudden changes in the pH of the buffer. The two types of resin have complementary properties and are often effectively used in sequence. Weak-acid cation exchangers can be useful for large or highly basic peptides.[226]

Ion-exchange celluloses (notably diethylaminoethylcellulose, carboxymethylcellulose, and phosphocellulose) have an exchange efficiency superior to that of ion-exchange resins[16a,227] and are very effective materials for the separation of peptides (see, e.g., Haylett and Swart[228] and Elzinga[229]). Ion-exchange materials based on dextrans and polyacrylamide are also available.

#### *1.6.1.1. Detection of Peptides in Effluents from Chromatographic Columns*

Larger peptides give relatively low color yields with ninhydrin.[230] Increased sensitivity can be attained by alkaline hydrolysis before reaction with ninhydrin.[231] Comparison of the colors produced with ninhydrin before and after hydrolysis often permits a rough estimate of the size of a peptide to be made. Automatic monitoring of peptide separations by splitting the effluent stream and allowing a fixed proportion to react with ninhydrin[232,233] saves much manipulative effort. Cystine-containing peptides have been detected by the reaction with phosphotungstic acid[234] and by a modification of the method of Zahler and Cleland[177]—reduction with 1,4-dithioerythritol and reaction with 5,5-dithiobis (2-nitrobenzoic acid) in the presence of arsenite.[177a,235]

Proteins may be labeled with radioactive isotopes by biosynthetic routes, or inactive proteins may be tritiated by homolytic reaction with tritiated hydrogen sulfide.[236] Peptides containing radioactive isotopes can be detected by passage of the column effluent through flow-type scintillation cells.

### 1.6.2. Desalting of Peptides

The presence of salt in peptide mixtures can cause interference in paper chromatographic or electrophoretic separations. Desalting may be effected by ion exchange[237] or gel filtration.[238]

### 1.6.3. Separation of Peptides on Paper

One-dimensional paper chromatography and paper electrophoresis are widely used for the preparative separation of peptides on a microscale. If the molecular weight of a peptide is known, the net charge can be deduced from electrophoretic mobility data,[239] and this is an aid in the differentiation of residues of the dicarboxylic acids and their amides. A combination of chromatography in one dimension with electrophoresis in a second dimension[(240)] (peptide mapping, fingerprinting) provides a rapid method of evaluating the complexity of peptide mixtures and of detecting small differences between related proteins. Reagent sprays specific for particular amino acid residues[(89,241)] and radioautography enable additional information to be obtained from peptide maps.

### 1.6.4. Diagonal Electrophoresis

This elegant method of selectively purifying certain peptides from a complex mixture has been extensively developed in recent years, and its application to the pairing of half-cystine residues in proteins[(242)] has been particularly successful.

The principle is as follows: (1) the peptide mixture is separated by electrophoresis; (2) the peptides are treated, on the paper, with a reagent that modifies certain amino acid residues in such a way that peptides containing them have altered electrophoretic mobilities (e.g., oxidation of a cystine peptide to a pair of cysteic acid peptides); and (3) a second electrophoretic separation is carried out at the same pH but in a direction at right angles to the first. The modified peptides lie off the 45° diagonal line determined by the unmodified peptides. Reviews of the method and its applications are given by Perham[(243)] and Hartley.[(244)] The technique has been applied by Cruickshank et al. to histidine-containing peptides.[(244a)]

## 1.7. Amino-Acid-Sequence Determination

While the subjects treated in earlier sections of this chapter form an important part of the technique of primary structure determination, the actual ordering of amino acid residues in peptides, large or small, depends heavily on the topological uniqueness of the N- and C-terminal residues of a polypeptide chain. Much ingenuity and hard work has been expended on the development of methods for their identification. Some of these methods are applicable to the determination of the terminal residues of intact proteins and their subunits: for this purpose, completeness of reaction and quantitative analytical procedures are essential. For work on peptides produced by fragmentation of larger structures, where the amount of material available is often small and not precisely known, sensitivity may be the paramount consideration.

### 1.7.1. N-Terminal Residues

In chemical methods of N-terminal-residue determination, the free *N*-amino group is labeled with a suitable substituent group. The modified N-terminal amino acid is then released from the peptide under such conditions that the bond formed in the labeling reaction is not broken. The modified N-terminal amino acid possesses properties different from those of amino acids; generally, it will not possess a dipolar structure, and the modifying group is usually chosen to confer characteristic optical properties or radioactivity; consequently, it is easily separated and identified.

The conditions required to break the peptide bond between the N-terminal residue and the next residue in the chain may be so drastic (e.g., acid hydrolysis) that all other peptide bonds in the molecule are broken; or, where cyclization or neighboring-group participation occurs, they may be sufficiently mild for the rest of the peptide to remain intact, thus permitting the labeling and cleavage reaction to be repeated on the new N-terminal amino acid (*sequential degradation*).

A complete scheme of N-terminal-residue determination thus requires the availability of the derivatives of all the protein amino acids and methods for their separation and identification and (preferably) quantitation. In addition to the $\alpha$-amino group, other nucleophiles in the protein or peptide that may react are the thiol group of cysteine, the $\epsilon$-amino group of lysine, the phenolic hydroxyl group of tyrosine, and the imidazole ring of histidine. The chromatographic separation of the derivatives involved in N-terminal-residue determination has been comprehensively reviewed by Rosmus and Deyl.[244b]

#### *1.7.1.1. 1-Fluoro-2,4-dinitrobenzene* (*FDNB*)

This reagent was introduced by Sanger[245] and used by him in the elucidation of the structure of insulin. Reaction with the unprotonated amino group gives a dinitrophenyl (DNP) peptide, which is then hydrolyzed to free amino acids and the DNP amino acid derived from the N-terminal residue. During the hydrolysis, the DNP amino acids themselves are destroyed to varying extents, and corrections for this must be applied in quantitative work. The method has played an important part in the investigation of protein structures by chemical means, and has been extensively studied and reviewed.[246–249] In recent years it has largely been replaced for qualitative purposes by the more sensitive DNS method (see below), but it still remains important for quantitative end-group determination. Automatic methods for the analysis of DNP amino acids have been developed.[250] The application of the DNP method to the determination of the end groups of structural

$$\text{2,4-}(O_2N)_2C_6H_3F + H_2NCHRCONHCHR'CO\text{—} \xrightarrow{\text{pH 8–9}} \text{2,4-}(O_2N)_2C_6H_3\text{—}NHCHRCONHCHR'CO\text{—}$$

$$\xrightarrow{5\,N\ \text{HCl, 4–16 hr}} \text{2,4-}(O_2N)_2C_6H_3\text{—}NHCHRCOOH + H_3\overset{\oplus}{N}CHR'COOH + \cdots$$

proteins has been described by Beyer and Schenk,[251,252] who summarize the contributions made by the Aachen school.

The use of 2-fluoro-3-nitropyridine[253] as a reagent for N-terminal residues typifies attempts to facilitate the release of the labeled N-terminal residue by neighboring-group participation. Hydrolysis for 30 min at 100°C in 0.2 *N* hydrochloric acid containing 20% of formic acid is sufficient to release the 3-nitro-2-pyridyl-amino acids, which can be separated and estimated by thin-layer chromatography.[254]

### *1.7.1.2. Cyanate Method*

The observation that ribonuclease undergoes slow inactivation in urea solutions led to a realization of the importance of the small amount of cyanate in equilibrium with urea,[254a] and subsequently to the exploitation of cyanate as a reagent for the determination of N-terminal residues of proteins and large peptides.[255] Carbamylation of the peptide or protein is carried out at pH 8 in 8 *M* urea or other denaturing solution. The reagents are removed by gel filtration or dialysis. The terminal *N*-carbamyl residues are cyclized to hydantoins by heating in acid:

$$NCO^- + \overset{+}{N}H_3CHRCONHCHR'CO\text{—} \xrightarrow{\text{pH 8}}$$

$$NH_2CONHCHRCONHCHR'CO\text{—} \xrightarrow{H^+} \begin{array}{c} CHR\text{—}CO \\ | \qquad\quad | \\ HN \qquad NH \\ \diagdown \quad \diagup \\ CO \end{array} + \overset{+}{N}H_3CHR'CO\text{—}$$

The hydantoins are separated chromatographically from the amino acids and peptides produced by the acid treatment, and converted to free amino acids by hydrolysis in either acid or alkali. The N-terminal-residue determination thus depends ultimately on an amino acid analysis. The yields of serine and threonine from the hydantoins are low, and correction factors must be applied. Nevertheless, the cyanate method has already found considerable use for quantitative end-group determination.

#### *1.7.1.3. 1-Dimethylaminonaphthalene-5-sulfonyl Chloride (Dansyl Chloride)*

The methodology based on the fluorescent amino acid derivatives of this reagent[256,257] offers a 100-fold increase in sensitivity over the DNP method, and it is at present the method of choice for qualitative N-terminal-residue determination. Liberation of the DNS amino acid from the labeled peptide requires complete acid hydrolysis of the peptide, but because of the small amount of material it requires, the DNS method can be used in conjunction with the Edman method for sequential degradation of peptides (see below). Effective separations of DNS amino acids can be achieved by chromatography on thin layers of polyamide.[244,258]

While the qualitative methods suffice for many applications of the DNS method in amino-acid-sequence work, the availability of quantitative procedures utilizing either fluorescence properties[259–261] or $^{14}C$-labeling[262] will be most valuable. The use of the DNS reaction in biochemical analysis has been reviewed by Seiler.[263]

#### *1.7.1.4. Edman Degradation*

This method, introduced by Edman,[264] permits the stepwise degradation of a protein or peptide from the N terminus. It is at present the most important method for amino-acid-sequence determination, and has been extensively applied and studied. The peptide is allowed to react with phenylisothiocyanate at pH 9–10 to form the phenylthiocarbamyl (PTC) peptide (Eq. 1-11). In anhydrous trifluoracetic acid, the PTC peptide is cleaved to liberate the N-terminal amino acid residue as the 2-anilino-5-thiazolinone derivative and a peptide lacking this amino acid residue (Eq. 1-12). The 2-amino-5-thiazolinones are too unstable for identification purposes, and are converted, via the PTC amino acids, to the isomeric 3-phenyl-2-thiohydantoins (Eq. 1-13). The shortened peptide liberated according to (Eq. 1-12) can then be subjected to further degradation cycles.

Because the cleavage reaction (Eq. 1-12) is fast, whereas the conversion reaction (Eq. 1-13) is slow, it is advantageous to separate the thiazolinone from the shortened peptide before converting it to the PTH, thereby minimizing exposure of the peptide to the hydrolytic conditions required for conversion. As the peptide becomes shorter, however, losses of peptide during

$$C_6H_5NCS + H_2NCHR_1CONHCHR_2CO\text{—} \xrightarrow{\text{pH 9–10}} C_6H_5NHCSNHCHR_1CONHCHR_2CO\text{—} \quad (1\text{-}11)$$

PTC peptide

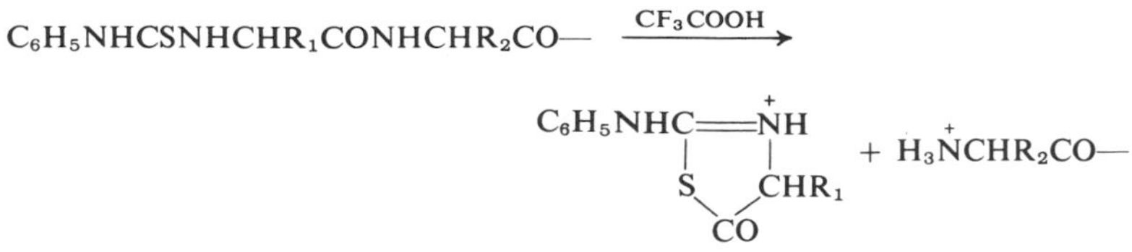

$$C_6H_5NHCSNHCHR_1CONHCHR_2CO\text{—} \xrightarrow{CF_3COOH} \underset{\text{2-Anilino-5-thiazolinone}}{C_6H_5NHC{=}\overset{+}{N}H\ (\text{ring: } S\text{—}CO\text{—}CHR_1)} + H_3\overset{+}{N}CHR_2CO\text{—} \quad (1\text{-}12)$$

2-Anilino-5-thiazolinone

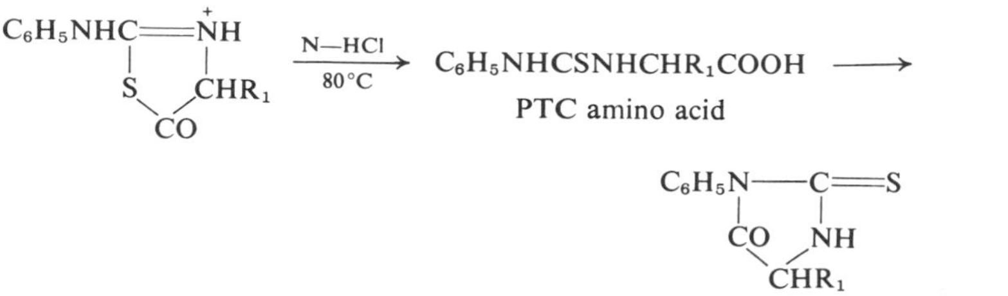

$$C_6H_5NHC{=}\overset{+}{N}H\ (\text{ring: } S\text{—}CO\text{—}CHR_1) \xrightarrow[80^\circ C]{N\text{—}HCl} C_6H_5NHCSNHCHR_1COOH \longrightarrow C_6H_5N\text{—}C{=}S\ (\text{ring: } CO\text{—}CHR_1\text{—}NH) \quad (1\text{-}13)$$

PTC amino acid

3-Phenyl-2-thiohydantoin (PTH) derivative of amino acid

extraction of the thiazolinone increase. Losses can be reduced by the use of a polymeric carrier.[264a] Rigorous exclusion of oxygen is desirable because the PTC group is easily desulfurized.

The amino acid PTHs liberated at each step are identified by chromatography on paper or silica gel. Increased sensitivity can be obtained by exposure to iodine vapor.[264b] Edman has continued to develop and refine the method over a period of 20 years[265] into an automated procedure in which the PTHs are delivered for identification at a rate of 15 stages per day, and an average yield per stage of 98.5%.[266,266a] Several commercial versions of this "protein sequenator" are now available. Another well-established version of the manual technique[267,267a] utilizes a paper strip as support for the peptide throughout the degradation. An ultramicro modification of the paper-strip technique has been described.[268]

The Edman degradation can be applied in other ways. In a straightforward subtractive version, the thiazolinone is removed and a portion of the shortened peptide is taken for amino acid analysis to identify the missing N-terminal residue by difference.[269] A powerful and productive modification[270] is the combination of the Edman degradation with the DNS method. The new N-terminal amino acid uncovered at each stage is identified as its

DNS derivative, as little as 0.005 $\mu$mol of peptide per stage being required. Gray and Smith[271] have modified the technique so that a pentapeptide can be sequenced in one day. All versions of the degradation require highly purified reagents and close attention to experimental detail.

The established versions of the Edman degradation are those described above. Because of its importance, several new modifications that promise increased sensitivity, speed, convenience, or reliability are under development. One approach is to use a thiocyanate reagent, such as fluoresceinisothiocyanate[272] or 1-naphthylisothiocyanate[273] which yields fluorescent thiohydantoins, thus making the sensitivity of the direct derivative method comparable with that of the DNS method. Increased sensitivity is also being sought by gas-chromatographic methods: separation procedures have been described for PTHs,[274] their TMS derivatives,[275] and the 3-methyl-2-thiohydantoins obtained by using methylisothiocyanate as reagent.[276–277b] These last derivatives have also been identified by mass spectrometry and quantitatively determined by isotope ratio assay after addition of $^{15}$N-enriched amino acids.[278] High-speed liquid chromatography has been applied to the analysis of amino acid PTHs.[278a] Further advances in technique will be necessary before the full potential sensitivity of these methods is realized.

Solid-phase modifications circumvent the difficulty of separating cyclized derivatives from short peptides by solvent extraction. Dowling and Stark[279] use a polystyrene matrix containing isothiocyanate groups, the shortened peptide being cleaved from the resin at each step. Degradation is followed by amino acid analysis or dansylation. In the modifications of Laursen[280] and Schellenberger et al.,[281] the peptide remains attached through its carboxy terminus to an aminated polystyrene resin. An automatic sequenator based on this principle has been described by Laursen.[282,283] Tritium labeling of the PTHs is employed.

Cyanomethyl dithiobenzoate has been suggested as a reagent for the sequential degradation of peptides.[284] In this case the primary cyclization product (a 2-phenylthiazolinone) is stable, in contradistinction to the 2-anilinothiazolinones of the Edman procedure.

#### *1.7.1.5. N-Terminal-Sequence Determination with Leucine Aminopeptidase*

Leucine aminopeptidase (LAP) from swine kidney catalyzes the hydrolysis of peptide bonds involving the carboxyl group of an amino acid residue whose amino group is free, the only exception being the imide bond to a proline residue. The rates of hydrolysis are highest for those amino acids with nonpolar aliphatic or aromatic side chains. Sequential liberation of amino acids from the N terminus of the peptide occurs, and the determination of the free amino acids present in the digest as a function of time often provides sufficient information for the N-terminal sequence of the peptide to be deduced.[(285)] Prolonged digestion with LAP results in complete hydrolysis of the peptide unless the process is terminated by an XPro— sequence. If complete hydrolysis of the peptide is the primary objective (e.g., for the location of amide groups), the more efficient ribosomal enzyme, aminopeptidase-M,[(286)] is to be preferred, although its action is limited by proline residues in the same way as that of LAP. With the availability of reliable chemical techniques, the use of LAP has declined, but it is still useful in cases where other methods give inconclusive results.

A sequencing strategy based on the use of dipeptidyl aminopeptidase has been described by Lindley.[(286a)]

#### *1.7.1.6. Blocked N-Terminal Residues*

The occurrence of a terminal *N*-acetyl group was first demonstrated by Narita[(287)] in tobacco mosaic virus protein; it was characterized as acethydrazide. Since then, it has been recognized that this blocking group is fairly common in globular and fibrous proteins. Well-authenticated *N*-acetyl terminal sequences have been reported in both high-sulfur[(288)] and low-sulfur[(289)] proteins from wool. For the quantitative determination of acetyl groups in wool, O'Donnell et al.[(290)] preferred to distill acetic acid from 12 *N* sulfuric acid and determine it by titration. Micromethods employing gas chromatography[(291)] and enzymatic analysis[(292,292a)] are available. Terminal *N*-formylproline has been detected in lamprey hemoglobin.[(293)]

Pyrrolid-2-one-5-carboxylic acid (PCA) has been identified as the N-terminal residue of several peptides and proteins (e.g., fibrinopeptides[(294)] and a low-sulfur protein fraction from wool[(295)]). A discussion of the occurrence of this group and a critical summary of the evidence that suggests that it may be an artifact derived from N-terminal glutamine is given by Blombäck.[(296)] Peptides with N-terminal PCA cannot be degraded by the Edman technique. An enzyme that specifically cleaves off this residue has been reported[(297)] and should prove a useful adjunct in sequence determination. Another approach that has been explored is the reduction of PCA to a prolyl residue.[(298)]

### 1.7.2. C-Terminal Residues

#### *1.7.2.1. Chemical Methods*

The reaction of hydrazine with proteins or peptides at 100°C is the most frequently used chemical method for the qualitative and quantitative determination of C-terminal amino acids.[(299)] In this reaction, transamidation converts all the amino acids whose carboxyl groups are involved in peptide-bond formation into their hydrazides, leaving the C-terminal residue unaltered:

$$\mathrm{H_2NCHR_1CONHCHR_2CO{-}NHCHR_nCOOH} \xrightarrow[\ 100°\mathrm{C}\ ]{\mathrm{H_2NNH_2}} \mathrm{H_2NCHR_1CONHNH_2 + H_2NCHR_2CONHNH_2 + \cdots + H_2NCHR_nCOOH}$$

The free amino acid can then be identified (e.g., by reaction with FDNB[300,301] or dansyl chloride[302,303]). Quantitative results can be obtained by using the numerous chromatographic methods available.[304,305]

A disadvantage is the low yield of C-terminal amino acids obtained and the large correction factors which have to be applied in quantitative work. Phillips[306] has attributed these low yields to the partial conversion of the C-terminal residue to its hydrazide, and in this connection a lower reaction temperature is probably desirable. Bradbury[307] used hydrazine sulfate to catalyze the reaction at 80°C, but the efficacy of this addition has never been established. Improved yields have been achieved by using ion-exchange resin as a catalyst.[305]

By treating proteins or polypeptides with acetic anhydride, the C-terminal residue is cyclized to give a peptidyl oxazolone. Matsuo et al.[308] tritiate the peptidyl oxazolone with $^3H_2O$ in pyridine:

$$\mathrm{{-}CONHCHR_{n-1}CONHCH(R_n)C({=}O)OH} \xrightarrow{\text{acetic anhydride}}$$

$$\mathrm{{-}CONHCHR_{n-1}{-}C{=}N{-}CH(R_n){-}C({=}O){-}O\ (ring)} \xrightarrow[\text{pyridine}]{^3\mathrm{H_2O}} \mathrm{{-}CONHCHR_{n-1}{-}C{=}N{-}C(R_n)(^3H){-}C({=}O){-}O\ (ring)} \xrightarrow[110°\mathrm{C}]{6\,N\,\mathrm{HCl}}$$

Peptidyloxazolone

$$\mathrm{H_2NCR_n \cdot {}^3HCOOH} + \text{inactive amino acids}$$

Acid hydrolysis and chromatographic separation of the hydrolysate, followed by detection of the radioactive residue, provides a sensitive method of identifying C-terminal amino acids. A critique of the method is given by Holcomb et al.,[309] who recommend that the reaction be carried out in an aqueous medium.

Under mild conditions, a mixture of ammonium thiocyanate and acetic anhydride reacts with the C-terminal residue of a peptide to form a peptidyl-thiohydantoin. Liberation of the thiohydantoin using acetohydroxamate yields the penultimate residue of the peptide as the new C-terminal amino acid:

$$\text{—CONHCHR}_{n-1}\text{CONHCHR}_n\text{COOH} \xrightarrow[\text{NH}_4\text{SCN}]{\text{acetic anhydride}}$$

```
                         Rn
                         |
                         HC——C=O             O
                         |    |              ||
  —CONHCHR(n-1)CO—N     NH       CH3CNHO-
                           \   /       ——————————>
                            C            or H+
                            ||
                            S
```

Peptidylthiohydantoin

```
                                    Rn
                                    |
                                    HC——C=O
                                    |    |
  —CONHCHR(n-1)COOH + HN     NH
                          \   /
                           C
                           ||
                           S
```

An examination of this reaction by Stark[310,310a] for the sequential removal of C-terminal residues from peptides showed that under favorable conditions six degradations were achieved, but application of the method was vitiated by the occurrence of C-terminal aspartic acid or proline. With proteins,[311] sequential removal of up to four residues has been achieved.[311a]

#### *1.7.2.2. Enzymatic Methods*

Two pancreatic exopeptidases having different specificities are available for C-terminal amino-acid-sequence determination. Carboxypeptidase A preferentially releases amino acids with aromatic or large aliphatic side chains from the C terminus of a peptide chain, whereas carboxypeptidase B preferentially releases the basic amino acids lysine and arginine. Ambler,[312] in a review of the use of the carboxypeptidases, has referred to the use of carboxypeptidase C, which is specific for proline and some other amino acids. For sequence determination, it is usually the practice to use a combination of carboxypeptidases A and B. From the kinetics of the release of amino acids from the C terminus, a partial sequence may frequently be deduced. Carboxypeptidase Y from bakers' yeast can remove most C-terminal amino acids, including proline.[312a]

### 1.7.3. Location of Disulfide Bridges

The pairing of half-cystine residues is usually the last step in the determination of the primary structure of a protein. It requires the isolation of peptides containing intact disulfide bonds, which are then cleaved, usually by oxidation, and the component cysteic acid-containing peptides separated, characterized, and located in the known primary structure. Most of the problems that arise, the most important being the danger of spurious results as a consequence of disulfide interchange, were dealt with in the classic studies of Ryle et al.[313] and Spackman et al.[234] An important advance was the introduction of the diagonal electrophoretic technique[242] for disulfide bridge location (Section 1.6.4). A procedure for the assignment of disulfide bridges which utilizes Ellman's reagent (Section 1.4.7) for the visualization of cystine peptides on peptide maps has been described by Glaser et al.[314]

Buchanan and Corfield[314a] have isolated cystine-containing peptides from a partial acid hydrolysate of wool, and Lindley and Cranston have discussed the difficult problem of characterizing the disulfide bonds in wool.[315]

### 1.7.4. Mass Spectrometry Applied to Amino-Acid-Sequence Determination

Given the availability of instrumentation and computing facilities, the use of mass spectrometry for the determination of the amino acid sequences of peptides offers high sensitivity ($<1$ $\mu$g of material for the actual analysis) and a tremendous reduction in demand on skilled man-hours. Considerable success has been achieved with naturally occurring peptide derivatives, but the problems that arise in routine application to peptides obtained by fragmentation of proteins are only now being overcome.[316] The subject is reviewed by Jones[316a] and Shemyakin et al.[317]

Because of the very low volatility of peptides, acylation of amino groups and esterification of carboxyl groups is necessary. Schier and Halpern form the ester by pyrolysis of the trifluoroacetylpeptide trimethylanilinium salt on the spectrometer probe.[317a] Functional side chains, notably that of arginine, may still limit volatility, and special artifices are necessary to deal with these. Permethylation of the peptide converts —CO—NH— to —CH—$NCH_3$— and eliminates the possibility of interchain hydrogen bonding between peptide links, another cause of involatility.

Winkler and Beckey,[318] using field desorption techniques, have shown that it is possible to obtain high intensities of molecular or quasi-molecular ions from peptides without recourse to derivatization.

Fragmentation of peptide derivatives takes place predominantly at the peptide link, proceeding from the C terminus in a sequential manner and producing "sequence ions." An alternative type of cleavage at the peptide link is also observed to a lesser extent, and certain amino acid side chains show characteristic fragmentation behavior. (Desulfurization of sulfur-containing peptides[318a] reduces the complexity of their mass spectra.) Isotopic labeling facilitates the recognition of sequence ions (e.g., the combined use of acetyl and trideuterioacetyl groups for N protection, which produces pairs of peaks separated by three mass units). With high-resolution instruments, such techniques are less important.

In an ingenious method for the determination of N-terminal sequences,[319] a protein is acylated with an equimolar mixture of acetic anhydride and trideuterioacetic anhydride and then digested with a suitable proteolytic enzyme. The mixture of peptides is permethylated, and the hydrophobic peptide derivatives from the N-terminal sequence are extracted into chloroform and the mass spectrum examined. Proteins containing a naturally acetylated N terminus give singlet peaks in place of the $n$, $n + 3$ doublets otherwise observed. Proteins containing N-terminal PCA also give singlet peaks, including an intense peak at $m/e$ 98 ($C_5H_8NO$). Reference has already been made (Section 1.7.1) to the use of mass spectrometry to identify amino acid derivatives produced in the Edman degradation.[278,320,320a]

The primary structures of proteins containing 300–500 amino acid residues are currently being determined. An annual publication[321] is devoted to sequence data and compilations are appearing in the literature.[321a] Nevertheless, an inexhaustible list of candidates presses for attention, and it seems that the demands of protein science as a whole for sequence determinations will only be met by the development and use of automatic and computational facilities, and, for the time being at least, an increase in the number of "occasional" sequencers. Two books[322,323] deal with all aspects of the subject, and Hartley[244] has given an inspiriting account of the important contributions that he and his colleagues have made.

## 1.8. Reagents Specific for Particular Amino Acids in Proteins

A particular reagent can generally react with more than one type of functional group in a protein. Toward fluoro-2,4-dinitrobenzene, for example, the nucleophiles thiol, amino, phenolic hydroxyl, imidazolyl, and aliphatic hydroxyl constitute a series of decreasing reactivity. For some purposes, such as the determination of N-terminal residues, concomitant reaction of other functional groups is not a serious drawback, but there are increasingly important applications for which limitation of modification to a single type of functional group is desirable. Such specifity, achieved by careful choice of reagent and conditions, enables the reactivity of individual residues of a particular amino acid to be studied (see, for example, Glick and Barnard[324]). In this way, deductions can be made about the environments of these residues and about the tertiary structure of the protein. Other applications include the labeling of specific sites with radioactive, chromophoric, or fluorescent groups and the preparation of heavy-metal derivatives for structural analysis.

Several examples of relatively specific reagents have been referred to in previous sections, and here only a few selected aspects can be touched upon. For a detailed account of the topic, and further references, the reader is referred to the reviews by Cohen,[325] who pays particular attention to the mechanistic aspect, and by Stark.[326] A convenient tabular summary of reagents and the groups with which they react is given by Vallee and Riordan.[327] A general method for studying the reactivity of individual functional groups in a protein, based on competition for trace amounts of a radioactive label, has been described by Kaplan et al.,[195] who used it to study the amino groups of porcine elastase.

### 1.8.1. Amino Groups

The $\epsilon$-amino groups of lysine residues are usually more reactive than the less nucleophilic $\alpha$-amino groups of proteins. It is therefore interesting to note that carbon disulfide and cyanate ion show some selectivity for $\alpha$-amino groups. In contrast, treatment with *O*-methylisourea converts lysine residues to homoarginine residues (Eq. 1-14) without much modification of $\alpha$-amino groups.[328] Amidination with imidic esters[329] (Eq. 1–15) is less specific for $\epsilon$-amino groups.

$$-(CH_2)_4NH_2 + CH_3OC(=NH)NH_2 \longrightarrow -(CH_2)_4NHC(=NH)NH_2 + CH_3OH \qquad (1\text{-}14)$$

*O*-Methylisourea

$$-(CH_2)_4NH_2 + CH_3OC(=NH)CH_3 \longrightarrow -(CH_2)_4NHC(=NH)CH_3 + CH_3OH \qquad (1\text{-}15)$$

Methyl acetimidate

The basic character of the side chain is retained, and many enzymes are not inactivated by these modifications. In both cases, however, the susceptibility of the lysine residues to trypsin-catalyzed hydrolysis at the neighboring peptide bond is lost. The use of methyl picolinimidate has also been proposed.[329a]

2,4,6-Trinitrobenzenesulfonic acid reacts with $\alpha$- and $\epsilon$-amino groups, but not with histidine nor tyrosine residues. In addition to being used for studying differential reactivity,[330] it is useful for determining lysine residues that have not reacted with other reagents. Reduced reactivity of certain proteins toward 2,4,6-trinitrobenzenesulfonic acid in concentrated urea solutions has been reported.[331]

Acrylonitrile, frequently used for blocking thiol groups (Section 1.4.6), reacts more slowly with $\epsilon$-amino groups and ultimately with terminal $\alpha$-amino groups[332,333] and histidine residues.[333a]

### 1.8.2. Tyrosine

Considerable selectivity for the acetylation of phenolic hydroxyl groups in preference to amino groups is shown by *N*-acetylimidazole.[334] Deacetylation can be effected with hydroxylamine. Tetranitromethane rapidly and selectively nitrates tyrosine in proteins at pH 8.[335,336] Beyer and Schenk[337] have studied the reaction between tetranitromethane and fibrous proteins. Diazonium-1H-tetrazole converts histidine and tyrosine residues in denatured proteins to the bisazo derivatives, which exhibit characteristic spectra.[338] The number of residues that has reacted can be determined by photometry at two wavelengths.

### 1.8.3. Carboxyl Groups

Esterification of carboxyl groups[339] is usually carried out by treatment with methanolic HCl, although derivatives of diazoacetic acid are sometimes suitable. There is evidence that methanolic HCl can cause some conversion of amide groups to ester groups[340] and also peptide-bond cleavage via $N \rightarrow O$ acyl shift.[341] A potentially useful reagent for the esterification of peptides and soluble proteins is triethyloxonium fluoroborate in aqueous sodium bicarbonate solution.[342]

The activation of carboxyl groups with water-soluble carbodiimides and their subsequent coupling with a nucleophile such as glycine methyl ester[343,343a] has been used for the determination of the total carboxyl content of proteins and in studies of the reactivity of individual carboxyl groups.[344] Frater[345] has shown that in reduced and alkylated proteins, the reactivity of carboxyl groups toward carbodiimides may be drastically modified, *S*-cyanoethyl derivatives being completely unreactive. The effect is attributed to the facilitation of electrostatic interactions involving carboxyl groups when the structure is no longer stabilized by disulfide bonds.

## 1.9. References

1. K. Linderstrøm-Lang, *Lane Medical Lectures*, Vol. 6, p. 58, Stanford University Press, Stanford, Calif. (1952).

1a. IUPAC-IUB Commission on Biochemical Nomenclature, *J. Mol. Biol.* **52**, 1 (1970).

2. B. Robson and R. H. Pain, *J. Mol. Biol.* **58**, 237 (1971).
2a. H. A. Scheraga, *Chem. Rev.* **71**, 195 (1971).
3. D. B. Wetlaufer and S. Ristow, *Ann. Rev. Biochem.* **42**, 135 (1973).
4. J. D. Bernal, *Discussions Faraday Soc.* **25**, 7 (1958).
5. I. M. Klotz, N. R. Langerman, and D. W. Darnall, *Ann. Rev. Biochem.* **39**, 25 (1970).
5a. C. Manwell and C. M. A. Baker, *Molecular Biology and the Origin of Species: Heterosis, Protein Polymorphism and Animal Breeding*, Sidgwick & Jackson Ltd., London (1969).
6. C. Tanford, *Advan. Protein Chem.* **23**, 121 (1968).
7. C. Tanford, *Advan. Protein Chem.* **24**, 2 (1970).
8. W. T. Astbury, *Trans. Faraday Soc.* **34**, 378 (1938).
9. S. Seifter and P. M. Gallop, in: *The Proteins* (H. Neurath, ed.), 2nd ed., Vol. 4, p. 155, Academic Press, New York (1966).
10. W. G. Crewther, R. D. B. Fraser, F. G. Lennox, and H. Lindley, *Advan. Protein Chem.* **20**, 191 (1965).
10a. J. H. Bradbury, *Advan. Protein Chem.* **27**, 1 (1973).
11. E. W. Davie and O. D. Ratnoff, in: *The Proteins* (H. Neurath, ed.), 2nd ed., Vol. 3, p. 360, Academic Press, New York (1966).
12. G. N. Ramachandran (ed.), *Treatise on Collagen, Chemistry of Collagen*, Academic Press, New York (1967).
12a. P. M. Gallop, O. O. Blumenfeld, and S. Seifter, *Ann. Rev. Biochem.* **41**, 617 (1972).
13. J. Steinhardt and S. Beychok, in: *The Proteins* (H. Neurath, ed.), 2nd ed., Vol. 2. p. 139, Academic Press, New York (1964).
14. S. Keller and R. J. Block, in: *Analytical Methods of Protein Chemistry* (P. Alexander and R. J. Block, eds.), Vol. 1, p. 2, Pergamon Press, Oxford (1960).
15. J. E. Eastoe and A. Courts, *Practical Analytical Methods for Connective Tissue Proteins*, pp. 71 ff., E. and F. N. Spon Ltd., London (1963).
16. H. A. Sober, R. W. Hartley, W. R. Carroll, and E. A. Peterson, in: *The Proteins* (H. Neurath, ed.), 2nd ed., Vol. 3, p. 1, Academic Press, New York (1965).
16a. E. A. Peterson, in: *Laboratory Techniques in Biochemistry and Molecular Biology* (T. S. Work and E. Work, eds.), Vol. 2, North-Holland Publishing Company, Amsterdam (1970).
17. L. Fischer, *An Introduction to Gel Chromatography*, North-Holland Publishing Company, Amsterdam (1969).
18. G. K. Ackers, *Advan. Protein Chem.* **24**, 343 (1970).
19. J. Porath and N. Fornstedt, *J. Chromatog.* **51**, 479 (1970).
19a. P. Cuatrecasas and C. B. Anfinsen, *Ann. Rev. Biochem.* **40**, 259 (1971).
20. L. G. Longsworth, in: *Electrophoresis* (M. Bier, ed.), p. 91, Academic Press, New York (1959).
21. H. Svensson, in: *A Laboratory Manual of Analytical Methods of Protein Chemistry, Including Polypeptides* (P. Alexander and R. J. Block, eds.), Vol. 1, p. 193, Pergamon Press, Oxford (1960).
22. O. Vesterberg and H. Svensson, *Acta Chem. Scand.* **20**, 820 (1966).
23. O. Vesterberg, *Biochim. Biophys. Acta* **257**, 11 (1972).
24. H. Haglund, *Science Tools* (*LKB Instr. J.*) **17**, 2 (1970).
24a. N. Catsimpoolas (ed.), *Ann. N.Y. Acad. Sci.* **209**, 1 (1973).
25. J. Porath and S. Hjerten, *Methods Biochem. Anal.* **9**, 193 (1962).
26. O. Smithies, *Advan. Protein Chem.* **14**, 65 (1959).

27. L. Ornstein, *Ann. N.Y. Acad. Sci.* **121**, 321 (1964).
28. A. Brownstone, *Anal. Biochem.* **27**, 25 (1969).
29. M. B. Mann and P. C. Huang, *Anal. Biochem.* **32**, 138 (1969).
30. K. Weber and M. Osborn, *J. Biol. Chem.* **244**, 4406 (1969).
30a. K. Weber, J. R. Pringle, and M. Osborn, *Methods Enzymol.* **26**, 3 (1972).
31. A. K. Dunker and R. R. Rueckert, *J. Biol. Chem.* **244**, 5074 (1969).
32. P. D. Jeffrey, *Aust. J. Biol. Sci.* **23**, 809 (1970).
33. C. J. O. R. Morris and P. Morris, *Biochem. J.* **224**, 517 (1971).
34. W. W. Fish, J. A. Reynolds, and C. Tanford, *J. Biol. Chem.* **245**, 5166 (1970).
34a. K. G. Mann and W. W. Fish, *Methods Enzymol.* **26**, 28 (1972).
35. A. Pusztai and W. B. Watt, *Biochim. Biophys. Acta* **214**, 463 (1970).
36. P. Andrews, *Methods Biochem. Anal.* **18**, 1 (1970).
37. R. F. Dicamelli, P. D. Holohan, S. F. Basinger, and J. Lebowitz, *Anal. Biochem.* **36**, 470 (1970).
38. J. W. Williams (ed.), *Ultracentrifugal Analysis*, Academic Press, New York (1963).
38a. T. J. Bowen, *An Introduction to Ultracentrifugation* John Wiley & Sons, Inc., New York (1970).
39. C. Tanford, *Physical Chemistry of Macromolecules*, John Wiley & Sons, Inc., New York (1961).
39a. J. R. Cann, *Interacting Macromolecules*, Academic Press, New York (1970).
40. K. S. McCarty, D. Stafford, and O. Brown, *Anal. Biochem.* **24**, 314 (1968).
41. J. B. Ifft and J. Vinograd, *J. Phys. Chem.* **70**, 2814 (1966).
41a. C. H. W. Hirs and S. G. Timasheff (ed.), *Methods Enzymol.* **27** (1973).
42. P. Alexander and R. J. Block (eds.), *A Laboratory Manual of Analytical Methods of Protein Chemistry*, Vol. 1, Pergamon Press, Oxford (1960).
43. P. Alexander and R. J. Block (eds.), *A Laboratory Manual of Analytical Methods of Protein Chemistry*, Vol. 3, Pergamon Press, Oxford (1961).
44. S. J. Leach (ed.), *Physical Principles and Techniques of Protein Chemistry*, Part A, Academic Press, New York (1969).
45. J. T. Edsall, in: *The Proteins* (H. Neurath and K. Bailey, eds.), 1st ed., Vol. 1B, p. 549, Academic Press, New York (1953).
46. I. Tinoco and C. R. Cantor, *Methods Biochem. Anal.* **18**, 81 (1970).
47. P. Urnes and P. Doty, *Advan. Protein Chem.* **16**, 401 (1961).
48. Y. T. Yang, in: *Conformations of Biopolymers*, Vol. 1, p. 173, Academic Press, New York (1967).
49. D. E. Goldsack, *Biopolymers* **7**, 299 (1969).
50. P. M. Bayley, in: *Amino Acids, Peptides and Proteins* (R. C. Sheppard, ed.), Vol. 5, p. 237, The Chemical Society, London (1974).
51. G. C. K. Roberts and O. Jardetzky, *Advan. Protein Chem.* **24**, 448 (1970).
52. B. Sheard and E. M. Bradbury, in: *Progress in Biophysics and Molecular Biology* (J. A. V. Butler and D. Nobel, eds.), p. 187, Pergamon Press, Oxford (1970).
52a. F. R. N. Gurd and P. Keim, *Methods Enzymol.* **27**, 836 (1973).
53. C. C. F. Blake, *Advan. Protein Chem.* **23**, 59 (1968).
54. F. H. C. Crick and J. C. Kendrew, *Advan. Protein Chem.* **12**, 134 (1957).
55. K. C. Holmes and D. M. Blow, *Methods Biochem. Anal.* **13**, 113 (1965).
56. R. E. Dickerson, in: *The Proteins* (H. Neurath, ed.), 2nd ed., Vol. 2, p. 603, Academic Press, New York (1964).
57. D. M. Blow and T. A. Steitz, *Ann. Rev. Biochem.* **39**, 63 (1970).

57a. T. L. Blundell, in: *Amino Acids, Peptides and Proteins* (R. C. Sheppard, ed.), Vol. 5, p. 145, The Chemical Society, London (1974).
58. R. D. B. Fraser and T. P. Macrae, *Conformation in Fibrous Proteins*, Academic Press, New York (1973).
58a. S. Moore and W. H. Stein, *Methods Enzymol.* **6**, 819 (1963).
59. F. Downs and W. Pigman, *Intern. J. Protein Res.* **1**, 181 (1969).
60. J. C. Fletcher and A. Robson, *Biochem. J.* **87**, 553 (1963).
61. A. S. Inglis and T. Y. Liu, *J. Biol. Chem.* **245**, 112 (1970).
62. F. Sanger and E. O. P. Thompson, *Biochim. Biophys. Acta* **71**, 468 (1963).
63. D. Roach and C. W. Gehrke, *J. Chromatog.* **52**, 393 (1970).
64. R. L. Hill and W. R. Schmidt, *J. Biol. Chem.* **237**, 389 (1962).
64a. H. P. J. Bennett, D. F. Elliot, B. E. Evans, P. J. Lowry, and C. McMartin, *Biochem. J.* **129**, 695 (1972).
65. B. Milligan, L. A. Holt, and J. B. Caldwell, *Appl. Polymer Symp.* **18**, 113 (1971).
66. M. Cole, J. C. Fletcher, K. L. Gardner, and M. C. Corfield, *Appl. Polymer Symp.* **18**, 147 (1971).
67. H. B. Bensusan, *Biochemistry* **8**, 4716 (1969).
67a. H. B. Bensusan, S. N. Dixit, and S. D. McKnight, *Biochim. Biophys. Acta* **251**, 100 (1971).
68. E. A. Noltmann, T. A. Mahawald, and S. A. Kuby, *J. Biol. Chem.* **237**, 1146 (1962).
68a. T. E. Hugli and S. Moore, *J. Biol. Chem.* **247**, 2828 (1972).
69. D. J. McCaldin, *Chem. Rev.* **60**, 39 (1960).
70. M. Friedman and C. W. Sigel, *Biochemistry* **5**, 478 (1966).
71. S. Moore and W. H. Stein, *J. Biol. Chem.* **176**, 367 (1948).
72. S. Moore and W. H. Stein, *J. Biol. Chem.* **211**, 907 (1954).
73. D. H. Spackman, W. H. Stein, and S. Moore, *Anal. Chem.* **30**, 1190 (1958).
74. S. Moore, D. H. Spackman, and W. H. Stein, *Anal. Chem.* **30**, 1185 (1958).
75. K. A. Piez and L. Morris, *Anal. Biochem.* **1**, 87 (1960).
76. P. B. Hamilton, *Anal. Chem.* **35**, 2055 (1963).
76a. B. C. Starcher, L. Y. Wenger, and L. D. Johnson, *J. Chromatog.* **54**, 425 (1971).
77. P. B. Hamilton, *Advan. Chromatog.* **2**, 3 (1966).
78. D. H. Spackman, *Methods Enzymol.* **11**, 3 (1967).
78a. J. W. Eveleigh and G. D. Winter, in: *Protein Sequence Determination* (S. B. Needleman, ed.), p. 91, Chapman & Hall Ltd., London; Springer-Verlag, Berlin (1970).
78b. A. J. Thomas, in: *Automation, Mechanization and Data Handling in Microbiology* (A. Baillie and R. J. Gilbert, ed.), p. 107, Academic Press, New York (1970).
79. S. Stein, P. Bøehlen, J. Stone, W. Dairman, and S. Udenfriend, *Arch. Biochem. Biophys.* **155**, 203 (1973).
80. B. Klein, J. E. Sheehan, and E. Grunberg, *Clin. Chem.* **20**, 272 (1974).
81. M. Weigele, S. L. de Bernardo and W. Leimgruber, *Biochem. Biophys. Res. Commun.* **50**, 352 (1973).
82. M. L. Johnson, E. A. Khairallah, and D. A. Yphantis, *Anal. Biochem.* **50**, 364 (1972).
82a. D. M. Kirschenbaum, *Anal. Biochem.* **44**, 159 (1971); **49**, 248 (1972); **52**, 234 (1973); **53**, 223 (1973).
83. H. Michl, *J. Chromatog.* **1**, 93 (1958).
84. D. Gross, *J. Chromatog.* **5**, 194 (1961).
85. W. J. Dreyer and E. Bynum, *Methods Enzymol.* **11**, 32 (1967).

86. S. Blackburn, in: *A Laboratory Manual of Analytical Methods of Protein Chemistry* (P. Alexander and H. P. Lundren, eds.), Vol. 4, p. 77, Pergamon Press, Oxford (1966).
87. M. C. Corfield and E. C. Simpson, *J. Chromatog.* **17**, 420 (1965).
88. R. Consden, A. H. Gordon, and A. J. P. Martin, *Biochem. J.* **38**, 224 (1944).
89. I. Smith (ed.), *Chromatographic and Electrophoretic Techniques*, 3rd ed., Vol. 1, William Heinemann Ltd., London (1969).
90. I. M. Hais and K. Macek, *Paper Chromatography: A Comprehensive Treatise*, Academic Press, New York (1963).
91. I. M. Hais and K. Macek, *Bibliography of Paper Chromatography, 1944–1956*, Academic Press, New York (1960).
92. I. M. Hais and K. Macek, *Bibliography of Paper Chromatography, 1957–1960*, Academic Press, New York (1962).
93. K. Macek, *Bibliography of Paper and Thin-Layer Chromatography, 1961–1965*, Elsevier Publishing Company, Amsterdam (1968).
94. A. L. Levy and D. Chung, *Anal. Chem.* **25**, 396 (1953).
95. G. Pataki, *Dünnschichtchromatographie in der Aminosaure- und Peptidchemie*, Walter de Gruyter & Co., Berlin (1965).
96. G. Pataki, *Techniques of Thin-Layer Chromatography in Amino Acid and Peptide Chemistry*, Ann Arbor–Humphrey Science Publishers, Ann Arbor, Mich. (1969).
97. M. Brenner and A. Niederwieser, *Methods Enzymol.* **11**, 39 (1967).
97a. A. Niederwieser, *Methods Enzymol.* **25**, 60 (1972).
98. J. G. Heathcote and C. Haworth, *Biochem. J.* **114**, 667 (1969).
98a. I. E. Bush, *J. Chromatog.* **29**, 157 (1967).
99. V. Pollak and A. A. Boulton, *J. Chromatog.* **45**, 200 (1969).
99a. S. Samuels, *J. Chromatog.* **96**, 1 (1974).
100. S. R. Snodgrass and L. L. Iversen, *Nature New Biol.* **241**, 154 (1973).
100a. J. P. Brown and R. N. Perham, *European J. Biochem.* **39**, 69 (1973).
101. B. Weinstein, *Methods Biochem. Anal.* **14**, 203 (1966).
102. K. Blau, in: *Biomedical Applications of Gas Chromatography* (H. A. Szymanski, ed.), Vol. 2, p. 1, Plenum Press, New York (1968).
103. A. Darbre and K. Blau, *J. Chromatog.* **29**, 49 (1967).
104. D. Roach and C. W. Gehrke, *J. Chromatog.* **43**, 303 (1969).
105. D. Roach, C. W. Gehrke, and R. W. Zumwalt, *J. Chromatog.* **43**, 311 (1969).
105a. C. W. Gehrke, K. Koo, and R. W. Zumwalt, *J. Chromatog.* **57**, 209 (1971).
106. D. J. Casagrande, *J. Chromatog.* **49**, 537 (1970).
106a. J. Jönsson, J. Eyem, and J. Sjoquist, *Anal. Biochem.* **51**, 204 (1973).
107. D. Roach and C. W. Gehrke, *J. Chromatog.* **44**, 269 (1969).
108. A. Islam and A. Darbre, *J. Chromatog.* **43**, 11 (1969).
109. C. W. Gehrke, H. Nakamoto, and R. W. Zumwalt, *J. Chromatog.* **45**, 24 (1969).
110. C. W. Gehrke and K. Leimer, *J. Chromatog.* **53**, 201 (1970).
111. F. Shahrokhi and C. W. Gehrke, *J. Chromatog.* **36**, 31 (1968).
112. E. D. Smith and K. L. Shewbart, *J. Chromatog. Sci.* **7**, 704 (1969).
113. W. Bandlow and A. Nordwig, *J. Chromatog.* **39**, 326 (1969).
114. R. C. Page, D. Jones, and R. Hansen, *Anal. Biochem.* **37**, 293 (1970).
115. A. D. Mitchell and I. E. P. Taylor, *Analyst (Lond.)* **95**, 1003 (1970).
116. J. R. Spies, *Anal. Chem.* **39**, 1412 (1967).
117. T. F. Spande and B. Witkop, *Methods Enzymol.* **11**, 498 (1967).
118. E. Scoffone and A. Fontana, in: *Protein Sequence Determination* (S. B. Needleman, ed.), p. 185, Chapman & Hall Ltd., London; Springer-Verlag, Berlin (1970).

118a. T-Y. Liu, *Methods Enzymol.* **25**, 44 (1972).
119. C. H. W. Hirs, *Methods Enzymol.* **11**, 59 (1967).
120. R. L. Spencer and F. Wold, *Anal. Biochem.* **32**, 185 (1969).
121. J. C. Fletcher, A. Robson, and J. Todd, *Biochem. J.* **87**, 560 (1963).
122. M. J. Horn, D. B. Jones, and S. J. Ringel, *J. Biol. Chem.* **138**, 141 (1941).
123. A. Patchornik and M. Sokolovsky, *J. Amer. Chem. Soc.* **86**, 1860 (1964).
124. Z. Bohak, *J. Biol. Chem.* **239**, 2878 (1964).
125. K. Ziegler, *J. Biol. Chem.* **239**, PC 2713 (1964).
126. F. P. Chinard, *J. Biol. Chem.* **199**, 91 (1952).
127. A. Robson, M. J. Williams, and J. M. Woodhouse, *J. Chromatog.* **31**, 284 (1967).
128. M. Tasdhomme, *Bull. Inst. Textile Fr.* **24**, 237 (1970).
129. L. M. Dowling and W. G. Crewther, *Anal. Biochem.* **8**, 244 (1964).
130. P. Miró and J. J. García-Domínguez, *Melliand Textilber.* **48**, 558 (1967).
131. J. Derminot, C. Belin, and M. Tasdhomme, *Bull. Inst. Textile Fr.* **22**, 785 (1968).
132. K. Ziegler, *Forschungsber. Landes Nordrhein-Westfalen* **1275** (1963).
133. M. Bauters, L. Lefebvre, and M. van Overbèke, *Bull. Inst. Textile Fr.* **21**, 425 (1967).
134. P. E. Wilcox, *Methods Enzymol.* **11**, 63 (1967).
135. S. J. Leach and E. M. J. Parkhill, *Proc. Intern. Wool Textile Res. Conf., Australia* C, 101 (1955).
136. R. Cecil and J. R. McPhee, *Advan. Protein Chem.* **14**, 256 (1959).
137. R. Cecil, in: *The Proteins* (H. Neurath, ed.), 2nd ed., Vol. 1, p. 380, Academic Press, New York (1963).
137a. M. Friedman, *The Chemistry and Biochemistry of the Sulfhydryl Group in Amino Acids, Peptides and Proteins*, Pergamon Press, Oxford (1973).
138. W. E. Savige and J. A. Maclaren, in: *The Chemistry of Organic Sulphur Compounds* (N. Kharasch, ed.), Vol. 2, p. 367, Pergamon Press, Oxford (1966).
139. C. H. W. Hirs, *J. Biol. Chem.* **219**, 613 (1956).
140. D. R. Goddard and L. Michaelis, *J. Biol. Chem.* **106**, 605 (1934).
141. D. R. Goddard and L. Michaelis, *J. Biol. Chem.* **112**, 361 (1935–1936).
142. E. O. P. Thompson and I. J. O'Donnell, *Aust. J. Biol. Sci.* **15**, 757 (1962).
143. A. M. Crestfield, S. Moore, and W. H. Stein, *J. Biol. Chem.* **238**, 622 (1963).
144. A. R. Williamson and B. A. Askonas, *Biochem. J.* **107**, 823 (1968).
145. W. W. Clelland, *Biochemistry* **3**, 480 (1964).
146. H. D. Weigmann, *J. Polymer Sci. A-1* **8**, 2237 (1968).
147. R. E. J. Mitchel, I. M. Chaiken, and E. L. Smith, *J. Biol. Chem.* **245**, 3485 (1970).
148. F. H. White, Jr., *Methods Enzymol.* **11**, 481 (1967).
149. B. J. Sweetman and J. A. Maclaren, *Aust. J. Chem.* **19**, 2347 (1966).
150. J. A. Maclaren and B. J. Sweetman, *Aust. J. Chem.* **19**, 2355 (1966).
151. S. J. Leach, A. Meschers, and O. A. Swanepoel, *Biochemistry* **4**, 23 (1965).
152. W. R. Middlebrook and H. Phillips, *Biochem. J.* **36**, 428 (1942).
153. R. Cecil and J. R. McPhee, *Biochem. J.* **60**, 496 (1955).
154. H. Lindley, in: *Sulphur in Proteins* (R. Benesch and R. E. Benesch, eds.), p. 33, Academic Press, New York (1959).
155. A. Massaglia, F. Pennisi, U. Rosas, S. Ronca-Testoni, and C. A. Rossi, *Biochem. J.* **108**, 247 (1968).
156. H. T. Clarke, *J. Biol. Chem.* **97**, 235 (1932).
157. L. J. Wolfram and D. L. Underwood, *Textile Res. J.* **36**, 947 (1966).
158. C. Wood, The Action of Sulphites on the Disulphide Bonds of Wool and Model Compounds, Ph.D. thesis, University of Leeds, 1966.
159. J. M. Swan, *Nature* (*Lond.*) **180**, 643 (1957).

160. W. W-C. Chan, *Biochemistry* **7**, 4247 (1968).
161. J. F. Pechère, G. H. Dixon, R. H. Maybury, and H. Neurath, *J. Biol. Chem.* **13**, 646 (1958).
162. J. L. Bailey and R. D. Cole, *J. Biol. Chem.* **234**, 1733 (1959).
163. H. Zahn, E. Drechsel, and W. Puls, *Hoppe-Seylers Z. Physiol. Chem.* **349**, 385 (1968).
164. N. Catsimpoolas and J. L. Wood, *J. Biol. Chem.* **241**, 1790 (1966).
165. G. R. Jacobson, M. H. Schaffer, G. R. Stark, and T. C. Vanaman, *J. Biol. Chem.* **248**, 6583 (1973).
166. W. H. Ward, *Textile Res. J.* **37**, 1085 (1967).
167. C. H. W. Hirs, *Methods Enzymol.* **11**, 199 (1967).
168. F. R. N. Gurd, *Methods Enzymol.* **25**, 424 (1972).
169. R. D. Cole, *Methods Enzymol.* **11**, 315 (1967).
169a. A. F. Bradbury and D. G. Smyth, *Biochem. J.* **131**, 637 (1973).
170. T. S. Seibles and L. Weil, *Methods Enzymol.* **11**, 204 (1967).
171. J. F. Cavins and M. Friedman, *J. Biol. Chem.* **243**, 3357 (1968).
172. J. F. Riordan and B. L. Vallee, *Methods Enzymol.* **11**, 541 (1967).
173. A. G. Weeds and B. S. Hartley, *Biochem. J.* **107**, 531 (1968).
174. S. J. Leach, in: *Analytical Methods of Protein Chemistry* (P. Alexander and H. P. Lundgren, eds.), Vol. 4, p. 3, Pergamon Press, Oxford (1966).
175. S. J. Leach, A. Meschers, and P. H. Springell, *Anal. Biochem.* **15**, 18 (1966).
176. D. Cavallini, M. T. Graziani, and S. Dupre, *Nature (Lond.)* **212**, 294 (1966).
177. W. L. Zahler and W. W. Clelland, *J. Biol. Chem.* **243**, 716 (1968).
177a. A. S. F. A. Habeeb, *Methods Enzymol.* **25**, 457 (1972).
178. D. R. Grassetti and J. F. Murray, *Arch. Biochem. Biophys.* **119**, 41 (1967).
179. L. H. Krull, D. E. Gibbs, and M. Friedman, *Anal. Biochem.* **40**, 80 (1971).
180. M. Friedman, L. H. Krull, and J. F. Cavins, *J. Biol. Chem.* **245**, 3868 (1970).
180a. M. Friedman and A. T. Noma, *Textile Res. J.* **40**, 1073 (1970).
180b. T-Y. Liu and A. S. Inglis, *Methods Enzymol.* **25**, 55 (1972).
181. A. Z. Budzynski and M. Stahl, *Biochim. Biophys. Acta* **175**, 282 (1969).
182. J. M. Gillespie and P. H. Springell, *Biochem. J.* **79**, 280 (1961).
183. P. H. Springell, J. M. Gillespie, A. S. Inglis, and J. A. Maclaren, *Biochem. J.* **91**, 17 (1964).
184. E. Chiancone, D. L. Currell, P. Vecchini, E. Antonini, and J. Wyman, *J. Biol. Chem.* **245**, 4105 (1970).
185. R. L. Hill, *Advan. Protein Chem.* **20**, 37 (1965).
186. W. G. Crewther and B. S. Harrap, *J. Biol. Chem.* **242**, 4310 (1967).
187. M. E. Hilburn, P. T. Speakman, and R. E. Yarwood, *Biochim. Biophys. Acta* **214**, 245 (1970).
188. E. Mihalyi and W. F. Harrington, *Biochim. Biophys. Acta* **36**, 447 (1959).
189. J. C. Bennett, *Methods Enzymol.* **11**, 211 (1967).
190. C. F. Jacobsen, J. Léonis, K. Linderstrøm-Lang, and M. Ottesen, *Methods Biochem. Anal.* **4**, 359 (1965).
191. B. Foltmann and B. S. Hartley, *Biochem. J.* **104**, 1064 (1967).
192. R. P. Ambler and R. J. Meadway, *Biochem. J.* **108**, 893 (1968).
193. M. A. Naughton and F. Sanger, *Biochem. J.* **78**, 156 (1961).
194. D. G. Smyth, *Methods Enzymol.* **11**, 214 (1967).
195. H. Kaplan, K. J. Stephenson, and B. S. Hartley, *Biochem. J.* **124**, 289 (1971).
196. A. D. Gounaris and G. E. Perlmann, *J. Biol. Chem.* **242**, 2739 (1967).

197. R. F. Goldberger, *Methods Enzymol.* **11**, 317 (1967).
198. M. Riley and R. N. Perham, *Biochem. J.* **118**, 733 (1970).
199. A. Marzotto, P. Pajetta, L. Galzigna, and E. Scoffone, *Biochim. Biophys. Acta* **154**, 450 (1968).
200. G. Braunitzer, K. Beyreuther, H. Fujiki, and B. Shrank, *Hoppe-Seylers Z. Physiol. Chem.* **349**, 265 (1968).
201. H. A. Itano and A. J. Gottlieb, *Biochem. Biophys. Res. Commun.* **12**, 405 (1963).
202. K. Toi, E. Bynum, E. Norris, and H. A. Itano, *J. Biol. Chem.* **242**, 1036 (1967).
202a. A. S. F. A. Habeeb and J. C. Bennett, *Biochim. Biophys. Acta* **251**, 181 (1971).
203. T. P. King, *Biochemistry* **5**, 3454 (1966).
204. A. L. Grossberg and D. Pressman, *Biochemistry* **7**, 272 (1968).
205. R. C. Davies and A. Neuberger, *Biochim. Biophys. Acta* **178**, 306 (1969).
206. J. A. Yankeelov, *Biochemistry* **9**, 2433 (1970).
207. K. Nakaya, H. Horinishi and K. Shibata, *J. Biochem.* (*Tokyo*) **61**, 345 (1967).
208. K. Takahashi, *J. Biol. Chem.* **243**, 6171 (1968).
208a. L. Patthy and E. L. Smith, *J. Biol. Chem.* **250**, 557 (1975).
209. B. Witkop, *Advan. Protein Chem.* **16**, 221 (1961).
210. B. Witkop, *Science* (*Wash., D.C.*) **162**, 318 (1968).
211. T. F. Spande, B. Witkop, Y. Degani, and A. Patchornik, *Advan. Protein Chem.* **24**, 98 (1970).
212. E. Gross, *Methods Enzymol.* **11**, 238 (1967).
213. A. S. Inglis and P. Edman, *Anal. Biochem.* **37**, 73 (1970).
214. R. S. Adelstein and W. M. Kuehl, *Biochemistry* **9**, 1355 (1970).
215. E. O. P. Thompson and I. J. O'Donnell, *Aust. J. Biol. Sci.* **20**, 1001 (1967).
216. L. K. Ramachandran and B. Witkop, *Methods Enzymol.* **11**, 283 (1967).
217. S. C. Rall and R. D. Cole, *J. Amer. Chem. Soc.* **92**, 1800 (1970).
218. K. Iwai and T. Ando, *Methods Enzymol.* **11**, 263 (1967).
219. J. Schultz, *Methods Enzymol.* **11**, 255 (1967).
220. C. M. Tsung and H. Fraenkel-Conrat, *Biochemistry* **4**, 793 (1965).
221. S. J. Leach, G. E. Rogers, and B. K. Filshie, *Arch. Biochem. Biophys.* **105**, 270 (1964).
222. M. A. Naughton, F. Sanger, B. S. Hartley, and D. C. Shaw, *Biochem. J.* **77**, 149 (1960).
223. B. Alexyev, V. Holeyšovsky and F. Sŏrm, *Collection Czech. Chem. Commun.* **29**, 1296 (1964).
224. W. A. Schroeder, *Methods Enzymol.* **11**, 351 (1967).
225. W. A. Schroeder, *Methods Enzymol.* **11**, 361 (1967).
226. A. B. Edmundson, *Methods Enzymol.* **11**, 369 (1967).
227. C. J. O. R. Morris and P. Morris, *Separation Methods in Biochemistry*, p. 322, Sir Isaac Pitman & Sons Ltd., London (1964).
228. T. Haylett and L. S. Swart, *Textile Res. J.* **39**, 917 (1969).
229. M. Elzinga, *Biochemistry* **9**, 1365 (1970).
230. E. Slobodian, G. Mechanic, and M. Levy, *Science* (*Wash., D.C.*) **135**, 441 (1962).
231. C. H. W. Hirs, *Methods Enzymol.* **11**, 325 (1967).
232. R. L. Hill and R. Delaney, *Methods Enzymol.* **11**, 339 (1967).
233. R. T. Jones, *Methods Biochem. Anal.* **18**, 205 (1970).
234. D. H. Spackman, S. Moore, and W. H. Stein, *J. Biol. Chem.* **235**, 648 (1960).
235. K. A. Walsh, R. M. McDonald, and R. A. Bradshaw, *Anal. Biochem.* **35**, 193 (1970).
236. F. H. White, Jr., B. Hank, K. Kon, and P. Riesz, *Anal. Biochem.* **30**, 295 (1969).

237. C. H. W. Hirs, *Methods Enzymol.* **11**, 386 (1967).
238. P. Flodin, *J. Chromatog.* **5**, 103 (1961).
239. R. E. Offord, *Nature* (*Lond.*), **211**, 591 (1966).
240. J. C. Bennett, *Methods Enzymol.* **11**, 330 (1967).
241. C. W. Easley, B. J. M. Zegers, and M. de Vijlder, *Biochim. Biophys. Acta* **175**, 211 (1969).
242. J. R. Brown and B. S. Hartley, *Biochem. J.* **101**, 214 (1966).
243. R. N. Perham, in: *Amino-acids, Peptides and Proteins* (G. T. Young, ed.), Vol. 1, p. 61, The Chemical Society, London (1969).
244. B. S. Hartley, *Biochem. J.* **119**, 805 (1970).
244a. W. H. Cruickshank, T. M. Radhakrishnan, and H. Kaplan, *Can. J. Biochem.* **49**, 1225 (1971).
244b. J. Rosmus and Z. Deyl, *Chromatog. Rev.* **13**, 163 (1971); *J. Chromatog.* **70**, 221 (1972).
245. F. Sanger, *Biochem. J.* **39**, 507 (1945).
246. H. Fraenkel-Conrat, J. I. Harris, and A. L. Levy, *Methods Biochem. Anal.* **2**, 383 (1955).
247. R. R. Porter, *Methods Enzymol.* **4**, 221 (1957).
248. G. Biserte, J. W. Holleman, J. Holleman-Dehove, and P. Sautière, *Chromatog. Rev.* **2**, 60 (1959).
249. R. L. Munier, *Z. Anal. Chem.* **236**, 358 (1968).
250. L. Kesner, E. Muntwyler, G. E. Griffin, and P. Quaranta, *Methods Enzymol.* **11**, 94 (1967).
251. H. Beyer and U. Schenk, *J. Chromatog.* **39**, 482 (1969).
252. H. Beyer and U. Schenk, *J. Chromatog.* **39**, 491 (1969).
253. A. Signor, L. Biondi, A. M. Tamburro, and E. Bordignon, *European J. Biochem.* **7**, 328 (1969).
254. E. Celon, L. Biondi, and E. Bordignon, *J. Chromatog.* **35**, 47 (1968).
254a. P. Hagel, J. J. T. Gerding, W. Fieggen, and H. Blomendal, *Biochim. Biophys. Acta* **243**, 366 (1971).
255. G. R. Stark, *Methods Enzymol.* **25**, 103 (1972).
256. W. R. Gray, *Methods Enzymol.* **25**, 121 (1972).
257. C. Gross and B. Labouesse, *European J. Biochem.* **7**, 463 (1969).
258. K. R. Woods and K. T. Wang, *Biochim. Biophys. Acta* **133**, 369 (1967).
259. V. V. Nesterov, B. G. Belenky, and L. G. Senyutenkova, *Biokhimiya* **34**, 824 (1969).
260. V. A. Spivak, V. M. Orlov, V. V. Shcherbukhin, and Ya. M. Varshavsky, *Anal. Biochem.* **35**, 227 (1970).
261. J. P. Zanetta, G. Vincendon, P. Mandel, and G. Gombos, *J. Chromatog.* **51**, 441 (1970).
262. C. Rapoport, M-F. Glatron, and M-M. Lecadet, *Compt. Rend.* **265**, 639 (1967).
263. N. Seiler, *Methods Biochem. Anal.* **18**, 259 (1970).
264. P. Edman, *Acta Chem. Scand.* **4**, 277, 283 (1950).
264a. H. D. Niall, J. W. Jacobs, I. Rietschoten, and G. W. Tregear, *FEBS Letters* **41**, 62 (1974).
264b. A. S. Inglis, P. W. Nicholls, and L. G. Sparrow, *J. Chromatog.* **90**, 362 (1974).
265. P. Edman, in: *Protein Sequence Determination* (S. B. Needleman, ed.), p. 211, Chapman & Hall Ltd., London: Springer-Verlag, Berlin (1970).
266. P. Edman and G. Begg, *European J. Biochem.* **1**, 80 (1967).
266a. H. D. Niall, *Methods Enzymol.* **27**, 942 (1973).
267. W. A. Schroeder, *Methods Enzymol.* **11**, 445 (1967).

267a. W. A. Schroeder, *Methods Enzymol.* **25**, 298 (1972).
268. J. M. Boigne, N. Boigne, and J. Rosa, *J. Chromatog.* **47**, 238 (1970).
269. W. Konigsberg, *Methods Enzymol.* **11**, 461 (1967).
270. W. R. Gray, *Methods Enzymol.* **11**, 469 (1967).
271. W. R. Gray and J. F. Smith, *Anal. Biochem.* **33**, 36 (1970).
272. H. Maeda, N. Ishida, H. Kawauchi, and K. Tuzimura, *J. Biochem.* (*Tokyo*) **65**, 777 (1969).
273. Z. Deyl, *J. Chromatog.* **48**, 231 (1970).
274. J. J. Pisano and T. J. Bronzert, *J. Biol. Chem.* **244**, 5597 (1969).
275. M. R. Guerin and W. D. Shults, *J. Chromatog. Sci.* **7**, 704 (1969).
276. M. Waterfield and E. Haber, *Biochemistry* **9**, 832 (1970).
277. D. E. Vance and D. S. Feingold, *Anal. Biochem.* **36**, 30 (1970).
277a. D. E. Vance and D. S. Feingold, *Nature* (*Lond.*) **229**, 121 (1971).
277b. M. D. Waterfield, C. Corbett, and E. Haber, *Anal. Biochem.* **38**, 475 (1970).
278. T. Fairwell, W. T. Barnes, F. F. Richards, and R. E. Lovins, *Biochemistry* **9**, 2260 (1970).
278a. C. L. Zimmerman, J. J. Pisano, and E. Appella, *Biochem. Biophys. Res. Commun.* **55**, 1220 (1973).
279. L. M. Dowling and G. R. Stark, *Biochemistry* **8**, 4728 (1969).
280. R. A. Laursen, *J. Amer. Chem. Soc.* **88**, 5344 (1966).
281. A. Schellenberger, H. Jeschkeit, R. Henkel, and H. Lehmann, *Z. Chem.* **7**, 191 (1967).
282. R. A. Laursen, *European J. Biochem.* **20**, 89 (1971).
283. R. A. Laursen, *Methods Enzymol.* **25**, 344 (1972).
284. A. Previero and J.-F. Pechère, *Biochem. Biophys. Res. Commun.* **40**, 549 (1970).
285. A. Light, *Methods Enzymol.* **25**, 253 (1972).
286. K. Hofmann, F. M. Finn, M. Limetti, J. Montibeller, and G. Zanetti, *J. Amer. Chem. Soc.* **88**, 3633 (1966).
286a. H. Lindley, *Biochem. J.* **126**, 683 (1972).
287. K. Narita, *Biochem. Biophys. Acta* **28**, 184 (1958).
288. T. Haylett and L. S. Swart, *Textile Res. J.* **39**, 917 (1969).
289. I. J. O'Donnell, *Aust. J. Biol. Sci.* **22**, 471 (1969).
290. I. J. O'Donnell, E. O. P. Thompson, and A. S. Inglis, *Aust. J. Biol. Sci.* **15**, 732 (1962).
291. G. R. Shepherd and B. J. Noland, *Anal. Biochem.* **26**, 325 (1968).
292. L. D. Stegink, *Anal. Biochem.* **20**, 502 (1967).
292a. S. C. Kuo and E. S. Younnathan, *Anal. Biochem.* **55**, 1 (1973).
293. H. Fujiki, G. Braunitzer, and V. Rudloff, *Hoppe-Seylers Z. Physiol. Chem.* **351**, 901 (1970).
294. B. Blombäck, M. Blombäck, P. Edman, and B. Hessel, *Biochim. Biophys. Acta* **115**, 371 (1966).
295. I. J. O'Donnell, *Aust. J. Biol. Sci.* **21**, 1327 (1968).
296. B. Blombäck, *Methods Enzymol.* **11**, 398 (1967).
297. R. F. Doolittle and R. W. Armentrout, *Biochemistry* **7**, 516 (1968).
298. S. Takahashi and L. A. Cohen, *Biochemistry* **8**, 864 (1969).
299. S. Akabori, K. Ohno, and K. Narita, *Bull. Chem. Soc. Japan* **25**, 214 (1952).
300. S. Akabori, K. Ohno, T. Ikenaka, H. Haruna, A. Tsugita, K. Sugae, and T. Matsushima, *Bull. Chem. Soc. Japan* **29**, 507 (1956).
301. Y. Kawanishi, K. Iwai, and T. Ando, *J. Biochem.* (*Tokyo*) **56**, 314 (1964).
302. P. Nedkov and N. Genov, *Biochim. Biophys. Acta* **126**, 544 (1966).

303. B. Mesrob and V. Holeyšovsky, *Collection Czech. Chem. Commun.* **32**, 1976 (1967).
304. H. Fraenkel-Conrat and C. M. Tsung, *Methods Enzymol.* **11**, 155 (1967).
305. W. A. Schroeder, *Methods Enzymol.* **25**, 138 (1972).
306. D. M. P. Phillips, *J. Chromatog.* **37**, 132 (1968).
307. J. H. Bradbury, *Biochem. J.* **68**, 482 (1958).
308. H. Matsuo, Y. Fujimoto, and T. Tatsuno, *Biochem. Biophys. Res. Commun.* **22**, 69 (1966).
309. G. N. Holcomb, S. A. James, and D. N. Ward, *Biochemistry* **7**, 1291 (1968).
310. G. R. Stark, *Biochemistry* **7**, 1796 (1968).
310a. G. R. Stark, *Methods Enzymol.* **25**, 369 (1972).
311. L. D. Cromwell and G. R. Stark, *Biochemistry* **8**, 4735 (1970).
311a. S. Yamashita, *Biochim. Biophys. Acta* **229**, 301 (1971).
312. R. P. Ambler, *Methods Enzymol.* **25**, 143 (1972).
312a. R. Hayashi, S. Moore, and W. H. Stein, *J. Biol. Chem.* **248**, 2296 (1973).
313. A. P. Ryle, F. Sanger, L. F. Smith, and R. Kitai, *Biochem. J.* **60**, 541 (1955).
314. C. B. Glaser, H. Maeda, and J. Meienhofer, *J. Chromatog.* **50**, 151 (1970).
314a. J. H. Buchanan and M. C. Corfield, *Appl. Polymer Symp.* **18**, 101 (1971).
315. H. Lindley and R. W. Cranston, *Biochem. J.* **139**, 515 (1974).
316. H. R. Morris, D. H. Williams, G. G. Midwinter, and B. S. Hartley, *Biochem. J.* **141**, 701 (1974).
316a. J. H. Jones, *Quart. Rev. Chem. Soc.* **22**, 302 (1968).
317. M. M. Shemyakin, Yu. A. Ovchinnikov, and A. A. Kiryushkin, in: *Mass Spectrometry: Techniques and Applications* (G. W. Milne, ed.), John Wiley & Sons, Inc., New York (1971).
317a. G. M. Schier and B. Halpern, *Aust. J. Chem.* **27**, 393 (1974).
318. H. V. Winkler and H. D. Beckey, *Biochem. Biophys. Res. Commun.* **46**, 391 (1972).
318a. M. A. Paz, A. Bernath, E. Henson, O. O. Blumenfeld, and P. M. Gallop, *Anal. Biochem.* **36**, 527 (1970).
319. W. R. Gray and U. E. Del Valle, *Biochemistry* **9**, 2134 (1970).
320. N. A. Aldanova, E. I. Vinogradova, S. A. Kasaryan, B. V. Rosinov, and M. M. Shemyakin, *Biokhimiya* **35**, 854 (1970).
320a. F. F. Richards and R. E. Lovins, *Methods Enzymol.* **25**, 314 (1972).
321. M. O. Dayhoff (ed.), *Atlas of Protein Sequence and Structure*, Vol. 5, National Biomedical Research Foundation, Silver Spring, Md. (1972); Suppl. 1 (1973).
321a. T. H. Jukes and R. Holmquist, *J. Mol. Evolution* **1**, 273 (1972); **2**, 343 (1973).
322. S. Blackburn, *Protein Sequence Determination*, Marcel Dekker, Inc., New York (1970).
323. S. B. Needleman, *Protein Sequence Determination*, Chapman & Hall Ltd., London; Springer-Verlag, Berlin (1970).
324. D. M. Glick and E. A. Barnard, *Biochem. Biophys. Acta* **214**, 326 (1970).
325. L. A. Cohen, *Ann. Rev. Biochem.* **37**, 695 (1968).
326. G. R. Stark, *Advan. Protein Chem.* **24**, 261 (1970).
327. B. L. Vallee and J. F. Riordan, *Ann. Rev. Biochem.* **38**, 733 (1969).
328. J. R. Kimmel, *Methods Enzymol.* **11**, 584 (1967).
329. M. J. Hunter and M. L. Ludwig, *Methods Enzymol.* **25**, 585 (1972).
329a. R. N. Perham and F. M. Richards, *J. Mol. Biol.* **33**, 795 (1968).
330. R. B. Freedman and G. K. Radda, *Biochem. J.* **114**, 611 (1969).
331. A. R. Goldfarb, *Biochim. Biophys. Acta* **200**, 1 (1970).
332. J. P. Riehm and H. A. Scheraga, *Biochemistry* **5**, 93 (1966).
333. J. F. Cavins and M. Friedman, *Biochemistry* **6**, 3766 (1967).

333a. H. R. Bosshard, K. H. Jorgensen, and R. E. Humbrel, *European J. Biochem.* **9**, 353 (1969).
334. J. F. Riordan and B. L. Vallee, *Methods Enzymol.* **25**, 494 (1972).
335. M. Sokolovsky, J. F. Riordan, and B. L. Vallee, *Biochemistry* **5**, 3582 (1966).
336. J. F. Riordan and B. L. Vallee, *Methods Enzymol.* **25**, 515 (1972).
337. H. Beyer and U. Schenk, *Kolloidzeitschrift* **233**, 890 (1969).
338. M. Sokolovsky and B. L. Vallee, *Biochemistry* **5**, 3574 (1966).
339. P. E. Wilcox, *Methods Enzymol.* **25**, 596 (1972).
340. D. J. Kilpatrick and J. A. Maclaren, *Textile Res. J.* **39**, 279 (1969).
341. D. Levy and F. H. Carpenter, *Biochemistry* **9**, 3215 (1970).
342. O. Yonemitsu, T. Hamada, and R. Kanaoka, *Tetrahedron Letters* 1819 (1969).
343. D. G. Hoare and D. E. Koshland, *J. Biol. Chem.* **242**, 2447 (1967).
343a. K. L. Carraway and D. E. Koshland, Jr., *Methods Enzymol.* **25**, 616 (1972).
344. A. Z. Budzynski and G. E. Means, *Biochem. Biophys. Acta* **236**, 767 (1971).
345. R. Frater, *FEBS Letters* **12**, 186 (1971).

2

# The Chemistry and Reactivity of Silk

## M. S. Otterburn

### 2.1. Introduction

Silk fibers have long caught the imagination of man from both scientific and technological viewpoints. The production of silk for textile purposes, *sericulture*, was practiced in the Far East over 4000 years ago. The first recorded commercial production was during the reign of the Chinese emperor Huang Ti in 2640 B.C. This emperor based the Chinese economy on silk production and the fiber had a vital role as a medium of exchange within the country, as well as being an important export commodity. The Chinese guarded their knowledge of sericulture and weaving techniques for many centuries, but, eventually, the techniques spread via Korea to Japan and India.

Silk came to Europe through the Byzantine Empire and sericulture was eventually established in Italy and the Rhône valley in France. By the fifteenth and sixteenth centuries the production of silk had become established in England. Sericulture and silk production were introduced into North America after the American Revolution.

The importance of silk as a textile fiber lay in its sheen, "handle," and draping qualities. With the advent of regenerated and synthetic fibers, these qualities could be matched at a far cheaper cost, and thus the importance of silk has declined. However, silk is still used for specialist and high-quality luxury items. Japan is by far the largest producer and consumer of the fiber.

---

**M. S. Otterburn** • Department of Industrial Chemistry, The Queen's University, Belfast, Northern Ireland

From a scientific viewpoint, commercial silk provided a readily available source of protein material for the work of Fischer et al.[1–3] and Abderhalden[4] in their pioneering studies of protein chemistry. The chemistry and technological properties of silk fibers have been studied by Lucas et al. and their review of 1958 deals with many aspects of the chemistry, reactivity, and structure of silk proteins.[5]

Because of the relatively simple composition of some silk proteins, certain silks have been used by biochemists interested in the method of biosynthesis of protein molecules as well as the genetic mechanisms involved in conformational transitions of proteins.[6–8]

## 2.2. Sources and Functions of Silk

Silk is an extracellular continuous filament produced by a wide variety of animals in the phylum Arthropoda, the most notable examples coming from the classes Insecta and Arachnida. The filaments fall into two categories: (a) those which are proteinaceous in composition and (b) those which are predominantly chitinous. In the chitinous silks, which are exemplified by those from the praying mantis, the silk is produced during hatching from the ootheca. The mantis is suspended from a pair of silk threads 5 cm long and 5 $\mu$m in diameter.[9] Rudall has identified these threads to be mainly chitinous in nature by X-ray diffraction techniques. Further work showed that the aminopolysaccharide has associated protein components similar in structure to the cuticle protein anthropodin[10] which has since been extensively studied by Hackmann.[11,12] The work on chitinous silks has been ably summarized by Rudall.[13]

Most silks produced in nature are proteinaceous in character, the main structural protein being fibroin and to a lesser extent collagen. As far as silks which are composed of collagen are concerned, the best example is that obtained from the gooseberry sawfly *Nematus ribesii*. This silk has been shown to give excellent X-ray data which proved to be characteristic of collagen.[13]

The majority of the silks produced by the classes Insecta and Arachnida are composed of fibroin, usually associated with the viscous protein sericin. The most common variety arising from the subclass Bombycidae is the species *Bombyx mori*, which provides the silk of commerce. The silkworms of *Bombyx mori*—which are in fact caterpillars, as they belong to the order *Lepidoptera*—are now entirely domesticated. In the practice of sericulture the silk moths lay eggs on leaves, the eggs coalescing together by virtue of a sticky exudate. When these hatch, the larvae are produced (1 g = 1500–2000 larvae). The larvae grow in length from 3 mm to approximately 8 cm in about 20 days, the main food being the leaves of the mulberry bush. By the fifth instar, which is the final period of larval life, the silk glands have enlarged and occupy a considerable proportion of the body mass. It is at this time that the silk filament is spun. The main function of the silk is to protect the pupa during metamorphosis. When spinning the cocoon the silk worm anchors a filament of silk to a twig or piece of straw and proceeds to spin an oval casing around itself by eccentric motions of its head. Each thread of silk consists of two filaments of fibroin cemented together by the protein sericin. Commercial silk yarn is obtained by drawing off the filaments of several cocoons.

The silk produced by representatives of the subfamily Saturniidae are exemplified by the species *Antherae*, which yields a wild silk known as tussah. Wild silk is obtained from *A. pernyi* and *A. mylitta* found in Mongolia and India, respectively. The difficulty with *Antherea* species is that they are not domesticated and consequently include much debris in their cocoons, which makes their commercial exploitation difficult. Finally, in this brief survey of the occurrence of silk, mention must be made of the silk produced by the class *Arachnida*, of which there are an enormous variety of species and types of silk. Some spiders produce more than one type of silk (e.g., the orb-web spinning spider produces six different types, each with a different function). This fascinating subject has been reviewed by Lucas[14] and by Lucas and Rudall.[15] Here mention should be made of *Nephila madagascariensis*, which was once the basis of a commercial silk industry in Madagascar.

## 2.3. Histology of Silk

Under microscopic examination raw silk exhibits an irregular surface structure. When the outer layer of sericin is removed, the fiber appears to be smooth and translucent. A cross section of a raw silk fiber shows it to be composed of two triangular components, called *brins* (fibroin), cemented together and surrounded by the protein sericin.

## 2.4. Chemical Composition of Silk

The two component proteins which together make up raw silk, fibroin and sericin, have been studied in detail by various workers over the past 60 years. This chemical work was initiated by Emil Fischer and extended by Abderhalden; their work has been summarized by Howitt.[16] However, since the development of the ion-exchange chromatographic methods of Moore and Stein, it has been possible to extend this early work in a truly quantitative manner.

Although microbiological methods for the analysis of fibroin hydrolysates have been used,[17] the easiest and most useful method of examining the protein is by acid hydrolysis followed by examination of the hydrolysates by ion-exchange chromatography. Lucas et al.[5,15] employed such an approach in their extensive and significant work. These workers characterized the amino acid composition of both sericin and fibroin from most known sources of these proteins.

The present chapter is concerned only with silk from the silkworm *Bombyx mori*. Table 2-1 gives the amino acid composition of both fibroin

**Table 2-1. Amino Acid Composition (Residues/1000) of Sericin and Fibroin from *Bombyx mori* Silk**

| Amino acid | Side chain | Sericin[15] | Fibroin[15,68] |
|---|---|---|---|
| Glycine | H— | 147 | 445 |
| Alanine | $CH_3$— | 43 | 293 |
| Leucine | $(CH_3)_2CHCH_2$— | 14 | 5 |
| Isoleucine | $CH_3CH_2CH(CH_3)$— | 7 | 7 |
| Valine | $(CH_3)_2CH$— | 36 | 22 |
| Phenylalanine | $C_6H_5CH_2$— | 3 | 6 |
| Serine | $CH_2OH$— | 373 | 121 |
| Threonine | $CH_3CH(OH)$— | 87 | 9 |
| Tyrosine | $HOC_6H_5CH_2$— | 26 | 52 |
| Aspartic acid | $HOOCCH_2$— | 148 | 13 |
| Glutamic acid | $HOOCCH_2CH_2$— | 34 | 10 |
| Arginine | $NH_2C(NH)NH(CH_2)_3$— | 36 | 5 |
| Cystine (half) | $(—S—CH_2)_2$— | 5 | 2 |
| Methionine | $CH_3SCH_2CH_2$— | | 1 |
| Lysine | $NH_2(CH_2)_4$— | 24 | 3 |
| Proline | $CH_2—CH_2$[a] / $CH_2$ $CH—COOH$ / NH (pyrrolidine ring) | 7 | 3 |
| Histidine | N—CH=C—$CH_2$— / CH—NH (imidazole ring) | 12 | 2 |
| Tryptophan | indole ring with —$CH_2$—, N—H | | 2 |
| $NH_3$ or amide | | 86 | |

[a] Complete formula of residue

and sericin from this source. The fibroin protein is composed of 15 or 16 α-amino acids linked together to form a biopolymer. The sericin, which is chemically separate from the fibroin, can be physically separated from the fibrous protein by extraction of the raw silk with aqueous soap solutions. The isolated fibroin is treated further by subsequent extractions with dilute alkalies, or even proteolytic enzymes, and finally extracted with organic solvents. By such means the fibrous protein can be purified from any residual sericin, lipids, waxes, or chemical impurities.

The most notable feature of the amino acid composition of the fibroin is the presence of large quantities of the more chemically simple amino acids (i.e., glycine, alanine, and serine). Also, the concentrations of cystine and the acidic amino acids—glutamic and aspartic—are low. Both glutamic and aspartic residues occur in the free-acid form rather than in the amide form, as is shown by the fact that little or no ammonia is liberated upon acid hydrolysis of the protein.

The total number of basic side chains in silk fibroin (i.e., lysine, histidine, and arginine) is relatively small; consequently, the quantity of acid or alkali that the fiber can absorb is also small: ≏0.15 equivalent/kg.[18,19] However the majority of the polar groups in the silk fiber are supplied by the hydroxyl-containing amino acids—serine, threonine, and tyrosine. Without doubt the largest component of the amino acid composition is derived from glycine and the aliphatic amino acid residues.

Because of the presence of such large quantities of glycine and amino acids with hydrocarbon side chains rather than residues with bulky aromatic or heterocyclic groups, close packing of the polypeptide chains in the protein is possible. Consequently, hydrogen bonding plays an important part in the conformation and structure of the fibroin—rather than depending on inter-chenic polypeptide crosslinks such as cystine. The influence of hydrogen bonding on the stability of the fibroin molecule can be shown by the ease of dissolution of the protein in known hydrogen-bond-breaking solvents (e.g., cupriethylenediamine or lithium thiocyanate).

## 2.5. Chemical Structure of Fibroin

Before an investigation of the chemical structure of silk fibers can take place, two problems have to be overcome: (1) the sericin and fibroin must be separated out, and (2) the fibroin must be solubilized with minimum degradation.

The standard methods of degumming silk (i.e., removing the sericin) involve the use of hot dilute soap solutions or superheated steam.[20] These methods cause little loss in tensile strength of the fiber, which is interpreted as reflecting little degradation of the protein. Enzymic methods to remove sericin have also been investigated, the preferred enzyme being papain, and this forms the basis of the technical process of producing Schappe silk. Once the sericin has been removed, the fibroin is extracted with water and organic solvents to remove chemical additives (e.g., soap) and residual fats and waxes.

The second problem, that of solubility of the fibroin, has to be overcome in order to apply the usual tests for purity and homogeneity. Hydrolysis may be accomplished by acids or alkalies, but these cause some degradation of the constituent amino acids. For molecular-weight determinations it has been found that the fibroin can be solubilized by using formic acid or cupriethylenediamine,[21] although even in these solvents some degradation does occur. The most successful solvent for fibroin appears to be concentrated solutions of lithium thiocyanate (50%),[22,23] and most recent work has been performed using this solvent.

### 2.5.1. Molecular-Weight Determination of Fibroin

Molecular-weight data of silk fibroin is somewhat inconclusive in that a variety of results have been obtained by the use of different techniques (e.g., 33,000 daltons by osmotic pressure measurements,[21] 84,000 daltons by ultracentrifugation,[24] 55,000 daltons by viscosity and streaming birefringence,[23] and by light scattering, values of several million daltons.[25] This wide range of values is not surprising in view of the fact that different molecular-weight averages are being measured.

Attempts have been made to use solutions of fibroin in cupriethylenediamine[26] and lithium isocyanate[27] in order to fractionate the protein. In both solvents two fractions were obtained, having the same amino acid composition, although they varied in viscosity, thus indicating heterogeneity of the protein.

Narita,[28] using cupriethylenediamine with 2-ethoxy-6,9-diaminoacridine lactate (Rivanol), was able to separate out two components from the fibroin solution: an insoluble fraction, which he termed silk plastin, and a more soluble fraction, which was only precipitated by the use of acetone.

### 2.5.2. N-Terminal Amino Acid Determination

When the chemical structure of the fibroin is considered, classical methods of protein chemistry have been employed. End-group analysis reveals, for example, that the N-terminal residues consist of glycine, alanine, and serine[28–30] and that the C-terminal residues are glycine, valine, and serine,[28] although other workers have shown tyrosine, valine, and proline to be C-terminal residues.[31] Discrepancies are bound to arise in such work, as there appears to be no consistency in the methods of extraction of sericin, nor is there any common solvent used for the solution work. Thus some methods will cause more fibroin degradation, while others will leave the fibroin contaminated by residual sericin.

### 2.5.3. Primary Structure of Fibroin as Determined by Acid Hydrolysis

Work on the peptide sequences of the fibroin molecule stems from the early German workers Fischer and Abderhalden, who, after subjecting the fibroin to partial acid hydrolysis, were able to identify the following peptides: Gly-Ala, Ala-Gly, and Gly-Tyr, as well as Gly-Ser-Pro-Tyr.[16,32] At first it was believed that long sequences of Ala and Gly, arranged in an alternating fashion, might exist in the protein chains. This idea was destroyed when Levy and Slobodian[33] showed that quantities of Ala-Gly were much greater than Gly-Ala rather than being in equimolar amounts. These workers also pointed out that the dipeptide Gly-Gly should be found in appreciable quantities if the residues glycine and alanine were joined in a random fashion. They were able to show, however, that less than 2.0% of this dipeptide occurred. This fact, together with the greater preponderance of Ala-Gly to Gly-Ala, leads to the suggestion that in fibroin the glycine and alanine residues are in the form

X-Ala-Gly-Ala-Gly-Y

where X and Y are amino acids other than alanine or glycine. This suggested structure is supported by the existence of the tri-peptide Gly-Ala-Gly[34] and by the fact that on hydrolysis twice as much Ala-Gly as Gly-Ala is returned.[35]

However, Kay and Schroeder[36] have shown the presence of the dipeptide Ala-Ala and suggest that the type of substituents in X and Y will determine, by hydrolytic rate effects, which peptides are released. This result would, in turn, have an effect on the yield of the dipeptide Ala-Gly.

Thus the evidence for the sequences of fibroin from chemical degradation of the protein gives some facts, but these are rather discontinuous. The alternative use of enzymic hydrolysis as a method of degrading the peptide chain will now be considered.

### 2.5.4. Primary Structure of Fibroin as Determined by Enzymic Hydrolysis

In the determination of the primary structure of silk fibroin, wide use has been made of proteolytic enzymes, mainly chymotrypsin and trypsin.

Lucas et al.[37] made use of chymotrypsin to digest fibroin and, after a short period of time, this digestion resulted in a solution and in a granular precipitate termed Cp. It was believed that the Cp fraction of the protein accounted for about 60% of the total fibroin. It was shown that the precipitate had a crystalline structure which was very similar to that of native fiber.[38] Lucas et al., from this early work, showed that the Cp fraction had an empirical formula of

$$\mathrm{Gly_{29} \cdot Ala_{20} \cdot Ser_{9} \cdot Tyr}$$

One conclusion from this work was that the Cp fraction was derived from the highly ordered crystalline regions of the fiber.

Lucas et al.,[15,37] using direct chemical methods of partial hydrolysis and dinitrophenylation, pursued a detailed study of the Cp fraction of fibroin of *Bombyx mori*. These results may be summarized as follows: (1) four dipeptides: Ala-Gly, Gly-Ala, Ser-Gly, and Ala-Ala were observed; (2) three DNP tripeptides: Ser-Gly-Ala, Gly-Ala-Gly, and Ala-Gly-Ala were found; (3) the dipeptides Ser-Ala and Gly-Gly could not be located.

The problem of the distribution of serine in the Cp fraction was attacked by Lucas et al. by use of the N → O peptidyl shift which is brought about in proteins under anhydrous acid conditions. In this case phosphoric acid at 40°C was chosen as the reagent. After the acid treatment FDNB was added and slowly the pH was increased. By this means it was possible to produce DNP serine before reversal to the *N*-acyl product. Although this procedure is not entirely satisfactory, owing to the reversibility of the reaction, it does give some indication of the amino acid residues next to serine in the polypeptide chain. Lucas et al. concluded from this work that 36% of the Cp fraction consisted of a hexapeptide of the structure Ser-Gly-Ala-Gly-Ala-Gly, which they believed to be a main element in the total structure of Cp. Despite some inefficiency of the method of isolation, they proposed that the Cp fraction of fibroin had the structure

$$\mathrm{Gly\text{-}Ala\text{-}Gly\text{-}Ala\text{-}Gly\text{-}[Ser\text{-}Gly\text{-}(Ala\text{-}Gly)_{\mathit{n}}]_{8}\text{-}Ser\text{-}Gly\text{-}Ala\text{-}Ala\text{-}Gly\text{-}Tyr}$$

where $n = 2$. Confirmation of this structure has been given by the synthetic work of Schnabel and Zahn, coupled with their analytical and X-ray work.[39–42] Similar but later work by Stewart et al.[43,44] also showed that the structure proposed by Lucas et al. was satisfactory.

Zahn et al.[45] isolated two fractions from the Cp fraction of fibroin which they found had the composition Ser-($Gly_6$, $Ala_4$, Ser) and Ser-($Gly_4$, $Ala_2$, or $Ala_3$, Ser). They claimed that these were the double repeating units of Lucas et al.,[37] namely [Ser-Gly-$(\mathrm{Ala\text{-}Gly})_n]_2$, for $n = 2$ and 1, respectively.

Thus the combination of enzymatic and chemical methods has succeeded in elucidating some considerable portion of the Cp fraction of fibroin. The remaining 40% of the protein, the soluble fraction of the chymotryptic digest, has also been investigated.[46] This soluble fraction has been designated by the symbol Cs. Peptides present in the solution, following treatment of the silk with the endopeptidase chymotrypsin, would be expected to terminate in residues of tyrosine, tryptophan, leucine, or phenylalanine. Peptides with only tyrosine and phenylalanine residues have been located by use of the hydrazine method.[46] Other workers, using a variety of analytical techniques, have located and identified mixtures of peptides from the Cs fraction of fibroin.[47–51] These peptides, which vary from two to eight residues in length, are shown in Table 2-2. In the work of Ziegler et al.[47,48] identification of the peptides was achieved by the use of countercurrent extraction and high-voltage electrophoresis. The latter technique proved extremely useful for the separation of peptides which contained acidic and basic residues.

The results of the work on the Cs fraction in the isolation of octa and tetra peptides, together with the 59 residue peptide identified in the Cp fraction, account for almost 75% of the total fibroin molecule. Use has been

**Table 2-2. Peptides Isolated from the Cs Fraction of *Bombyx mori* Fibroin**

| Reference | Composition |
|---|---|
| Lucas et al.[46] | Gly-Ala-Gly-Tyr |
| | Gly-($Gly_3$, $Ala_2$, Val)-Tyr |
| | Gly-Ala-Gly-Ala-Gly-Ala-Gly-Tyr |
| Lucas[49] | Gly-Pro-Tyr |
| | Ser-Gly-Tyr |
| | Gly-Val-Gly-Tyr |
| | Val-($Gly_2$, Ala, Asp)-Tyr |
| Ziegler et al.[47,48] | Ala-Trp |
| | Glu-Tyr |
| | Val-Lys-Phe |
| | Leu-(Glu, Lys)-Phe |
| | Ser-Gly-Glu-Tyr |
| | Ile-Thr-Ala-His |
| | Ser-Glu-Asp-Tyr |
| | Lys-Gly-(Arg, Glu)-Lys |
| | Val-Ala-Gly-His-Gly-Tyr |
| | Val-Ala-Gly-Asp-Gly-Tyr |
| Zuber[50] | ($Gly_4$, $Val_2$, Tyr) |
| | ($Gly_5$, $Ala_3$, Val, Tyr) |
| | ($Gly_4$, $Ala_3$, Tyr) |
| | ($Gly_{10-15}$, $Ala_{10-15}$, Tyr) |
| Cebra[51] | Gly-(Gly, Val)-Tyr |
| | Gly-(Gly, Ala)-Tyr |
| | Ala-($Gly_2$, Val, Asp)-Tyr |
| | Gly-($Gly_3$, $Ala_2$, Val)-Tyr |
| | Gly-($Gly_3$, $Ala_3$)-Tyr |

made of trypsin digests to determine the mode of connecting the known peptides of both fractions together in the total protein.[46,50,52] The results of the trypsin digest work suggest that the octa- and tetrapeptides of the Cs fraction are directly connected to the 59 residue peptide of the Cp fraction. The latter portion, in turn, forms the major part of the ordered crystalline regions of the fibroin molecule.

#### *2.5.4.1. Existence of Cystine in Silk Fibroin*

The primary structures of the various fractions of fibroin have been well studied, and much progress has been made in elucidating the overall picture of the composition of the crystalline and amorphous regions of the protein. However, factors such as molecular weight, numbers of polypeptide chains, and the arrangement of the chains into the overall structure of the protein is somewhat less clear. At the present time there is no evidence to indicate the presence of covalent interchenic crosslinks in the fibroin molecule, unlike other fibrous proteins, such as the keratins, collagens, elastin, and fibrin,[53] which have a variety of natural covalent crosslinks. Fibroin apparently relies solely on hydrogen bonding to maintain the structural integrity of the protein.

One of the main deficiencies of the early studies as far as covalent crosslinks are concerned was the apparent absence of that ubiquitous amino acid cystine; in all early analyses of the composition of fibroin no mention was made of this residue. The lack of evidence of cystine was probably due to two factors: (1) the techniques being used were laborious and insensitive, and (2) the amounts of amino acid present were very small. Confusion also arose from the fact that cystine was known to be present in the sericin molecule, and when reports of the finding of cystine in fibroin were published, the results were dismissed as arising from artifacts or by contamination from the sericin. In 1955, however, Schroeder and Kay[54] reported the presence of cystine in the fiber. These findings were confirmed by Zuber et al.[55] Later, Earland and Raven[56] isolated from acid-extracted fiber the 2,4-dinitrophenyl cysteic acid derivative which conclusively substantiated the presence of cystine in fibroin. It is now accepted that *Bombyx mori* silk fibroin contains approximately 0.2% cystine in its composition. However, no evidence was produced that the moiety was acting as an interchenic crosslink.

The potential importance of the presence of cystine in the fibroin molecule, albeit in small quantities, was very great. Such residues strategically placed in the primary structure of the protein could affect the reactivity, secondary structure, and molecular weight of the molecule. Consequently, it was of the greatest importance to isolate fragments of the protein which contained cystine or cysteic acid residues.

Earland and Robins succeeded in isolating from the chymotryptic digests of acid-washed silk fibroin an octapeptide containing cystine.[57] To this peptide they assigned the sequence

Arg-Ala-Leu-Pro-Cys-Asp($NH_2$)-Val-Cys

This structure has one of the smallest cystine containing ring structures known but has been shown to be stereochemically feasible by use of space-filling molecular models.[57,58] The octapeptide isolated by Earland and Robins is believed to account for 33% of the total cystine present in the fibroin.[58]

Robson et al. have examined fibroin from glands of *Bombyx mori* silkworms fed on mulberry leaves coated with $^{35}S$ cysteine.[59] The unspun silk was digested with enzymes and these workers isolated a 13-residue peptide containing cysteic acid to which they assigned the structure

Gly-Ala-Gly-Ala-Gly-Cys-Asx-Ser-Ala-Val-Cys-(Pro, Leu)

These workers found that this peptide accounted for approximately 75% of the cysteic acid present in the silk fibroin. A fragment similar to that of Robson et al. has also been isolated by Schade et al.[60]

The two cystine-containing peptides do differ significantly in their primary structure but are not self-exclusive, as they were extracted from different materials by different techniques. Neither of the sequences account for 100% of the total cystine in the fiber. Consequently, it is quite possible that both sequences could occur in separate regions of the fiber.

## 2.6. Integrated Picture of the Fibroin Molecule

In order to establish a total picture of the structure of the silk fiber, account must be taken of the secondary structure as elucidated by X-ray diffraction work, and this must be combined with the primary and tertiary structures derived from chemical investigations.

Silk proteins, because of their fibrous nature, have been the subject of intense examination by X-ray techniques. Most silks have been found to possess $\beta$ structures, although they exhibit $\alpha$-helical, polyglycine II, and collagen conformations. As with all polymers, whether synthetic or natural, the fibroins are not 100% crystalline but do possess local regions of disorder.

The work of Warwicker,[61] and later of Lucas and Rudall,[62] classed silk proteins as antiparallel $\beta$ structures. These were subdivided into six groups, all of which were pleated $\beta$ sheets. Group 1 comprised *Bombyx mori* fibroin, and the other five groups were given over to the remaining silks (see

Table 2-3. Values for Dimensions *c* of the Parallel-β-Pleated Fibroin Unit Cell[a,b]

| X-ray group | Example | *c* (Å) |
|---|---|---|
| 1 | *Bombyx mori* | 9.3 |
| 2a | *Anaphe moloneyi* | 10.0 |
| 2b | *Canephora asiatica* | 10.0 |
| 3a | *Antheraea pernyi* | 10.6 |
| 3b | *Cricula andrei* | 10.6 |
| 4 | *Digelansinus diversipes* | 13.8 |
| 5 | *Thaumetopoea pityocampa* | 15.0 |
| 6 | *Nephila senegalensis* | 15.7 |

[a] Data from Warwicker[61] and Lucas and Rudall.[15,62]
[b] The values for dimensions *a* and *b* are 9.44 and 6.95 ± 0.05 Å, respectively.

Table 2-3). All six groups exhibited a common periodicity of approximately 6.95 Å, which corresponds to direction *b* (Fig. 2-1), along the fiber axis. A repeat distance of 9.44 Å was found for *Bombyx mori* in the hydrogen-bonded direction. The value of dimension *c*, which corresponds to the separation of the pleated sheets, was found to be 9.3 Å. This value for the separation

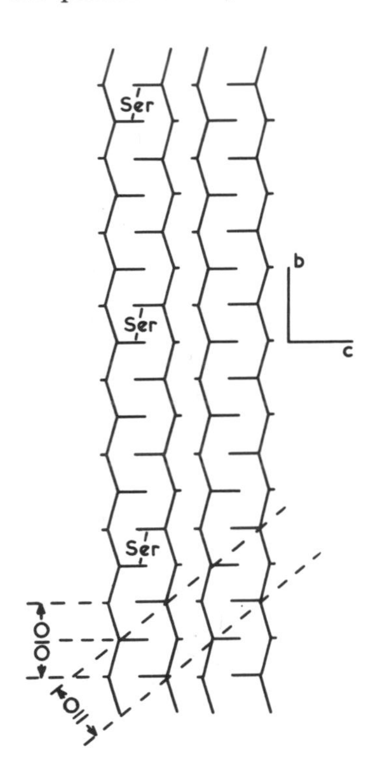


Fig. 2-1. Arrangement of chain for *Bombyx mori* fibroin with the sequence of hexapeptides (Gly-Ala)$_2$Gly-Ser. Reflection 010 is absent. Axial period at 21 Å.

distance of the pleated sheets appears to represent subtle changes in the primary sequences of the different types of fibroin—as this distance $c$ is very characteristic of the different X-ray groups of Warwicker[61] and of Lucas and Rudall.[62]

This very general X-ray information can be related to the chemical structure of *Bombyx mori* fibroin. Shaw[52] has postulated that fibroin consists of a multichain arrangement of polypeptide chains. These chains may be linked together by cystine or ester interchenic links. Such ester links were first postulated by Zuber.[50] The presence of cystine as an interchenic link has yet to be established. Also, the presence of ester-type crosslinks would be difficult to substantiate by normal chemical analysis, as they would be acid or alkali labile. The presence of other interchenic links, such as the isopeptides $N^{\epsilon}$-($\gamma$-L-glutamyl)-L-lysine or $N^{\epsilon}$-($\beta$-L-aspartyl)-L-lysine, is another possibility.[63,64] These latter links, which are also acid-labile, were first isolated by enzymic methods. Possibly the presence of ester-type crosslinks in fibroin could be investigated by similar methods. It is interesting to note, in the context of covalent crosslinkages in fibroin, that Raven et al.[65] have identified dityrosine in tussah silk fibroin. While these workers did not claim that this compound was a crosslink, it is worth noting that such moieties have been shown to act as interchenic links in the elastic protein, resilin, which is obtained from the cuticle of insects.[66,67]

Shaw[52] has divided the primary structure of fibroin into three phases. Phase I consists of repeating sequences of the type

$$\text{Gly-Ala-Gly-Ala-Gly-[Ser-Gly-(Ala-Gly)}_n]_8\text{-Ser-Gly-Ala-Ala-Gly-Tyr}$$

where $n = 2$ or has a mean value of 2. This segment of the molecule consists of 60% of the protein. The characteristic X-ray pattern of fibroin almost certainly originates from this highly oriented and crystalline region. Phase II of the primary structure consists of tetra and octa peptides in which the residues glycine, alanine, tyrosine, and valine predominate. This phase consists of approximately 30% of the total protein. The remaining 10% of the molecule, phase III, which consists of the truly amorphous part of the protein, is made up of amino acid residues with ionic or bulky side groups. Presumably in such a model for silk fibroin, any cystine present in the protein would reside in phase III.

The molecular weight of the proteins in *Bombyx mori* fibroin has been the subject of much work and argument in the past. Values from as low as 16,000 daltons up to several million daltons have been determined by different techniques. Lucas[68] claimed that the fibroin molecule was made up of four polypeptide chains held together by four cystine disulfide bridges. From this he estimated the molecular weight of the fibroin to be 400,000 daltons. However, more recent work on the cystine content of the fiber by Robson et al.[59]

and by Earland and Robins[58] has established the molecular weight of fibroin to be 103,000 and 100,000 daltons, respectively.

In conclusion, it may be fairly stated that, owing to the work of various research groups, especially that of Lucas and his co-workers, a tremendous amount of knowledge concerning the primary and secondary structure of the fibroin proteins is available. One area in which there is a lack of information is in the interbonding of the polypeptide chains. Other than hydrogen bonds and secondary valence bonding, little or no evidence is available to support the concept of covalent bonding of the polypeptide chains.

## 2.7. Studies on the Chemical Reactivity of *Bombyx mori* Silk Fibroin

In sharp contrast to the large amount of work that has been carried out on the primary and secondary structures of various silks, comparatively little work has been done on their chemical reactivity. The work that has been carried out tends to be as offshoots of industrial investigations. This lack of interest in the reactivity of silk fibroin may well be due to the decline in commercial importance of the fiber. Another reason may be the relative chemical inertness of the constituent amino acids present in the protein (see Table 2-1).

Most information concerning the fiber's reactivity has been derived from technical investigations into the chemical and physical properties of modified silk, prior to or during industrial processing. Thus hydrolysis, the action of oxidizing agents, and the use of crosslinking agents, resins, and dyestuffs with fibrous silk have been studied by a variety of authors. Early surveys of silk processing were made by Huber[69] and by Lucas et al.[5]

### 2.7.1. Hydrolysis of Silk Fibroin

Hydrolysis of fibroin can be achieved to some extent by boiling water or steam for extended periods of time. Such degradation is relatively slow but may be significant in aqueous dyeing and finishing processes.

Hydrolytic action, caused by mineral acids and alkalies on silk fibers, causes main-chain fission of the polypeptide chains. The efficiency of this hydrolysis is interrelated with the pH of the hydrolytic agent, temperature, and physical agitation.[70] In general it has been found that the action of mineral acids is more efficient than that of alkalies.[71,72] In the case of the acids, peptide fission occurs in a random manner throughout the protein, whereas in the case of alkalis, attack seems to take place initially at the ends of the peptide chains. Korchagin[72] has studied the effects of the addition of various inorganic salts on the efficacy of hydrolysis of silk by dilute acids and alkalis. He found, in the case of alkaline hydrolysis, that the nature of the cation of the salt had a marked effect on the degree of hydrolysis. However, in the case of acid hydrolysis the anion had the more significant effect. As well as causing fission of the peptide bonds in the fiber, the action of hot alkali causes excessive loss of hydroxyamino acid residues such as serine and threonine; this also causes an increase in the amino nitrogen content of the hydrolysate. It has been suggested by some authors[73] that the hydroxyamino acids show increased sensitivity to alkaline attack when they are incorporated in the polypeptide chains.

Degradative studies of acids and alkalis on silk fibers are linked with viscosity measurements and solubility tests to determine the degree and efficiency of degradation by the hydrolytic agent. Most work has been carried out using hot hydrochloric acid as the main hydrolyzing agent. Other mineral acids have, however, been studied. Hot concentrated sulfuric acid has similar hydrolytic properties as hydrochloric acid, but sulfonation of the tyrosine residues occurs under extended time conditions.

The action of concentrated or anhydrous mineral acids on silk fibroin and other proteins has shown that the peptide bonds involving the amino groups of serine and threonine are extremely labile. Such reactions have usually been carried out at low temperatures for extended periods of time.

```
    |                     |                    |
    NH                    NH                   NH
    |                     |                    |
 O=C                   O=C                  O=C
    |                     |                    |
    CHCH2OH  ----->       CHCH2X  ----->       CHCH2X  ----->
    |                     |                    |
    NH                    NH                   N
    |                     |                    |
    C=O                   C=O                  C—OH
    |                     |                    |

                      |                     |
                      NH                    NH
                      |                     |
                   O=C                   O=C
                      |                     |
                      CH—CH2  ----->        CH—CH2
                      |    |                |    |
                      N    O                NH3+ O
                       \\  /                     |
                         C                  X-   C=O
                         |                       |
```

X = ester intermediate

The action of such acids has considerable technical importance in textile chemistry and is the basis of the carbonizing process for wool.[74]

One important consequence of the action of the concentrated acids on proteins is the rearrangement reaction, termed the N → O peptidyl or acyl shift. This was first proposed by Bergmann et al.[75] and shown to occur for *N*-acyl-*β*-hydroxyamino acids. The reaction was monitored by the increase in amide nitrogen of the reactants by the Van Slyke method. The mechanism of the N → O acyl shift occurs through the formation of an ester from the hydroxyl groups of serine or threonine with the acid. This is followed by enolization, cyclization, and finally elimination of the acid. The reaction scheme is as shown. More recent work on the mechanism of the N → O acyl rearrangement has shown that it is not essential for the rearrangement to proceed via an ester–oxazoline intermediate, as it may go through a hydroxyoxazolidine.[76]

```
 |                      |                      |
 CH—CH2     H+          CH—CH2     H+          CH——CH2
 |   |    <====>        |   |    <====>        |     |
 NH  OH     OH-         NH  O      OH-         NH2   O
  \                      \ /                        /
   CO                     C                       CO
   |                     / \                      |
   R                    R   OH                    R
```

Direct proof of the N → O acyl shift, which has great significance in protein chemistry (see Section 2.5.3), is somewhat sparse. Most results are based on Van Slyke analyses of the solutions. This method of monitoring the reaction has, however, proved suspect and both Elliott[77] and Narita[78] have emphasized caution in the interpretation of the results using this method. Elliott[77] used concentrated sulfuric acid on silk fibroin and found, by the Van Slyke method, that 62% acyl rearrangement of serine peptide bonds had occurred. However, on examination of the treated fibroin he was able, using the FDNB blocking technique, to show that only 23% of the available serine was converted to the DNP derivative. Also, Elliott was unable to show the presence of any DNP threonine. These surprising results may be due to either the N → O acyl rearrangement being incomplete or, alternatively, the pH conditions necessary for reacting FDNB causing the reverse O → N acyl rearrangement. The work of Sakakibara et al.[79] has shown, using glycylserine with hydrogen fluoride, that 93% conversion to the *O*-glycylserine product occurred. However, when concentrated sulfuric acid was used only 36% conversion was recorded.

Nitric acid not only degrades the fiber by bringing about peptide fission but also causes nitration of phenyl residues in the protein. This side reaction of the hydrolytic process also causes discoloration in partially degraded silk, which is still fibrous.

Organic acids, which have little effect when used in dilute solution at ambient temperatures, can cause dissolution of the fiber at higher concentrations (e.g., 98% formic acid) with some associated degradation of the protein.

Enzyme proteolysis of fibrous silk is extremely difficult to achieve. This is probably due to the inability of the enzymes to penetrate into the highly ordered regions of the fiber. Mercer[80] found that it took up to 6 months to even partially hydrolyze fibroin with the enzyme trypsin. Thus, to facilitate any degree of enzymic hydrolysis of silk fibroin, it is necessary to swell and dissolve the fiber in a solvent prior to enzyme treatment.

### 2.7.2. Chemical Modification of Silk by Alkalis

It has been known for many years that the action of alkaline solutions on proteins causes the formation of moieties which are capable of acting as crosslinks in the protein structure.[81–83] The subject of crosslink formation in alkali-treated proteins has received a great deal of attention, especially in the high-cystine-containing proteins such as the keratins.[84–86] However, in the case of silk fibroin, the presence or absence of such crosslinks has received little attention. This may be due to the fact that the formation of moieties (e.g., lanthionine) demanded a mechanism of formation based on cystine. It was not until cystine was located in fibroin that much attention was drawn to potential alkali-induced crosslinks in the fiber. In 1964, Bohak isolated from alkaline-treated ribonuclease an amino acid, $N^{\epsilon}$-(2-amino-2-carboxyethyl)-lysine, to which he assigned the trivial name lysinoalanine.[87] He was also able to show that in the protein phosphovitin, which does not contain cystine, it was possible to form lysinoalanine. In fact, he claimed that up to 80% of the lysyl residues were incorporated into lysinoalanine in this protein. Bohak suggested that dehydroalanine was being formed via serine-*O*-phosphate by a $\beta$-elimination reaction due to the alkaline conditions. Ziegler, who discovered lysinoalanine in alkali-treated keratin, also suggested that seryl residues as well as half-cystine were involved in the formation of dehydroalanine.[88,89] The possibility of lysinoalanine formation in silk fibroin was first proposed and investigated by Mellet and Louw.[90] This resulted in the detection of 20 $\mu$mol/g of the moiety in alkali-treated silk fibroin. These workers claimed that lysinoalanine was formed totally from dehydroalanine-derived serine. This work was later substantially confirmed by Robson and Zaidi,[91] who showed that only 10% of the lysinoalanine formed in the fibroin had its origins in cystine. The possible routes for the formation of lysinoalanine are as shown.

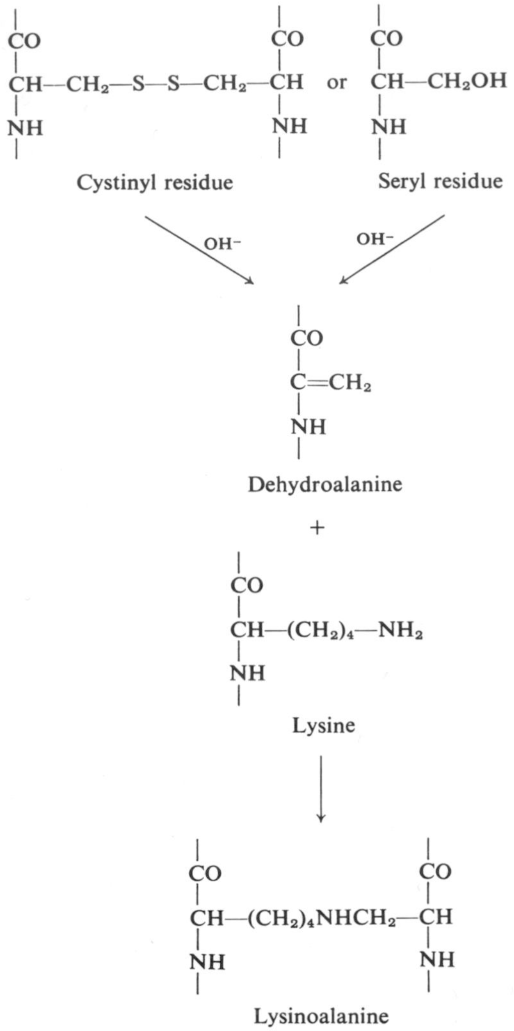

$$\begin{array}{l} | \\ CO \\ | \\ CH-CH_2-S-S-CH_2-CH \\ | \qquad\qquad\qquad\qquad\quad | \\ NH \qquad\qquad\qquad\qquad NH \\ | \qquad\qquad\qquad\qquad\quad | \end{array} \quad \text{or} \quad \begin{array}{l} | \\ CO \\ | \\ CH-CH_2OH \\ | \\ NH \\ | \end{array}$$

Cystinyl residue Seryl residue

$\xrightarrow{OH^-}$ $\xleftarrow{OH^-}$

$$\begin{array}{l} | \\ CO \\ | \\ C{=}CH_2 \\ | \\ NH \\ | \end{array}$$

Dehydroalanine

+

$$\begin{array}{l} | \\ CO \\ | \\ CH-(CH_2)_4-NH_2 \\ | \\ NH \\ | \end{array}$$

Lysine

$\downarrow$

$$\begin{array}{l} | \qquad\qquad\qquad\qquad\qquad\quad | \\ CO \qquad\qquad\qquad\qquad\quad CO \\ | \qquad\qquad\qquad\qquad\qquad\quad | \\ CH-(CH_2)_4NHCH_2-CH \\ | \qquad\qquad\qquad\qquad\qquad\quad | \\ NH \qquad\qquad\qquad\qquad\quad NH \\ | \qquad\qquad\qquad\qquad\qquad\quad | \end{array}$$

Lysinoalanine

In 1967 Ziegler[92] isolated another amino acid, formed by the action of alkali on silk proteins. This amino acid, $N^{\delta}$-(-2-amino-2-carboxyethyl)-ornithine, which he named ornithinoalanine, was only found in the sericin protein of silk. This compound is believed to be formed by the reaction of the $\gamma$-amino groups of ornithine with the dehydroalanyl residues. Ornithine is a degradation product of arginine, and it is therefore not surprising that this compound has not been located in the fibroin fraction of silk, as no arginine is present in fibroin. The probable reaction scheme for the formation of orninithoalanine is shown on page 71.

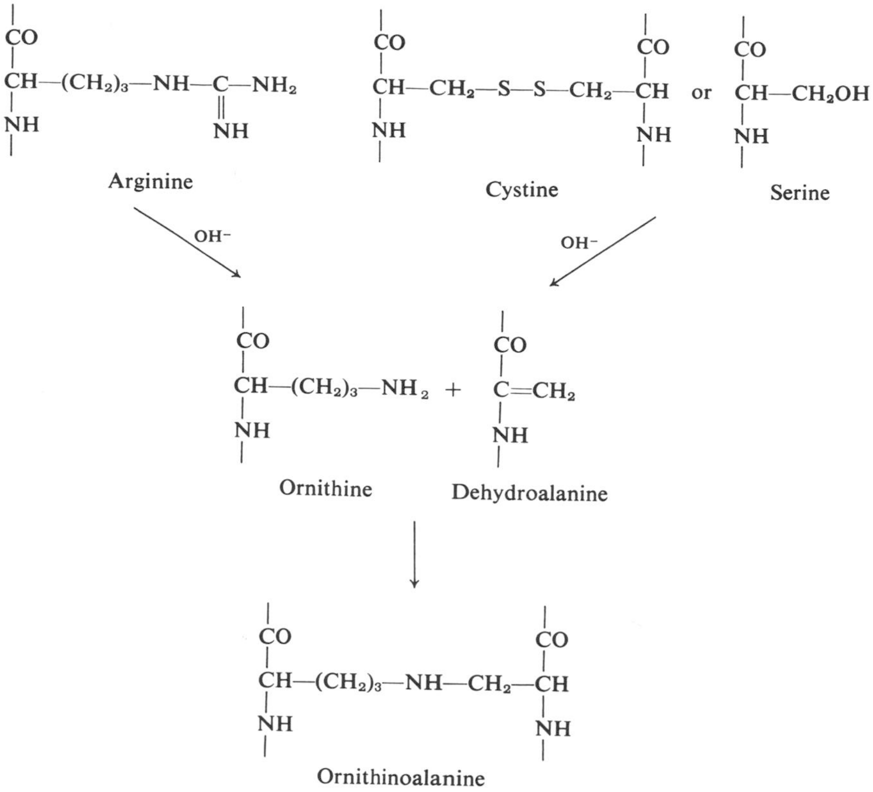

$$\begin{array}{l} | \\ CO \\ | \\ CH{-}(CH_2)_3{-}NH{-}C{-}NH_2 \\ | \qquad\qquad\qquad\quad \| \\ NH \qquad\qquad\qquad NH \\ | \end{array} \qquad \begin{array}{l} | \qquad\qquad\qquad\qquad\quad | \\ CO \qquad\qquad\qquad\quad CO \\ | \qquad\qquad\qquad\qquad\quad | \\ CH{-}CH_2{-}S{-}S{-}CH_2{-}CH \\ | \qquad\qquad\qquad\qquad\quad | \\ NH \qquad\qquad\qquad\quad NH \\ | \qquad\qquad\qquad\qquad\quad | \end{array} \quad \text{or} \quad \begin{array}{l} | \\ CO \\ | \\ CH{-}CH_2OH \\ | \\ NH \\ | \end{array}$$

Arginine Cystine Serine

$OH^-$ $OH^-$

$$\begin{array}{l} | \\ CO \\ | \\ CH{-}(CH_2)_3{-}NH_2 \\ | \\ NH \\ | \end{array} \quad + \quad \begin{array}{l} | \\ CO \\ | \\ C{=}CH_2 \\ | \\ NH \\ | \end{array}$$

Ornithine Dehydroalanine

$$\downarrow$$

$$\begin{array}{l} | \qquad\qquad\qquad\qquad\qquad\quad | \\ CO \qquad\qquad\qquad\qquad\quad CO \\ | \qquad\qquad\qquad\qquad\qquad\quad | \\ CH{-}(CH_2)_3{-}NH{-}CH_2{-}CH \\ | \qquad\qquad\qquad\qquad\qquad\quad | \\ NH \qquad\qquad\qquad\qquad\quad NH \\ | \qquad\qquad\qquad\qquad\qquad\quad | \end{array}$$

Ornithinoalanine

### 2.7.3. Oxidation and Induced Crosslinking of Silk Fibroin

In general, protein chemists have placed great emphasis, when considering the oxidation of proteins, on the reactions of cystine. This labile and reactive residue occurs only to a small extent in silk fibroin, and it is therefore not surprising that the more usual products of oxidation are absent in this protein. The mechanisms of oxidation of the fiber are very complex, and various workers have postulated different reaction mechanisms for the observed results. Oxidation of silk fibroin can take place at any of three sites: side chains, N-terminal residues, and the polypeptide backbone.

Numerous oxidizing agents have been used, among them hydrogen peroxide, which is absorbed by the fiber and is believed to form complexes with the amino groups and peptide bonds within the molecule.[93] Great emphasis has been placed by some researchers on the stability and reactivity of tyrosine. Thus, Nakanishi and Kobayashi[94] showed that the tyrosine content of silk fibers was diminished and that the peptide links involving tyrosine were fissioned by the action of hydrogen peroxide.

Quite early on in the study of oxidation reactions on silk it was found that colors were produced by the action of the oxidizing agent in the fiber. These colors varied from yellow to pink to brown. Such color reactions were assumed to be associated with the oxidation of tyrosine residues, with the formation of quinones.(95,96) However, the presence of *o*-quinones has not been established.(97)

Oxidative attack by potassium permanganate has been studied.(98,99) From this work it has been concluded that an overall loss in strength of the fiber occurs together with a gradual dissolution of part of the protein. In 1955 Howitt,(100) from solubility studies, proposed that new crosslinks were formed in oxidized silk fibers. The crosslinks he proposed were simple ether links formed between two tyrosine residues on adjacent polypeptide chains. Although such ether links were subsequently shown by Asquith et al.(101) not to exist, the possibility of crosslinks being formed via an oxidative mechanism was pursued. In this area of research, Earland et al. proposed a crosslink based on the reaction of a *p*-quinone with a free amino group.(102,103) Later, these workers also proposed that the insolubilization of oxidized fibroin was due to the formation of melamine-type crosslinks, again indicating tyrosine as the main crosslinking moiety.(104) The proposed crosslinks are as shown.

Another aspect of inducing crosslinkages into silk fibers is to react the protein with known crosslinking agents such as formaldehyde.(105,106) Such treatments increase the molecular weight and thus decrease the solubility of the silk. The reaction of 1-fluoro-2,4-dinitrobenzene with fibroin also depresses the protein's solubility. Zahn and Würz(30) found, from this work some extremely interesting information concerning the availability and positioning of various amino acid residues in the fibroin molecule. They were able to show that only 79% of the ε-amino groups of lysine were available for reaction with FDNB. They also found that 96–99% of the hydroxyl groups of tyrosine were available for reaction. This led these workers to the conclusion that both these amino acid residues were in the amorphous region of the protein. This important concept of availability of amino acids has been utilized by Carpenter and Booth[107,108] and others[109] to estimate the availability of amino acid residues in other proteins.

The fact that tyrosine and lysine were readily available in the amorphous region of the fiber allowed Zahn and Zuber[110,111] to make use of a series of molecules as "molecular calipers" in an attempt to determine the distances between adjacent polypeptide chains in the amorphous region of the protein. The molecules they synthesized and used were bifunctional fluorine-containing reagents with molecular lengths of 3–14 Å. After reacting the fiber with such compounds they were able to establish that bifunctional reaction had taken place. Maximum crosslinking occurred in the protein when the reactive molecule was about 10 Å in length. Thus, with 4,4′-difluoro-3,3′-dinitrodiphenylsulfone, 100% bifunctional reaction with tyrosine residues was achieved.

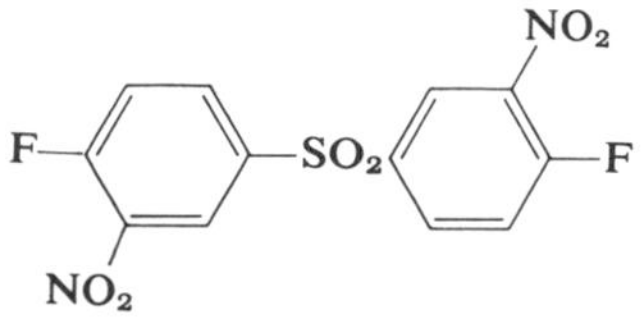

$NO_2$

F— —$SO_2$— —F

$NO_2$

The introduction of chemically stable foreign crosslinks into the protein gave valuable information concerning the molecular structure of the amorphous regions of silk fibroin. Also, from a technological viewpoint, the possibility of introducing crosslinks that would increase the physical and chemical resistance of the fiber was established.

## 2.8. Technological Processes Involved in Silk Dyeing and Finishing

Silk fibroin is not only an interesting fibrous protein but also a valuable and useful textile material. Many of the facts learned by protein chemists concerning the reactivity, stability, and physical properties of the fiber have been known by dyers and finishers for centuries, albeit in empirical ways. As with most industries that have developed via empiricism, dyeing has a vocabulary and a mystique all its own, which varies from fiber to fiber. The success in the preparation, dyeing, and finishing of expensive silk items shows a wealth of knowledge and feeling for what is essentially a delicate biopolymer.

### 2.8.1. Degumming of Silk

The first stage in the preparation of silk for dyeing is to separate the fiber from the silk gum (fibroin from sericin). This has been referred to earlier in this chapter; methods such as boiling water or steam have been used in the past. The use of enzymic methods have also been employed successfully. However, the preferred method is to use dilute soap solutions. The action of the soap is thought to be due to the formation of a readily dispersed alkali derivative of the sericin, by the action of the alkali formed on the hydrolysis of the soap. Thus to use nonhydrolyzable synthetic detergents is not a good alternative. The action of the soap may be increased by adding alkali to give solutions of pH 10.0.

In practice, raw silk may be degummed by boiling the fiber for 2 h at 95°C in 0.5% soap solution. The silk is usually placed in linen mesh bags to prevent entanglement and reduce abrasion. The degree of degumming desired will dictate whether or not a second degumming process is required. The sericin which is removed will be in suspension in the liquor. This is known as "boiled-off liquor" and is used extensively in the dyeing of silk. Recently Machon et al.[112] have introduced a degumming process that utilizes sodium oleate, sodium hexametaphosphate, and an alkali. This system has proved excellent for degumming as well as for scouring the fiber.

### 2.8.2. Bleaching of Silk

The next step in the processing of silk fibers is to bleach the fibers to remove any residual color present. Sulfur stoving of silk has been used in the past but has been superseded by aqueous chemical methods. The most common method used today is the hydrogen peroxide method. In this process the silk is treated with hydrogen peroxide (100 vol), sodium silicate, and a soap solution. The liquor is raised to 90°C and allowed to cool overnight. The fibers are then washed and rinsed.

### 2.8.3. Weighting of Silk

After silk has been degummed, a loss in weight occurs as a result of the removal of the sericin. In practice this weight loss, as well as loss in "handle" and "drape" of the fiber (or fabric), is regained by the process of weighting. Basically, weighting is a three-stage process in which the silk is allowed to absorb, in sequence, various inorganic salts. First, the fiber is treated with anhydrous stannic chloride, which is subsequently hydrolyzed to stannic oxide by water. This salt is insoluble and remains within the fiber. Next, the silk is allowed to absorb sodium phosphate. The phosphate ion complexes with the tin. The process can be repeated to increase the amount of salts within the fiber. Finally, the treated silk is boiled in sodium silicate to adjust the exact amounts of weighting materials present in the fiber.

### 2.8.4. Dyeing of Silk

As silk fibers are proteinaceous and possess a variety of functional groups, it is not surprising that the fibers are relatively easy to dye. In fact, silk has a remarkable affinity for most classes of commercial dyestuffs. Thus, most dyestuffs which are applicable to wool can be applied to silk (e.g., acid, basic, metal-complex, vats, and reactive dyestuffs have been used, as well as the older, natural colors, such as logwood). Silk textiles are usually found in the "luxury trade" (e.g., evening wear, braids, hatbands, tapestry threads). Most of these items demand clear, bright shades but on the whole do not require high fastness properties, particularly to light. Consequently, a dye may be chosen for its color rather than placing too much emphasis on its other properties. For a more detailed exposition of dye selection for silk goods, other, more specialized texts should be consulted.[113]

Acid dyestuffs are widely used in silk dyeing. With this class of dyestuff a wide range of shades with good fastness properties can be produced. It is found in practice that acid dyestuffs tend to have low covering power on the fiber, and if very deep shades are used, "bronzing" occurs. However, to produce deep rich shades a basic dyestuff may be added to the acid-dyed fiber. When dyeing with acid dyestuffs, it is necessary to use a retarding agent to reduce the rate of dyeing and thus give a level distribution of the dyestuff on the fiber. To this end, an anionic detergent may be used or "boiled-off liquor" (i.e., soap solution containing sericin from the degumming process). Both equalizing acid dyestuffs, using an acid dye bath, and milling dyestuffs, using neutral conditions with 20% sodium sulfate, can be used on the fiber.

Basic dyestuffs are extremely useful for application to silk where brightness and clarity of shade are required, but where fastness properties are of secondary importance. Again, it is necessary to use a retarding agent in the form of the "boiled-off liquor" or acetic acid in order to promote levelness. The wet-fastness properties of the basic dyeings can be improved by after-treatment with tannic acid solutions.[114] More recently, some workers[115] have used a fusion product of phenol, sulfur, and sodium hydroxide as an after-treatment for basic dyed silk. The resulting finish did not affect the physical characteristics of the fiber but did improve the wet-fastness properties.

Neutral dyeing metal-complex dyestuffs have also been used for silk. These can be applied at the boil from solutions containing 20% sodium sulfate. To decrease the dyeing time and enable lower temperatures to be used,[116] use has also been made of organic solvents such as benzyl alcohol or *n*-butanol.

Reactive dyestuffs for cellulosic and wool fibers can be applied quite satisfactorily to silk by both acid and alkaline dyeing procedures. The latter method is preferred, as it gives more level and reproducible results. Bakker and Johnson[117] studied the dyeing of silk with monochlorotriazinyl dyestuffs and concluded that dye–fiber hydrolysis occurred during dyeing and recommended that fixation of the dyestuff on the fiber could be improved by the use of tertiary base catalysts.

Mordant dyestuffs were used extensively on silk prior to the advent of synthetic organic dyestuffs. In recent times their importance has declined, principally because of the tediously long dyeing process involved. Chrome mordanting of silk differs from that of wool in that the silk has no affinity for the chromic acid. Silk also has a much lower capacity for dissociating metallic salts than does wool.[118] Thus, to mordant silk, concentrated solutions of basic salts have to be used, and precipitation of the basic salt or metallic hydroxide is achieved by treatment with sodium silicate. After mordanting, the fiber is dyed in an acid dyebath with boiled-off liquor or detergent.

Very little work has been carried out on the mechanism of reaction or the physical chemistry of the dyeing processes of silk fibers. Work that has been reported usually appears as an addendum to the work on keratin. This probably reflects the lack of commercial interest today in silk fibers.

### 2.8.5. Finishing of Silk

As silk, owing to its high cost, tends to have limited usage, and goods made of silk are not normally subjected to harsh treatment, little work has been carried out on surface finishes. Occasionally in the past such compounds as gums, dextrin, and starch have been used to give stiff finishes to the fabric.[119] Generally, recent developments in polymer and resin chemistry have led to the application of such compounds to textile substrates. In all cases the use of a synthetic resin should enhance the natural properties of the textile, and in no way detract from them. Murase and Shiozaki[120] treated silk fibers with methylolacrylamide and stannic chloride by a pad–dry–bake process to impart a wash-and-wear finish to silk. By this process good wet and dry crease-recovery properties were attained. More recently, Shiozaki and Tanaka[121,122] have carried out work on the reaction of mono- and multifunctional epoxides on silk. This work showed that the reaction rates of the compounds with the protein were governed by the type of salt used in the process. The finished fabric exhibited good moisture-regain and flexibility, and excellent crease-resistant, properties.

Attempts have also been made to improve the strength and abrasion resistance of silk fibers. In this process the silk fibers were treated with aqueous solutions of a dialdehyde polysaccharide and their tensile strengths were improved.[123] If the fiber was treated with an aqueous solution of silica and an organic polymer containing an isocyanate group, then increased resistance to abrasion and creasing resulted.

Finally, attempts have been made to impart durable flame and soil resistance to fabrics[124] by treating silk with aqueous solutions of condensed phosphoric acid and a quaternary ammonium salt. The finish so produced is claimed to be effective and durable.

## 2.9. References


1. E. Fischer, *Chem. Ber.* **34**, 433 (1901).
2. E. Fischer and A. Skita, *Z. Physiol. Chem.* **33**, 177 (1901).
3. E. Fischer and E. Abderhalden, *Chem. Ber.* **40**, 3544 (1907).
4. E. Abderhalden, *Z. Physiol. Chem.* **120**, 207 (1922).
5. F. Lucas, J. T. B. Shaw, and S. G. Smith, *Advan. Protein Chem.* **13**, 107 (1958).
6. M. Florkin and C. Jeuniaux, *Arch. Intern. Physiol. Biochim.* **66**, 552 (1958).
7. M. Florkin, S. Bricteux-Grégorie, and A. Dewandre, *Biochim. Z.* **333**, 370 (1960).
8. M. Florkin, *Bull. Acad. Roy. Belg.* **4**, 441 (1965).
9. Anonymous, *Turtox News* **26**, No. 8 (1948).
10. G. Fraenkel and K. M. Rudall, *Proc. Roy. Soc. (Lond.)* **B134**, 111 (1947).
11. R. H. Hackmann and M. Goldberg, *J. Insect Physiol.* **2**, 221 (1958).
12. R. H. Hackmann, *Proc. 4th Intern. Congr. Biochem., Vienna* (1958).
13. K. M. Rudall, in: *Comparative Biochemistry* (M. Florkin and H. S. Mason, eds.) Vol. IVB, Chap. 9, p. 397, Academic Press, New York (1962).
14. F. Lucas, *Discovery* **25**, 20 (1964).
15. F. Lucas and K. M. Rudall, in: *Comprehensive Biochemistry* (M. Florkin and H. Stotz, eds.), Vol. 26B, Chap. VII, Elsevier Publishing Company, Amsterdam (1968).
16. F. O. Howitt, *Bibliography of the Technical Literature on Silk*, The Hutchinson Publishing Group Ltd., London (1946).
17. J. Kirimura, *Protein Chemistry*, Vol. 5, Kyonitsu Shuppan, Tokyo (1957).
18. D. J. Lloyd and P. B. Bidder, *Trans. Faraday Soc.* **31**, 864 (1935).
19. M. C. Corfield, F. O. Howitt, and A. Robson, *Nature (Lond.)* **174**, 603 (1954).
20. M. S. Dunn, M. N. Camien, L. S. Rockland, S. Shanckman, and S. C. Goldberg, *J. Biol. Chem.* **155**, 591 (1944).
21. D. Coleman and F. O. Howitt, *Symposium on Fibrous Proteins*, Society of Dyers and Colourists, p. 144, Bradford, England (1946); D. Coleman and F. O. Howitt, *Proc. Roy. Soc. (Lond.)* **109A**, 145 (1947).

22. E. Waldschmidt-Leitz and O. Zeiss, *Z. Physiol. Chem.* **300**, 49 (1955).
23. R. Signer and R. Strässle, *Helv. Chim. Acta* **30**, 155 (1947).
24. F. H. Holmes and D. I. Smith, *Nature* (*Lond.*) **169**, 193 (1952).
25. E. H. Mercer, *Textile Res. J.* **24**, 135 (1954).
26. B. Drucker and S. G. Smith, *Nature* (*Lond.*) **65**, 196 (1950).
27. R. Signer and R. Glanzmann, *Makromol. Chem.* **5**, 257 (1951).
28. K. Narita, *J. Chem. Soc.* (*Japan*) **75**, 1005 (1954).
29. G. Braunitzer and D. Wolff, *Z. Naturforsch.* **106**, 404 (1955).
30. H. Zahn and A. Würz, *Biochem. Z.* **322**, 327 (1952).
31. J. T. B. Shaw and S. G. Smith, *J. Textile Inst.* **45**, 934T (1954).
32. E. Abderhalden and A. Bahm, *Z. Physiol. Chem.* **215**, 246 (1933); **219**, 72 (1933).
33. M. Levy and E. Slobodian, *Cold Spring Harbor Symp. Quant. Biol.* **14**, 113 (1949); M. Levy and E. Slobodian, *J. Biol. Chem.* **199**, 563 (1952).
34. H. G. Ioffe, *Biokhimiya* **19**, 495 (1954).
35. E. Slobodian and M. Levy, *Fed. Proc.* **11**, 288 (1952).
36. L. M. Kay and W. A. Schroeder, *J. Amer. Chem. Soc.* **76**, 3564 (1957).
37. F. Lucas, J. T. B. Shaw, and S. G. Smith, *Biochem. J.* **66**, 468 (1957).
38. J. O. Warwicker, *Acta Cryst.* **7**, 565 (1954).
39. H. Zahn and E. Schnabel, *Ann. Chem.* **604**, 62 (1957).
40. E. Schnabel, *Ann. Chem.* **615**, 165 (1968).
41. E. Schnabel and H. Zahn, *Ann. Chem.* **614**, 141 (1958).
42. E. Schnabel, *Ann. Chem.* **615**, 173 (1958).
43. F. H. C. Stewart, *Aust. J. Chem.* **19**, 489 (1966).
44. R. D. Fraser, T. P. MacRae, and F. H. C. Stewart, *J. Mol. Biol.* **19**, 580 (1966).
45. H. Zahn, W. Schade, and K. Ziegler, *Biochem. J.* **104**, 1019 (1967).
46. F. Lucas, J. T. B. Shaw, and S. G. Smith, *Biochem. J.* **83**, 164 (1962).
47. K. Ziegler and H. Spoor, *Biochim. Biophys. Acta* **33**, 138 (1959).
48. K. Ziegler and N. H. LaFrance, *Z. Physiol. Chem.* **322**, 21 (1960).
49. F. Lucas, unpublished work reported in *Comprehensive Biochemistry* (M. Florkin and E. H. Stotz, eds.), Vol. 26b, Chap. VII, Elsevier Publishing Company, Amsterdam (1968).
50. H. Zuber, *Kolloid-Z.* **179**, 100 (1961).
51. J. J. Cebra, *J. Immunol.* **86**, 197 (1961).
52. J. T. B. Shaw, *Biochem. J.* **93**, 45 (1964).
53. R. S. Asquith, M. S. Otterburn, and W. J. Sinclair, *Angew. Chem. Intern. Ed.* **13**, 514 (1974).
54. W. A. Schroeder and L. M. Kay, *J. Amer. Chem. Soc.* **77**, 3908 (1955).
55. H. Zuber, K. Ziegler, and H. Zahn, *Z. Naturforsch.* **12b**, 734 (1957).
56. C. Earland and D. J. Raven, *Nature* (*Lond.*) **192**, 1185 (1961).
57. C. Earland and S. P. Robins, *Experientia* **25**, 905 (1969).
58. C. Earland and S. P. Robins, *Intern. J. Peptide Protein Res.* **5**, 327 (1973).
59. A. Robson, J. M. Woodhouse, and Z. H. Zaidi, *Intern. J. Protein Res.* **2**, 181 (1970).
60. W. Schade, I. Liensenfeld, and K. Ziegler, *Kolloid-Z.* **242**, 1161 (1970).
61. J. O. Warwicker, *J. Mol. Biol.* **2**, 350 (1960).
62. F. Lucas and K. M. Rudall, in: *Symposium on Fibrous Proteins, Australia, 1967* (W. G. Crewther, ed.), Butterworth & Co. (Australia) Ltd., Sydney (1968).
63. R. S. Asquith, M. S. Otterburn, J. H. Buchanan, M. Cole, J. C. Fletcher, and K. L. Gardner, *Biochim. Biophys. Acta* **221**, 342 (1970).
64. R. S. Asquith, M. S. Otterburn, and K. L. Gardner, *Experientia* **27**, 1388 (1971).
65. D. J. Raven, C. Earland, and M. Little, *Biochim. Biophys. Acta* **251**, 96 (1971).

66. T. Weis-Fogh, *J. Exptl. Biol.* **37**, 889 (1960).
67. S. O. Andersen, *Biochim. Biophys. Acta* **93**, 213 (1964).
68. F. Lucas, *Nature* (*Lond.*) **210**, 952 (1966).
69. C. J. Huber, in: *Matthews' Textile Fibers* (H. R. Mauersberger, ed.), 5th ed., John Wiley & Sons, Inc., New York (1947).
70. M. S. Otterburn and W. J. Sinclair, *J. Sci. Food Agr.* **24**, 929 (1973).
71. A. S. Tweedie, *Can. J. Res.* **16**, 134 (1938).
72. M. V. Korchagin, *Zh. Prikl. Khim.* **25**, 212 (1952).
73. B. H. Nicolet and L. A. Shinn, *J. Biol. Chem.* **140**, 685 (1941).
74. K. L. Gardner, Ph.D. thesis, University of Leeds (1973).
75. M. Bergmann, E. Brand, and F. Weinmann, *Z. Physiol. Chem.* **131**, 1 (1923).
76. K. Iwai and T. Ando, *Methods Enzymol.* **11**, 263 (1967).
77. D. F. Elliott, *Biochem. J.* **50**, 542 (1952).
78. K. Narita, *J. Amer. Chem. Soc.* **81**, 1751 (1959).
79. S. Sakakibara, H. H. Shin, and G. F. Hess, *J. Amer. Chem. Soc.* **84**, 4921 (1962).
80. E. Mercer, *Aust. J. Sci. Res.* **5**, 365 (1952).
81. J. B. Speakman and C. S. Whewell, *J. Soc. Dyers Colourists* **52**, 380 (1936).
82. M. J. Horn, D. B. Jones, and S. J. Rignel, *J. Biol. Chem.* **138**, 141 (1938).
83. M. C. Corfield, C. Wood, A. Robson, M. J. Williams, and J. M. Woodhouse, *Biochem. J.* **103**, 15c (1967).
84. R. S. Asquith, A. K. Booth, and D. J. Skinner, *Biochim. Biophys. Acta* **181**, 164 (1969).
85. R. S. Asquith and J. J. García-Domínguez, *J. Soc. Dyers Colourists* **84**, 155 (1968).
86. R. S. Asquith and P. Carthew, *Biochim. Biophys. Acta* **278**, 346 (1972).
87. Z. Bohak, *J. Biol. Chem.* **239**, 2878 (1964).
88. K. Ziegler, *J. Biol. Chem.* **239**, PC2713 (1964).
89. K. Ziegler, *Proc. 3rd Intern. Wool Textile Res. Conf., Paris* **2**, 312 (1965).
90. P. Mellet and D. F. Louw, *Chem. Commun.* **17**, 396 (1965).
91. A. Robson and Z. H. Zaidi, *J. Textile Inst. Trans.* **58**, 267 (1967).
92. K. Ziegler, *Nature* (*Lond.*) **214**, 404 (1967).
93. P. Alexander, D. Carter, and C. Earland, *Biochem. J.* **47**, 251 (1950).
94. M. Nakanishi and K. Kobayashi, *J. Soc. Textile Cellulose Ind. Japan* **10**, 128, 131 (1954).
95. D. B. Das and J. B. Speakman, *J. Soc. Dyers Colourists* **66**, 583 (1950).
96. C. Schrile and J. Meybeck, *Compt. Rend.* **232**, 732 (1951).
97. D. A. Sitch and S. G. Smith, *J. Textile Inst. Trans.* **48**, 341 (1957).
98. S. Akune, *Bull. Fac. Agr. Kagoshima Univ.* **2**, 91, 97 (1953).
99. S. Akune and K. Koga, *Bull. Fac. Agr. Kagoshima Univ.* **2**, 103 (1953).
100. F. O. Howitt, *Textile Res. J.* **25**, 242 (1955).
101. R. S. Asquith, I. Bridgeman, and A. J. Smith, *Proc. 3rd Intern. Wool Textile Res. Conf., Paris* **2**, 385 (1965).
102. C. Earland and J. G. P. Stell, *Biochim. Biophys. Acta* **23**, 97 (1957).
103. C. Earland, J. G. P. Stell, and A. Wiseman, *J. Textile Inst. Trans.* **51**, 817 (1960).
104. C. Earland and J. G. P. Stell, *Polymer* **7**, 549 (1966).
105. D. French and J. T. Edsall, *Advan. Protein Chem.* **2**, 277 (1945).
106. H. Fraenkel-Conrat and H. S. Olcott, *J. Amer. Chem. Soc.* **68**, 34 (1946).
107. K. J. Carpenter, *Biochem. J.* **77**, 604 (1960).
108. K. J. Carpenter and V. H. Booth, *Nutritional Abstr. Rev.* **43**, 424 (1973).
109. R. S. Asquith, D. K. Chan, and M. S. Otterburn, *J. Chromatog.* **43**, 382 (1969).
110. H. Zahn and H. Zuber, *Textil-Rundschau* **9**, 119 (1954).

111. H. Zuber, Dissertation, University of Heidelburg (1953).
112. R. Machon, J. Fléchet, and E. Hugo, Brit. Pat. 1,278,707 (1971).
113. C. C. Wilcock and J. L. Ashworth, *Whittakers-Dyeing with Coal-Tar Dyestuffs*, 6th ed., Baillière, Tindall and Cox, London (1964).
114. C. H. Giles, *A Laboratory Course in Dyeing*, Society of Dyers and Colourists, Bradford, England (1971).
115. A. N. Filippov and Z. A. Nabatnikova, *Sb. Nauchn.-Issled. Rabot Khim. i Khim. Tekhnol. Vysokomolekul. Soedin., Tashkentsktekstil'n Inst.* **1**, 299 (1964).
116. H. R. Chipalkatti, C. H. Giles, and D. G. M. Vallance, *J. Chem. Soc.* 4375 (1954).
117. P. G. H. Bakker and A. Johnson, *J. Soc. Dyers Colourists* **89**, 203 (1973).
118. C. H. Giles, *J. Soc. Dyers Colourists* **60**, 312 (1944).
119. J. T. Marsh, *An Introduction to Textile Finishing*, Chapman & Hall Ltd., London (1957).
120. P. Murase and H. Shiozaki, *Bull. Textile Res. Inst. Japan* **79**, 17 (1966).
121. H. Shiozaki and Y. Tanaka, *Bull. Res. Inst. Polymers Textiles Japan* **94**, 15 (1971).
122. H. Shiozaki and Y. Tanaka, *Bull. Res. Inst. Polymers Textiles Japan* **99**, 27 (1972).
123. Miles Laboratories, U.S. Pat. 3,479,128 (1965).
124. Asaki Chemical Co. Ltd., Brit. Pat. 1,069,946 (1967).

# 3

# The Histology of Keratin Fibers

## J. A. Swift

### 3.1. Introduction

*Keratin* (Greek, horn + in) is the generic term applied to the resilient structures such as hair, horn, nail, feather, and skin which comprise the integument and appendages of the higher vertebrates and whose prime function is to protect the animals from their environment. They are composed mainly of protein and are characterized by their high cystine content[1] and birefringence[2] and by the fact that they yield an $\alpha$-type X-ray diffraction pattern which reverts to a $\beta$ type on stretching.[3]

The keratins have been subdivided into two groups, "hard" and "soft," according to their chemical and physical properties but more particularly according to the amount of cystine they contain.[4,5] On this basis, skin with a sulfur content of less than 3% is defined as a soft keratin, and the considerably tougher materials, such as horn, hoof, and hair, which contain more than 3% sulfur, are classified as hard keratins. The present chapter is concerned with the microscopic structure of keratin fibers (hair and wool) in the latter group of materials. Particular emphasis will be placed on the chemical properties of the various microscopic components of these fibers.

Although keratin fibers have an overall cylindrical form with an external overlapping scale pattern, there are many subtle variations in shape and surface architecture not only in fibers from different mammalian species (Figs. 3-1 to 3-4) but also within a single species, as for example in man (Figs. 3-5 to 3-8).

Microscopic studies of the internal structure of keratin fibers compel us

---

**J. A. Swift** • Unilever Research, Isleworth Laboratory, Isleworth, Middlesex, England

to recognize that they are multicomponent and that, like their external form, there are many generalizations of structure with superimposed fine variations both between and within the animal species. Similar variations also occur in the chemical composition of keratin fibers.[6] For these reasons it is clear that "keratin" cannot be considered as a material of clearly defined composition, and the term should be used in only a generalized way and certainly with caution.[7,8]

## 3.2. Microscopy Methods

Microscopy studies of keratin fibers are carried out in many diverse disciplines, covering such fields as textiles, criminology, archeology, dermatology, and cosmetics. It is pertinent, therefore, to mention here some of the

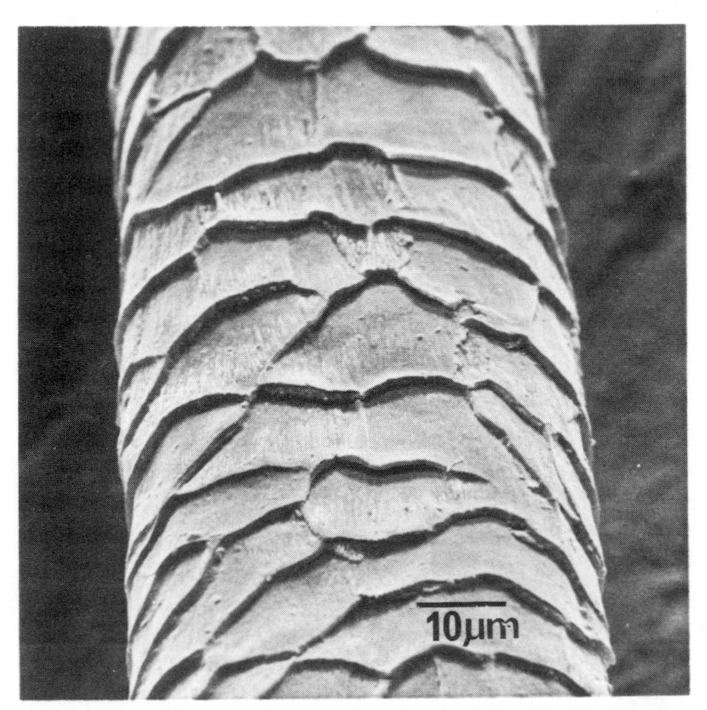


Figs. 3-1–3-4. Scanning electron micrographs of hairs from (1) 64's Merino wool, (2) barbary sheep, (3) bat-eared fox, and (4) otter. The following abbreviations are used on the electron micrographs in this chapter: A, A layer; Co, cortex; CoCM, cortical cell membrane complex; CuCM, cuticle cell membrane complex; Cy, cell cytoplasm; D, delta band; E, endocuticle; Ex, exocuticle; I, intercellular space; Im, intermacrofibrillar matrix; In, inner layer; M, melanin pigment granule; Ma, macrofibril; Me, medulla; Mi, microfibril; N, nucleus; NR, nuclear remnant; Nu, nucleolus; O, orthocortex; P, paracortex.

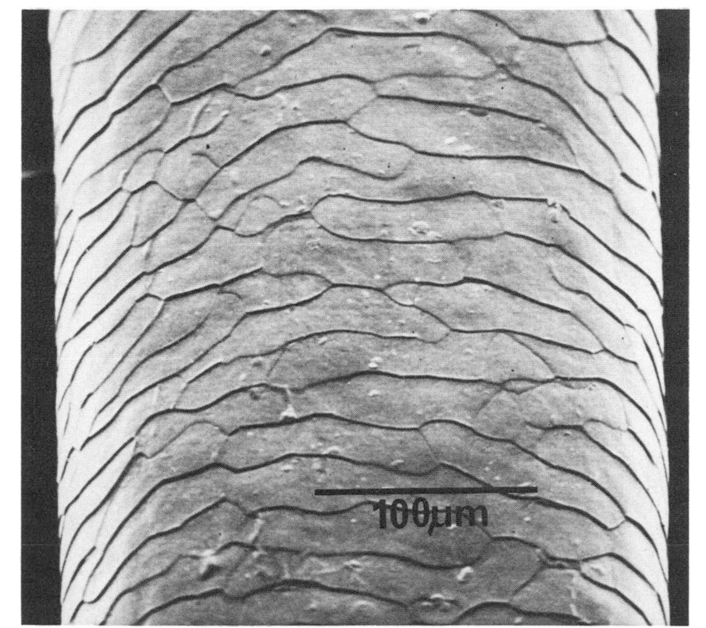


Fig. 3-2

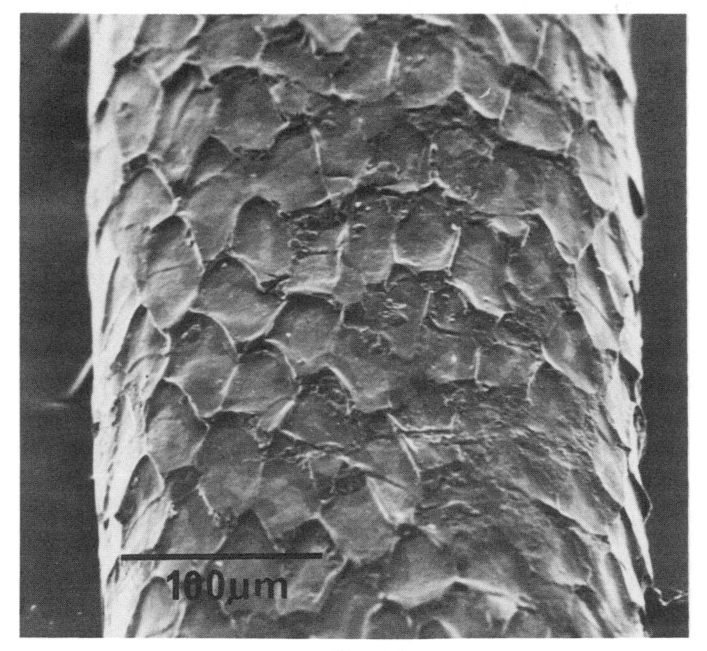


Fig. 3-3

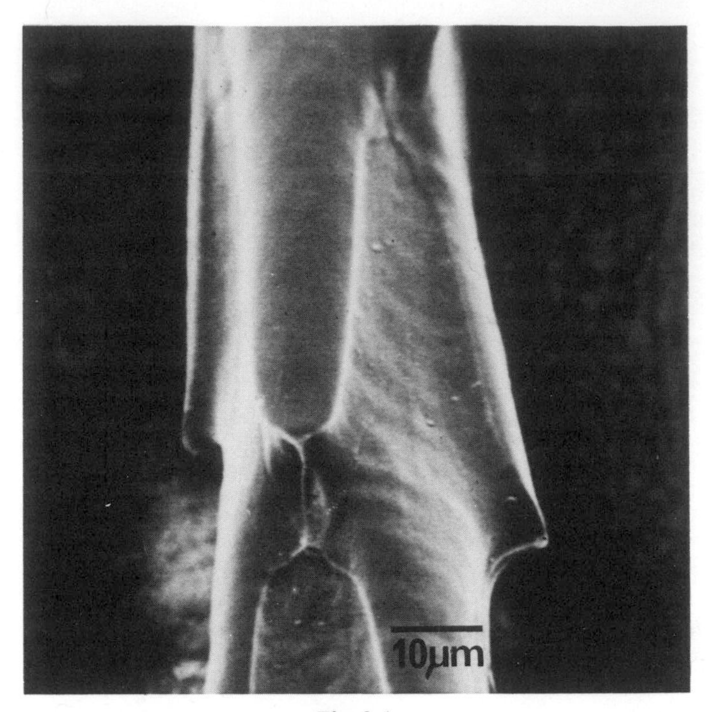


Fig. 3-4

microscopic methods available, although for detailed descriptions and variations the reader is referred to more extensive reviews.[9–13]

### 3.2.1. Optical Microscope Methods

The optical (or light) microscope has been by far the most widely used instrument in the microscopic study of keratin fibers, and indeed some of the earliest studies of wool and hair using this instrument were recorded by Hooke in the *Micrographia*.[14]

For simple studies of the diameter, shape, color, and to some extent scale pattern and internal structure, keratin fibers are simply mounted and examined by transmitted light. The high refractive index of the fibers tends to make the direct study of their surface architecture difficult, although this may be overcome to some extent, and the scale margins revealed with better clarity, by mounting the fibers in a medium of refractive index sufficiently removed from that of the fibers themselves.[10] The vacuum evaporation of certain metals onto the fiber surface is particularly useful for the subsequent study of surface topography with the optical microscope.[15]

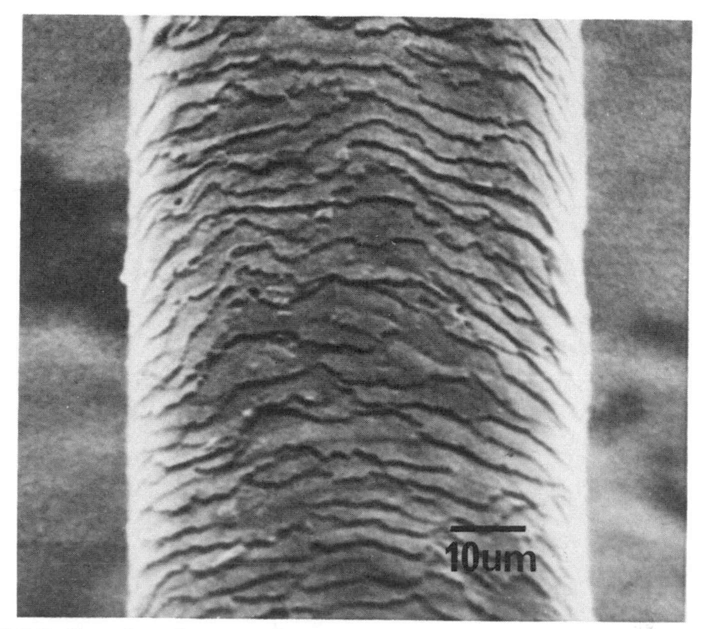


Figs. 3-5–3-8. Scanning electron micrographs of Caucasian human (5) head hair, (6) axillary hair, (7) pubic hair, and (8) moustache.

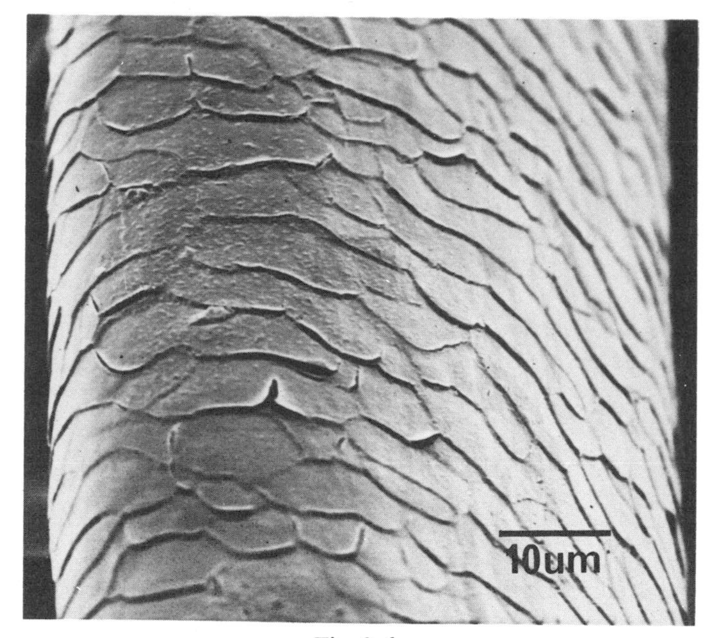


Fig. 3-6

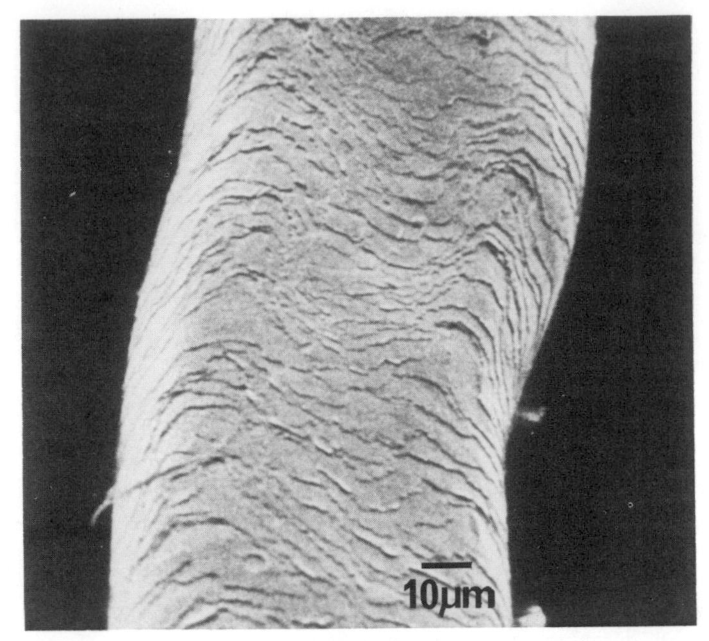


Fig. 3-7

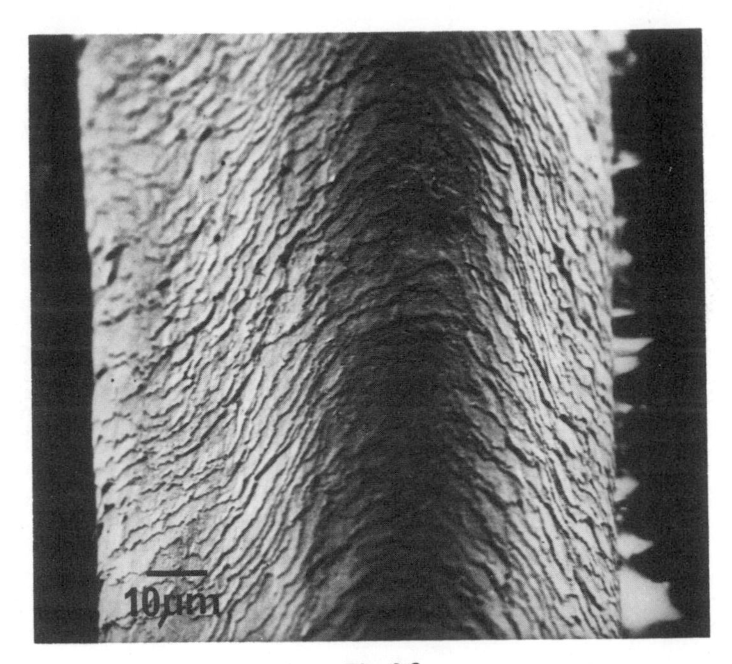


Fig. 3-8

Another quick method for examining fiber scale patterns with the optical microscope is to use a cast or impression of the fiber in a suitable plastic material.[16,17] In an extension of this approach, the impressions of a "rolled" fiber permit examination of the scale patterns around the whole circumference of the fiber.[10] This technique has been used extensively in the study of keratin fibers from many different animal species.[10]

A very sensitive method for observing minute surface irregularities is by interference microscopy. This technique has been pioneered particularly by Tolansky[18] and extended for the observation of fiber surfaces by Howell and Mazur[19] and by Skertchly.[20] It does, however, suffer the disadvantage that only minute areas of an individual fiber surface can be studied at one time. In Skertchly's method the fiber to be examined is coated with a thin layer of Canada balsam to give a smooth outer surface and sandwiched between two sheets of optically flat glass. This specimen is illuminated by monochromatic sodium light which is reflected into a compound microscope. Interference occurs between light reflected from the outer, smooth balsam surface and that reflected from the inner balsam surface, which conforms to the shape of the fiber. The interference patterns so observed in the compound microscope relate to the minute irregularities of the fiber surface.

For the detailed study of the internal structure of keratin fibers with the optical microscope, it is necessary to examine slices of the fibers. Although transverse sections of fibers embedded in a suitable supporting medium can be cut on a conventional microtome, many of the simple hand microtomes, [21–23] particularly the Hardy microtome,[24] are extremely convenient for the rapid preparation of fiber sections of reasonable quality. Such sections can be used for examining cross-sectional shape and size, the distribution of natural pigments, for assessing the extent of medullation, and for determining the extent of dyestuff penetration. By treating the cut sections with various colored stains or dyes, some of the fine structure within the fiber sections may be rendered visible, and by applying techniques of histochemistry, in which the attachment of a stain can be made to occur at a specific chemical site,[25] the distribution of various chemical components within the fiber section can be determined.

The techniques required for the optical microscope study of such soft tissues as hair follicles and skin differ somewhat from those for mature fibers. In general, to preserve the spatial arrangement of the various morphological components, the piece of tissue is treated with a chemical reagent, such as formalin, which will introduce crosslinks (this step is known as fixation), and all the water is replaced by a suitable supporting medium, such as paraffin wax.[25] The embedded specimen is appropriately oriented and trimmed and sections of the order of 5 $\mu$m in thickness are cut on a mechanical microtome so as to contain the portion of tissue of interest. After mounting on glass microscope slides the paraffin is removed from the sections with a suitable solvent, such as xylene (this step is known as "clearing"), leaving the tissue proper attached to the slide. Before examination in the optical microscope several general colored dyes or stains are used to treat the sections so that the components in the sections will be rendered visible.

Particular mention should be made of the stain Sacpic, which was developed by Auber[26] for use with hair-follicle sections. This stain has been used widely in the study of follicle structure and is particularly valuable for revealing the various stages in the formation of the fully hardened fiber. In addition, a wide variety of histochemical procedures[25,27] are available for examining the distribution of chemical components among the morphological structures observed under the optical microscope. Also, by applying histochemical methods usually to frozen sections of the original tissue, the distribution of various enzyme systems can be studied, although the internal structure within these frozen sections tends not to be so well preserved.[25]

Phase-contrast microscopy is a particularly powerful optical microscope technique for enhancing image contrast. Using this method it is possible to observe structure such as cell boundaries of sectioned keratin fibers which are normally invisible using the optical microscope in its conventional mode of operation. Appleyard and Dymoke[28] have used it to particular advantage for observing the boundaries of cortical cells in cross sections of keratin fibers mounted in *o*-chlorophenol.

Polarization microscopy is an additional method which is useful in the microscopic study of keratin fibers. It can be used for determining refractive index and, more particularly, the birefringence of fibers[29] (i.e., the numerical difference between the refractive indexes parallel and perpendicular to the fiber axis), a physical phenomenon that reflects the oriented internal structure of fibers.[30] Examination of polarization colors is also a valuable aid in the identification of fibers.[12,31]

The microscopic examination of the visible light emitted from specimens which are illuminated with ultraviolet irradiation (i.e., fluorescence microscopy) has also been used in the study of keratin fibers. For example, by this technique yellow autofluorescence of fungoid growth of mildewed wool has been observed[12] and for wool stained with 1% benzopurpurin 10B, mechanical damage in the wool appears pink, acid damage bluish pink, alkali damage yellowish pink, and the intact wool colorless.[12]

Fluorescence microscopy can also be used for identifying the location of antibody–antigen interaction in tissues.[25] Indeed, Kemp and Rogers[32] have used methods of immunofluorescence for determining the location of sulfur-rich proteins in the hair follicle.

Autoradiography is another powerful adjunct to optical microscopy and has been used to investigate the incorporation of various radio-labeled compounds into the hair follicle. The method usually involves feeding the animal concerned with the radio-labeled compound, cutting sections of the follicles, and overlaying these with a photographic emulsion which is exposed and then processed with developer and fixed in the usual manner. Under the optical microscope, blackening of the photographic emulsion layer due to the presence of the label within the tissue can be seen with reference to the overlying structure of the tissue section. Such techniques have been used, for example, to study the migration of cells in the hair follicle (using tritiated thymidine),[33] to examine the incorporation of cystine into the follicle (using $^{35}$S-cystine),[34] and to study the site of melanin synthesis and associated tyrosinase activity in the hair follicle (using $^{14}$C-tyrosine).[35]

### 3.2.2. Conventional Electron Microscope Methods

Whereas the optical microscope is limited in resolution to about 0.2 $\mu$m and has an exceedingly narrow depth of focus (about 0.2 $\mu$m for microscopes of high numerical aperture), the conventional transmission electron microscope is capable of high resolution (down to 2 nm for most biological materials) and has a depth-of-image focus of the order of 2 $\mu$m, which is consideraby greater than the normal specimen thickness (about 100 nm).

The earliest electron microscope studies of keratin were made in the early 1940s by Zahn[36] and Hock and McMurdie,[37] who examined wool fragments obtained by different physical and chemical treatments. Since that time, advances have been made in both the performance of the commercial instrument and in specimen preparation, so that the technique is now widely used in many branches of scientific research. This does not mean that the conventional electron microscope is the panacea of microscopy, for it presents its own problems of specimen preparation (particularly for animal fibers) and interpretation of the actual images obtained. Certainly it is difficult to carry out the speedy, routine examination of fibers that one normally would carry out with the optical microscope.

In the simplest form of specimen preparation the specimen is supported directly on a mesh electron microscope grid and observed in the microscope by transmission. Such methods were used in the early studies of keratin fiber fragments[36,37] and have been used more recently to obtain profiles or silhouettes of wool fibers.[38] Nowadays metal "shadowing"[39] or "negative staining"[40] are more commonly used for enhancing the image contrast of fiber and follicle fragments supported on a relatively electron-transparent thin plastic film overlaying the mesh electron microscope grid.

As with the optical microscope, the use of sections is essential if one is to carry out a detailed study of the internal structure of fibers or follicles with the conventional electron microscope. To limit degradation of the electron image by chromatic aberration caused by the change in energy of the electrons as they pass through the specimen, it is essential to use sections that are extremely thin and of the order of 40–100 nm thick. The fully hardened fiber is therefore supported in a suitable polymer (e.g., Araldite), and ultrathin sections are cut with the aid of an ultramicrotome and mounted on an electron microscope support grid.[41] For the examination of soft tissues such as the hair follicle, it is necessary to preserve the structural integrity of the specimen by chemical fixation, then to replace the tissue water with a suitable solvent, such as ethanol, followed by infiltration of a suitable resin, which is subsequently hardened.[42] From such embedded tissues, ultrathin sections are cut and mounted on the specimen support grid.

Since fiber and follicle sections are composed mainly of low-atomic-number elements, there is little differential scattering of electrons in the electron microscope from one fiber component to another, resulting in insufficient image contrast to permit the discrimination of many of the substructures present in the section. To facilitate differential scattering of electrons by the various components which will enable structure to be seen, the sections are usually impregnated with one of a variety of heavy-metal compounds. Many such "electron stains" enable one to identify the fine structural components of the section,[42] but because of the complexities of interaction between many of the stains with the specimen, it is often not possible to specify that staining of specific chemical groups has occurred.

Although the interpretation of transmission electron micrographs of sectioned keratin fibers is relatively straightforward, at high magnification interpretations must be made with caution and certainly with due regard to the physical interaction of the incident electrons with the specimen.[43] More recently, a few methods have been devised for attaching heavy elements to specific chemical sites in tissues or their sections for electron microscope identification. This aspect will be covered in more detail in Section 3.4.4. In addition, a few electron histochemical techniques are available for identifying the location of enzyme systems[44] and antibody–antigen interactions,[45] in tissues but few of these have been applied in the electron microscope study of keratin fibers and follicles.

By overlaying electron microscope sections with photographic emulsions autoradiographic studies can be made of specimens that have been radioactively labeled.[46] This method permits the identification of the radioactive emissions of the attached labeling substance with respect to the normal morphological components of the tissue as they are normally seen under the electron microscope. The resolution of the technique is limited to about 0.1–0.2 $\mu$m at best, and it is necessary to choose labeling atoms that have very weak $\beta$ emission; in this respect, tritium is the commonly used label. One disadvantage of the technique is that migration of the labeled substance may occur from its original position during the procedures leading up to the preparation of the thin section, which is overlayed with emulsion. This has been overcome to some extent more recently with the development of techniques for cutting thin frozen tissue sections.[47] Such sections are mounted on a cold electron microscope grid and then overlayed with the emulsion.

Except by reflection electron microscopy,[48] which is a comparatively difficult method to use, it is not possible to examine directly the surface structure of animal fibers in the conventional electron microscope. For the examination of fiber surfaces with this microscope it is necessary to use thin-film-replica techniques. A variety of these methods have been developed which permit the identification of fiber surface detail at quite high magnification,[39] but considerable care has to be taken to avoid artifactual observations. Except at very high magnification, the examination of fiber surface structure is now normally accomplished with the aid of the scanning electron microscope, and this is outlined in detail in the next section.

Recently, the transmission electron microscope has been used to obtain diffraction patterns from sectioned keratin fibers.[40] This technique is particularly powerful, for it enables one to select a small area from the section for diffraction and permits one to examine long-range periodicity. By use of this method, Dobb[49] has shown differences in the organization of the ortho- and paracortical segments of wool.

### 3.2.3. Scanning-Electron-Microscope Methods

The scanning electron microscope (SEM), the first commercial model of which became available in 1965,[50] is extremely diverse in its modes of operation and can provide a wealth of microscopic information about surface architecture, elemental composition, crystalline makeup, and electrical and magnetic properties of specimens. In this microscope a beam of electrons derived from a heated tungsten filament and accelerated through a potential difference of from 1 to 50 kV is brought to focus at the surface of the specimen under study. This beam is scanned in a raster about the specimen surface by means of electromagnetic scanning coils. Interaction of the electron beam results in some of the electrons being backscattered with little or no loss of energy, in the emission of low energy (secondary) and Auger electrons, and in the emission of X rays and light. In addition, a current will pass through the specimen, and if the specimen is sufficiently thin, some of the electrons will be transmitted. Images relating to various properties of the specimen are obtained by detecting one or other of these secondary effects and using the resulting rapidly varying signal to modulate the intensity of brightness of a cathode ray tube which is scanned in synchronism with the electron beam scanning of the specimen surface. Magnification in the final image is governed by the ratio of the size of the cathode ray display tube to the size of the area actually scanned out by the electron beam on the specimen and is normally in the range from about $\times 20$ up to $\times 100{,}000$.

In the most common mode of operation of the scanning electron microscope, secondary electron emission from the specimen is used to yield images that relate to the surface architecture of the specimen. The high depth of image focus of this instrument (typically 150 $\mu$m at $100\times$ and 10 $\mu$m at $10{,}000\times$), its wide magnification range, and its maximum resolving power of about 10 nm have made it valuable in the study of animal fibers, particularly for assessing the effects of various textile processes on the surface structure of wool.[51–53] In the most general method for the preparaton of specimens for surface examination in the SEM, the specimen (e.g., keratin fiber) is glued to a small metal mount and vacuum-coated with a thin layer of a suitable metal. Several metals have been used, but silver is particularly useful, since it is known that even for very thick evaporated deposits, the top surface of the silver very accurately conforms to the underlying specimen surface.[54] Such evaporated metal layers are usually necessary to conduct away the heat generated during impact by the electron beam and to prevent electrostatic charging of the specimen, which would otherwise seriously affect the quality of the subsequent SEM image. Good results have been reported by omitting the metallic coating and, instead, spraying the specimen with a textile anti-static compound.[55] In some cases by operating the scanning electron microscope at low accelerating voltage (e.g., 2.5 kV) keratin fibers have been examined satisfactorily without any sort of coating.[56,57] Using these low accelerating voltages Swift and Brown[56,57] have devised a technique for the SEM so that the same part of the fiber surface can be examined in detail both before and after treatment. This has enabled them to observe minute surface changes due to cosmetic treatments of human hair.

Considerable work has now been carried out on the direct examination of keratin fiber surfaces in the SEM, but by variations in the methods of specimen preparation, additional information about the internal structure of the fibers can be gleaned. Thus Anderson and Lipson[58] have used razor-cut transverse surfaces of wool fibers for assessing the extent of damage in the fibers caused by various chemical treatments, and Swift and co-workers[59] have used the SEM in its secondary electron emissive mode for studying the internal structure of metal-impregnated thick keratin fiber slices. The SEM has also been used in the fluorescence mode (i.e., the detection light emitted from the specimen) for observing the disposition of textile fluorescers on wool fibers.[60]

The SEM has been used to advantage recently in the transmission mode for the study of keratin fiber sections.[61-65] Fiber sections somewhat thicker, larger, and easier to cut than those normally used in conventional electron microscopy can be examined in the SEM in a wide range of direct magnifications, extending from ×20 to a useful maximum of about ×50,000 and with a maximum resolution of about 10 nm. Both metal impregnation and enzyme digestion of the sections has been used to yield the necessary image contrast for observing structural detail. The methods are generally amenable to the routine examination of keratin fiber internal structure at quite a moderate resolution.

Recent developments in methods of electron-probe X-ray microanalysis with the SEM[66] and of methods for carrying out dynamic experiments within this microscope[67] have extended still further the experimental scope of the SEM. They can be expected in the future to contribute yet further to our knowledge of the keratin fiber and the effect of various processes on it.

## 3.3. The Hair Follicle

### 3.3.1. Genesis

Mammalian hair fibers emerge from the skin from follicles, which are embryonically formed down-growths of the basal layers of the epidermis into the dermis.[68] The first sign of formation of the follicle is the localized mitosis of the basal cells of the epidermis, producing a column of cells advancing obliquely into the dermis but still separated from the dermis by the basement membrane. In the human embryo, for example, such events occur during the late second and early third month in the region of the eyebrows, upper lips, and chin.[69] With this advance the dermis responds with the approach of a number of cells, including potential pigment-producing cells of neural crest origin, which position themselves in a recess at the tip of the advancing column. As this hemispherical shaped and recessed unit known as the dermal papilla is formed, the downward column of cells ceases to advance and mitosis in the epidermal cells is restricted to those cells around the apex of the newly formed dermal papilla. During these stages cellular differentiation occurs on the side of the follicle, giving outgrowths which become the sebaceous gland and erector pili muscle; the latter extend from the side of the follicle obliquely through the dermis to a point of attachment at the dermo-epidermal junction.[68]

In the next stage, cell division of the epidermal cells around the apex of the dermal papilla yields a column of cells arranged in the form of concentric cylinders which move upward through the epidermal cells of the initial down-growth. The outer cylinder, identified as the cells of the inner root sheath, initially covers the tip of the column and rapidly undergoes hardening to provide the necessary resilience for the upward thrust. At the level of the opening of the sebaceous gland onto the follicle, the cells of the inner root sheath disintegrate, permitting the continued upward movement of the now-hardening cells of the innermost parts of the cylinder to their point of emergence at the skin surface in the form of a fully hardened hair. The outer part of the follicle, through which the inner root sheath and inner elements progressed upward, is continuous with the basal cells of the epidermis and is identified as the outer root sheath. The cells of the inner root sheath evidently provide the rigid constricting net within which the initial softer elements of the hair are molded. The cells of the hair proper, contained within the inner root sheath as concentric cylinders, undergo differentiation in the part of the hair-follicle bulb matrix cells level with the apex of the dermal papilla. Progressing inward from the inner root sheath, these differentiated cell layers are the cuticle, cortex, and medulla (see Section 3.4). The whole growing fiber is fed by blood vessels in the papilla and around the lower part of the follicle.[70]

### 3.3.2. Hair Growth Cycle

All mammalian keratin fibers grow cyclically in periods of active growth alternated by periods of shedding and resting of the follicle and accompanied by dramatic structural and metabolic changes. There are considerable variations in the cycle among different animal species. In the mouse the hairs have a cycle of 15 to 20 days, and phased growth occurs in well-defined regions of the coat, progressing steadily to other regions (i.e., adjacent follicles tend to behave in the same manner).[70] In man the cycle is measured in years, and the phases of the cycle of each hair are independent.[71] Merino sheep have virtually no cyclic activity, whereas cattle tend to molt seasonally but with at least some of the follicles remaining in an active growth phase.[72] The elephant seal molts annually but with the shedding of large sheets of epidermis.[73] The rate of growth of keratin fibers and the time periods for active growth and resting are determined by many factors, including nutrition,[74] hormones,[75–77] and even light.[78]

It was Dry who first recognized, in 1925–1926,[79] the detailed phases of the hair growth cycle in the coat of the mouse. Although much work has been contributed since then to the finer details of the hair growth cycle, his nomenclature for the various stages is still generally applied to all mammalian keratin fibers.[79] The various stages of the cycle are as follows:

1. *Telogen*. This is the resting phase of the follicle. The base of the hair

shaft is attached by a brushlike end in a plug of mitotically inert cells at about the level of the sebaceous gland. In this situation the hair is readily dislodged and shed as "molt." Beneath the inert plug the dermal papilla exists as a small, spherically shaped zone of cells.

2. *Anagen*. This is the period for redevelopment of the follicle and active formation of the fiber shaft. Several stages have been identified.[80] In anagen I mitotic activity commences from the inert cellular plug proliferating downward so as to surround the dermal papilla. During anagen II and III, cellular differentiation occurs to give a new inner root sheath and the cells of the hair proper; the tip of this new hair remains static as the follicle and its dermal papilla move downward into the dermis. This development proceeds until the follicle is at its full penetration into the dermis and some six times longer than it was during the resting (telogen) phase. In anagen IV and V, the matrix

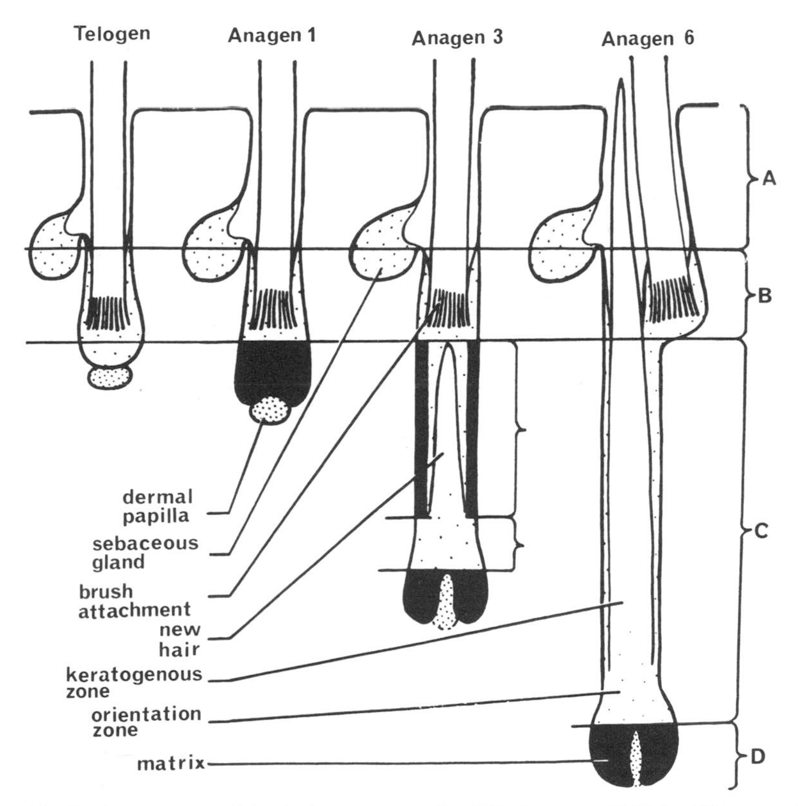


Fig. 3-9. Various stages of the hair growth cycle: (A) the upper follicle, (B) zone of no mitosis, (C) zone in which mitoses are transiently present during follicle elongation, (D) the matrix. Areas of high mitotic activity are shaded black. (From Bullough and Laurence.[80])

cells immediately above the dermal papilla region renew mitotic activity, and by differentiation a new hair is thrust upward and to the side of the old hair (if it remains), to emerge at the skin surface.

3. *Catagen*. In this phase, active formation of the hair shaft ceases and the follicle reverts to its dormant (telogen) state. Initially, the matrix cells of the hair bulb cease to differentiate into the normally concentric cylindrical layers, but instead yield a column of germ cells which push the lower end of the hair shaft upward, to the level of the sebaceous gland. At this stage, the lowest part of the hair has modified to produce a sheath of connecting club cells attaching it to the germ cells. The lower column of cells then undergoes extensive degeneration, leaving finally a short hair club set in germ cells, beneath which is a small ball of dermal papilla cells (i.e., the telogen phase).

The various stages of growth of the hair follicle are summarized in Fig. 3-9.

### 3.3.3. Structure of the Hair Follicle

The optical microscopic histology of the hair follicles of a wide range of animals has been reported extensively, and such studies are frequently an integral part of wider investigations of the hair follicle.[82] It is not our intention here to examine the optical microscopic structure of the follicle, but to highlight transmission electron microscope studies which have served to amplify our knowledge of follicle morphology. Although there are fine variations in follicle structure from one animal species to another and indeed from one fiber type to another within an individual animal species, general cellular organization of the follicle into multiple concentric cylinders is evident in all keratin fibers. The earliest electron microscope descriptions of follicle fine structure have been noteworthy for their detail.[83–86] Subsequent observations of structure[87–111] have confirmed this earlier work and provided yet more detail. The following subsections are concerned with the generalized structure, formation, and fate of these cells layers. Figure 3-10 shows the spatial arrangement of the various subcomponents of the follicle.

#### *3.3.3.1. Matrix*

These cells are roughly spherical to cuboidal and are distinguished by their large nuclei (Fig. 3-11). Coiled nucleoli are often found close to the nuclear membrane and mitotic figures are frequently seen. The cytoplasm of each cell is packed with numerous ribosomes, which are often grouped together in the form of "rosettes." A well-developed Golgi apparatus is present, and there are numerous, though small, mitochondria, but only a

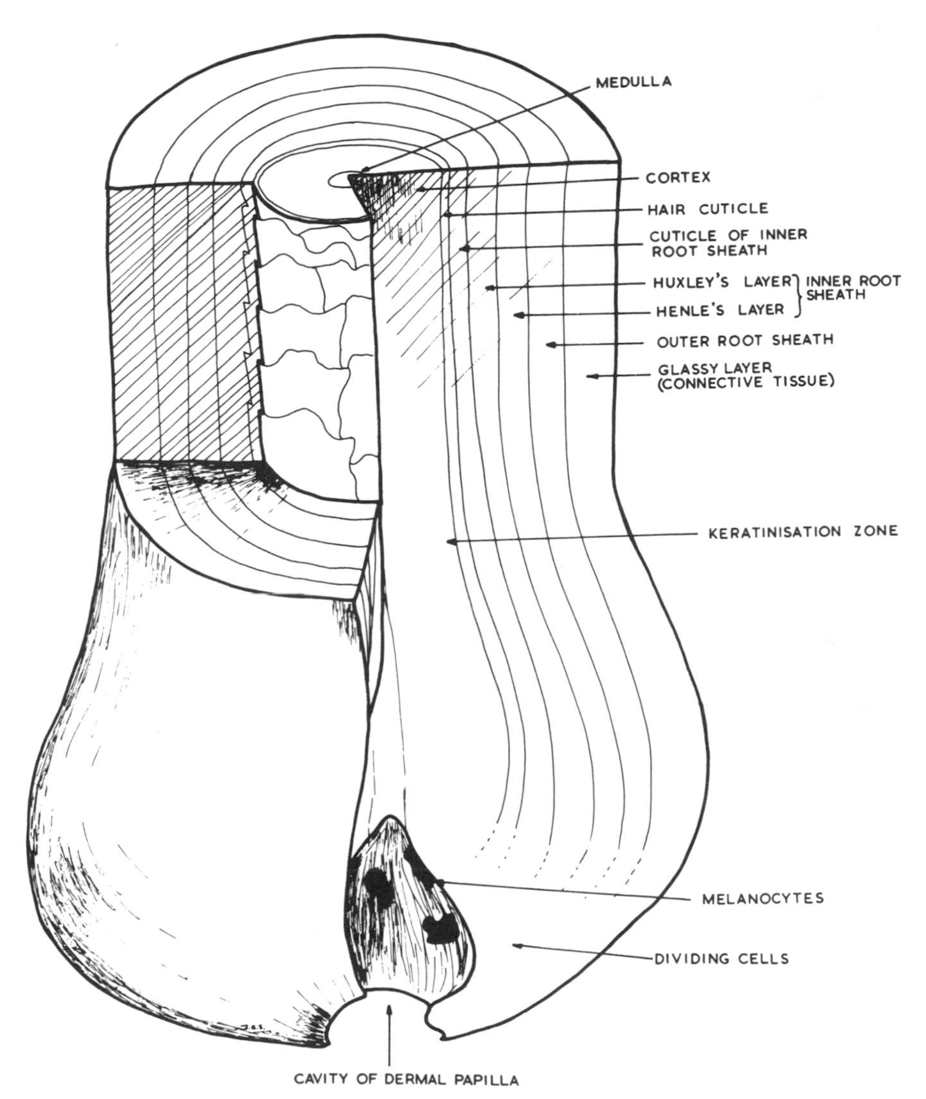


Fig. 3-10. Schematic diagram of a hair follicle, illustrating the radial distribution of cells derived from the hair follicle. (From Montagna and Van Scott.[81])

sparse endoplasmic reticulum is encountered. The cells are generally closely apposed, except where separated by melanocytes or melanocytic dendrites, and are occasionally cemented together by small desmosomal plaques similar in structure to those normally encountered in the epidermis.[88] A few tonofibrils cemented at one end into the desmosomal plaques extend into the cell cytoplasm. No attachment devices are seen at the interface between the basal

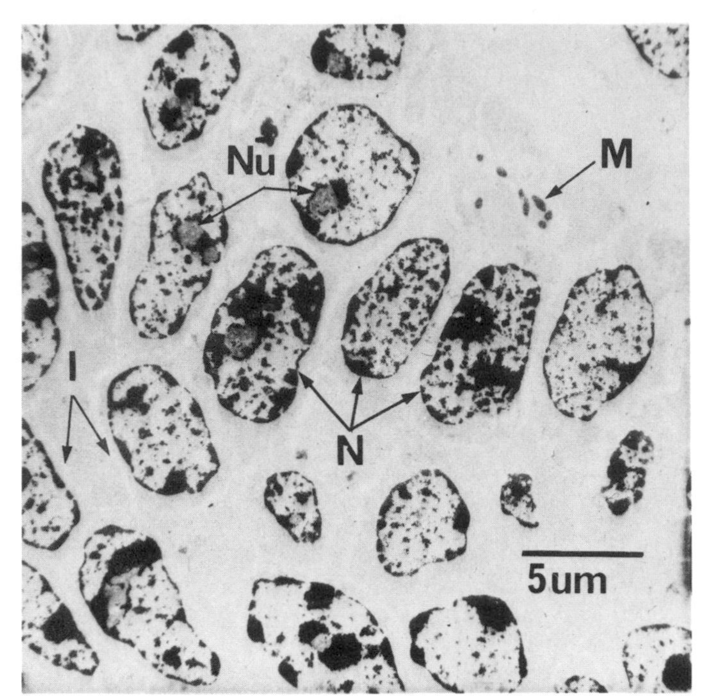


Fig. 3-11. Longitudinal section of guinea pig hair follicle showing follicle matrix cells. This section was stained with silver–methenamine reagent.

cells and the prominent dermoepidermal junction. Many of the structural features of the follicle basal cells are associated with cells which are actively engaged in protein synthesis and which retain their products of synthesis.[12]

*3.3.3.2. Melanocytes*

Melanocytes, with their main body abuting onto the basement membrane at the apex of the dermal papilla, are interspersed among the follicle matrix cells. These cells are responsible for the synthesis of all the natural pigments of the fiber shaft. On the whole, the ultrastructure of the hair follicle melanocyte has been considered apart from that of the rest of the hair follicle.[113–202]

The melanocyte is principally dendritic, with its dendrites extending upward between the follicle matrix cells. It contains a well-developed rough-surfaced endoplasmic reticulum and a clearly defined Golgi apparatus consisting of characteristic flattened sacs, vacuoles, and small vesicles. In any single section of a follicular melanocyte, melanin granules are seen in various stages of formation (Fig. 3-12). In the initial stages of synthesis in the Golgi

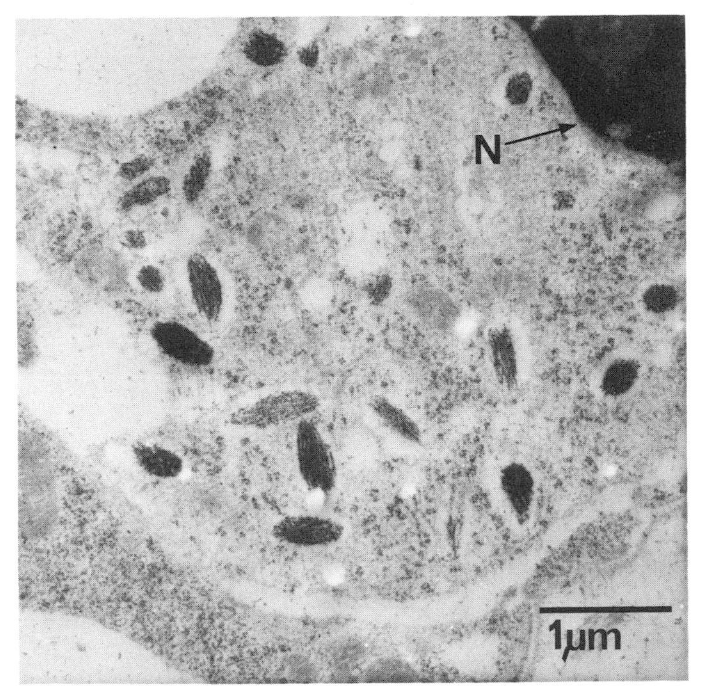


Fig. 3-12. Same section as for Fig. 3-11 but showing melanin granules at different stages of formation within a melanocyte.

zone, small vesicles some 5 nm in diameter grow in size by lateral fusion almost to the dimension of the mature granule. Simultaneously, another membrane is laid down at the periphery of the granule and the whole sac becomes filled with fibers extending between the poles of the oval-shaped structure. In albino animals this represents the last stage of synthesis of the granule, but for normal pigmented systems, progressive melanin deposition occurs on the fibers. For black hairs, deposition of melanin within the granule carries on until the whole unit is more or less homogeneously dense. For less intensely pigmented hairs, and particularly for blonde hair, the granules are incompletely melanized and exhibit a curious moth-eaten appearance. During these various stages of granule growth the granules move out along the dendrites of the cell.

The mechanism of transfer of melanin granules from the melanocyte to the matrix and differentiating cells of the hair follicle has been the subject of much speculation. Whereas it was thought earlier that the dendrites penetrated the cell membrane of the recipient cell to inoculate it with melanin,[123] there is now good evidence for believing that melanin transfer involves phagocytosis by the recipient cell.[121,124] In this process the melanin-laden dendrite of the melanocyte is engulfed by the recipient matrix cell membrane, pinched off and drawn into the cell cytoplasm. These bits are then partially digested, releasing the granules into the cytoplasm of the matrix cells. Whereas the pigment-laden matrix cells undergo differentiation, moving upward with progressive cell division to form the hair shaft, the melanocyte usually remains attached at the apex of the dermal papilla.

#### *3.3.3.3. Cortex*

The first signs of the differentiation of matrix cells into hair cortex are seen in the upper bulb region of the follicle. At this site the cells in this stream have by lateral compression elongated quite considerably into a spindle form. Fibrils some 7 to 8 nm in diameter aligned parallel to the long axis of the cell and attached to the desmosomes begin to aggregate, initially at the periphery of the cell. Simultaneously, an electron dense material appears between the fibrils so that discrete bundles are seen. The formation of fibrillar bundles is soon seen throughout the cytoplasm of the cell, and the space between the bundles is densely packed with ribosomes, mitochondria, and a few cytoplasmic membranes. Proceeding upward through the neck of the follicle these cells continue to elongate and the growth of fibril aggregates continues to fill nearly the whole of the cell, to the exclusion of the other cell components, which either disappear or are compressed into an amorphous residue between the fibril bundles. The cell nucleus during these processes undergoes considerable distortions, finally becoming roughly spindle-shaped and stellate in transverse section, and occupying the central portion of the cell.

The cell membrane also undergoes characteristic changes. In the matrix cells the outer membranes are about 9 nm thick and are separated by an unstained intercellular space of about 100 nm. As the cell cytoplasm undergoes transformation, an amorphous deposit is laid down in the intercellular space. This deposit is irregular and varies in thickness up to 20 nm.

#### *3.3.3.4. Cuticle*

The first signs of differentiation of matrix cells into cuticle occur in the upper bulb region of the follicle. Initially, electron-opaque granules 30 to 40 nm in diameter are deposited around the periphery of the cell and then these grow in size on the side of the cell immediately beneath the outermost cell membranes. This granular layer continues to grow until it occupies approximately half of the cell, compressing the nucleus and other cytoplasmic constituents into an amorphous layer occupying the other half of the cell. The granular layer then abruptly transforms into a layer composed of curious large blocks, and then a further rapid transition converts it into two zones, identified with the A and exocuticle layers of the mature fiber cuticle. As the single layer of transforming cells approaches the neck of the hair bulb the cells progressively elongate, flatten, and then tilt and slide over each other to emerge as an overlapping layer of cells. This latter process has been likened by Mercer[86] to the action of a zip.

As with the cortex characteristic changes occur in the vicinity of the cell membranes. The early differentiated cuticle cell membranes are about 9 nm thick and separated by an intercellular space of 8 nm thickness, but as keratinization proceeds, an amorphous material is deposited in the intercellular space to form a very regular layer about 20 nm in thickness.

#### *3.3.3.5. Medulla*

Medullary cells in those follicles producing medullated fibers are elaborated from the matrix cells close to the apex of the dermal papilla, and from this site they differentiate to form a central core of the hair shaft. In the first observable stages there is an increase in the amount of rough-surfaced endoplasmic reticulum, and the Golgi apparatus becomes very prominent. Arising from the Golgi zone dense spherical granules about 30 to 50 nm in diameter develop and grow in size as the cell rises up the follicle. Simultaneously, a number of fibrils appear randomly in the cell cytoplasm, but their significance in terms of the total changes occurring in the cell is not known. The medullary granules coalesce to give a relatively structureless single granule, which occupies a large proportion of the original cell. At the same time the other cell organelles, such as nucleus, endoplasmic reticulum, Golgi apparatus, and mitochondria, begin to disintegrate, giving rise to small, apparently empty vacuoles. With coalescence the vacuoles yield large empty spaces within the cell. Dehydration would appear to take place in these later stages, and finally the medullary cell is seen as a sparse entity wedged between cortical cells in ladderlike fashion down the central axis of the hair shaft.

#### *3.3.3.6. Inner Root Sheath*

There are three layers of the inner root sheath, which arise as separate cylindrical layers from the matrix cells of the hair follicle. Immediately adjacent to and closely adhering to the hair cuticle is the inner root sheath cuticle. Proceeding outward from the hair are two further layers, Huxley's layer and Henle's layer. All these cells undergo similar transformations into hardened cells, but the changes occur at different rates in each layer, and indeed the order of hardening of the layers differs between species.[102] In the first stage the matrix cells acquire numerous fibrillar bundles 7–8 nm in diameter, and these are associated with small electron-opaque granules (trichohyalin), from which the fibrils appear to emanate. These cells undergo flattening as they pass through the upper reaches of the hair bulb. The trichohyalin droplets increase in size so that they occupy a good proportion of the cell cytoplasm, and the other cell organelles, such as nuclei and mitochondria, disintegrate. Quite dramatically, and usually within the length of one cell, the whole cytoplasm is transformed to contain a well-organized array of fibrils with their long axes parallel to the cell length. Some believe that the fibrillar material arises from the transformation of trichohyalin,[84,87] whereas others believe that the trichohyalin is merely transformed into an interfibrillar matrix.[94]

The prime function of the inner root sheath is believed to be a mold within which the hair shaft is elaborated.[125] At the complete hardening of the hair shaft the function of the inner root sheath is spent, and it is progressively degraded at about the level of the sebaceous gland, to leave the hair to emerge freely at the skin surface. In the wool follicle Gemmell and Chapman[102] have shown that this breakdown is initiated at the surface of Henle's layer with the appearance of protrusions and the loss of cytoplasmic material. Such breakdown is probably mediated by enzymes secreted by the outer root sheath cells or more likely derived from the sebaceous gland.

#### *3.3.3.7. Outer Root Sheath*

At the lower part of the hair bulb proper and forming the outermost layer of the follicle is a single layer of differentiating cells identified as the outer root sheath. This sheath progresses up the follicle with a multiplication of its layers, and, at the level of the sebaceous gland orifice, these layers appear to be continuous with the various cell layers of the epidermis. Indeed, the development and fate of the outer root sheath are very similar to those of the epidermis. They contain the organelles associated with active protein synthesis, such as numerous mitochondria, agranular membranes, and ribosomes.[112] The cells flatten as they proliferate and rise upward. They are attached to each other and to the cells of Henle's layer by desmosomes. Although no specialized attachments are seen between the outer root sheath cells and the basement membrane in the lower reaches of the follicle, hemidesmosomes appear close to the sebaceous gland orifice. There are also increases in the amounts of fibrils deposited in the cells, but these have no specific orientation other than being parallel to the place of the cellular sheet. Dense granules also appear in association with these fibrils and grow in size up to 400 nm in diameter. At the level of the sebaceous gland these granules resemble the keratohyalin granules of the epidermis, and indeed at this point the cells harden into layers identical with the stratum corneum and are sloughed into the hair canal.

As early as 1927 Pinkus[126] described a line of morphologically distinct cells of the outer root sheath immediately adjacent to Henle's layer. Recently, Orwin,[104] from his studies of the Romney wool follicle, has shown that this line of cells, for which he has suggested the term "companion layer," is formed from follicle matrix cells near the base of the dermal papilla. They elongate and flatten in shape, are tightly apposed to keratinized Henle's layer cells, and tend to accumulate large quantities of glycogen. Their function is not at all clear, but Orwin suggests that they may have a role in the differentiation and breakdown of the Henle layer cells.

### 3.3.4. Histochemistry

Three primary methods have been used to determine the chemical nature of the various morphological components of the hair follicle: histochemistry, autoradiography, and the analysis of materials extracted from specific sites in the follicle. Histochemistry for the optical microscope involves the chemical treatment of a section of the follicle so that a colored substance is introduced by specific reaction with the tissue chemical group of interest. Electron histochemistry is similar in principle but requires the specific deposition in the tissue of an agent that will preferentially scatter electrons.

In autoradiography the living tissue is inoculated with a radio-labeled trace of the substance whose incorporation one wishes to follow. Sections of this tissue are then overlayed with a photographic emulsion and after a period of time that is dependent upon the specific activity of the label, the emulsion is developed and fixed. Blackening of the emulsion layer due to the presence of the radio-label can then be observed in relation to the underlying morphological structure of the tissue. Optical microscope autoradiography usually makes use of $^{14}C$-labeled materials, but in specific instances tritium or $^{35}S$ will be used. Only tritium-labeled tracers are satisfactory for electron microscope autoradiography.

The third general method is much less specific in terms of structure but can lead to a more detailed examination of chemical composition. Chemical analyses are usually straightforward, but careful histological examinations are required to determine precisely which components are extracted. This method can yield extremely valuable information.

#### *3.3.4.1. Cystine and Cysteine*

The presence of cysteine sulfydryl groups in the hair follicle and cystine disulfide crosslinks in the fully keratinized fiber are perhaps the most important aspects of keratin fiber chemistry. Using a nitroprusside staining reaction for sulfydryl groups, Giroud and Bulliard[127] demonstrated the presence of high concentrations of this chemical group in keratinizing tissues, and observed an inverse relationship between the intensity of staining and hardening of the tissues. Those structures with a high intensity of staining were classified[4] as hard keratins (e.g., hair and nails) and those with a lower intensity of staining were classified as soft keratins (e.g., epidermis). These observations have been largely confirmed by other workers using more specific histochemical stains.

A further significant finding, which occurred with the development of optical histochemical methods for disulfide group staining, was that decreases in sulfhydryl group concentration and increased hardening in the hair follicle are accompanied by increases in disulfide concentration. This stabilization and hardening, occurring in the upper reaches of the hair follicle bulb, is due to the oxidation of cysteine-containing (—SH) proteins into proteins highly crosslinked by cystine (—S—S—) groups. The presence of —SH and —S—S— groups is confined mainly to the hair cortex and cuticle, although small amounts of —SH have been seen in the outer and inner root sheaths. The dermal papilla is devoid of any demonstrable —SH.

Recently, the present author has devised an electron histochemical procedure for examining the distribution of —S—S— groups with respect to the ultrastructure of the hair follicle.[141] In the hair cortex the appearance of protein containing —S—S— occurs initially as curious small blocks distributed along fibril bundles. Thereafter, cystine appears to be distributed evenly throughout the fibril bundles, although there is some evidence that the cystine reaction occurs predominantly with the interfibrillar matrix rather than with fibrils themselves. In the hair cuticle, cystine is first observed in association with 40-nm-diameter globules deposited around the entire periphery of the cell cytoplasm. These cystine-laden globules increase steadily in size, particularly in that part of the cell farther from the follicle axis, so the whole cell is divided into two main zones, one half where keratinization is proceeding and the other half in which the cell constituents (nucleus, endoplasmic reticulum, etc.) are compressed. The cystine-rich layer then undergoes rapid transition, first into a layer of curious cuboidal blocks and then into two final layers identifiable as the A layer and exocuticle of the fully keratinized fiber. The only other site in the follicle where cystine has been observed using this electron histochemical procedure is Henle's layer. Cystine is observed immediately after fibrillation of this cell layer as a thin layer approximately 16 nm thick immediately adjacent to the cell membrane. The significance of cystine at this site is unknown, but it may be associated with the separation of this cell layer at the sebaceous gland orifice.

Stabilization and hardening in the hair follicle are accompanied by increases in the birefringence of the cortex.[142,29] In addition, whereas X-ray diffraction patterns of the lower bulb show an unoriented α-pattern, keratinization is accompanied by the appearance of an oriented α-type pattern identical to that of the final hair.[29] Various workers have shown that the appearance of the α-type pattern actually precedes the formation of disulfide-rich proteins, and corresponds to the appearance of the fibrillar bundles in the cortex which later acquire the cystine-rich proteins.[29,143–145]

The kinetics of keratin synthesis in the follicle has been examined extensively using techniques of autoradiography,[34,146–159] but the precise mechanism of the incorporation of proteins, particularly high-sulfur proteins, remains obscure. $^{35}$S-labeled cystine appears to concentrate in the upper reaches of the bulb, a process which many have suggested indicates that the incorporation of cystine is not in phase with normal protein biosynthesis in the follicle. It is now generally thought that α-keratin-like sulfur-poor fibrils (microfibrils) are synthesized first, and that the synthesis and deposition of a sulfur-rich amorphous interfibrillar matrix occurs at a later stage. Certainly this mechanism is in keeping with the established ultrastructural development of the cortex and with the electron histochemical observations that cystine appears in the cortex after the formation of the fibrillar bundles. Downes et al.[148] have examined the $^{35}$S activity in the *S*-carboxymethyl keratines extracted from various levels of the follicle following the administration of $^{35}$S-cystine. They interpreted their results in terms of a two-stage process of biosynthesis in the cortex. More recently, Fraser[150] has obtained evidence indicating that there is simultaneous synthesis of high- and low-sulfur-containing proteins, and he suggests that whereas the concentration of the low-sulfur protein increases steadily in the developing cortex, the concentration of sulfur-rich protein increases rapidly during the later stages of maturation. Also, in keeping with this general thesis, Wilson et al.,[46] using tritiated methionine, have found that there is no specific intracellular site for low-sulfur protein synthesis, and they suggest that a large proportion of the low-sulfur proteins formed within the developing cortical cell ultimately contribute to the keratin microfibrils.

### *3.3.4.2. Nucleic Acids*

The common method for the optical histochemical demonstration of DNA is the Feulgen–Schiff reaction.[25,159] It involves the mild acid hydrolysis of tissue sections to release aldehyde groups from the deoxypentose sugar of DNA and coupling with fuchsin-sulfurous acid (Schiff's reagent) to produce a purple coloration. RNA staining is achieved either by basophilic staining with toluidine blue or by the methyl green–pyronin method devised by Brachet.[161] The dividing cells of the hair-follicle matrix stain intensely by the Feulgen–Schiff reaction. The other matrix cells stain almost as intensely, but the intensity diminishes with differentiation of the cells, and the nuclei completely lose this stainability as keratinization proceeds.[140,162,163] The nuclei of the papilla cells are strongly positive.[162] Silver–methenemine solution has been used for the electron histochemical staining of DNA[164] and has been used in investigations of the follicle[60,141]; these results largely confirm optical histochemical observations and show that there is complete elimination of DNA staining in the nuclear remnants of the developing fiber cortex.

The basal cells of the follicle possess the greatest quantity of RNA.[162] The differentiating hair cortical cells contain large quantities of RNA,[162] in keeping with ultrastructural observations of large numbers of ribosomes between the fibrillar bundles. RNA staining ceases abruptly as the cortical cells completely keratinize.[162]

The chemical fate of nuclear DNA and ribosomal RNA in the fully keratinized fiber is somewhat of a mystery. Mercer[5] has suggested that resorption of phosphorus occurs in the follicle.

### *3.3.4.3. Carbohydrates*

Glycogen is the main carbohydrate reserve store in animal tissues, and there are many optical histochemical procedures for staining it.[25] It can usually be removed from tissue sections by saliva digestion. Under the electron microscope it has a very characteristic appearance in sites usually corresponding to those seen under the optical microscope.[165]

Whereas glycogen is absent from the matrix cells of the hair follicle, all the fully differentiated cells contain glycogen but the amount gradually diminishes as the various cell layers undergo hardening.[162,163] Shipman et al.[166] have observed interesting changes in the glycogen of the murine hair follicle with the various phases of the hair growth cycle. Whereas glycogen is present in most anagen phases, it disappears abruptly with catagen.

Acid mucopolysaccharides are visualized histochemically by a variety of methods, including Alcian blue staining, Hale's reaction, metachromasia, and the PAS (periodic acid–Schiff) reaction.[44] An intense PAS reaction is produced by the connective tissue sheath of the hair follicle[167] and strong metachromatic staining occurs in the dermal papilla.[163,168] Sylven[168] has suggested that these acid mucopolysaccharides are important in supplying sulfur for keratinization. Metachromatic staining, mostly in the peripheral cell layers of the outer root sheath, has been reported,[162] but there would appear to be very little acid mucopolysaccharide associated with the cells of the hair proper. An electron histochemical procedure has been devised[169] which yields results analogous to the PAS reaction of optical histochemistry.[170] Rambourg and Leblond[171] have found, using this method, that many mammalian cells are coated with a polysaccharide-rich layer. Following the observations of Mercer et al.[172] that epidermal cells are similarly coated, Orwin[106] has found polysaccharide-rich layers around the matrix cells of the hair follicle, particularly at the desmosomal points of cell contact. There is some uncertainty as to the fate of these cell coats in the hair cuticle, cortex, and medulla, but the coat remains in the inner root sheath layer.[106] Orwin[106] believes that the retention of this polysaccharide coat may have significance in the breakdown of the inner root sheath higher in the follicle.

#### *3.3.4.4. Lipids*

The histochemical disposition of lipids in the hair follicle has received very little attention.[162] The cells of the hair follicle contain only traces of sudanophilic lipids, and usually only in the undifferentiated matrix cells.[173] Lipid granules are found in the early differentiated inner root sheath, but these disappear as trichohyalin is deposited in the cells.[163,174] Phospholipid is also found in the matrix cells and the inner root sheath, the amount decreasing as hardening proceeds.[162]

#### *3.3.4.5. Arginine and Citrulline*

The Sakaguchi staining method[25,175] is a highly specific optical histochemical procedure for arginine and involves treatment of tissue sections with $\alpha$-naphthol in alkaline hypochlorite or hypobromite solution to yield a red color. When applied to the hair follicle, intense staining of arginine is observed in the trichohyalin droplets of the developing medulla and inner root sheath.[162,176,177] The intensity of this staining reaction diminishes as these cells undergo hardening. It is noteworthy that the trichohyalin droplets also have a high affinity for acidic and basic dyes,[26,178] indicating that the droplets are rich in polar groups. Rogers[87] has devoted considerable attention to the biochemistry of trichohyalin. He has used a wax-sheet stripping method for isolating inner root sheath-rich components from the hair follicle[179] and a trypsin digestion method for isolating medullary proteins.[120] These procedures have revealed that both inner root sheath and medullary proteins are rich in acidic and basic amino acids, contain the unusual amino acid citrulline, and are almost totally devoid of cystine.[87] Rogers[87] has proposed that the trichohalin droplets of the inner root sheath and medulla contain arginine-rich proteins (he refers to this material as "arginine-trichohyalin") and that by desimidation of the arginine, this is transferred into the citrulline-rich hardened proteins which subsequently fill the cells ("citrulline-trichohyalin").

*3.3.4.6. Enzymes*

Optical histochemical procedures have been used extensively for determining the distribution of various enzymes within the morphological components of the hair follicle, notable contributions having been made by Brown–Falco.[162] These enzymes have included phosphorylase,[162,181–184] aldolase,[162] succinic dehydrogenase,[162,185–187] cytochrome oxidase,[188] alkaline phosphatase,[140,162,163,189–191] acid phosphate,[190,192] glucose 6-phosphatase,[182,183,193] esterases,[163,187,194,195] carbonic anhydrase,[162,196] aminopeptidase,[194] β-glucuronidase,[197] and arginase.[198] In addition, the distribution of tyrosinase and dopa oxidase activity within the melanocytes of pigmented hair fibers has been examined with reference to the various stages of the hair growth cycle by autoradiographic methods.[35] More recently, the distribution of certain enzymes in the follicle has been examined by electron histochemical and autoradiographic procedures. Thus Wong[199] has demonstrated alkaline phosphatase activity along the basement membrane and around the periphery of the cells of the dermal papilla; Wolff and Schreiner[200] have demonstrated acid phosphatase activity associated with the premelanosomes, Golgi apparatus, and endoplasmic reticulum of the melanocyte; and Seiji and Iwashita[201] have examined the detailed localization of tyrosinase activity in the melanocyte.

Two recent papers by Adachi, Uno, and Montagna,[182,183] concerned with possible enzymatic differences between normal follicles and the quiescent vellus follicles of bald scalp, are notable for their detail. Although these studies did not yield any significant information as to why baldness occurs, considerable information was gleaned about carbohydrate metabolism within the follicle.

## 3.4. The Fully Keratinized Hair

This section is divided for convenience into two main parts, considering the morphological structure of keratin fibers in the first part and the chemical nature of some of these morphological components in the second part.

### 3.4.1. General Structure

The main structural component of all fully hardened mammalian keratin fibers is the *cortex*, and this consists of a cylindrical array of closely packed interdigitating spindle-shaped cells whose long axes are aligned parallel with the fiber axis. This core is covered by a *cuticle*, and this is composed of sheet-like cells that overlay each other from root to tip of the fiber. The third main component of the keratin fiber is the *medulla*, and this consists of a series of specialized and often air-filled cells arranged along the axis of the fiber.

It is reasonable to generalize on the structure of all keratin fibers by saying that they are composed of cortex, cuticle, and medulla, but on the other hand, there are fine variations in the arrangements of these different cell types in hairs not only from one animal species to another, but also from one body part to another (Fig. 3-1 to 3-8). Thus, whereas the wool fiber is generally small in diameter (10–30 $\mu$m) and is covered by a layer of cuticle cells that overlap singly, human head hair is some 60–100 $\mu$m in diameter and is covered by 6–10 layers of overlapping cuticle cells and pigs' bristle, which is often 0.25 mm in diameter and has as many as 35 layers of cuticle cells. Also, whereas dark brown or black human head hair contains only a sparse medulla that is filled with a low-sulfur protein, most rodent hairs are so highly medullated that the medulla comprises some three-fourths of the total cross-sectional area of the fiber and is longitudinally segmented by ladderlike zones of low-sulfur protein separating large air-filled spaces. It is not possible within the scope of the present chapter to consider all variations in keratin-fiber morphology, but Fig. 3-13, relating to human hair, will serve as a useful reference point for most animal fibers.

### 3.4.2. Surface Structure

The scale surfaces of a wide variety of animal fibers have been examined by optical microscopy,[10,11,202,203] particularly from casts of the fibers prepared in soft, settable plastic materials. Many different types of scale pattern have been encountered, and in some cases attempts have been made to classify the various patterns. Some descriptions[10,202] have considered the type of pattern (e.g., mosaic, waved, shallow, chevron, pectinate, or petal), the form of scale margins (e.g., smooth, crenate, rippled, or scalloped), and the distance apart of successive scale margins (e.g., close, distant, or near). Kassenbeck,[203] using diagrammatic descriptions of scale-pattern type showing an overall progression from seal fibers through to human and alpaca hairs, has attempted to classify scale pattern with respect to such factors as animal species, fiber growth rate, and the physical properties of the fibers. It is clear that scale-pattern descriptions are a valuable aid in the identification of unknown fiber types, but the extreme variability of these patterns is such that only tentative species assignments can be made, and certainly it is not possible to classify according to breed (race)[10] or individual animal.

More recently, the scanning electron microscope has aided in the detailed

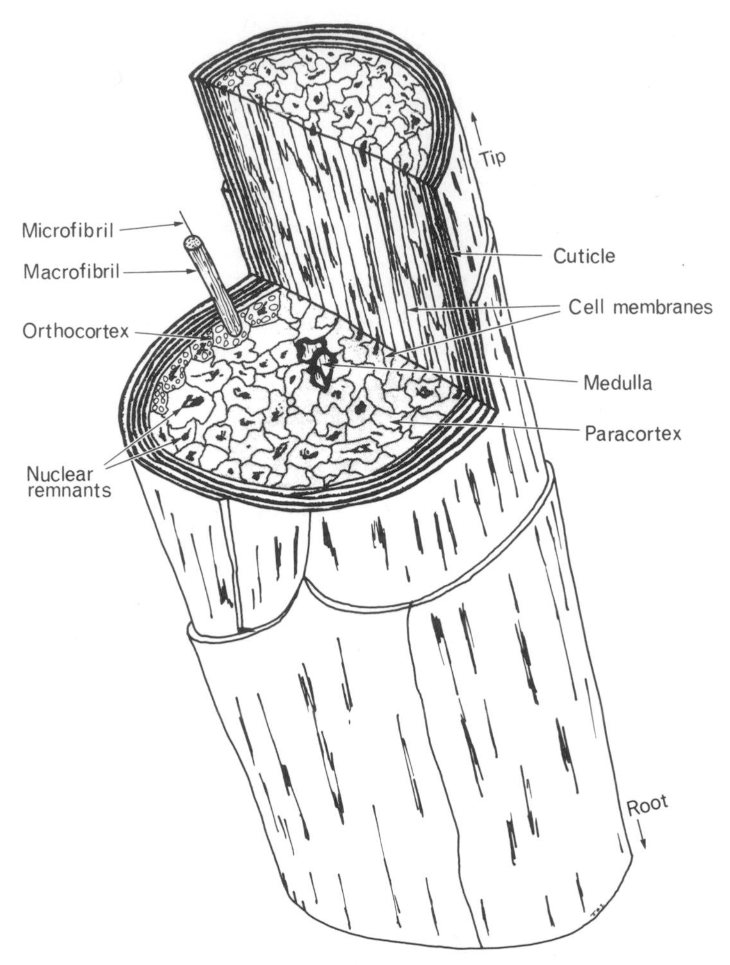


Fig. 3-13. Schematic diagram of human hair structure. (Redrawn by the author from a similar diagram for wool published by Dobb et al.[205])

examination of fiber surfaces. Bottoms et al.[204] and Swift and Brown[57] have shown that in human head hair there is considerable variation in the scale pattern from root to tip, and that these variations are due to the natural wearing processes of the fibers. In the examination of hair surfaces, considerable care must therefore be exercised that the right conclusions are drawn about the causes of scale-pattern variation. Since the root ends of animal fibers are relatively free from externally promoted defects,[57] this part of the hair, where it emerges from the skin, is perhaps the most suitable for comparative studies both between and within different animal species. The nature of the fiber surface at the root end is also a good reference point for examining the effects of various chemical or physical treatments on the fiber surface. A particularly powerful procedure for assessing the fine changes in surface architecture due to any sort of treatment is that in which the same part of the same fiber is examined before and after treatment. Swift and Brown[57] have used this scanning electron miscroscope technique to show that the combing of human hair gradually chips away small pieces from the distal edges of the surface cuticle cells.

Whereas photographs frequently have the quality of conveying "obvious" information, it is often difficult to apply verbal descriptions for this information or to assess the information quantitatively. One is often therefore forced to define specific features in such a way that the nature of the feature is readily grasped by all, and then to note the appropriate variations in this feature from one photograph (or experiment) to another. Such verbal descriptions for the overall scale patterns of animal fibers have been used in optical microscopy studies,[10,202,203] but more recently the detailed surface architecture of wool fibers seen in electron micrographs has been subjected to literary descriptions.[205] Such descriptions of specific features and the variations in them have proved valuable in assessing the effects of various chemical treatments on the wool fiber surface.[52]

There has been considerable interest in the nature of the keratin fiber surface in relation to felting and shrink proofing.[206] Keratin fibers have two coefficients of friction, according to whether the impinging surface moves along the fiber from root to tip (i.e., "with-scales" friction) or from tip to root (i.e., "against-scales" friction and usually having the greater magnitude). It is this differential friction effect (DFE) which causes individual fibers within a mat to move preferentially in one direction, thereby becoming locked within the fiber mat. In shrinkproofing the main aim is to modify the fiber surfaces either by chemical treatment or polymer deposition so that the DFE is minimized or eliminated altogether.[206] In the case of chemical shrinkproofing treatments, Makinson[207] believes that the material inside the surface cuticle cells is attacked and that the whole surface becomes much softer than that in the untreated fiber.

Many mechanisms have been proposed for keratin fiber differential friction, the most common being the ratchet mechanism and the ploughing mechanism.[206] There has not yet been direct experimental proof that any particular one of these mechanisms is the important one. It is clear, however, that recent advances in scanning electron microscope technique for observing the area of contact between two moving fiber surfaces,[67,208] coupled with the simultaneous measurement of the frictional forces involved,[209] might enable us to answer this question of differential friction.

Whereas the cuticle surfaces of untreated human hair are usually quite smooth,[57] in many wool fibers the scale surface is longitudinally fluted[210] (see Fig. 3-1). Wolfram[211] has recently examined sections of such wool under the transmission electron microscope and has shown that the flutings form an integral part of the overall structure of the cuticle, the projecting ridges being composed entirely of exocuticle. The origin or possible function of these flutings has not been established, although Wolfram[211] has proposed that they assist in spreading of sebum over the fiber surface.

### 3.4.3. Internal Structure

#### *3.4.3.1. Cuticle*

All mammalian fiber cuticle cells are thin, hard cellular sheets that overlay the main core of the fiber. They are 0.2–0.4 μm thick and, when flattened out, are approximately square (sides 30–50 μm) with rounded corners. On the fiber the cuticle cells are in the form of a segment from a truncated conical shell and overlay each other from root to tip of the fiber in a way analogous to that in a stack of disposable conical plastic cups. The free length of each cell visible at the surface of the fiber is to some extent dependent upon the overall diameter of the fiber.[211a–213] In fibers of small

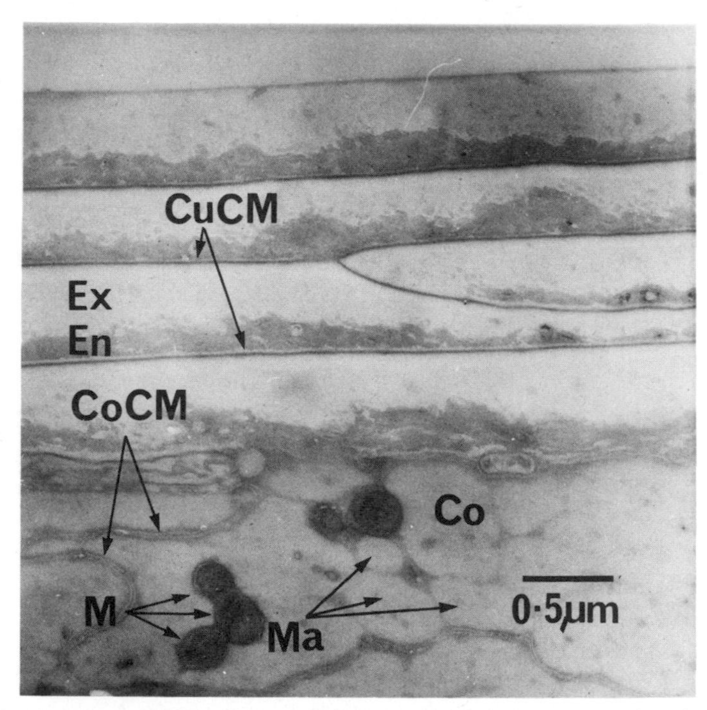


Fig. 3-14. Transverse section of human hair stained with dodecatungstophosphoric acid.

diameter such as wool or human vellus hair, the amount of scale overlap is relatively small (i.e., some three-fourths of the cell length is exposed), and the distance between each scale margin at the fiber surface is relatively large. A transverse section of such fibers usually shows only one or two thicknesses of cuticle.[214,205] In human hair, on the other hand, the scales overlap to a much greater extent, so only about one fourth of their length is visible at the fiber surface, and in transverse section 6–10 cuticle cell layers are seen (Fig. 3-14).[215,216] At the other extreme, in pig's bristle the scale margins at the fiber surface are quite close together, the proportion of each scale exposed at the fiber surface is small (about one tenth), and as many as 35 cuticle cell layers have been seen in transverse section[60,217] (Fig. 3-15).

Most animal fibers have cuticle cells which in longitudinal or transverse section are more or less of constant thickness. The cuticle cells of wool fibers, on the other hand, tend to be somewhat irregular in thickness, are frequently fluted on their exposed outer surfaces,[210,211a] and at the exposed distal margins of the cells there is frequently a ridge of about twice the thickness of the main bulk of the cells.[214,205]

The junction between adjacent cuticle cells and between the cuticle and

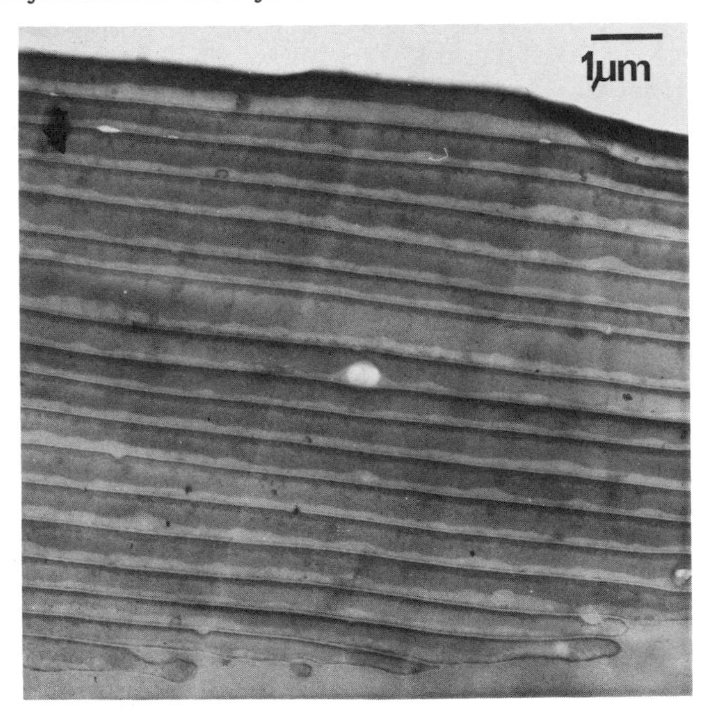


Fig. 3-15. Longitudinal section of pig's bristle illustrating the numerous layers of cuticle. This section was stained with ammoniacal silver nitrate.

cortex is usually relatively planar (Fig. 3-14). Sometimes "ball-and-socket" interconnections are seen between the cells and these probably contribute to the mechanical strength of the scale layer.[218] Some fibers and particularly human nostril,[60] vicuna, swamp wallaby, and guanaco fibers[219,220] have large numbers of these ball-and-socket interconnections. Piper[219,220] believes that these may have been formed in the follicle by a process of partial phagocytosis in which one cell attempts to pinch off and ingest a portion of an adjacent cell.

Each cuticle cell is composed of lamellar subcomponents, and these can readily be seen in the transmission electron microscope from metal-stained transverse or longitudinal sections of the fiber. Phosphotungstic acid staining of human head hair sections (Fig. 3-14) reveals that each cuticle cell is divided into two principal components. The lamina that is found at the greater radial distance from the fiber axis is known as the *exocuticle*, and the one at the lesser radius is known as the *endocuticle*. On average in human hair the exocuticle is about 0.2 $\mu$m thick and the endocuticle about 0.1 $\mu$m thick, the junction between the two layers being rather irregular. On staining sections by reduction and treatment with osmium or neutral silver salts or with complex alkaline silver solutions, yet additional laminae are revealed within each cuticle cell (Fig. 3-16). The *A* layer, which stains very intensely, is of constant thickness in each cell (about 40 nm) and is located at the periphery of the cell on the outer-facing aspect. The *inner layer* is rather irregular, up to 20 nm thick and located at the periphery of the cell on the inner-facing aspect.

The contents of the cuticle cell generally lack fine structure when observed under the transmission electron microscope. Orfanos and Ruska[216] claim to have observed fine structure in the exocuticle of human hair that consists of interwoven threadlike osmophilic units about 3 nm in diameter. Interpretations of high-magnification transmission electron micrographs such as those used by Orfanos must, however, be made with extreme caution and certainly with due regard to the phase- and amplitude-contrast components of the image[221]; otherwise, one risks the danger of drawing erroneous conclusions about specimen structure.

The endocuticle usually shows a disorganized substructure of sparse membranelike elements. These membranes are probably the remnants of effete cytoplasmic structures, such as the cell nucleus, endoplasmic reticulum, mitochondria, and so on, which are compressed to one side of the cell as formation and keratinization of the exocuticle proceeds in the follicle (cf. Section 3.3.3.4). Recently, Dobb et al.,[222] from observations of sections of wool which have been quantitively reduced and alkylated with methyl mercuric iodide, have observed an additional layer approximately 10 nm thick along the outer edge of the peripheral cuticle cells.

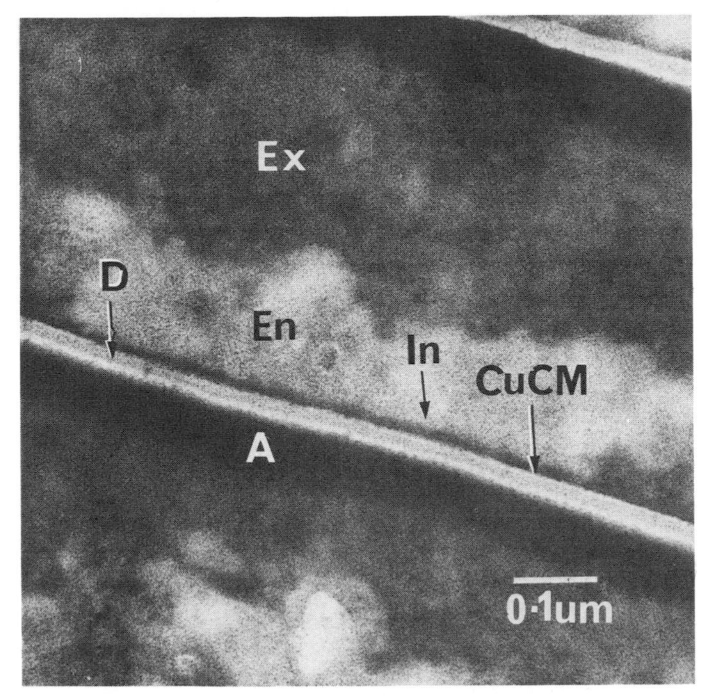


Fig. 3-16. Transverse section of human hair cuticle. The section was stained with osmium (after reducing the section), uranyl acetate, and lead citrate.

Between adjacent cuticle cells there is a *cell membrane complex*, the structure of which can readily be seen under the electron microscope in fiber sections stained with dodecatungstophosphoric acid (Fig. 3-17).(124,224) The overall thickness of this lamellar complex is about 28.0 nm. It consists of a central stainable layer of very constant thickness (about 18.0 nm) generally referred to as the *delta band*(214) or intercellular cement. This is covered on both sides with a layer of constant thickness (about 2.5 nm) which cannot be stained by any of the normal heavy-metal staining procedures for electron microscopy, assumed to represent the lipid components of the protoplasmic *membrane* which normally bounds each cell. A further layer which does stain and is about 2.5 nm thick is found as the outermost lamina between which the other cell membrane components are sandwiched.

Since each cell of the keratin fiber is bounded by a protoplasmic membrane, it is reasonable to expect that the surface of the fiber will be covered with membranes associated with the underlying cells. This membrane is normally very difficult to observe under the electron microscope from thin sections because of the lack of image contrast between the surface membrane layer and the surrounding epoxy embedding medium of the section. By

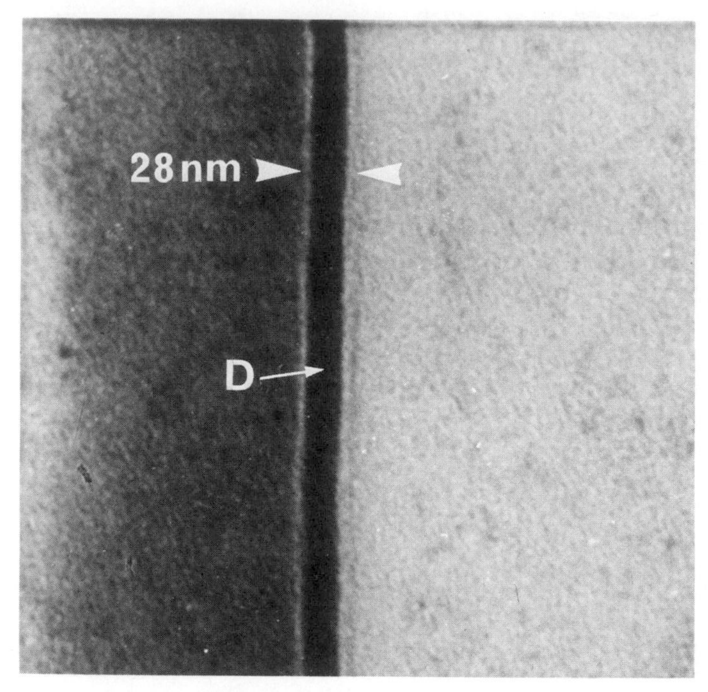


Fig. 3-17. Cuticle cell membrane complex revealed by dodecatungstophosphoric acid staining of a transverse human hair section.

evaporating a thin metal coating onto the hair surface and staining the sectioned fibers, sufficient image contrast can be obtained to observe a relatively electron-transparent layer approximately 2.5 nm thick immediately at the surface of the fiber.[223] This surface layer, which is presumed to be the lipid component of the protoplasmic membrane, is removed from the surface of human hair by various cosmetic and chemical treatments.[223]

### *3.4.3.2. Epicuticle*

This component is observed when animal fibers are treated with certain oxidizing agents such as chlorine water,[225–227] bromine water,[228,229] or permonosulfuric acid.[60] The phenomenon was first described by Allwörden,[225] who showed that small bubbles were produced at the surface of keratin fibers immersed in chlorine water. These bubbles involve small areas of each surface cuticle cell in the case of chlorine water treatments, but bromine water frequently causes eruptions which emanate from a depth corresponding to several cuticle layers in coarse fibers.[229] The bubbles are believed to arise through an osmotic process in which the oxidizing agent diffuses through a surface semipermeable membrane and dissolves part of the underlying protein. The solution of this protein within the cell creates an osmotic imbalance across the surface membrane such that water flows into the cell, increasing the internal pressure and raising the surface membrane as a bubble in the process.[230] King and Bradbury[231] have isolated fragments of epicuticle by the chlorine and bromine water methods and have shown them to be thin sheets of 3.2 and 14.0 nm thickness, respectively.

In terms of the normal histological structure of a fiber section in the electron microscope, the origin of the epicuticle has not been clearly defined. The most likely explanation is that it is a surface membrane, consisting of the cytoplasmic membrane and or the A layer, to which remnants of protein from the exocuticle are attached.

### 3.4.3.3. *Cortex*

This is the component which constitutes the main bulk of all keratin fibers (cf. Fig. 3-18) and largely determines the main mechanical properties

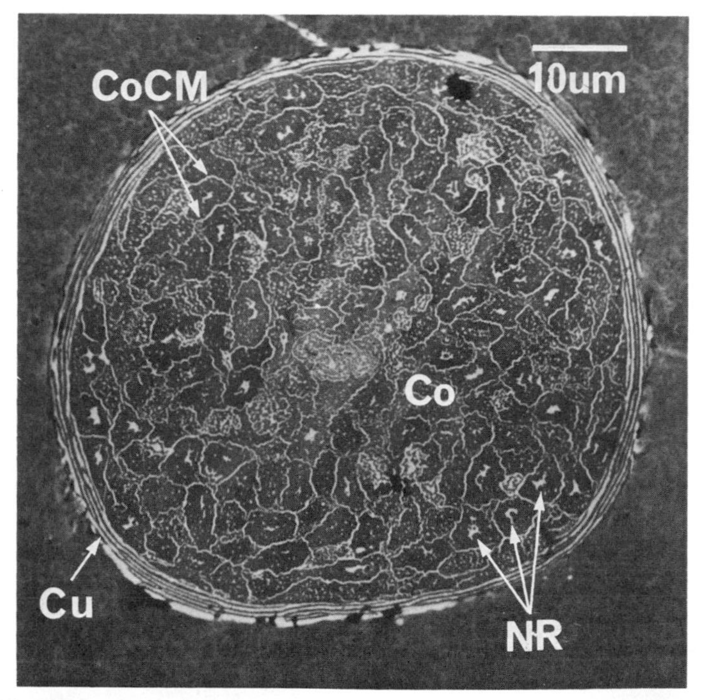


Fig. 3-18. Transverse section of yak hair. The section has been partially digested with a proteolytic enzyme and then examined unstained by transmission scanning electron microscopy.

of the fiber. The cortex consists of "spindle"-shaped cells (cortical cells) some 50–100 μm long and 3–6 μm in diameter, closely packed, with their long axes parallel to the fiber axis. Each cell is bounded by a cell membrane complex and, although the term "spindle"-shaped is frequently used, in reality the cell is also longitudinally fluted and its extremities separated into fingerlike units (Fig. 3-19). This complex shape has evidently arisen as the partially formed cortical cells are laterally compressed as the hair passes through the neck of the hair follicle. In the fully hardened fiber adjacent cells are closely interlocked at their fluted surfaces and interdigitate with each other at their extremities. In transverse section the boundary between adjacent cells appears to be quite irregular, and small pockets of membrane-bounded cortical material are often observed, apparently within the boundaries of a larger sectioned cell.

Each cortical cell contains a prominent *nuclear remnant*. This is stellate in transverse section (cf. Fig. 3-18), 0.5–1 μm in overall diameter and about 40 μm long, lying along the axis of each cell. No fine structure has been observed within the nuclear remnant.

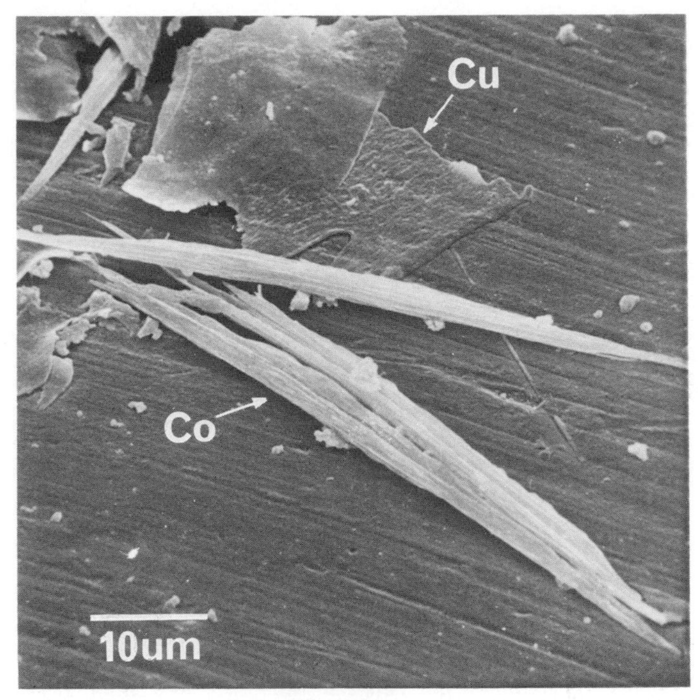


Fig. 3-19. Cuticle (Cu) and cortical (Co) cell fragments released by the partial hydrolysis of human hair with papain/bisulfite.

The major proportion of each cortical cell is occupied by close-packed *macrofibrils*. These are approximately solid cylindrical units 0.05–0.2 μm in diameter and of variable length but often stretching the full length of each cortical cell. Between each macrofibril is a variable amount of *intermacrofibrillar matrix* and, in the case of pigmented fibers, melanin granules. This matrix is composed of those cytoplasmic fragments, such as endoplasmic reticulum, Golgi apparatus, and mitochondria, which have been squeezed together as the macrofibrils were forming in the follicle. In many respects this matrix will probably be similar in composition to the endocuticle. Fraser et al. have suggested that 4.7 nm equatorial arcs frequently seen in X-ray diffraction photographs of keratin fibers are due to crystalline lipids and believe that some of this may be present in the intermacrofibrillar matrix.[232]

Each cortical cell, as mentioned earlier, is separated from its adjacent cells by a *cell membrane complex* (Fig. 3-20). The structure of this is certainly not as regular as that normally encountered in the cuticle, for it irregularly contours the surface of each cell, but it nevertheless consists of membrane–cement–membrane laminations. The membrane component associated with each cell does not normally appear stained in electron microscope sections, is

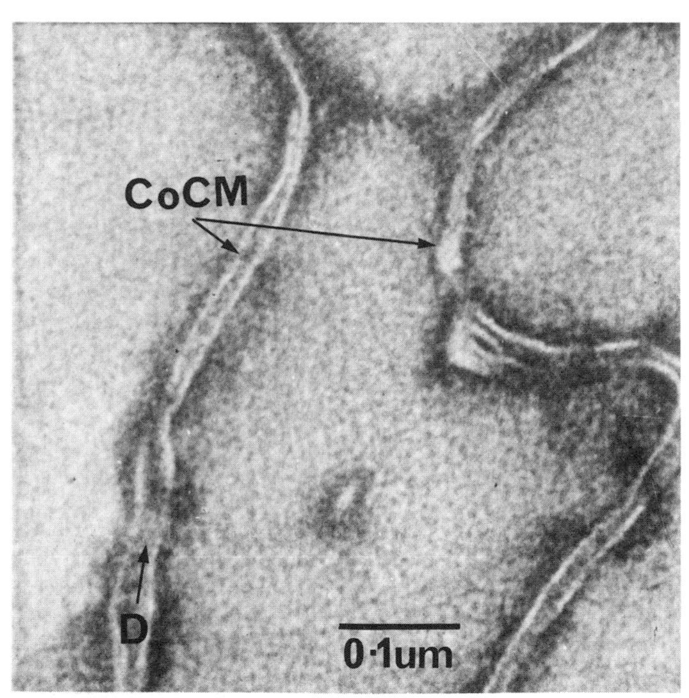


Fig. 3-20. Cortical cell membrane complex revealed by dodecatungstophosphoric acid staining of a transverse section of human hair.

of more or less constant thickness (about 2.5 nm), and is probably composed of lipids. The amount of cement between the cell membranes is at any given point quite variable. In some cases it is impossible to distinguish any cement, so the membranes are apparently in close apposition and in other places the membranes are separated by up to a 15-nm-thick layers of cement.[87] On the cytoplasmic side of the membrane there is usually an irregular layer up to 20 nm thick, extending as far as the peripheral macrofibrils of the cell, which stains somewhat differently from other cell components and which appears to be associated in some way with the external membrane.[60]

*3.4.3.3a. Orthocortex, Paracortex, and Fiber Crimp.* In some cortical cells the macrofibrils are so closely packed, with the intervention of little intermacrofibrillar matrix, that it is difficult, from the electron microscopy of sections, to delineate the individual macrofibrils; this is a major histological feature of the *paracortical cell.* In other cells there is sufficient matrix to enable one to discriminate each macrofibril, and this is the prominent feature of the *orthocortical cell*[214] (these two major cell types can be seen in Fig. 3-21). A third type of cell, the *mesocortical cell,* has also been described as

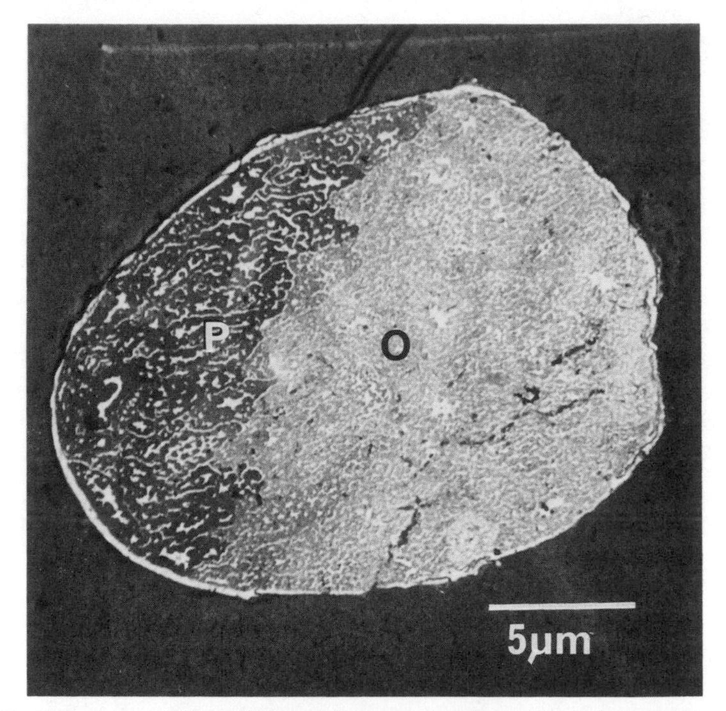


Fig. 3-21. Transverse section of Shropshire wool showing ortho and para cortex (O and P, respectively). The section has been partially digested with a proteolytic enzyme and then examined unstained by transmission scanning electron microscopy.

containing both types of macrofibrillar arrangement.[233] The orthocortical cell is generally more accessible to staining and is more susceptible to chemical attack than the paracortical cell, and it was such properties as these which led to the initial discovery that there are two major cortical types within the wool fiber and that in many wool fibers the whole cortex is transversely segmented.[234–239] It is now generally accepted that in wool, segmentation of the cortex is associated with the natural crimping of the fiber. Indeed, the orthocortical segment is always found on the outside of the crimp curl, and this is achieved by the twisting of the two cortical segments in phase with the crimping of the fiber.[240] In some fibers, such as Lincoln wool or mohair, which are only lightly crimped, a radial distribution of cortical cell types has been observed with a central core of orthocortex.[238,241]

Various workers have described the human hair cortex as being of the "para type."[237,239] Using various staining and digestion methods, Spearman and Barnicot[242] have carried out a detailed optical microscope examination of Bushman, Hottentot, and Negro hair and compared the structure of these fibers with European and Chinese hair. They concluded that there was no bilateral difference in the cortex of the spiralized hairs, that the cortex was intermediate in character between orthocortex and paracortex (presumably in comparison with wool), but that there was marked asymmetry in the distribution of pigment in the Negro and Bushman hair which was not present in the other hairs. The transmission electron microscopy of fiber sections is at present the only reliable method for discriminating between various cortical cell types within human hair, and certainly the difference observed by this method between the ortho and para types of cells is not as marked as that in wool.[60] It is not surprising, therefore, that optical microscope methods have failed to reveal bilaterality in human hair. The careful examination of fiber sections in the transmission electron microscope has now revealed that Mongolian head hair (i.e., straight hair) contains all paracortical cells,[60] Caucasian hair (curly) contains a central core of para-type cells,[64] and Negro hair (woolly) is bilaterally segmented into the two cortical regions with the ortho-type segment on the outside of the crimp curl.[60]

*3.4.3.3b. Substructure of the Macrofibril.* The internal structure of the cortical macrofibril has been most effectively studied by the transmission electron microscopy of fiber sections appropriately loaded with heavy-metal salts. In the paracortex of wool, and to some extent in the para-type cells of human hair, the whole transverse section of the macrofibril appears to be composed of transversely cut rodlike elements approximately 7 nm in diameter (*microfibrils*) in pseudo-hexagonal array and embedded in a matrix of structureless material (the *intermacrofibrillar matrix*), which usually contains most of the metal of the heavy-metal staining procedure (Fig. 3-22). A general microfibril/matrix organization is also observed in the orthocortical

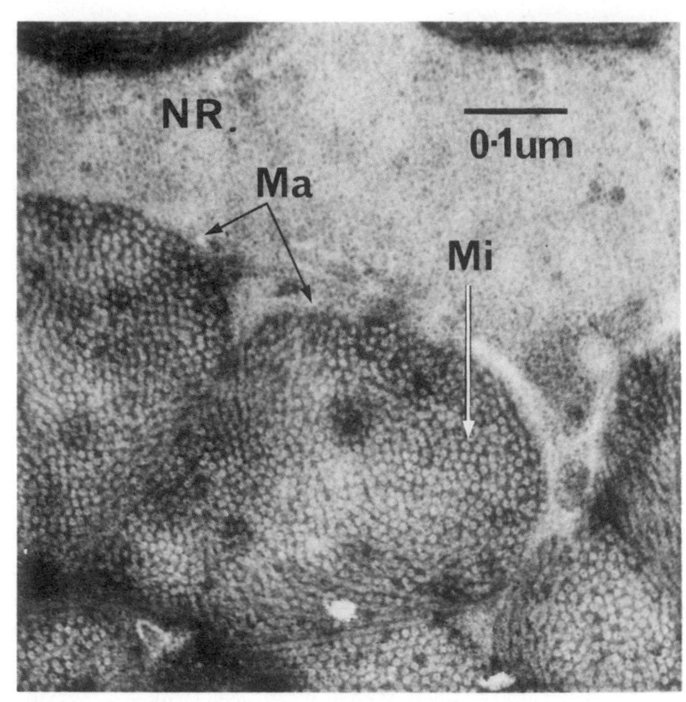


Fig. 3-22. Transverse section of paracortical macrofibrils of human hair. The section has been stained with osmium (after reduction), uranyl acetate, and lead citrate.

macrofibril, but the ends of the microfibrils are only seen in pseudo-hexagonal array about the center of the macrofibril, the remainder consisting of concentric circular dark and light zones and the whole rather resembling a thumbprint (Fig. 3-23).[214] Interpretation of these microfibril patterns can be made by considering that the electron microscope image represents a simple geometric projection section structure and that image contrast is due to the differential electron scattering between the metal-rich and metal-poor components (amplitude contrast).[243,248] On this basis the microfibrils of the paracortex would appear to be all parallel (and within the section, all parallel to the electron optical axis of the microscope to achieve circular "end-on" microfibril images). In the orthocortex, on the other hand, it would appear that the microfibrils at the center of the macrofibril are parallel to the electron beam and imaged as circular dots, but that radially the microfibrils are tilted relative to the electron beam so that overlapping oblique imaging is obtained and concentric light and dark bands seen. Johnson and Sikorski[244] and Dobb et al.[245] have examined the microfibrillar pattern in specific parts of a transverse fiber section after tilting the section at various angles in the electron

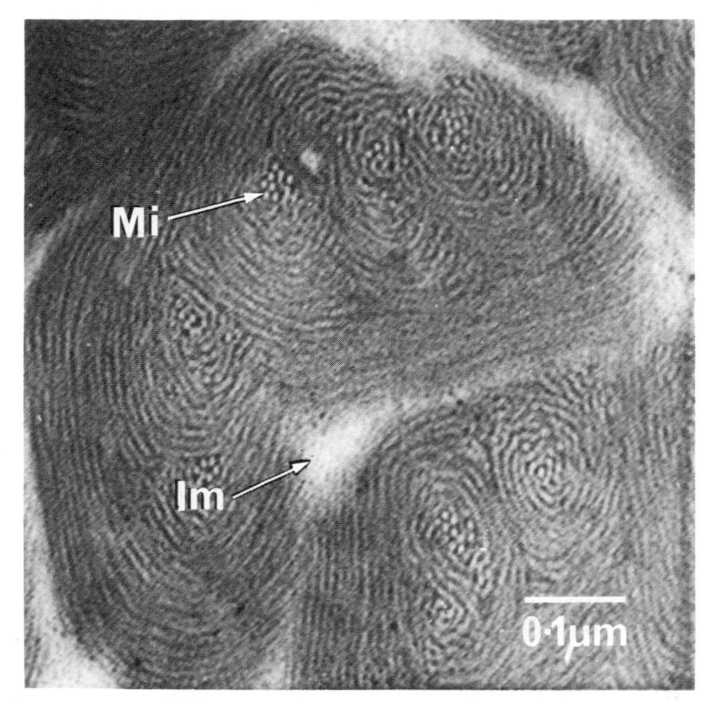


Fig. 3-23. Transverse section of orthocortical macrofibrils of human hair. The section has been stained with ammoniacal silver nitrate.

microscope. They found that microfibrils were tilted by as much as 40° to each other in the section assuming simple projection imaging. This is at variance with a maximum angular dispersion of only 20° measured from X-ray fiber diffraction photographs.[243] A rather surprising observation was that even in sections where from geometric considerations "end-on" microfibril imaging should disappear after tilting the section through 3°, the "end-on" image remained even after 16° tilt. The latter evidence by this group of workers has been taken to indicate that high-magnification electron micrographs of transversely sectioned keratin fibers cannot be interpreted solely in simple projections terms. They believe that materials such as keratin, which are known from X-ray diffraction studies to be paracrystalline, will also, through the regular disposition of diffracting sites within a thin section, give rise to electron diffraction contributions to the electron microscope image and pseudo "end-on" microfibril imaging.

A further disturbing aspect concerning the physical identity of the microfibril is that although fibrils of predominantly 2.0 nm diameter are found in ultrasonicated fragments from keratin fibers, units of 7.5 nm diameter, consistent with the dimensions of microfibrils seen in transverse section, are absent.[246] This has been interpreted[246] to mean that either the forces between adjacent microfibrils are similar to those between their internal components, or the microfibril may be considered as a fringe–fibrillar system involving the meadering of internal components, so that the apparent filamentous nature of microfibril represents an average mode of aggregation.

From the examination of sectioned wool fibers it has been noted that the microfibrils of the paracortex are separated by more intermicrofibrillar matrix than they are in the orthocortex.[247] Using low-angle selected area electron diffraction techniques, Dobb et al.[49,248,249] have recently obtained discrete equatorial reflections from metal-stained longitudinal sections of merino wool and found that these diffraction spacings for paracortical regions (8.80–10.60 nm) are significantly greater than those for orthocortical material (7.05–7.28 nm). They have determined the average separation of microfibrils in the orthocortex and paracortex from analyses of their electron micrographs with a convolution camera and shown that there is good correspondence between these average separations and those calculated from the electron diffraction data, assuming that the equatorial spots were first-order diffraction reflections from a hexagonally packed system of cylinders. By taking the average diameter of the microfibrils measured from electron micrographs as 7.0 nm they were able to calculate that the volume occupied by microfibrils in the para- and ortho-like cells of merino wool are approximately 33–48% and 67–70%, respectively.

High-resolution electron micrographs of longitudinally sectioned metal-stained wool show discontinuous striations.[248,250] The most common lateral periodicity observed is 7.55 nm, with a few at 4.37 nm and very occasionally 2.77 nm.[250] These are consistent with the 010, 110, and 120 spacings, respectively, for a hexagonally packed system of microfibrils within the plane of the section in which the center-to-center distances between adjacent microfibrils is 8.58 nm.[248,250] Consistent with this model, tilting of the longitudinal sections through 30° about an axis parallel to the direction of the striations results in some of the 7.55-nm spacings being modified to 4.37 nm, and vice versa.[250]

One striking feature of longitudinal sections of keratin fiber cortex is that the microfibrillar elements are terminated axially at various intervals[250] and indeed the equatorial reflections in low-angle electron diffraction patterns from these areas are split parallel to the equator into bands with spacings up to 200 nm.[248] Dobb and Sikorski[248] consider that these apparent discontinuities may be due to axial changes in azimuthal orientation of twisted pseudo-hexagonal arrays of microfibrils. Such twisting has already been proposed by Rogers[214] to explain the "thumbprint" appearance of transversely sectioned macrofibrils. On the other hand, Dobb and Sikorski[248] believe that the relative frequency of the 010, 110, and 120 spacings seen in the longitudinal sections may indicate that change in azimuthal orientation is abrupt rather than continuous. They point out that such a system of adjacent but continuous segments of supramolecular aggregates twisted with respect to each other is likely to profoundly affect the torsional and elastic properties of this component of the keratin fibers.

*3.4.3.3c. Substructure of the Microfibril.* In the previous section some emphasis was placed on the problems of determining macrofibril substructure from transmission electron micrographs of sectioned keratin fibers. This problem becomes even more acute as attempts are made to determine whether there is any specific arrangement of subunits within the microfibril. Several claims have been made to the existence of substructure,[251,252] and in the most notable of these, Filshie and Rogers[251] proposed that the microfibril is composed of nine 20-Å-diameter *protofibrils* arranged in an annular ring with two fibrils forming a core (i.e., "9 + 2"). Filshie and Rogers[251] point out that this type of structural organization does occur in nature in cilia and flagella (although admittedly at a different dimensional level), but this obviously cannot be taken as proof that this organization occurs in keratin fibers.

The problems of reaching workable conclusions about the organizational substructure of the microfibril from the electron microscopy of sections has been highlighted in a series of papers by Johnson and coworkers.[43,244,245,253,254] They have shown that the internal structure of the microfibril, which one apparently sees in a micrograph, is markedly dependent upon the operating conditions used with the electron microscope (e.g., position of focus, accelerating potential aperture size, etc.), and they claim that the image contains contributions from amplitude contrast, phase contrast, and probably also diffraction which are difficult to separate from each other. This group of workers maintains that there is only evidence for an indefinite ring-core structure.[244]

Support for the existence of a ring-core structure present within electron micrographs of sectioned microfibrils has been obtained by Fraser et al.[255,256] from averaged microfibril images derived from electron micrographs in which phase structure was minimized close to image focus. They have also pointed out that this ring-core structure is in accord with the rotationally averaged electron density distributions calculated from low-angle X-ray diffraction patterns of various keratins.[256,257] The validity of this approach has been hotly disputed on the grounds that the procedure involves assumptions about the extent and disposition of axial repeats in the microfibril within the thickness of the fiber section.[258,259]

Another approach to determining the histological makeup of the microfibril has been to examine fragments of keratin fibers in the electron microscope using negative staining and heavy-metal staining procedures.[40,246,248,260–263] Dobb and Sikorski[248] have shown that the debris obtained by taking merino wool, exhaustively converting the cystine in it to *S*-carboxymethyl cysteine, and grinding and ultrasonicating it consists predominantly of sheets of 2.0-nm-diameter filaments (presumably protofibrils). In similar preparations Dobb and Rogers[40] have occasionally encountered rodlike fragments of approximately 7.5 nm diameter which were probably microfibrils and as many as nine 2.0-nm-diameter fibrils were counted at the frayed ends of the microfibrils. This evidence, taken with that of Whitmore,[264] who has isolated 7.5-nm-diameter fibrils from hair roots after treatment with chymotrypsin, would seem to establish the structural identity of the microfibril but not reliably establish the number nor the organization of the protofibrillar subunits.

In fragment preparations from keratin fibers single, separate 2.0-nm-diameter filaments have been observed, and these are presumed to be single protofibrils.[40,260–262] Further fraying of some of these into 1.0-nm fibrils has been reported,[40,260–262] and these smaller units are believed to be single-protein chains. Indeed, this is in accord with the X-ray diffraction evidence of Fraser et al.[265] that protofibrils are two- or three-strand protein ropes of the type of coiled-coil rope originally described by Crick.[266] That the fibrillar units apparently observed in fragmented keratin are protein fibers has been questioned,[258,259] but the recent electron diffraction evidence of Dobb and Sikorski[248] that the material at least contains denatured protein would seem to relieve us of doubts in this matter.

#### *3.4.3.4. Medulla*

In terms of morphological substructure, this has been the least studied of keratin fiber components. The gross form of the medulla varies widely from animal to animal and has been classified into different types by Wildman.[10] His classification subdivides medulla into two main types according to whether it forms a broken or an unbroken column along the axis of the fibers. Further subdivisions are made into "latticed" and "simple" for unbroken medulla, and "fragmented" and "ladder" for broken medulla. In most rodent hairs, for example, the proportion of the transverse section occupied by the medulla can be as much as 80%, and the fibers are segmented by a series of ladderlike structures (Fig. 3-24). The rungs of the ladder are formed by annular thickenings of cortexlike material at regular intervals along the inner surface of the cylindrical shell of cortex.[60] The space between these "cortical" rungs contains a close-fitting spherical shell of amorphous material, the center of which is occupied by a large air-filled space. The air-filled spaces are thereby arranged in columns along

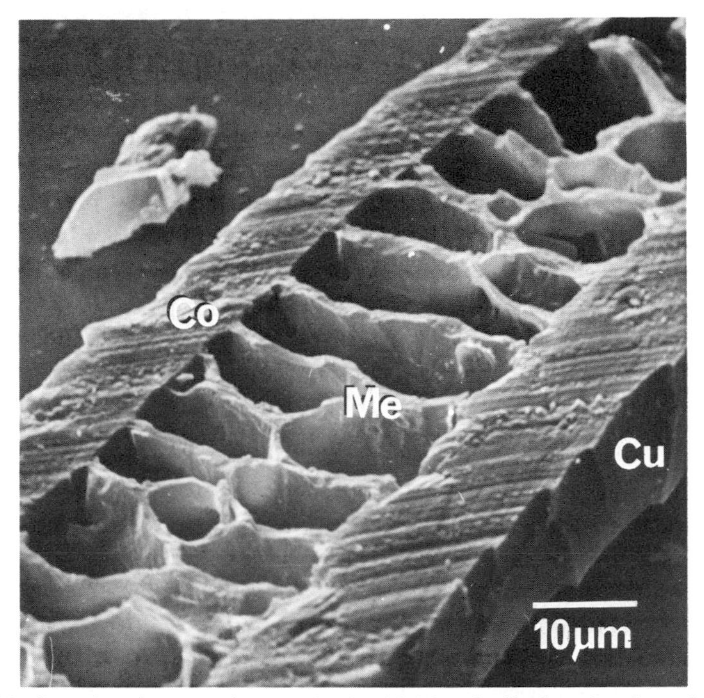


Fig. 3-24. Scanning electron micrograph of a thick longitudinal section through a guinea pig hair.

the axis of the fiber and separated by thin transverse discs of the amorphous material.

The presence of a cortexlike framework supporting thin shells of amorphous material bounding air spaces seems to be a generalized feature of the medulla of all keratin fibers. In human hair the medulla is not as regular as that of rodent hair, but nevertheless contains an irregular girderlike framework of cortical material over which is deposited an amorphous layer of irregular thickness giving a system of air-filled spaces of variable size.[267] The girderlike network contains macrofibrillar elements indistinguishable in structure from those found in the fiber cortex but aligned preferentially with respect to the girder rather than parallel to the fiber axis.[267] The material covering this framework is completely devoid of structure, tends to stain readily with uranyl salts (indicating the presence of reasonable quantities of acidic groups[60]), and contains very little cystine, as judged by an electron histochemical staining technique for this amino acid.[215] These observations are in accord with amino acid analysis of medulla separated from porcupine quill.[87]

The medulla of keratin fibers is quite resistant to various chemicals used, as keratinolytic solvents, and indeed alkali metal hydroxides, have been used to effect the isolation of medulla for subsequent chemical analysis.[178,268] In the absence of cystine crosslinks, the reason for the insolubility of medulla has been puzzling. It has been recently established, however, that the high insolubility is due to the presence of amide crosslinks between the protein chains.[269] These crosslinks are formed between lysyl and glutamyl or aspartyl residues on adjacent protein chains and are resistant to degradation by proteolytic enzymes. Although the amorphous component of the medulla is rapidly dissolved by enzymes such as trypsin, and indeed this has been used to obtain amino acid analysis for medulla,[87] the dipeptide at the crosslink is unaffected. This has been convenient not only for establishing the presence of the crosslink in medulla, but also for obtaining quantitative estimates of the extent of crosslinking by determining the amount of the dipeptide among the individual amino acids.[269]

#### *3.4.3.5. Hair Pigment*

Mammalian keratin fibers contain one of three principal types of pigment elaborated in the form of small granules within specialized dendritic cells (melanocytes) in the hair bulb matrix at the apex of the dermal papilla. The granules are "passed on" to the cells which will form the hair fiber proper and are fixed in position as keratinization and hardening of the hair takes place. The most important and widespread of these pigments is *eumelanin* (also known as *tyrosine melanin*). This dense black polymer is synthesized by a series of enzymatic oxidation steps from tyrosine and is responsible for the natural pigmentation of all black, brown, and blonde hairs.[35] *Pheomelanin* and *trichosiderin* are the pigments of yellow and red (auburn) hairs, respectively. Very little is known about their chemical makeup except that pheomelanin may resemble the ommochrome pigments of insect eyes[35] and that the hydrochloric acid–soluble red pigments contain iron.[270]

The tyrosine melanin pigment granules of black mammalian hairs (generally referred to as *melanin granules*) are ellipsoidal in shape, 0.8–1.0 $\mu$m in length, 0.3–0.4 $\mu$m in diameter,[271] and are generally situated in the intermacrofibrillar matrix of the cortex with their long axis parallel to the length of the fiber. In thin transverse sections of black hairs examined under the transmission electron microscope, the pigment granules are inherently electron-dense and, despite the use of various heavy-metal staining techniques, they only show a fairly granular and irregular internal structure. Progressing to fibers of lighter color the granules are less electron-opaque and in the case of blonde hairs, the granules shows a transverse internal structure consisting of 20- to 25-nm-thick folded sheets of dense material[272] (Fig. 3-25). The main difference between fibers of different color, ranging from blonde to black, is

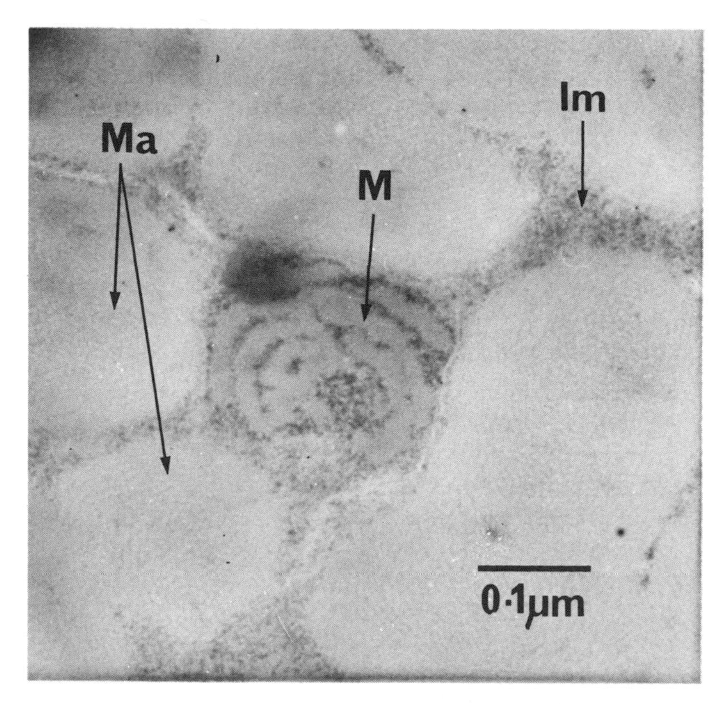


Fig. 3-25. Transverse section through a melanin granule of blonde hair. Osmium tetroxide staining.

due to the actual amount of the dense tyrosine melanin polymer within each granule rather than the physical number of granules in the fiber.[271]

Although melanin granules are distributed throughout the cortex of all pigmented mammalian hairs, in human hair, for example, their numerical concentration tends to be less toward the center of the hair than at the periphery.[60] Also in highly crimped pigmented wool fibers, there are more granules in the paracortex than in the orthocortex,[273] and this appears to be the consequence of the asymmetric distribution of melanocytes in the follicle.[273] Melanin granules are only found occasionally in the fiber cuticle, but when they are present, they reside in the endocuticle and tend to cause a minor irregularity in the parallel arrangement of the cuticle cells at the point where they occur. The presence of melanin granules within the cuticle has been found particularly in human nostril,[60] guanaco, vicuna, and swamp wallaby hairs[219,220] and appears to be associated with a predominance of interlocking devices not only between individual cortical cells but between cuticle and cortex.[219] Piper[220] has proposed that these unusual structures are due to the higher-than-normal phagocytic activity of the follicular cuticle cells, leading in one case to the ingestion of pigment and in the other to partial phagocytosis of one cuticle cell by another. From detailed examinations of thin transverse sections of human hair stained with dodecatungstophosphoric acid, the present author found that for many of the melanin granules in the cortex, there was a corresponding abutment of cortical cell membrane material up to the granule. In some cases a thin membrane resembling the cortical cell membrane actually encompassed some of the granules. It is believed that this may be the result of a process of phagocytosis whereby the early cortical cell of the hair matrix has engulfed portions of melanocytes or single granules present in the intercellular space, drawing them into the cell cytoplasm. The arrangement of membranes seen in sections of the mature fibers is consistent with this process, having been "frozen in" by the keratinization of the cortical cell.

The melanin granules of black hairs are exceedingly hard, as judged by the difficulty of sectioning them at the ultramicrotone. In addition, they are abnormally dense (about 1.72 g/ml),[271] are highly crystalline, as judged by the X-ray powder diffraction patterns which have been obtained,[271,274] and exhibit a high refractive index. Such properties as these are at variance with a random polymerization of indole-5,6-quinone (the immediate precursor of melanin) units, as has been proposed for the melanin macromolecule.[275,276] On the other hand, successive Diels–Alder condensations of indole-5,6-quinone would lead to a polymer containing extensive conjugation, a quinonoid terminus, and would be likely to form a graphitic-like three-dimensional structure, exhibiting properties similar to natural tyrosine melanin.[277] In addition, this polymerization process would lead to the evolution of carbon dioxide in the correct amounts from positions 5 and 6 of the original indole quinone (there has been no previous satisfactory explanation for the loss of carbon dioxide from these positions).

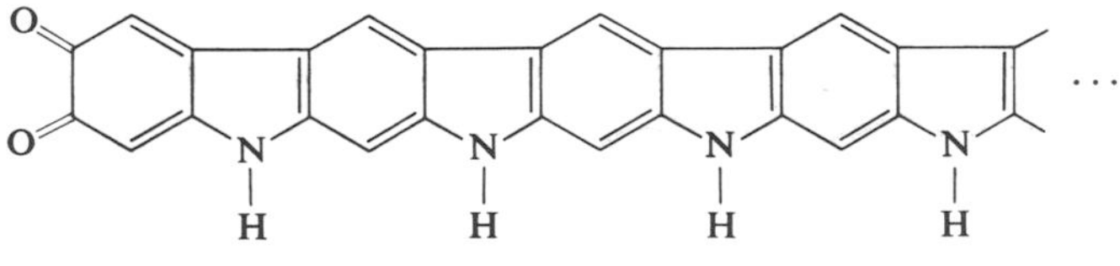

### 3.4.4. Histochemistry

There have been three main approaches to elucidating the chemical composition of the various morphological components of mature keratin fibers. In one method fractions have been obtained by the dissolution of the fiber and then, after the event, attempts have been made to assign the morphological origin of these fractions in terms of the known microscopic structure of the fiber. In the case of the separation of major components such as those obtained by peracetic acid oxidation, dissolution in ammonia, and differential precipitation, it is generally believed that the low-sulfur fraction ($\alpha$-keratose) is derived from the microfibrils, the insoluble fraction ($\beta$-keratose) comes from the cell membrane complex, and the high-sulfur fraction ($\gamma$-keratose) originates from the intermicrofibrillar matrix.[278] In a similar way, it is considered that the two major proteins separated at different pHs from fibers after reduction and *S*-carboxymethylation are obtained primarily from the microfibril (i.e., the low-sulfur, *S*-carboxymethyl kerateine A) and the intermicrofibrillar matrix (i.e., the high-sulfur *S*-carboxymethyl kerateine B).[279,280]

A more widely used and reliable method has been to use microscopic techniques to establish which morphological components dissolve or are separated with given treatments. The validity of subsequent analysis is, of course, dependent upon clean morphological separation of components, and to this end selective enzymatic digestion procedures are valuable.[281–283] This type of method has been used for obtaining analysis of cuticle,[284–286] orthocortex and paracortex,[281] medulla,[268] endocuticle and exocuticle,[282,287,288] macrofibrils and intermacrofibrillar matrix,[283,289,290] and various fractions obtained from the cell membrane complex.[285,291]

The third method, which we will consider in more detail here, is that of histochemistry, that is, the selective staining of specific chemical groups so that the distribution of these groups with respect to the morphological substructure of the fibers can be observed under a microscope. In the present section only those techniques applicable to the transmission electron microscope (i.e., electron histochemistry) will be described. It is a requirement that the product of an electron histochemical procedure will scatter electrons more effectively than the original organic components of the hair, so that sufficient image contrast will be available to determine the distribution of the stain in a section of the fiber.[292] To this end it is necessary to seek those chemical treatments that will culminate in the deposition of substantial quantities of a heavy metal at the appropriate chemical sites.

#### *3.4.4.1. Cystine and Cysteine*

Many metal-staining procedures give rise to intense staining of the exocuticle and A layer and staining of the intermicrofibrillar matrix but not the microfibrils. These have included osmium tetroxide,[251,295,296] silver nitrate,[41,293,294] ammoniacal silver salts,[297] lead hydroxide,[251,298] sodium plumbite,[217] and potassium permanganate,[298] used either singly or in various combinations and applied to untreated or reduced keratin fibers. Many of these stains are valuable for revealing the detailed internal structure of keratin fibers visible in thin sections examined under the transmission electron microscope, but, although it is generally assumed that the metals bind to the cystine residues within the fibers, absolute specificity for cystine has not been established. There are at present three staining methods for which the staining specificity for cystine has been reasonably established.

*3.4.4.1a. Organomercurial Halide Method.* This relies upon the reaction between organomercurial halides and the sulfhydryl groups of reduced keratins.

$$\text{hair—S—S—hair} \xrightarrow{\text{reduction}} \text{hair—SH} \xrightarrow{\text{RHgX}} \text{hair—S—Hg—R}$$

Methyl mercuric iodide,[295] *p*-chloromercuribenzoic acid,[299] and phenyl mercuric chloride[300] have been used as alternatives in this electron histochemical procedure. More recently, Dobb et al.,[222] using *p*-chloromercuribenzoate and methyl mercuric iodide, have shown that there is near-stoichiometric uptake of mercury by the sulfhydryl groups of reduced wool fibers. Although their claims for high specificity of sulfhydryl staining are reasonably valid, their quantitative work is insufficiently accurate to take account of minor constituents which could react with the organomercurial but may not contain sulfhydryl groups. In this respect it is interesting to note that Levy[301] has demonstrated that organomercuric halides result in the chlormercuration of the amino groups of insulin.

*3.4.4.1b. Silver–Methenamine Method.* In this technique keratin fiber sections are treated with a complex solution of silver nitrate and hexamethylene tetramine (i.e., silver–methenamine reagent) under precisely controlled conditions of time, temperature, and pH.[215] Two forms of silver are deposited in the sections. One is a morphological stain which is readily removed by subsequent treatment of the section with sodium thiosulfate, and the other is a globular form of silver which resists dissolution by thiosulfate. It is this latter form of silver which it is claimed demonstrates the sites of cystine-containing proteins (except for the globular silver found over sectioned melanin granules).[215] The reasons for claiming high specificity are that (1) there is considerably less globular silver in sections which have been reduced and then stained,[215] (2) no silver globules are produced in sections which have been exhaustively reduced and alkylated,[215] and (3) if sections are reduced and alkylated with 1-iodo-2,3-dihydroxy propane, no silver globules are obtained with subsequent staining but are produced at the same sites as in untreated sections if these particular alkylated sections are oxidized with periodate.[60] This scheme is summarized as shown (Ag/meth refers to silver–methenamine reagent). In addition to this evidence it has been shown that for a series of cellulosic films to which various amino acid residues have been attached, only the model containing cystine yielded silver precipitation with

$$\text{hair—S—S—hair} \xrightarrow{\text{Ag/meth}} \text{silver globules}$$

$$\Big\downarrow \text{reduction}$$

$$\text{hair—SH} \xrightarrow{\text{Ag/meth}} \text{only a few silver globules}$$

$$\Big\downarrow ICH_2CH(OH)—CH_2(OH)$$

$$\text{hair—S—}CH_2\text{—}CH(OH)\text{—}CH_2(OH) \xrightarrow{\text{Ag/meth}} \text{no globular silver}$$

$$\Big\downarrow HIO_4$$

$$\text{hair—S—}CH_2\text{—CHO} \xrightarrow{\text{Ag/meth}} \text{silver globules}$$

silver–methanamine reagent, which resisted removal by sodium thiosulfate.[141] Silver deposition was also less after reduction of the cystine-containing model and eliminated altogether by exhaustive reduction and alkylation.

*3.4.4.1c. S-Carboxymethylation Uranyl Method.* There is strong evidence for believing that when solutions of uranyl salts are used for staining tissue sections there is binding between divalent uranyl ions and the carboxyl groups of the tissue (see Section 3.4.4.3). In addition, it is well known that subsequent treatment of these sections with alkaline lead salts gives intensification of the original uranium staining. Kassenbeck[302] has used this staining procedure as the basis for an electron histochemical demonstration of cystine. Although there is minimal staining of untreated hair sections, there is a dramatic increase in the amount of staining in sections which have been reduced and then alkylated with iodoacetic acid. Since there is no such increase in staining in sections which are reduced and alkylated with *N*-ethylmaleimide, this would

$$\text{hair—S—S—hair}$$

$$\Big\downarrow \text{reduction}$$

$$\text{hair—SH}$$

$$\Big\downarrow ICH_2COOH$$

$$\text{hair—S—}CH_2\text{—COOH}$$

$$\Big\downarrow UO_2(Ac)_2$$

$$(\text{hair—S—}CH_2\text{—COO})_2UO_2$$

seem to be clear-cut evidence that the increased staining of the *S*-carboxymethylated sections was occurring specifically at those sites originally containing cystine. This is summarized as shown.

The three electron histochemical procedures described above have been used, of course, to define the detailed distribution of cystine in mammalian keratin fibers. The highest concentration of cystine is in the exocuticle and A layer, and the present author believes that it could account for at least one amino acid residue in three of the proteins in these components. Such a high concentration of cystine, with associated intermolecular crosslinking, is likely to confer some interesting and unusual mechanical properties on these components. Little or no cystine is found in the endocuticle, the cell membrane complexes, the amorphous medullary protein, the nuclear remnants, and intermacrofibrillar matrix. Cystine is present, of course, in the macrofibrils but is distributed in the intermicrofibrillar matrix and not in the microfibrils. The latter distribution is in keeping with the normally accepted two-phase structure for the major components of all mammalian keratin fibers.

### *3.4.4.2. Free Amino and Basic Groups*

*3.4.4.2a. PTA Staining Technique.* When thin sections of keratin fibers are treated with concentrated aqueous solutions of dodecatungstophosphoric acid [i.e., phosphotungstic acid (PTA)] and examined under the transmission electron microscope, the staining which is observed is almost complementary to that obtained with stains for cystine.(124,297,303) Kuhn et al.,(304) in their work on the PTA staining of collagen, suggested that PTA bound at the basic sites and particularly at the free amino groups within the proteins. Kuhn(305) went on to show that solutions of PTA contained a trivalent anion of the type $O{=}P\,(W_2O_7H)_3$ and proposed that it was this species which bound to the cationic groups of the protein in a manner analogous to an acid dye. That the binding of PTA in keratin sections is due to the presence of free amino groups was established independently by Bonés(306) and Kassenbeck and Hagege,(297) who found that dinitrophenylation of the amino groups in the fibers eliminated PTA staining.

*3.4.4.2b. Kendall–Barnard Technique.* In 1963 Kendall and Barnard published an elegant electron histochemical procedure for staining the free amino groups of proteins.(307) The technique is dependent upon the acylation of amino groups by a complex of silver nitrate and *N*-acetyl homocysteine thiolactone (AHCTL), silver being covalently bound as a mercaptide with the opening of the thiolactone ring, as shown. Insofar as specificity is concerned, Kendall and Barnard showed that the loss of amino groups in rat liver and rat sperm resulted in an equivalent uptake of silver in the tissues.

$$Ag^+ + \underset{\displaystyle \underset{}{CH_2}\!-\!\underset{\displaystyle NHCOCH_3}{CH}}{CH_2 \overset{S}{\frown} CO} + NH_2\text{—protein} \longrightarrow \begin{array}{l} S\text{—Ag} \\ | \\ CH_2 \\ | \\ CH_2 \\ | \\ CHCONH\text{—protein} \\ | \\ NH \\ | \\ COCH_3 \end{array}$$

This staining procedure has been used by the present author on sections of human hair and, although the intensity of staining is not as good as with PTA, precisely the same components are stained.[60] In addition, staining can be eliminated completely by the prior dinitrophenylation of the amino groups in the hair sections.

Using both the PTA and the Kendall and Barnard techniques, the keratin fiber components which stain and evidently contain higher-than-average concentrations of free amino groups are the endocuticle, the intercellular membrane cements of the cuticle and cortex (i.e., the delta band), an irregular layer adjacent to all the cell membranes, the nuclear remnants of the cortex, and the intermacrofibrillar matrix. In contrast to this, the exocuticle, A layer, and macrofibrils remain unstained. Some staining within the cortical macrofibrils has been reported,[297] but since in this case staining was carried out in boiling-PTA solutions, it seems reasonable to think that the PTA is merely binding at the amino end of the hydrolyzed proteins.

### *3.4.4.3. Carboxyl Groups*

At present the only electron histochemical procedure for staining carboxyl groups relies upon the use of uranyl acetate acting as a basic dye. The main evidence for carboxyl group binding is that uranyl acetate does not stain collagen in which the carboxyl groups have been eliminated by methylation.[308] Further evidence for carboxyl group binding is the very high uptake of uranyl acetate in *S*-carboxymethylated keratin fiber sections.[302] The components of normal keratin fiber sections which stain with uranyl acetate and which evidently contain higher than average concentrations of carboxyl groups are the endocuticle, delta band, nuclear remnants, and intermacrofibrillar matrix. Although these are the most intensely stained components, there are also increases in the electron opacity of the macrofibrils, indicating the presence of a lower concentration of carboxyl groups in these components. Using uranyl acetate and combinations of uranyl acetate and lead salts Kassenbeck[302] has shown that within the macrofibril, staining occurs preferentially within the microfibrillar rather than in the intermicrofibrillar matrix.

### *3.4.4.4. Tyrosine*

In 1964 DeDeurwaerder et al.[309] reported the isolation of a new class of high-glycine-tyrosine proteins from keratin fibers. These proteins were obtained by the extraction of reduced wool with tris(diethylamino)phosphine in formamide, and from examinations of sections of the extracted fibers under the electron microscope it appeared that the proteins originated from the cell membrane complex. Since 1964 there has been considerable interest in the high-glycine-tyrosine proteins, which have been extracted from keratin fibers under a variety of different conditions.[310–313] The amount of these proteins varies widely from one animal species to another, and for example Gillespie[313] has shown that whereas there is less than 1% in Lincoln wool, there is as much as 30% in echidna quill.

The histological origin of the high-glycine-tyrosine proteins is of considerable interest. Although DeDeurwaerder et al.[309] demonstrated that their material is obtained from the cell membrane complex, this can hardly be the site for all the high-tyrosine proteins in echidna quill. Using the Millon mercury method for optical microscopy, Stoves[314] has shown that whereas the tyrosine content of cuticle and cortex is relatively low, there are high concentrations of tyrosine in the medulla. However, the medulla is unlikely to be the origin of the high-glycine-tyrosine proteins investigated by Gillespie,[313] because his extraction procedures would not have tended to dissolve the medulla. The only evidence for the histological origin of these new proteins is that the center-to-center separation of microfibrils appears to increase with the high-glycine-tyrosine content of the fibers. These proteins therefore probably contribute to the bulk of the intermicrofibrillar matrix.[315]

From the foregoing it is clear that a specific electron histochemical procedure for tyrosine would be desirable. Recently, the present author has devised such a technique.[60] This relies upon the conversion of tyrosyl residues to a diazonium chloride, with subsequent coupling to α-naphthol. The product contains a coordination site for the binding of divalent heavy metals:

OH ⟶ HO, $N_2Cl$ ⟶ N=N, HO···$M^{2+}$···HO

The specificity of this technique for tyrosine has been evaluated by using a series of model compounds containing different amino acids covalently linked to cellulosic films. Under the conditions used for staining sections, only the tyrosine-containing model gave rise to binding of platinum, indicating the high specificity of staining for tyrosine among all the amino acids normally

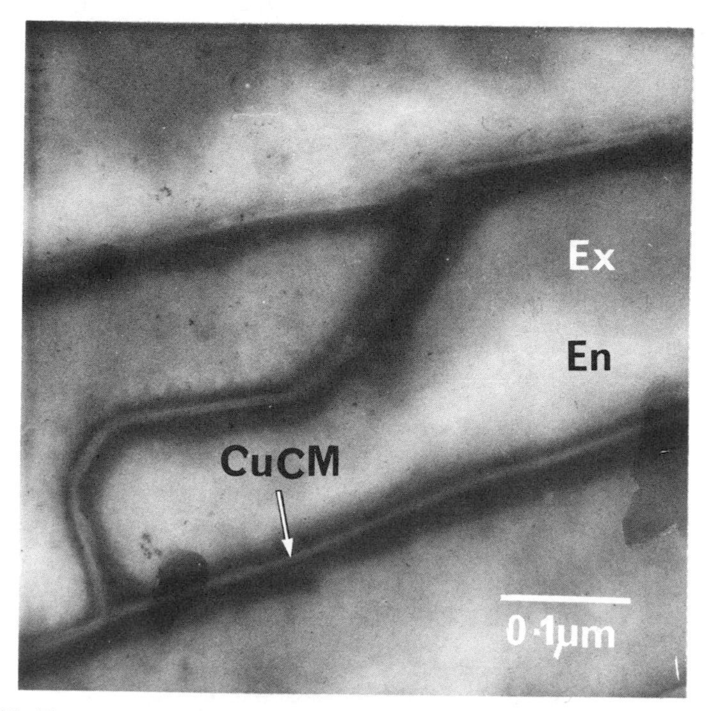


Fig. 3-26. Transverse section through human hair cuticle. A stain has been used to reveal the sites of tyrosine-containing proteins.

encountered in keratin fibers. The procedure is not without its difficulties, one of which is the problem of treating thin keratin sections with nitrous acid and maintaining them intact. For sections of human hair there is staining for tyrosine in thin layers on either side of the cuticle cell membrane complex (Fig. 3-26). Further work will be required with this staining procedure to establish the histological origin of the high-tyrosine-containing proteins in, for example, echidna quill.

## 3.5. References

1. W. Ward and H. P. Lundgren, *Advan. Protein Chem.* **9**, 243 (1954).
2. R. J. Barnes, Studies in the Optical Properties of Wool, Hair, and Related Fibers, Ph.D. thesis, University of Leeds (1933).
3. W. T. Astbury and F. O. Bell, *Tabulae Biol.* **17**, Pt. 1, 90 (1939).
4. A. Giroud, H. Bulliard, and C. P. Leblond, *Bull. Histol. Tech. Microscopy* **11**, 129 (1934).
5. E. H. Mercer, *Keratin and Keratinisation*, Pergamon Press, Oxford (1961).

6. J. M. Gillespie, in: *Biology of the Skin and Hair Growth* (A. G. Lyne and B. F. Short, eds.), p. 377, Angus & Robertson Ltd., Sydney (1965).
7. E. H. Mercer, in: *Advances in Biology of Skin* (W. Montagna and R. L. Dobson, eds.), Vol. 9, *Hair Growth* p. 556, Pergamon Press, Oxford (1969).
8. A. G. Matoltsy, in: *Advances in Biology of Skin* (W. Montagna and R. L. Dobson, eds.), Vol. 9, *Hair Growth*, p. 559, Pergamon Press, Oxford (1969).
9. G. L. Clark (ed.), *The Encyclopedia of Microscopy*, Reinhold Publishing Corp., New York (1961).
10. A. B. Wildman, *The Microscopy of Animal Textile Fibres*, Wool Industries Research Association, Leeds (1954).
11. F. Kidd, *Brushmaking Materials*, British Brush Manufacturers Research Association, Leeds (1957).
12. J. L. Stoves, *Fibre Microscopy*, National Trade Press, London (1957).
13. J. W. S. Hearle and R. H. Peters (eds.), *Fibre Structure*, Butterworth & Co Ltd. and the Textile Institute, London (1963).
14. R. Hooke, *Micrographia of Some Physiological Descriptions of Minute Bodies Made by Magnifying Glasses, with Observations and Enquiries Thereupon*, John Martyn, London (1665).
15. R. D. B. Fraser and G. E. Rogers, *Aust. J. Biol. Sci.* **8**, 129 (1955).
16. J. I. Hardy and T. M. Plitt, *U.S. Dept. Interior*, *Wildlife Circ.* **7** (1940).
17. L. Auber and H. M. Appleyard, *Nature* (*Lond.*) **168**, 736 (1951).
18. S. Tolansky, *Multiple Beam Interferometry*, Clarendon Press, Oxford (1948).
19. J. Howell and R. Mazur, *J. Textile Inst.* **44**, T59 (1953).
20. A. Skertchly, *J. Textile Inst.* **45**, T78 (1954).
21. J. M. Preston, *Modern Textile Microscopy*, Emmott & Co. Ltd., London (1933).
22. F. A. Mennerich, *Rayon Textile Monthly* **22**, 68 (1941).
23. P. Kassenbeck, J. Jacquemart, and R. Monrocq, *3ᵉ Congr. Intern. Recherche Textile Lainière, Paris* **1**, 209 (1965).
24. J. I. Hardy, *U.S. Dept. Agr. Circ. 378* (1933).
25. A. G. E. Pearse, *Histochemistry: Theoretical and Applied*, Vol. 1, J. & A. Churchill Ltd., London (1968).
26. L. Auber, *Trans. Roy. Soc. Edinburgh* **62**, 191 (1950–1951).
27. E. J. Holmes, *J. Histochem. Cytochem.* **16**, 428 (1968).
28. H. M. Appleyard and C. M. Dymoke, *J. Textile Inst.* **45**, T480 (1954).
29. E. H. Mercer, *Biochim. Biophys. Acta* **3**, 161 (1949).
30. W. J. Schmidt, *Die Bausteine des Tierskorpers in polarisierten Lichte*, Friedrich Cohen, Bonn (1924).
31. A. N. J. Heyn, *Textile Res. J.* **22**, 513 (1952).
32. D. J. Kemp and G. E. Rogers, *J. Cell Sci.* **7**, 273 (1970).
33. W. L. Epstein and H. I. Maibach, in: *Advances in Biology of Skin* (W. Montagna and R. L. Dobson, eds.), Vol. 9, *Hair Growth*, p. 83, Pergamon Press, Oxford (1969).
34. D. R. Harkness and H. A. Bern, *Acta Anat.* **31**, 35 (1957).
35. T. B. Fitzpatrick, P. Brunet, and A. Kukita, in: *The Biology of Hair Growth* (W. Montagna and R. A. Ellis, eds.), p. 255, Academic Press, New York (1958).
36. H. Zahn, *Melliand Textilber.* **21**, 505 (1940).
37. C. W. Hock and H. F. McMurdie, *J. Res. Natl. Bur. Std.* **31**, 299 (1943).
38. E. H. Mercer and A. L. G. Rees, *Aust. J. Exptl. Biol.* **24**, 147 (1946).
39. G. M. Jeffrey, J. Sikorski, and H. J. Woods, *Proc. Intern. Wool Textile Res. Conf., Australia* **F**, 130 (1955).

40. M. G. Dobb and G. E. Rogers, in: *Symposium on Fibrous Proteins* (W. G. Crewther, ed.), p. 267, Butterworth & Co. (Australia) Ltd., Sydney (1968).
41. M. G. Dobb, J. A. Nott, and J. Sikorski, *Proc. European Regional Conf. Electron Microscopy, Delft* **2**, 664 (1960).
42. F. S. Sjöstrand, *Electron Microscopy of Cells and Tissues*, Vol. 1, Academic Press, New York (1967).
43. D. J. Johnson and J. Sikorski, *Nature (Lond.)* **194**, 31 (1962).
44. A. G. E. Pearse, *J. Roy. Microscopy Soc.* **81**, 107 (1963).
45. L. A. Sternberger, *J. Histochem. Cytochem.* **15**, 139 (1967).
46. P. A. Wilson, R. C. Henrikson, and A. M. Downes, *J. Cell Sci.* **8**, 489 (1971).
47. A. K. Christensen, *J. Cell Biol.* **51**, 772 (1971).
48. J. A. Chapman and J. W. Menter, *Proc. Roy. Soc. (Lond).* **A226**, 400 (1954).
49. M. G. Dobb, *J. Textile Inst.* **61**, 232 (1970).
50. A. D. G. Stewart and M. A. Snelling, *Proc. 3rd European Regional Conf. Electron Microscopy, Prague* **A**, 55 (1964).
51. *Wool Sci. Rev.* **35**, 2 (1969).
52. A. Hepworth, J. Sikorski, D. J. Tucker, and C. S. Whewell, *J. Textile Inst.* **60**, 513 (1969).
53. C. A. Anderson, H. J. Katz, M. Lipson, and G. F. Wood, *Textile Res. J.* **40**, 29 (1970).
54. S. Tolansky, *Vacuum* (publ. Edwards High Vacuum), **4**, 456 (1954).
55. J. Sikorski and W. Sprenkmann, *Melliand Textilber.* **49**, 471 (1968).
56. A. C. Brown and J. A. Swift, *Proc. 5th European Regional Conf. Electron Microscopy, Manchester*, 386 (1972).
57. J. A. Swift and A. C. Brown, *J. Soc. Cosmetic Chem.* **23**, 695 (1972).
58. C. A. Anderson and M. Lipson, *Textile Res. J.* **40**, 88 (1970).
59. A. C. Brown and J. A. Swift, *Beitr. Elektronenmikroskop. Direktabb. Oberfl., Berlin* **3**, 299 (1970).
60. J. A. Swift, unpublished observations.
61. J. A. Swift, A. C. Brown, and C. A. Saxton, *J. Phys. E. (Sci. Instrum.)* **2**, 744 (1969).
62. J. A. Swift and A. C. Brown, *Proc. 3rd Ann. SEM Symp., Chicago*, 113 (1970).
63. J. A. Swift, *J. Textile Inst.* **63**, 64 (1972).
64. J. A. Swift, *J. Textile Inst.* **63**, 129 (1972).
65. J. A. Swift, *Appl. Polymer Symp.* **18**, 185 (1971).
66. J. C. Russ, *Proc. 5th Ann. Scanning electron Microscopy Symp., Chicago*, 73 (1972).
67. N. E. Bywater, T. Buckley, A. Hepworth, and J. Sikorksi, *Proc. 5th European Conf. Regional Electron Microscopy, Manchester*, 384 (1972).
68. H. Pinkus, in: *The Biology of Hair Growth* (W. Montagna and R. A. Ellis, eds.), p. 1, Academic Press, New York (1958).
69. W. E. Straile, in: *Biology of the Skin and Hair Growth* (A. G. Lyne and B. F. Short, eds.), p. 35, Angus & Robertson Ltd., Sydney (1965).
70. F. J. Ebling, in: *Biology of the Skin and Hair Growth* (A. G. Lyne and B. F. Short, eds.), p. 507, Angus & Robertson Ltd., Sydney (1965).
71. A. M. Kligman, *J. Invest. Dermatol.* **33**, 307 (1959).
72. D. F. Dowling and T. Nay, *Aust. J. Agr. Sci.* **11**, 1064 (1960).
73. J. K. Ling, in: *Biology of the Skin and Hair Growth* (A. G. Lyne and B. F. Short, eds.), p. 525, Angus & Robertson Ltd., Sydney (1965).
74. R. H. Hayman, in: *Biology of the Skin and Hair Growth* (A. G. Lyne and B. F. Short, eds.), p. 575, Angus & Robertson Ltd., Sydney (1965).

75. A. B. Houssay, C. E. Epper, and J. H. Pazo, in: *Biology of the Skin and Hair Growth* (A. G. Lyne and B. F. Short, eds.), p. 641, Angus & Robertson Ltd., Sydney (1965).
76. K. A. Ferguson, A. L. C. Wallace, and H. R. Lindner, in: *Biology of the Skin and Hair Growth* (A. G. Lyne and B. F. Short, eds.), p. 655, Angus & Robertson Ltd., Sydney (1965).
77. A. M. Downes and A. L. C. Wallace, in: *Biology of the Skin and Hair Growth* (A. G. Lyne and B. F. Short, eds.), p. 679, Angus & Robertson Ltd., Sydney (1965).
78. J. C. Hutchinson, in: *Biology of the Skin and Hair Growth* (A. G. Lyne and B. F. Short, eds.), p. 565, Angus & Robertson Ltd., Sydney (1965).
79. F. W. Dry, *J. Genet.* **16**, 287 (1925–1926).
80. W. S. Bullough and E. B. Laurence, in: *The Biology of Hair Growth* (W. Montagna and R. A. Ellis, eds.), p. 171, Academic Press, New York (1958).
81. W. Montagna and E. J. van Scott, in: *The Biology of Hair Growth* (W. Montagna and R. A. Ellis, eds.), p. 39, Academic Press, New York (1958).
82. W. Montagna and R. L. Dobson (eds.), *Advances in Biology of Skin*, Vol. 9, *Hair Growth*, Pergamon Press, Oxford (1969).
83. M. S. C. Birbeck and E. H. Mercer, *J. Biophys. Biochem. Cytol.* **3**, 203 (1957).
84. M. S. C. Birbeck and E. H. Mercer, *J. Biophys. Biochem. Cytol.* **3**, 215 (1957).
85. M. S. C. Birbeck and E. H. Mercer, *J. Biophys. Biochem. Cytol.* **3**, 227 (1957).
86. E. H. Mercer, in: *The Biology of Hair Growth* (W. Montagna and R. A. Ellis, eds.), p. 91, Academic Press, New York (1958).
87. G. E. Rogers, in: *The Epidermis* (W. Montagna and W. C. Lobitz, Jr., eds.), p. 179, Academic Press, New York (1964).
88. S. I. Roth and W. H. Clark, in: *The Epidermis* (W. Montagna and W. C. Lobitz, Jr., eds.), p. 303, Academic Press, New York (1964).
89. P. F. Parakkal, in: *Advances in Biology of Skin* (W. Montagna and R. L. Dobson, eds.), Vol. 9, *Hair Growth*, p. 441, Pergamon Press, Oxford (1969).
90. M. Bell, in: *Advances in Biology of Skin* (W. Montagna and R. L. Dobson, eds.), p. 61, Pergamon Press, Oxford (1969).
91. A. Charles, *Exptl. Cell Res.* **18**, 138 (1959).
92. F. Happey and A. G. Johnson, *J. Ultrastructure Res.* **7**, 316 (1962).
93. P. F. Parakkal, *J. Ultrastructure Res.* **14**, 133 (1966).
94. P. F. Parakkal and A. G. Matoltsy, *J. Invest. Dermatol.* **43**, 23 (1964).
95. G. E. Rogers, *Exptl. Cell Res.* **13**, 521 (1957).
96. S. I. Roth and E. B. Helwig, *J. Ultrastructure Res.* **11**, 33 (1964).
97. S. I. Roth and E. B. Helwig, *J. Ultrastructure Res.* **11**, 52 (1964).
98. S. I. Roth, in: *Ultrastructure of Normal and Abnormal Skin* (A. S. Zelickson, ed.), p. 105, Henry Kimpton, London (1967).
99. R. D. B. Fraser, T. P. MacRae, and G. E. Rogers, *Keratins*, Charles C Thomas, Springfield, Ill. (1972).
100. B. Forslind and G. Swanbeck, *Exptl. Cell Res.* **43**, 191 (1966).
101. R. E. Chapman and R. T. Gemmell, *J. Ultrastructure Res.* **36**, 342 (1971).
102. R. T. Gemmell and R. E. Chapman, **36**, 355 (1971).
103. K. Hashimoto, *Arch. Klin. Exptl. Dermatol.* **238**, 333 (1970).
104. D. F. G. Orwin, *Aust. J. Biol. Sci.* **24**, 989 (1971).
105. P. F. Parakkal, *Z. Zellforsch.* **107**, 174 (1970).
106. D. F. G. Orwin, *Aust. J. Biol. Sci.* **23**, 623 (1970).
107. V. A. Puccinelli, R. Caputo, and B. Ceccarelli, *Giorn. Ital. Dermatol.* **108**, 1, 1967.
108. V. A. Puccinelli and T. Cainelli, *Giorn. Ital. Dermatol.* **107**, 1147 (1966).

109. V. A. Puccinelli, R. Caputo, and B. Ceccarelli, *Arch. Klin. Exptl. Dermatol.* **233**, 172 (1968).
110. P. F. Parakkal, *J. Ultrastructure Res.* **29**, 210 (1969).
111. S. I. Roth, in: *Biology of the Skin and Hair Growth* (A. G. Lyne and B. F. Short, eds.), p. 233, Angus & Robertson Ltd., Sydney (1965).
112. M. S. C. Birbeck and E. H. Mercer, *Nature* (*Lond.*) **189**, 558 (1961).
113. N. A. Barnicot and M. S. C. Birbeck, in: *The Biology of Hair Growth* (W. Montagna and R. A. Ellis, eds.), p. 239, Academic Press, New York (1958).
114. A. S. Zelickson, in: *Ultrastructure of Normal and Abnormal Skin* (A. S. Zelickson, ed.), p. 163, Henry Kimpton, London (1967).
115. M. Seiji, in: *Ultrastructure of Normal and Abnormal Skin* (A. S. Zelickson, ed.), p. 183, Henry Kimpton, London (1967).
116. J. H. Mottaz and A. S. Zelickson, in: *Advances in Biology of Skin* (W. Montagna and R. L. Dobson, eds.), Vol. 9, *Hair Growth*, p. 471, Pergamon Press, Oxford (1969).
117. M. S. C. Birbeck and N. A. Barnicot, in: *Pigment Cell Biology* (M. Gordon, ed.), p. 549, Academic Press, New York (1959).
118. M. S. C. Birbeck, E. H. Mercer, and N. A. Barnicot, *Exptl. Cell Res.* **10**, 505 (1956).
119. H. Braunsteiner, F. Mlczock, and F. Pakesch, *Klin. Wochensch.* **36**, 262 (1958).
120. H. M. Hirsch, A. S. Zelickson, and J. F. Hartmann, *Z. Zellforsch.* **65**, 409 (1965).
121. J. H. Mottaz and A. S. Zelickson, *J. Invest. Dermatol.* **49**, 605 (1967).
122. A. S. Zelickson and J. F. Hartmann, *J. Invest. Dermatol.* **36**, 23 (1961).
123. A. Charles and J. T. Ingram, *J. Biophys. Biochem. Cytol.* **6**, 41 (1959).
124. J. A. Swift, *Nature* (*Lond.*) **203**, 976 (1964).
125. W. E. Straile, in: *Biology of the Skin and Hair Growth* (A. G. Lyne and B. F. Short, eds.), p. 35, Angus & Robertson Ltd., Sydney (1965).
126. F. Pinkus, in: *Handbuch der Haut- und Geschlechtskrankheiten* (J. Jadassohn, ed.), p. 85, Springer-Verlag, Berlin (1927).
127. A. Giroud and H. Bulliard, *Arch. Morphol.* **29**, 8 (1930).
128. A. Giroud and C. P. Leblond, *Ann. N.Y. Acad. Sci.* **53**, 613 (1951).
129. M. Chevremont and J. Frederic, *Arch. Biol. Liège* **54**, 589 (1943).
130. K. Rudall, *Advan. Protein Chem.* **7**, 253 (1952).
131. R. J. Barrnett, *J. Natl. Cancer Inst.* **13**, 905 (1953).
132. R. J. Barrnett and A. M. Seligman, *J. Natl. Cancer Inst.* **13**, 215 (1952).
133. R. J. Barrnett and A. M. Seligman, *J. Natl. Cancer Inst.* **14**, 769 (1954).
134. R. J. Barrnett and A. M. Seligman, *Science* (*Wash., D.C.*) **116**, 323 (1952).
135. A. Z. Eisen, W. Montagna, and H. B. Chase, *J. Natl. Cancer Inst.* **14**, 341 (1953).
136. W. Montagna, A. Z. Eisen, A. H. Rademacher, and H. B. Chase, *J. Invest. Dermatol.* **23** (1954).
137. G. F. Odland, *J. Invest. Dermatol.* **21**, 305 (1953).
138. A. E. Foraker and W. J. Wingo, *Amer. Med. Assoc. Arch. Dermatol.* **72**, 1 (1955).
139. R. J. Barrnett and R. F. Sognnaes, in: *Fundamentals of Keratinisation* (E. O. Butcher and R. F. Sognnaes, eds.), p. 27, American Association for the Advancement of Science Publication 70 (1962).
140. M. H. Hardy, *Amer. J. Anat.* **90**, 285 (1952).
141. J. A. Swift, *Histochemie* **19**, 88 (1969).
142. R. J. Barnes, Studies in the Optical Properties of Wool, Hair and Related Fibres, Ph.D. thesis, University of Leeds (1933).
143. J. C. Derksen, G. C. Heringa, and A. Weidinger, *Acta Neerl. Morph. Norm. Pathol.* **1**, 31 (1937).

144. A. Giroud and G. Champetier, *Bull. Soc. Chim. Biol.* **18**, 656 (1936).
145. K. M. Rudall, in: *Fibrous Proteins*, p. 15, Charley and Pickersgill, London (1946).
146. M. L. Ryder, *Proc. Roy. Soc. Edinburgh* **367**, 65 (1958).
147. R. T. Sims, *J. Cell Biol.* **185**, 157 (1964).
148. A. M. Downes, L. F. Sharry, and G. E. Rogers, *Nature* (*Lond.*) **199**, 1059 (1963).
149. A. M. Downes, K. A. Ferguson, J. M. Gillespie, and B. S. Harrap, *Aust. J. Biol. Sci.* **19**, 319 (1966).
150. I. E. B. Fraser, *Aust. J. Biol. Sci.* **22**, 231 (1969).
151. B. Forslind, *Acta Dermatovener* (*Stockholm*) **51**, 9 (1971).
152. B. Forslind, *Acta Dermatovener* (*Stockholm*) **51**, 1 (1971).
153. B. Forslind, B. Lindstrom, and G. Swanbeck, *Acta Dermatovener* (*Stockholm*) **51**, 81 (1971).
154. R. T. Sims, *J. Anat.* **103**, 507 (1968).
155. R. E. Chapman, *J. Cell Sci.* **9**, 791 (1971).
156. M. L. Ryder, *J. Histochem. Cytochem.* **7**, 133 (1959).
157. A. M. Downes and A. G. Lyne, *Aust. J. Biol. Sci.* **14**, 120 (1966).
158. A. M. Downes, *Aust. J. Biol. Sci.* **14**, 109 (1966).
159. R. Feulgen and H. Rossenbeck, *Z. Phys. Chem.* **165**, 215 (1927).
160. H. Hermann, J. S. Nicholas, and J. K. Boricious, *J. Biol. Chem.* **184**, 231 (1950).
161. J. Brachet, *Arch. Biol.* (*Paris*) **53**, 207 (1942).
162. O. Braun-Falco, in: *The Biology of Hair Growth* (W. Montagna and R. A. Ellis, eds.), p. 65, Academic Press, New York (1958).
163. W. Montagna, *The Structure and Function of Skin*, Academic Press, New York (1956).
164. C. de Martino and L. Zamboni, *J. Ultrastructure Res.* **19**, 273 (1967).
165. W. Bondareff, *Anat. Res.* **129**, 97 (1957).
166. M. Shipman, H. B. Chase, and W. Montagna, *Proc. Soc. Exptl. Biol. Med.* **88**, 449 (1955).
167. O. Braun-Falco, *Arch. Klin. Exptl. Dermatol.* **201**, 521 (1955).
168. B. Sylven, *Exptl. Cell Res.* **1**, 582 (1950).
169. N. Dettmer and D. Schwarz, *Z. Wiss. Mikroskopie* **61**, 423 (1954).
170. J. A. Swift and C. A. Saxton, *J. Ultrastructure Res.* **17**, 23 (1967).
171. A. Rambourg and C. P. Leblond, *J. Cell Biol.* **32**, 27 (1967).
172. E. H. Mercer, R. A. Jahn, and H. I. Maibach, *J. Invest. Dermatol.* **51**, 204 (1968).
173. W. Montagna, H. B. Chase, and J. B. Hamilton, *J. Invest. Dermatol.* **17**, 147 (1951).
174. W. Montagna, *Quart. J. Microscopical Sci.* **91**, 205 (1950).
175. S. Sakaguchi, *J. Biochem.* (*Tokyo*) **5**, 25 (1925).
176. G. E. Rogers, *J. Histochem. Cytochem.* **11**, 700 (1963).
177. M. L. Ryder, *Quart. J. Microscopical Soc.* **100**, 1 (1959).
178. A. G. Matoltsy, *Exptl. Cell Res.* **5**, 98 (1953).
179. G. E. Rogers, *Exptl. Cell Res.* **33**, 264 (1964).
180. G. E. Rogers, *Nature* (*Lond.*) **194**, 1149 (1962).
181. O. Braun-Falco, *Arch. Klin. Exptl. Dermatol.* **202**, 163 (1956).
182. H. Uno, K. Adachi, and W. Montagna, in: *Advances in Biology of Skin* (W. Montagna and R. L. Dobson, eds.), Vol. 9, *Hair Growth*, p. 221, Pergamon Press, Oxford (1969).
183. K. Adachi and H. Uno, in: *Advances in Biology of Skin* (W. Montagna and R. L. Dobson, eds.), Vol. 9, *Hair Growth*, p. 511, Pergamon Press, Oxford (1969).
184. R. A. Ellis and W. Montagna, *J. Histochem. Cytochem.* **6**, 201 (1958).
185. V. R. Formisano and W. Montagna, *Anat. Record* **120**, 893 (1954).

186. O. Braun-Falco and B. Rathjens, *Dermatol. Wochschr.* **130**, 1271 (1954).
187. T. S. Argyris, *Anat. Record* **125**, 105 (1956).
188. G. E. Rogers, *Quart. J. Microscopical Sci.* **94**, 253 (1953).
189. E. O. Butcher, *Ann. N.Y. Acad. Sci.* **53**, 508 (1951).
190. H. W. Spier and K. Martin, *Arch. Klin. Exptl. Dermatol.* **202**, 120 (1956).
191. P. L. Johnson and G. Bevelander, *Anat. Record* **95**, 193 (1946).
192. G. Moretti and H. Mescon, *J. Histochem. Cytochem.* **4**, 247 (1956).
193. O. Braun-Falco, *Dermatol. Wochschr.* **134**, 1252 (1956).
194. O. Braun-Falco, *Dermatol. Wochschr.* **134**, 1341 (1956).
195. O. Braun-Falco, *Arch. Klin. Exptl. Dermatol.* **202**, 153 (1956).
196. O. Braun-Falco and B. Rathjens, *Acta Histochem.* **1**, 82 (1954).
197. O. Braun-Falco, *Arch. Klin. Exptl. Dermatol.* **203**, 61 (1956).
198. R. G. Crounse and S. Rothberg, *J. Invest. Dermatol.* **35**, 107 (1960).
199. C. K. Wong, *Sapporo Med. J.* **34**, 1 (1968).
200. K. Wolff and E. Schreiner, *Arch. Dermatol. Forsch.* **241**, 255 (1971).
201. M. Seiji and S. Iwashita, *J. Invest. Dermatol.* **45**, 305 (1965).
202. H. M. Appleyard, *Guide to the Identification of Animal Fibres*, Wool Industries Research Association, Leeds (1960).
203. P. Kassenbeck, in: *Structure de la laine*, Textes et discussions du colloque de l' Institut Textile de France, p. 51 (July 1961).
204. E. Bottoms, E. Wyatt, and S. Comaish, *Brit. J. Dermatol.* **86**, 379 (1972).
205. M. G. Dobb, F. R. Johnston, J. A. Nott, L. Oster, J. Sikorski, and W. S. Simpson, *J. Textile Inst.* **52**, T153 (1961).
206. *Wool Sci. Rev.* **35**, 2 (1972).
207. K. R. Makinson, *Appl. Polymer Symp.* **18**, 1083 (1971).
208. A. Hepworth, T. Buckley, and J. Sikorski, *J. Phys. E.* (*Sci. Instrum.*) **2**, 789 (1969).
209. J. S. Halliday and P. H. Newman, *J. Phys. E.* (*Sci. Instrum.*) **2**, 444 (1969).
210. N. Ramanathan, J. Sikorksi, and H. J. Woods, *Proc. Intern. Wool Textile Res. Conf., Australia* **F**, 92 (1955).
211. L. J. Wolfram, *Textile Res. J.* **42**, 252 (1972).
211a. K. M. Rudall, *Proc. Leeds Phil. Lit. Soc.* **4**, 13 (1941).
212. H. M. Appleyard and C. M. Greville, *Nature* (*Lond.*) **166**, 1031 (1950).
213. J. H. Bradbury and J. D. Leeder, *Aust. J. Biol. Sci.* **23**, 843 (1970).
214. G. E. Rogers, *J. Ultrastruct. Res.* **2**, 309 (1959).
215. J. A. Swift, *J. Roy. Microscopical Soc.* **88**, 449 (1967).
216. C. Orfanos and H. Ruska, *Arch. Klin. Exptl. Dermatol.* **231**, 97 (1968).
217. J. Sikorski and W. S. Simpson, *Nature* (*Lond.*) **182**, 1235 (1958).
218. M. G. Dobb, Ph.D. thesis, University of Leeds (1963).
219. L. P. S. Piper, *J. Textile Inst.* **57**, T185 (1966).
220. L. P. S. Piper, *Nature* (*Lond.*) **213**, 596 (1967).
221. D. J. Johnson, *J. Roy. Microscopical Soc.* **88**, 39 (1967).
222. M. G. Dobb, R. Murray, and J. Sikorski, *J. Microscopy* **96**, 285 (1972).
223. J. A. Swift and A. W. Holmes, *Textile Res. J.* **35**, 1014 (1965).
224. D. F. G. Orwin and R. W. Thomson, *J. Cell Sci.* **11**, 205 (1972).
225. K. von Allwörden, *Z. Angew. Chem.* **29**, 77 (1916).
226. J. Lindberg, B. Philip, and N. Gralen, *Nature* (*Lond.*) **162**, 458 (1948).
227. C. Muller, *Z. Zellforsch. Mikrosk. Anat.* **29**, 1 (1939).
228. W. Herbig, *Z. Angew. Chem.* **32**, 120 (1919).
229. R. D. B. Fraser and G. E. Rogers, *Biochem. Biophys. Acta* **16**, 307 (1955).
230. J. Ames, *J. Textile Inst.* **43**, T262 (1952).

231. N. L. R. King and J. H. Bradbury, *Aust. J. Biol. Sci.* **21**, 375 (1968).
232. R. D. B. Fraser, T. P. MacRae, G. E. Rogers, and B. K. Filshie, *J. Mol. Biol.* **7**, 90 (1963).
233. R. M. Bonés and J. Sikorski, *J. Textile Inst.* **58**, 521 (1967).
234. M. Horio and T. Kondo, *Textile Res. J.* **23**, 373 (1953).
235. E. H. Mercer, *Textile Res. J.* **23**, 388 (1953).
236. R. D. B. Fraser, H. Lindley, and G. E. Rogers, *Biochim. Biophys. Acta* **13**, 295 (1954).
237. J. H. Dusenbury and A. B. Coe, *Textile Res. J.* **25**, 354 (1955).
238. R. D. B. Fraser and G. E. Rogers, *Aust. J. Biol. Sci.* **8**, 288 (1955).
239. J. Menkart and A. B. Coe, *Textile Res. J.* **28**, 218 (1958).
240. E. H. Mercer, *Textile Res. J.* **24**, 39 (1954).
241. R. D. B. Fraser and T. P. MacRae, *Textile Res. J.* **29**, 618 (1959).
242. R. I. Spearman and N. A. Barnicot, *Amer. J. Phys. Anthropol.* **18**, 91 (1960).
243. *Wool Sci. Rev.* **29**, 25 (1966).
244. D. J. Johnson and J. Sikorski, *Proc. 3rd Intern. Wool Textile Res. Conf., Paris* **1**, 147 (1965).
245. M. G. Dobb, D. J. Johnson, and J. Sikorski, *Proc. 5th Intern. Conf. Electron Microscopy, Philadelphia*, T-4 (1962).
246. M. G. Dobb, J. Sikorski, and P. G. Whitmore, *Proc. 4th European Regional Conf. Electron Microscopy, Rome*, 553 (1968).
247. G. E. Rogers and B. K. Filshie, in: *Ultrastructure of Protein Fibres* (R. Borasky, ed.), p. 123, Academic Press, New York (1963).
248. M. G. Dobb and J. Sikorski, *Appl. Polymer Symp.* **18**, 743 (1971).
249. M. G. Dobb and J. Sikorski, *Philips Bull.* (Sept. 1969).
250. G. R. Millward and J. Sikorski, *Proc. 4th European Regional Conf. Electron Microscopy, Rome*, 551 (1968).
251. B. K. Filshie and G. E. Rogers, *J. Mol. Biol.* **3**, 784 (1961).
252. P. Kassenbeck, *Proc. 3rd Intern. Wool Textile Res. Conf., Paris* **1**, 115 (1965).
253. D. J. Johnson and J. Sikorski, *Proc. 3rd European Regional Conf. Electron Microscopy, Prague*, 363 (1964).
254. D. J. Johnson and J. Sikorski, *Nature (Lond.)* **205**, 266 (1965).
255. R. D. B. Fraser, T. P. MacRae, and G. R. Mullward, *J. Textile Inst.* **60**, 343 (1969).
256. T. P. MacRae, G. R. Millward, D. A. D. Parry, E. Suzuki, and P. A. Tulloch, *Appl. Polymer Symp.* **18**, 65 (1971).
257. R. D. B. Fraser, T. P. MacRae, and D. A. D. Parry, in: *Symposium on Fibrous Proteins, Australia, 1967*, p. 279, Butterworth & Co. (Australia) Ltd., Sydney (1968).
258. M. G. Dobb and J. Sikorski, *J. Textile Inst.* **60**, 497 (1969).
259. R. D. B. Fraser, T. P. MacRae, and G. R. Millward, *J. Textile Inst.* **60**, 498 (1969).
260. M. G. Dobb, *J. Mol. Biol.* **10**, 156 (1964).
261. M. G. Dobb, *Nature (Lond.)* **207**, 293 (1965).
262. M. G. Dobb, *J. Ultrastructure Res.* **14**, 294 (1966).
263. G. R. Millward, *J. Ultrastructure Res.* **31**, 349 (1970).
264. P. G. Whitmore, *Aust. J. Biol. Sci.* **25**, 1373 (1972).
265. R. D. B. Fraser, T. P. MacRae, and A. Miller, *Nature (Lond.)* **203**, 1231 (1964).
266. F. H. C. Crick, *Acta Cryst.* **6**, 685 (1953).
267. G. Mahrle and C. E. Orfanos, *Arch. Dermatol. Forsch.* **241**, 305 (1971).
268. J. H. Bradbury and J. M. O'Shea, *Aust. J. Biol. Sci.* **22**, 1205 (1969).
269. H. W. J. Harding and G. E. Rogers, *Biochemistry* **10**, 624 (1971).
270. P. Soldt, *Naturwiss.* **51**, 365 (1964).

271. J. A. Swift, Ph.D. thesis, University of Leeds (1963).
272. C. Orfanos and H. Ruska, *Arch. Klin. Exptl. Dermatol.* **231**, 279 (1968).
273. G. Laxer, C. S. Whewell, and H. J. Woods, *J. Textile Inst.* **45**, T482 (1954).
274. G. Laxer, J. Sikorski, C. S. Whewell, and H. J. Woods, *Biochim. Biophys. Acta* **15**, 174 (1954).
275. R. A. Nicolaus, *Rass. Med. Sperimentale, Suppl. 1* (1962).
276. K. Hempel, in: *The Biologic Effects of UV Radiation* (F. Urbach, ed.), p. 305, Pergamon Press, Oxford (1969).
277. J. D. Dance and J. A. Swift, unpublished speculations.
278. P. Alexander and R. F. Hudson, *Wool: Its Chemistry and Physics*, Chapman & Hall Ltd., London (1954).
279. W. G. Crewther, R. D. B. Fraser, F. G. Lennox, and H. Lindley, *Advan. Protein Chem.* **20**, 191 (1965).
280. J. M. Gillespie, *J. Polymer Sci.* **C20**, 201 (1967).
281. V. G. Kulkarni, R. M. Robson, and A. Robson, *Appl. Polymer Symp.* **18** 1, 127 (1971).
282. J. H. Bradbury and K. F. Ley, *Aust. J. Biol. Sci.* **25**, 1235 (1972).
283. D. E. Peters and J. H. Bradbury, *Aust. J. Biol. Sci.* **25**, 1225 (1972).
284. J. H. Bradbury, G. V. Chapman, N. Hambly, and N. L. R. King, *Nature (Lond.)* **210**, 1333 (1966).
285. J. H. Bradbury, G. V. Chapman, and N. L. R. King, *Proc. 3rd Intern. Wool Textile Res. Conf., Paris* **1**, 359 (1965).
286. J. A. Swift and B. Bews, *J. Soc. Cosmetic Chem.* **25**, 13 (1974).
287. J. A. Swift and B. Bews, *J. Soc. Cosmetic Chem.* **25**, 355 (1974).
288. J. H. Bradbury and D. E. Petrs, *Textile Res. J.* **42**, 471 (1972).
289. S. J. Leach, G. E. Rogers, and B. K. Filshie, *Arch. Biochem. Biophys.* **105**, 270 (1964).
290. R. DeDeurwaeder, M. G. Dobb, L. A. Holt, and S. J. Leach, *Arch. Biochem. Biophys.* **120**, 249 (1967).
291. R. A. DeDeurwaerder, M. G. Dobb, and B. J. Sweetman, *Nature (Lond.)* **203**, 48 (1964).
292. E. Zeitler and G. F. Bahr, *Exptl. Cell Res.* **12**, 44 (1957).
293. W. S. Simpson and H. J. Woods, *Nature (Lond.)* **185**, 157 (1960).
294. J. Sikorski and H. J. Woods, *J. Textile Inst.* **51**, T506 (1960).
295. G. E. Rogers, *Ann. N.Y. Acad. Sci.* **83**, 378 (1959).
296. G. E. Rogers, *J. Ultrastructure Res.* **2**, 309 (1959).
297. P. Kassenbeck and R. Hagege, *Proc. 3rd Intern. Wool Textile Res. Conf., Paris* **1**, 245 (1965).
298. G. E. Rogers and B. K. Filshie, *Proc. 5th Intern. Congr. Electron Microscopy, Philadelphia* **2**, 1 (1962).
299. R. M. Bones and J. Sikorski, *Abstr. Symp. Electron Microsc. Cytochem., Leiden*, p. 145 (1966).
300. C. Bailey, Ph.D. thesis, University of Leeds (1967).
301. D. Levy, *Biochem. Biophys. Acta* **317**, 473 (1973).
302. P. Kassenbeck, *J. Z. Polymer Sci.* **C20**, 49 (1967).
303. R. M. Bones and J. Sikorski, *J. Textile Inst.* **58**, 521 (1967).
304. K. Kuhn, W. Grassman, and U. Hofman, *Z. Naturforsch.* **136**, 154 (1959).
305. K. Kuhn, *Kolloid-Z. Z. Polymere* **182**, 12 (1962).
306. R. M. Bonés, Ph.D. thesis, University of Leeds (1966).
307. P. A. Kendall and E. A. Barnard, *J. Roy. Microscopical Soc.* **81**, 203 (1963).

308. E. Gebhardt and K. Kuhn, *Z. Anorg. Chem.* **320**, 71 (1963).
309. R. A. DeDeurwaerder, M. G. Dobb, and B. J. Sweetman, *Nature* (*Lond.*) **203**, 48 (1964).
310. H. Zahn and M. Biela, *Textile-Praxis* **23**, 103 (1967).
311. H. Zahn and M. Biela, *Eur. J. Biochem.* **5**, 567 (1968).
312. B. Kingham, S. P. Riley, and A. Robson, *Biochem. J.* **122**, 60P (1971).
313. J. M. Gillespie, *Comp. Biochem. Physiol.* **41B**, 723 (1972).
314. J. L. Stoves, *Nature* (*Lond.*) **151**, 304 (1943).
315. R. D. B. Fraser, personal communication.

4

# The Chemical Composition and Structure of Wool

H. Lindley

*All the facts of Nature are so indissolubly linked together that it is only by first grasping the simpler ones that we can even begin to understand the superb craftsmanship that finds expression in the structure of living things.*

*W. T. Astbury*
*Fundamentals of Fibre Structure*

## 4.1. Introduction

The first two amino acids to be isolated from protein hydrolysates were glycine and leucine, both by Braconnot and both in 1820.[1] Of these two, leucine was isolated from wool, which can thus lay claim to a long history in protein chemistry. Other early investigators attempted to use the techniques of elementary analysis to derive an empirical formula for wool, but it was speedily recognized that the molecule was far too complex to be handled in this way. With the establishment in 1901 by Fischer of the polypeptide theory of protein structure, it was realized that the first step in the elucidation of the chemical structure of proteins was the determination of the relative amounts of the differing amino acids in a hydrolysate of the protein, and interest in elementary analysis became restricted to nitrogen and sulfur

---

**H. Lindley** • Division of Protein Chemistry, CSIRO, Parkville (Melbourne), Victoria 3052, Australia

estimations, which permitted estimates of the total recovery of amino acids and sulfur-containing amino acids to be made.

As early as 1907 an amino acid analysis of wool accounting for 33% of the total nitrogen was published by Abderhalden and Voitinovici.[2] The estimation of total sulfur led to one of the earliest controversies in wool chemistry, in which Barritt and King[3] maintained that not only did the sulfur content of wool vary between wool types, but also that it varied along the length of the fiber. Marston,[4] on the other hand, maintained that the sulfur content of wool was invariant at about 3.5% sulfur, but there is now no doubt that Barritt and King were correct. Viewed with hindsight their finding was the first piece of chemical evidence that wool is not a single compound but a mixture of proteins. However, in the absence of any technique of solubilizing wool and because techniques of protein fractionation and characterization were in any case very undeveloped, this line of thought was not pursued, and wool tended to be treated and thought of as a single protein.

Despite this oversimplification, the decade from 1930 witnessed great strides in our understanding of wool composition and structure, and the names of Astbury and Speakman will always be associated with this period. Looking back now some 40 years later it seems an act of incredible temerity on Astbury's part to attempt to solve the problem of wool structure by X-ray diffraction techniques, and if one looks at reproductions of his actual photographs,[5] one is left speechless. Yet Astbury was not the complete pioneer in the field. Silk had been studied by Brill as far back as 1923,[6] and a structure which even today must be regarded as essentially correct had been proposed in 1928.[7] This silk structure can be described quite simply as that of a fully extended peptide chain. Astbury's contribution was to discover that the unimpressive X-ray diffraction picture given by wool was changed by stretching the fiber into a different picture, equally unimpressive, but showing strong analogies with the picture given by silk. Astbury and Woods[8,9] then made the assumption that the $\alpha$ photograph of unstretched wool was due to regularly folded peptide chains which were converted on stretching the fiber to fully extended peptide chains giving a silklike $\beta$ photograph. To this concept of regularly folded chains Speakman and Hirst[10] added the additional postulate of interchain disulfide bonds and ionic interactions ("salt linkages") rather like rungs of a ladder. Crude though this picture may seem today, it formed the basis of much of the work of the Leeds group for a number of years and was successfully used to explain aspects of wool technology and to suggest novel techniques for such processes as setting. Indeed, it led directly to the formulation of a new industry—the permanent waving of human hair—as a result of what Astbury described as "crystallography invading the tonsorial parlour."[11]

More importantly, however, Astbury's ideas, especially as he extended them to proteins other than wool, sparked off a great interest in proteins and protein structure and led to a period of intense speculation. The cyclol hypothesis of protein structure of Wrinch[12] obviously stems directly from Astbury and Woods' postulated $\alpha$ structure. It would also seem likely that the Bergmann–Niemann[13] hypothesis of amino acids repeating at regular intervals along the peptide chain owed much to Astbury's ideas—certainly the hypothesis appealed strongly to him and he became one of its most active proponents. Today, of course, these theories are largely forgotten, and yet the fact that much of *Bombyx mori* silk can be represented as $(\text{gly ala})_n$ and much of collagen sequence as $(\text{gly pro X})_n$ show that Bergmann and Niemann's speculations came somewhere near to the truth as far as these fibrous proteins are concerned.

The ultimate truth or falsity of a scientific hypothesis is, however, not the only criterion of its merit; the amount of interest and experiment the hypothesis provokes can be of major importance. The many speculations on protein structure of this period were incredibly fruitful in this latter sense. An enormous effort—within the context of the period—went into attempts to improve techniques of amino acid analysis and obtain reliable analytical data on proteins. This culminated in the development by Martin and Synge of partition chromatography on silica gel for estimating mono amino acids[14] and was followed in rapid succession by the development of paper chromatography,[15] its use for sequence determination of small peptides,[16] chromatography on ion-exchange resins,[17] and electrophoresis in silica gel.[18] It is not widely recognized that the foundations of many of the techniques currently used in amino acid analysis and chemical structure of proteins were laid in a wool laboratory in the years during and immediately following World War II.

## 4.2. Chemical Studies on Wool Proteins

### 4.2.1. Oxidized Wool Proteins—Keratoses

The observation by Toennies and Homiller[19] that of the naturally occurring amino acids, only cystine, methionine, and tryptophan are oxidized by performic acid, was brilliantly used by Sanger[20,21] to separate the A and B chains of insulin as a primary step in the work which was to lead him to the elucidation of the first complete covalent structure of a protein. The success of this work led to studies in which wool was also oxidized by performic or the closely related peracetic acid to give products designated as keratoses.

However, whereas insulin contains neither tryptophan nor methionine, wool contains both these amino acids, and looking back there is little doubt that some of the conclusions drawn in the earlier work are of doubtful validity because of unrecognized peptide bond breakage and lack of specificity in the reagent. Peracetic acid seems to be especially unsatisfactory, since not only does it cause peptide-bond fission, but it also may not oxidize cystine completely.[22] Thus Corfield et al.[23] showed that there is a loss of 14% histidine, 6% phenylalanine, and 1% lysine when wool is oxidized with peracetic acid. In reading literature of the early 1950s, it is important to realize that we had the situation of workers using reagents, of whose lack of specificity they were unaware, on what they imagined was a relatively simple protein mixture but which in fact we now recognize to be a far more complex mixture than could possibly have been envisaged at that time.

Early work by Alexander and Earland[25] and Alexander et al.[26] established that the keratose obtained by oxidation of wool with peracetic could be separated on a solubility basis into three quite distinct fractions, the $\alpha$, $\beta$, and $\gamma$ keratoses. These were further differentiated on the basis of molecular weight and sulfur content, $\alpha$-keratose having the highest molecular weight and lowest sulfur content and the $\gamma$-keratose being especially rich in sulfur. Wool was said to be made up of 60% $\alpha$-keratose [subdivided into $\alpha_I$ high-molecular-weight fractions (45%) and $\alpha_{II}$ lower molecular weight (15%)], 30% $\gamma$-keratose of very low molecular weight, and 10% of the $\beta$ fraction, which was actually the insoluble residue remaining after the peracetic acid oxidized wool is extracted with dilute ammonia. Typical analyses of these fractions are given in Table 4-1.

Later work by O'Donnell and Thompson[27,28] and Thompson and O'Donnell[29] showed that both $\alpha$- and $\gamma$-keratose prepared using performic acid under carefully controlled conditions were heterogeneous on the basis of separations achieved by chromatography on DEAE cellulose columns, the reality of the separations being confirmed by amino acid analysis of the fractions and also by starch gel electrophoresis. Heterogeneity of the $\alpha$-keratose prepared by peracetic acid oxidation has also been shown by Corfield[30,31] using chromatography on hydroxyapatite. Whereas Thompson and O'Donnell took the view that the homogeneity was largely due to the nature of wool (i.e., that wool is made up of many proteins), Corfield took the other view—that the heterogeneity was an artifact of the isolation procedure and due to peptide bond breakdown. Heterogeneity of $\gamma$-keratose obtained by performic oxidation was also demonstrated by Haylett[32] using both moving-boundary electrophoresis and gel filtration. In order to remove one possible source of artifactual heterogeneity, these workers used extraction with pH 6 buffer instead of the more usual 0.2 $N$ ammonia.

Despite the known heterogeneity of $\alpha$-keratose, work has been carried

**Table 4-1. Amino Acid Composition (μmol/g) of Keratose Fractions from Merino Wool**

| | Peracetic acid oxidation[23] | | | Performic acid oxidation[24] | | |
|---|---|---|---|---|---|---|
| Amino acid | α-keratose | γ-keratose | β-keratose | α-keratose | γ-keratose | β-keratose |
| Alanine | 601 | 301 | 606 | 564 | 317 | 576 |
| Ammonia | 1275 | 1288 | 1078 | 919 | 740 | 815 |
| Arginine | 647 | 554 | 554 | 656 | 609 | 824 |
| Aspartic acid | 778 | 209 | 594 | 696 | 272 | 541 |
| Cysteic acid | 463 | 1690 | 551 | 673 | 1601 | 846 |
| Glutamic acid | 1356 | 684 | 916 | 1163 | 718 | 1037 |
| Glycine | 642 | 579 | 817 | 851 | 592 | 659 |
| Histidine | 51 | 61 | 110 | 63 | 70 | 138 |
| Isoleucine | 310 | 249 | 357 | 257 | 250 | 260 |
| Leucine | 908 | 297 | 772 | 822 | 362 | 687 |
| Lysine | 286 | 60 | 402 | 287 | 54 | 514 |
| Phenylalanine | 241 | 134 | 294 | 291 | 176 | 190 |
| Proline | 335 | 1148 | 572 | 366 | 1089 | 585 |
| Serine | 834 | 1131 | 1000 | 795 | 1134 | 928 |
| Threonine | 429 | 870 | 551 | 387 | 820 | 425 |
| Tyrosine | 304 | 164 | 282 | 403 | 169 | 153 |
| Valine | 495 | 484 | 552 | 429 | 444 | 308 |

**Table 4-2. Tryptic Peptides from α-Keratose[33,34]**

| | |
|---|---|
| Dipeptides | Phe-Arg Ser-Arg Ser-Lys Gly-Arg Ala-Lys<br>Ala-Arg Lys-Lys Val-Arg Leu-Arg |
| Tripeptides | Ser-Ser-Arg Gly-Ser-Arg (Asp, Val)-Arg<br>Glu-Ile-Lys Cya-Cya[a]-Glu |
| Tetrapeptide | (Thr, Glu, Leu)-Arg |
| Pentapeptides | Ala-(Thr, Val, Ile)-Arg Leu-(Thr, Glu, Leu)-Lys<br>(Asp, Leu, Glu, Ile)-Lys |
| Hexapeptides | (Arg, $Gln_2$, Val, Ile)-Arg Phe-(Asp, $Glu_2$, Leu)-Lys<br>Ser-(Asp, Glu, Gly, Leu)-Arg Phe-Leu-(Asn, Gln, Glu)-Lys<br>Ala-(Asp, Thr, Glu, Ile)-Lys Asp-(Val, Cya, Ala, Leu)-Arg<br>(Asp, Ser, Glu, Gly, Leu)-Arg |
| Heptapeptide | Leu-(Ser, Glu, Ala, Leu, Tyr)-Lys |
| Octapeptides | Ala-($Asp_2$, Ser, Ile, Tyr, His)-Lys<br>Ala-(Asp, Ser, $Glu_2$, Val, Leu)-Arg<br>Gly-(Glu, $Gly_4$, Ile)-Arg<br>Asp-Glu-Asp-(Val, Cya, Ala, Leu)-Arg |
| Pentadecapeptide | Glu-Ala-Glu-Cya-Ala-(Glu, Ser, Asp, Ala, Glu)-Glu-Ala-Ser-Gly-Arg |

[a] Cya, cysteic acid.

out on the amino acid sequence of this fraction both by Fell et al.[33,34] in Germany and by Blackburn, Corfield, and co-workers at Leeds.[35,36] The sequences established by Fell et al. are shown in Table 4-2. This work was confined to an examination of low-molecular-weight tryptic peptides. A much more ambitious task was undertaken by the Leeds group. These workers have isolated an α-keratose fraction which they designated U.S.3. They recognize that this fraction is not a single component but a mixture of polypeptides, but nevertheless they have done an enormous amount of work establishing the composition and some sequences from this fraction.[37] Tables 4-3 and 4-4 summarize their findings.

**Table 4-3. Tryptic Peptides from Wool Fraction U.S.3[35] [a]**

| | | |
|---|---|---|
| Dipeptides | Cya[b]-Lys Cya-Arg Thr-Lys Ser-Arg<br>Glx-Arg[c] Ala-Arg Val-Arg | |
| Tripeptide | Ala-Gly-Ser | |
| Tetrapeptides | Ser-Glx-Ser-Arg Ile-Leu-Glx-Arg | |
| Pentapeptides | Glx-Ser-Ala-Asx-Arg<br>Asx-Glx-Ser-Ala-Arg | Ile-Leu-Cya-Ala-Lys<br>Ile-Glx-Val-Glx-Arg |
| Hexapeptides | Glx-Thr-Glx-Leu-Asx-Lys<br>Asx-Val-Cya-Ala-Leu-Arg<br>Ala-Thr-Val-Tyr-Ile-Arg<br>Gly-Ile-Cya-Ser-Tyr-Arg<br>(Glx, His)-Leu-Thr-Gly-Arg | Ala-Gly-Ser-Cya-Gly-Arg<br>Ser-Glx-Leu-Gly-Asx-Arg<br>Phe-Leu-Glx-Glx-Asx-Lys<br>Leu-Glx-Tyr-Glx-Ile-Arg<br>Thr-Lys-Tyr-Ser-Glx-Arg |
| Heptapeptides | Ile-Leu-Leu-Cya-Gly-Gly-Lys<br>Ala-Ala-Glx-Glx-Leu-Ser-Arg<br>Leu-Leu-Glx-Gly-Glx-Glx-Arg | Ser-Lys-Cya-Glx-Glx-Ile-Lys<br>Thr-Glx-Leu-Ile-Glx-Glx-Lys<br>Leu-Ala-Ser-Tyr-Leu-Glx-Lys |
| Nonapeptides | Glx-Thr-Met-Glx-Phe-Leu-Asx-Asx-Arg<br>Glx-Ala-Asx-Cya-Glx-Ala-Ser-Gly-Arg<br>Ser-Ala-Ser-Phe-Ser-Cya-Gly-Ser-Arg-<br>Leu-Val-Val-Glx-Asx-Ile-Asx-Ala-Lys<br>Ile-(Glx, Ser)-(Asx, Glx, Leu)-Thr-Tyr-Arg | |
| Decapeptides | Glx-(Ala$_2$, Asx, Cya, Glx, Val)-Ser-Gly-Arg<br>Ala-Lys-Glx-Asx-Met-Ala-Leu-Cya-Leu-Lys<br>Ala-Glx-Asx-Tyr-Asx-Leu-Ala-Ile-Ser-Arg<br>(Asx, Phe)-Glx-Leu-Thr-Ala-Ala-Glx-Val-Lys | |
| Undecapeptide | (Leu, Thr)-Asx-Leu-Gly-Asx-Ser-Thr-Glx-Ala-Arg | |
| Heptadecapeptide | Leu-Asx-Ala-Pro-Thr-Val-Glx-Leu-Asx-Val-Asx-Glx-Ala-Val-Leu-Asx-Arg | |

[a] Some corrections given in ref. 37 have been incorporated.
[b] Cya, cysteic acid.
[c] Since glutamic acid and glutamine and aspartic acid and asparagine are not differentiated, Glx and Asx are used throughout.

**Table 4-4. Chymotryptic Peptides from Wool Fraction U.S.3**[36]

| | |
|---|---|
| Dipeptides | Thr-Leu Thr-Phe Gly-Tyr Arg-Leu |
| Tripeptides | Gly-Leu-Leu Ala-Leu-Leu Ala-Ala-Phe<br>Ser-Thr-Tyr Ala-Ser-Tyr Lys-Leu-Leu |
| Tetrapeptides | Ser-Ser-Glx-Leu Ile-Glx-Thr-Leu<br>Thr-(Leu, Pro)-Leu Gly-Ser-Gly-Phe<br>Thr-Gly-Gly-Phe Leu-(Glx, Lys)-Tyr |
| Pentapeptides | Ala-Ser-Asx-Asx-Phe Ala-Ala-Asx-Asx-Phe<br>Glx-Glx-Lys-Ile-Leu Ala-Ser-Gly-Ser-Tyr<br>Glx-Glx-Asx-Lys-Leu Leu-Glx-Thr-Lys-Leu<br>Lys-Asx-Ser-Lys-Leu |
| Hexapeptides | Arg-Gly-Ile-Ser-Cya[a]-Tyr Glx-Lys-Val-Arg-Glx-Leu |
| Heptapeptides | Glx-Asx-Glx-Arg-Ala-Glx-Leu Asx-Glx-Thr-Arg-Ala-Glx-Tyr<br>Cya-(Glx, $Leu_2$, Pro)-Ser-Phe Leu-Glx-Lys-Val-Arg-Glx-Leu<br>Leu-Glx-Glx-Glx-Asx-Lys-Leu Ala-Leu-Glx-(Glx, Gly, Ser)-Leu<br>Val-Asx-Leu-Asx-Arg-Val-Leu |
| Octapeptides | Glx-Glx-Lys-(Asp, Cya, Glx, Ile)-Leu<br>Glx-Arg-($Glx_{2-3}$, Pro, Ser)-Leu<br>Lys-Lys-Asx-Val-Asx-Cya-Ala-Tyr<br>Leu-Asx-Asx-Arg-Leu-Ala-Ser-Tyr<br>Arg-Ala-Thr-Ala-Glx-Asx-Glx-Phe<br>Ala-Gly-Ser-Cya-Gly-Arg-Ser-Phe |
| Nonapeptides | Gly-Leu-Asx-Ile-Glx-(Ala, Ile)-Thr-Tyr<br>Gly-Ser-Arg-Ser-Val-(Cya, Gly)-Gly-Phe<br>Cya-Ala-Lys-Ser-Glx-Asx-Ala-Arg-Leu |
| Decapeptides | Ala-Leu-Glx-(Ala, $Glx_3$, His, Val)-Leu<br>Glx-(Asx, Glx, Gly, Lys)-Asx-Phe-Glx-Thr-Tyr |
| Duodecapeptides | Thr-Ala-Ala-(Glx, Val)-Lys-Cya-(Asp, Glx, Ser)-Lys-Leu<br>Thr-($Ala_2$, Glx, Val)-Lys-Cya-Glx-Asx-Ser-Lys-Leu<br>[(Cya, Leu)-Lys, Glx-Lys, (Ala, Asx, $Glx_2$)-Arg]-Ser-Leu |
| Tridecapeptides | Ala-Glx-Leu-Arg-Ser-Glx-(Asx, Leu)-Arg-Asx-Glx-Glx-Tyr |
| Tetradecapeptide | Glx-($Ala_2$, Asx, Cya, Glx, $Leu_2$, $Lys_2$, Met)-Lys-Glx-Tyr |

[a] Cya, cysteic acid.

### 4.2.2. *S*-Carboxymethylated Wool Proteins

The concept of solubilizing wool by the two-stage process of reduction of the disulfide crosslinks to thiol groups and subsequent stabilization of the thiol protein against atmospheric oxidation by alkylation dates back to the work of Goddard and Michaelis in 1934 and 1935.[38,39] These workers not only introduced the use of alkylation with iodoacetate, still the most widely used alkylating agent in protein chemistry, but also fractionated the *S*-carboxymethylkerateine from wool into low- and high-sulfur fractions, which they designated as SCMKA and SCMKB, respectively. Furthermore, within the limits of the techniques available at that time they showed that the only chemical reactions occurring were reduction of disulfides to thiol groups and reaction of iodoacetate with thiol groups. Later work has fully confirmed these findings, and the conversion of cystine residues to *S*-carboxymethyl cysteine residues has now become a routine procedure in protein chemistry. It is possible that in the case of some of the very early work on wool, both the pH and temperature of reduction may have been sufficiently high to introduce the possibility of deamidation of amide side chains or peptide-bond breakage. However, this problem was recognized early in the work and avoided by, for example, extraction at lower pH values in the presence of urea solutions. Certainly no such objection can be taken to the most recent techniques involving reduction with tributyl phosphine,[40,41] where the chemistry can be carried out under what amounts virtually to physiological conditions of pH. The accord between the results obtained under these extremely mild conditions and the earlier work suggests that even under the more drastic conditions, little if any peptide-bond breakage occurred during the reducing procedures. Some specific problems associated with the use of thioglycollic acid as a reducing agent have been discussed[42,43] but have probably been circumvented in most work on wool using this reagent.

Many procedures have been published for the reduction and alkylation of wool and the subsequent extraction of a soluble protein fraction. The results of all these procedures, however, are consistent with the view that soluble proteins derived from wool are divisible into a high- and a low-sulfur fraction, and that each of these fractions is extremely heterogeneous. Earlier disagreements as to whether this heterogeneity is real or artifactual[44,45] seem to have been settled in favor of the former view by the determination of the complete amino acid sequence of some proteins (see Section 4.2.2.2).

#### *4.2.2.1. Low-Sulfur Proteins*

Table 4-5 gives the amino acid composition in micromoles per gram of two fractions SCMKA and SCMKA-2 obtained in early attempts to fractionate the low-sulfur proteins and, for comparison, two high-sulfur fractions SCMKB-1 and SCMKB-2 are also quoted. The very significant differences between high- and low-sulfur proteins are evident from this table, which compares closely with Table 4-1, giving similar data for $\alpha$- and $\gamma$-keratoses. On the other hand, SCMKA-2 (a specific fraction obtained from the whole low-sulfur fraction SCMKA) shows only minor differences in amino acid composition from its parent, and the isolation of completely homogeneous

**Table 4-5. Amino Acid Compositions (μmol/g) of Merino 64's Wool and Wool SCMK**

| Amino acid | Whole wool[a] (28) | SCMKA[b] (46) | SCMKA-2 (47) | SCMKB-1[b] (48) | SCMKB-2[a] (49) |
|---|---|---|---|---|---|
| Alanine | 417 | 518 | 591 | 238 | 275 |
| Ammonia | 887 | 1024 | 1076 | 888 | 1086 |
| Arginine | 602 | 585 | 632 | 398 | 248 |
| Aspartic acid | 503 | 655 | 764 | 60 | 82 |
| Half-cystine[c] | 943 | 546 | 578 | 1859 | 1734 |
| Glutamic acid | 1020 | 1138 | 1290 | 772 | 905 |
| Glycine | 688 | 709 | 542 | 497 | 702 |
| Histidine | 58 | 53 | 49 | 45 | 1 |
| Isoleucine | 234 | 295 | 303 | 215 | 330 |
| Leucine | 583 | 826 | 855 | 144 | 151 |
| Lysine | 193 | 326 | 320 | 38 | 1 |
| Methionine | 37 | 44 | 41 | Nil | Nil |
| Phenylalanine | 208 | 243 | 213 | 50 | 103 |
| Proline | 633 | 342 | 279 | 969 | 853 |
| Serine | 860 | 588 | 706 | 1163 | 1100 |
| Threonine | 547 | 354 | 394 | 893 | 832 |
| Tyrosine | 353 | 345 | 271 | 164 | 151 |
| Valine | 423 | 477 | 578 | 331 | 317 |

[a] It is known that wool also contains about 35 μmol/g of tryptophan and SCMKB-2 about 50 μmol/g.
[b] SCMKA and SCMKB-1 both contain about 30 μmol/g of tryptophan.
[c] For half-cystine, read *S*-carboxymethyl cysteine, where appropriate.

proteins from the low-sulfur fraction has proved to be a very intractable problem. All available techniques such as starch[50] and acrylamide[51] gel electrophoresis, immunodiffusion,[52] isoelectric focusing,[53] chromatography on ion-exchange celluloses, and molecular sieve gels have all been used with some measure of success, but to date no low-sulfur protein fraction has been prepared which can be claimed to be homogeneous by all criteria. Early work with starch gel electrophoresis[50] showed two components—the so-called components 7 and 8—to comprise almost 50% of the entire wool (see Fig. 4-1). However, hopes that these components were unique homogeneous proteins have proved illusory. Analyses of the *S*-carboxymethyl[54] and *S*-cyanoethyl[55] derivatives of components 7 and 8 are given in Table 4-6, and these clearly demonstrate significant differences between the two components. Components 7 and 8 both have molecular weights in the range 40,000–50,000 and acetylated N termini.[56] Anyone wishing to get some idea of the difficulties and complexities involved in wool in this area should read papers by Thompson and O'Donnell in which they attempted to characterize and

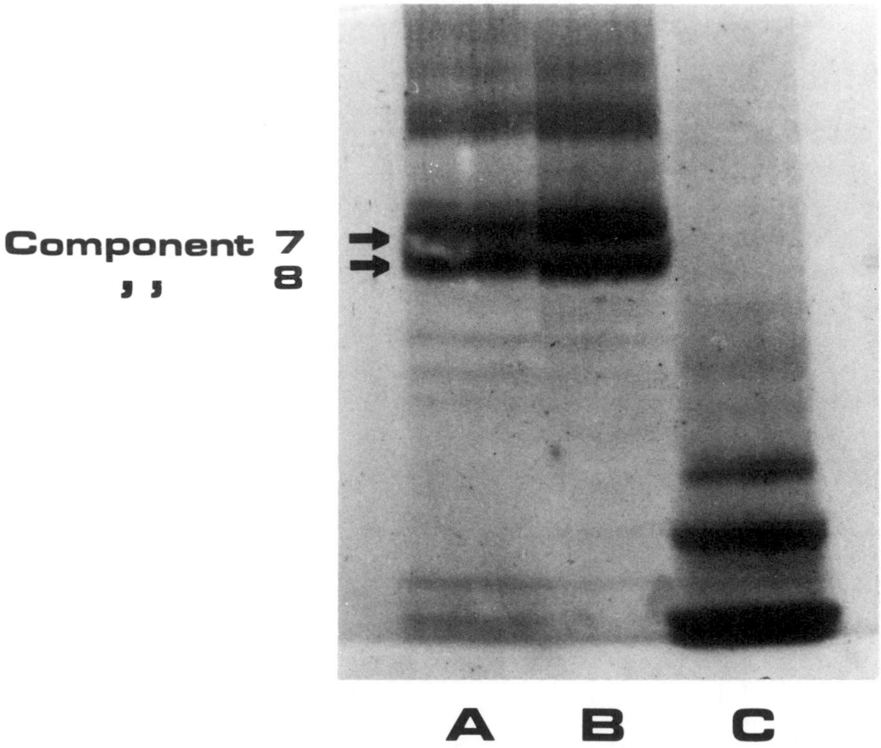

Fig. 4-1. Starch gel electrophoresis patterns of wool protein derivatives in pH 8.6 buffer containing 8 *M* urea.[50] A, SCMKA; B, SCMKA-2; C, SCMKB. (Courtesy *Aust. J. Biol. Sci.*)

**Table 4-6. Amino Acid Composition (μmol/g) of Components 7 and 8 Isolated as Carboxymethyl and Cyanoethyl Derivatives[54,55]**

| Amino acid | Component 7 | | Component 8 | |
|---|---|---|---|---|
| | SCM | SCNE | SCM | SCNE |
| Alanine | 667 | 727 | 503 | 552 |
| Ammonia | 1160 | 1445 | 1282 | 1436 |
| Arginine | 657 | 669 | 659 | 645 |
| Aspartic acid | 771 | 728 | 909 | 916 |
| Carboxymethyl cysteine[a] | 534 | 519 | 505 | 464 |
| Glutamic acid | 1340 | 1266 | 1533 | 1475 |
| Glycine | 640 | 727 | 351 | 370 |
| Histidine | 46 | 50 | 52 | 57 |
| Isoleucine | 334 | 315 | 318 | 303 |
| Leucine | 849 | 788 | 996 | 995 |
| Lysine | 375 | 385 | 251 | 267 |
| Methionine | 53 | 58 | 26 | 29 |
| Phenylalanine | 209 | 226 | 176 | 176 |
| Proline | 251 | 269 | 311 | 309 |
| Serine | 710 | 619 | 641 | 631 |
| Threonine | 375 | 379 | 443 | 407 |
| Tyrosine | 257 | 252 | 219 | 243 |
| Valine | 555 | 594 | 532 | 546 |

[a] Carboxyethyl cysteine in case of cyanoethyl derivatives.

sequence the peptides obtained from a highly purified component 8 after specific fission at methionine residues by reaction with cyanogen bromide.[57–60] Some sequence data were obtained, but the authors' main conclusion was that their starting material was not homogeneous or contained a number of closely related proteins. There are indications in the literature[61,67] that improved procedures for fractionation of components 7 and 8 have been developed, but detailed reports are not yet available.

Although progress with isolating a pure low-sulfur protein has been disappointingly slow, remarkable advances have nevertheless been made in the study and characterization of the α-helical portions of the low-sulfur fraction. The initial observations in this field were by Crewther and Harrap,[63] who used partial proteolysis with Pronase P of crude SCMKA and separated a highly helical fraction from the digest. However, although this fraction behaved as elongated particles of molecular weight 41,000 in borate buffer, in 8 *M* urea these dissociated into peptides with a wide variety of chain lengths, the smallest having molecular weights of 1000 or less. Later work by Crewther and Dowling[61] showed that digestion did not proceed so far when chymotrypsin rather than Pronase was used to digest the SCMKA. The initial α-helical fragment as isolated is similar to that obtained by Pronase digestion but dissociates in 8 *M* urea mainly to two fractions of molecular weight 12,500 and 26,000. Although the 12,500 fragment is reasonably homogeneous with respect to molecular weight, it can be further fractionated on a DEAE cellulose column into a number of components which fall into two distinct families, and these have been designated Type I and Type II peptides. All Type I peptides show considerable homologies with each other as demonstrated initially by peptide mapping[61] and later by sequence determination.[64,65] All Type II peptides similarly show strong familial resemblances to each other but little or no homology with Type I peptides. Initially, both Type I and Type II peptides were thought to be very heterogeneous, but it now appears that some of this apparent heterogeneity is artifactual and arises from carbamylation reaction in the strong urea solutions, and also cyclization of N-terminal glutamic or glutamine residues.[65] Present evidence suggests that both Type I and Type II families each have only three or four major components.

The larger (26,000 mol wt) peptides appear to be Type I and Type II peptides joined together end to end, possibly with an intervening length of nonhelical peptide chain. Components have been identified which can be described as Type I–Type I, Type I–Type II, and Type II–Type II.[66] No evidence of a Type II–Type I molecule has so far been uncovered. Much of this work has so far only been published in abstract, but, owing to the kindness of Dr. Crewther and his collaborators, it is possible to include

Type I

L C P N Y Q S Y F R T I E E
L Q Q K I L C A K S E N S R
L V I E I D N A A L A T D D
F R T K Y E S E R

Type II

Q N R Q C C E S N L E P L F
S G Y I E T L R R E A E C V
E A D S G R L S S E L N L V
Q E V E E C Y E R R Y E E E
I A L R A T A E N E F V A L
K K D V D C A Y L R K S D L
Q A N V E A L I Q E T D F L
R R L Y E E E I R V L

Fig. 4-2. Partial sequences of the Types I and II helical units. Every seventh residue commencing from residue 1 (Type I) and residue 7 (Type II) has been underlined to highlight the occurrence of hydrophobic side chains with this periodicity. There are four exceptions to this generalization in the 23 sites underlined. Curiously, the aberrant amino acid is always glutamic acid (E). The one letter amino acid code is as follows: A, alanine; C, ½ cystine (actually determined as *S*-carboxymethylcysteine); D, aspartic acid; E, glutamic acid; F, phenylalanine; G, glycine; H, histidine; I, isoleucine; K, lysine; L, leucine; M, methionine; N, asparagine; P, proline; Q, glutamine; R, arginine; S, serine; T, threonine; V, valine; W, tryptophan; and Y, tyrosine.

amino-acid-sequence information on the Type I and II peptides prior to publication. These data are given in Fig. 4-2. This and subsequent sequence data have been presented using the single-letter code for the amino acids. This saves considerable space and also facilitates comparison of related sequences. A key to the one-letter code is given as a footnote to Table 4-2.

These sequences show no evidence whatever of short-period-repeating sequences and hence cast grave doubts on theories of keratin structure which rely on postulated periodicities (see Section 4.3.2).

On the other hand, the sequences do show that if the appropriate starting residue is chosen, there is a strong probability that every subsequent seventh residue has a hydrophobic side chain. This finding is an important one for the coiled-coil hypothesis of keratin structure (Section 4.3.2).

#### *4.2.2.2. High-Sulfur Proteins*

It was recognized very early that the SCMKB proteins are heterogeneous. Because they display relatively little tendency to aggregate in solution it was a comparatively simple matter to demonstrate this by hydrodynamic methods, and in 1962 Gillespie could state that "there are at least eight proteins present."[67] At that time this seemed a fairly formidable degree of complexity, but as will appear later, this statement in fact probably understates the amount of heterogeneity by at least an order of magnitude.

Early attempts at fractionation relied on classical salting-out procedures from protein chemistry, and one fraction obtained in this way, SCMKB-2, appeared to be homogeneous for some time. The more recent developments with regard to this fraction will be considered later. In the overall field of fractionation of the high-sulfur proteins, however, every worthwhile technique has had to be employed, and in general each new technique has uncovered another previously unsuspected degree of complexity. Thus by gel filtration, followed by chromatography on DEAE-cellulose, Joubert and Burns[68] separated SCMKB into 26 fractions, and by prefacing this procedure with column electrophoresis, Joubert et al.[69] increased the number of such fractions to 33. Moving-boundary and acrylamide gel electrophoresis of the subfraction suggested that few of these fractions were homogeneous and that there were probably at least 70 components present. However, it was suggested that these probably represented families of closely related proteins.

Evidence confirming the extreme heterogeneity but extending the concept of families of proteins has been obtained for Merino wool by Darskus et al.,[70] who looked at one-dimensional peptide maps of subfractions of SCMKB prepared by chromatography on DEAE-cellulose. The results strongly support the idea of families of proteins containing common structural features. In an extension of this work to wool from other breeds of sheep

(Lustre Mutant Merino, English Leicester Merino crosses, Lincoln, Border Leicester, Southdown, and Corridale), it was found that the SCMKB components of the different wools gave qualitatively similar results when chromatographed on DEAE-cellulose, but the relative proportions of the subfractions were very variable.[71] This variability was not only dependent on breed but was strongly influenced by diet and even varied between individual animals of the same breed on identical diets. An exception to these generalizations was provided by SCMKB from wool of the primitive Soay sheep, which was extremely anomalous in giving very poor resolution into components.

First success in the isolation and determination of the complete amino acid sequence of a wool protein went to the South African group, who concentrated their effort on the smallest molecular-weight group (mol wt ≑ 11,500) of the high-sulfur proteins.[72] A complete sequence for one protein was published in 1969[73] and for a further two very closely related proteins in 1971.[74,75] These sequences are shown in Fig. 4-3. The two main conclusions which can be drawn from these sequences are (1) the great similarity between them, so that they can obviously be derived by a point-mutation process, and (2) that they can be arbitrarily divided near the midpoint of the chain into high- and low-sulfur portions.

The next group of proteins to be sequenced were the SCMKB-2 family by the Australian group. A fraction labeled SCMK-B2 was first isolated by Gillespie in 1962, which seemed to have considerable claim to homogeneity.[76] Early attempts at sequencing began very promisingly but ultimately became bogged down, and final success had to await the recognition that the

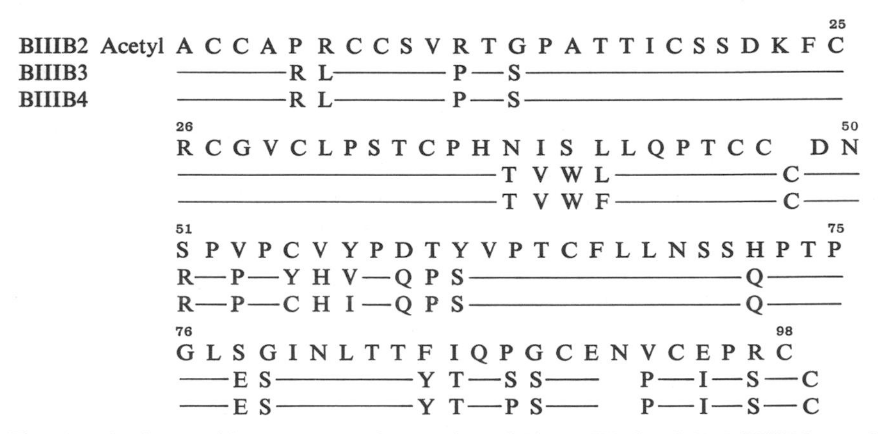


Fig. 4-3. Amino acid sequence of proteins SCMKBIIIB2, SCMKBIIIB3, and SCMKBIIB4. The sequence of IIIB2 is given in full; for IIIB3 and IIIB4, a continuous line indicates identity of sequence; code letters are given for differences in sequence. "Gaps" are indicated as such (i.e., there is no residue at position 48 in IIIB2 corresponding to C at this position for IIIB3 and IIIB4; cf. position 92).

fraction was heterogeneous and the devising of techniques for isolation of the four components.[77] Complete sequences of three components (SCMKB-2A, SCMKB-2B, and SCMKB-2C) and also that of a minor variant of the A component have now been published[78–80] and are shown in Fig. 4-4.

These data show one extremely interesting feature which is not shown by the earlier sequences—the presence of repeating decapeptide sequences. It would seem probable, in fact, that the proteins evolved by repeated partial gene duplications involving all or part of a decapeptide unit, and it is this structural feature which was responsible for the earlier difficulties in isolating, characterizing, and sequencing these proteins. It also provides an explanation of some of the results discussed by Lindley et al.[45] which had earlier led Gillespie[67] to propose a mechanism for the synthesis of high-sulfur proteins involving the stepwise addition of peptides to a precursor protein.

Swart and Haylett[81] have since sequenced another two high-sulfur proteins intermediate in size between their IIIB protein and the B2 group. They designate these proteins IIIA3 and IIIA3A, and the amino acid sequences are given in Fig. 4-5. They differ from all other wool proteins which have been sequenced so far in having threonine as the N-terminal amino acid. These proteins also show evidence of repeating sequence, and there has been some disputation as to whether the data for these (and the B2 group) are best explained on the basis of a repeating deca-[80] or pentapeptide unit.[82] Figure 4-6 shows the way in which the sequences of the two proteins SCMKB-2B and IIIA3 can be written so as to show maximum homology base on a decapeptide repeat.[83] A proposed repeating unit for each protein is shown and a common origin for the two proposed decapeptides is obvious from their homology.

In passing it may be mentioned that on the basis of some characteristic features of the peptide map of the SCMK-B2 family of proteins, it is a straightforward matter to look for proteins of this type in other keratins. In this way, "B2-like" proteins have been shown to occur in a variety of wool types, bovine hair, mohair, sheep horn, and sheep hoof.[84] On the other hand, they do not seem to occur in the hair of the yak, bison, antelope, camel, llama, or guinea pig, although the latter animals also fall into the genus artiodactyla.[85]

### 4.2.3. Ultra-High-Sulfur Proteins

Reis and Schinckel[86,87] showed that when sheep are fed a high-protein diet, or especially when cysteine or methionine is infused directly into the abomasum, they produce wool of increased sulfur content. In some cases this may result in the sulfur content increasing from 3% to 4.4%. Gillespie

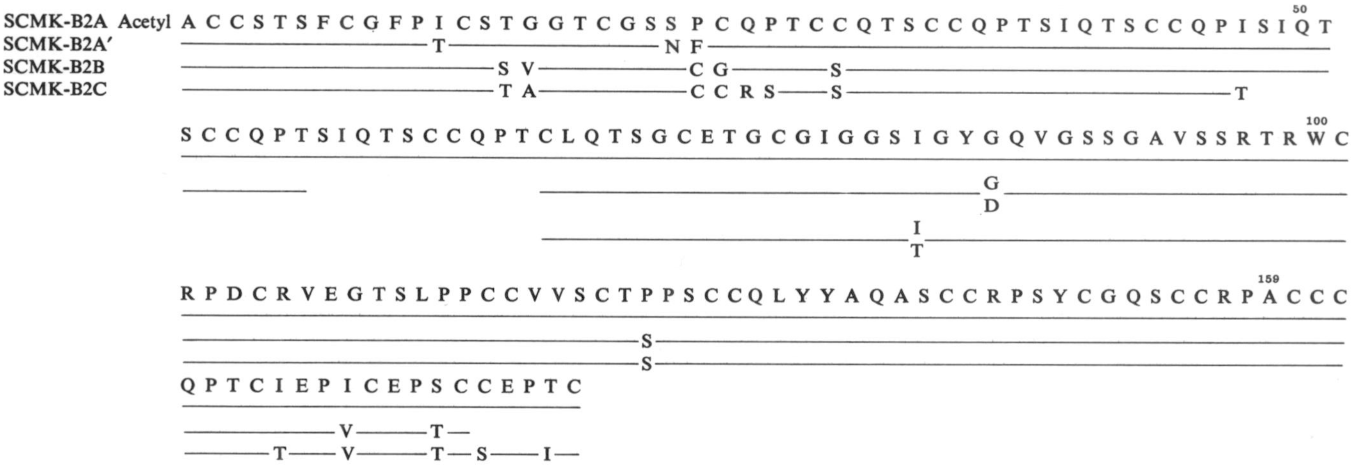

```
                                                                                            50
SCMK-B2A  Acetyl A C C S T S F C G F P I C S T G G T C G S S P C Q P T C C Q T S C C Q P T S I Q T S C C Q P I S I Q T
SCMK-B2A'        ---------------------T--------------------N F-------------------------------------------------------
SCMK-B2B         -------------------------S V---------------C G---------S-------------------------------------------
SCMK-B2C         -------------------------T A---------------C C R S-----S--------------------------------------T-----

                                                                                                        100
                 S C C Q P T S I Q T S C C Q P T C L Q T S G C E T G C G I G G S I G Y G Q V G S S G A V S S R T R W C
                 ------------------------------------------------------------------------------------------------------
                 -------             ------------------------------------------------G---------------------------------
                                                                                     D
                                                                                I
                                     -------------------------------------------T-------------------------------------

                                                                                                       159
                 R P D C R V E G T S L P P C C V V S C T P P S C C Q L Y Y A Q A S C C R P S Y C G Q S C C R P A C C C
                 ------------------------------------------------------------------------------------------------------
                 ---------------------------------------S-----------------------------------------------------------
                 ---------------------------------------S-----------------------------------------------------------

                 Q P T C I E P I C E P S C C E P T C
                 -----------------------------------
                 -------------V--------T--
                 --------T----V--------T--S-----I--
```

Fig. 4-4. Amino acid sequence of the proteins SCMK-B2A, SCMK-B2A′ (a minor variant of B2A), SCMK-B2B, and SCMK-B2C. A well-established minor variant involving only a single amino acid residue of each of the two latter proteins is also indicated.

```
SCMK-BIIIA3  T G S C C G P T F S S L S C G G G C L Q P R Y Y R D P C C C R P V S C Q   T V S R P V T F V P R C T
SCMK-BIIIA3A ———————————————————————————————————————————C C————————————————————S——T——————————————————S————

             R P I C E P C R R P V C C D P C S L Q E G C C R P I T C C P T S C Q A V V C R P C C N A T T C C Q P
             ———————————————————————————————————————————————————————————————————————————————————————————————————

             V S V Q C P C C R P T S C Q P   A P C S R T F R T S P C C
             ———————————————————————————  —S——   —————————————————
```

Fig. 4-5. Amino acid sequence of proteins SCMKBIIIA3 and a minor variant, SCMKBIIIA3A.

```
               A C C S T S F C G F P I G S S V C T
                       C G S S C G Q P T C S Q
                             T S C C Q P T S I Q
                             T S C C Q P I S I Q
                             T S C C Q P T C L Q
                             T S G C E T G C G I G G S I G Y G Q V
G S S G A V S S R T R W C R P D C R V E G T
                       S L P P C C V V S C T S
                             P S C G Q L Y Y A Q
                             A S C C R P S Y C G
                             Q S C C R P A C
                                 C C Q P T C I E
                               P V C E P T C

                                     Q     C
B2 Repeat                    T S C C   P T   I Q
                                     R     S
```

```
                       T G S C C G P T F S S L S C G G G C
L Q P R Y Y R D P C C G R P V S C Q
                         T V S R P V T F V
                       P R C T R P I C
                     [ E P C R R P V
                             C C D P C S L Q ] Fifteen
                     [ E G C C R P I             Residue
                             T C C P T S C Q ] Repeat
                       A V V C R P C C W A
                       T T C C Q P V S V Q
                       C P C C R P T S C Q
                       P A P C S R T T C R
               T F R T S P C C

                                   V
IIIA3 Repeat           D P C C R P   S C Q
                                   T
```

Fig. 4-6. Amino acid sequence of proteins SCMKB2B and IIIA3 arranged to show maximum homology based on a decapeptide repeat. Proposed repeating units for each protein are shown, and homologous residues are underlined.

et al.[88] showed that this increase was mainly due to the *de novo* synthesis of a group of proteins in which on the average one residue in three is half-cystine and which have been referred to as the ultra-high-sulfur proteins. They can be extracted from the wool of sheep on a high-sulfur diet with urea–thioglycollate, and after alkylation with iodoacetate can be separated from other high-sulfur proteins by chromatography on DEAE cellulose at pH 4.5 and QEA Sephadex at pH 11.[89,90]

Since the derived proteins have on the average one *S*-carboxymethyl group in every three residues, they are extremely acidic and hence exhibit, in an even more acute form, all the difficulties of fractionation of the normal SCMKB proteins. Some success has been achieved with similar techniques to those used for SCMKB-2 (i.e., chromatography at or below the isoelectric point), and there seems little doubt that this group of proteins is extremely heterogeneous.[89] The molecular weights of different fractions have been found to range from less than 10,000 to greater than 40,000. The *S*-carboxymethyl cysteine content remains remarkably constant from fraction to fraction, but there are large variations in other amino acids. Peptide maps have provided further evidence of heterogeneity, but again there is a suggestion of three or more familial groups of proteins. This class of proteins also occurs in animal fibers from other artiodactyla and has been found, for instance, in hair from goat, cattle, yak, bison, deer, camel, and pig.[90] No sequence studies have so far been attempted on these proteins.

### 4.2.4. High-Glycine-Tyrosine Proteins

In addition to the high- and low-sulfur fractions present in alkaline thioglycollate extracts of wool, Simmonds and Stall[91] found a minor fraction noteworthy for containing unusually large amounts of glycine and tyrosine. Similar fractions have since been obtained by acid extraction of either reduced or oxidized wool and also by prolonged formic acid extraction of untreated wool.

Zahn and Biela[92] gave a summary of these earlier findings in a paper in which they describe their own attempts to extract and purify high-glycine-tyrosine proteins (Table 4-7). It is obvious from their work that these proteins are, like every other group of wool proteins, extremely heterogeneous. Gillespie produced clear evidence of this heterogeneity using starch gel electrophoresis[98] and with Darskus[99] was able to show that the variable tyrosine content of different wools was due to, and also directly related to, the amount of high-glycine-tyrosine protein in the wool sample. This relationship, in fact, extends to other keratins, so that in the case of echidna quill, some 30% of the proteins are of this high-glycine-tyrosine type. Related proteins seem also

**Table 4-7. Amino Acid (μmol/g) Composition of High-Glycine-Tyrosine Fractions from Wool[92]**

| | | Source of protein fraction | | | | | | |
|---|---|---|---|---|---|---|---|---|
| Amino acid | Untreated Merino wool | Alkaline thioglycollate extraction[91] | Performic oxidation keratose-X[93] | Acetic acid extraction of reduced wool[94] | Minor fraction of reduced wool[95] | α-Keratose | Reduced wool[96] | Formic acid extraction of untreated wool[97] |
| Alanine | 456 | 293 | 146 | 400 | 201 | 253 | 216 | 576 |
| Arginine | 563 | 391 | 489 | 370 | 384 | 426 | 350 | 570 |
| Aspartic acid | 544 | 309 | 195 | 319 | 265 | 447 | 308 | 674 |
| *S*-Carboxymethyl cysteine | 0 | 0 | 0 | 0 | 352 | 0 | 0 | 0 |
| Cysteic acid | 27 | 0 | 989 | 19 | 0 | 559 | 228 | 110 |
| Cystine | 874 | 698 | 0 | 275 | 0 | 0 | 114 | 363 |
| Glutamic acid | 1226 | 702 | 373 | 476 | 147 | 780 | 86 | 1124 |
| Glycine | 763 | 1287 | 1796 | 1868 | 2193 | 1483 | 2432 | 1329 |
| Histidine | 81 | 66 | 36 | 62 | 114 | 46 | 50 | 200 |
| Isoleucine | 249 | 209 | 137 | 97 | 43 | 135 | 21 | 361 |
| Leucine | 647 | 418 | 402 | 440 | 437 | 569 | 405 | 760 |
| Lysine | 233 | 60 | 29 | 75 | 32 | 114 | 15 | 432 |
| Methionine | 22 | 0 | 0 | 0 | 0 | 0 | 18 | 100 |
| Phenylalanine | 206 | 255 | 224 | 464 | 833 | 327 | 766 | 401 |
| Proline | 558 | 693 | 585 | 966 | 402 | 329 | 478 | 395 |
| Serine | 992 | 1070 | 1025 | 961 | 970 | 832 | 1095 | 770 |
| Threonine | 560 | 592 | 458 | 377 | 276 | 267 | 282 | 450 |
| Tyrosine | 287 | 478 | 693 | 859 | 1054 | 622 | 1199 | 573 |
| Valine | 461 | 369 | 254 | 263 | 288 | 248 | 189 | 507 |

to occur in avian beaks and reptilian claws. In wool itself the amount of these proteins may vary from near zero for Lincoln up to 12% by weight in some samples of Merino wool. The actual amount present in a given wool sample seems to depend on both genetic and dietary factors.[100]

By the use of chromatography on quaternary ammonium ethyl cellulose at pH 10.5,[101] followed by chromatography on DEAE-cellulose at pH 8, Frenkel et al.[102] were able to isolate a protein fraction (component 0.62) which seemed to be homogeneous and had a molecular weight of 7000. Its amino acid analysis showed that aromatic amino acids, glycine, and serine accounted for more than 70% of the molecule. Its composition is also unusual in that it contains no lysine, histidine, glutamic acid, methionine, or isoleucine. Final confirmation of the homogeneity of the component came with the determination of its complete sequence by Dopheide[103] and this is given in Fig. 4-7.

While a cursory examination of this sequence shows no very obvious regularities, it is possible, by introducing one gap somewhere between residues 4 and 9 and regarding one amino acid between residues 32 and 40 and one between residues 60 and 65 as insertions, to write the sequence as a series of decapeptide repeats in which positions 4 and 10 are uniformly aromatic residues (Fig. 4-8). If correct, this view would relate the high-glycine-tyrosine proteins to the high-sulfur proteins (see Section 4.4), but it should be pointed out that conclusions based on amino acid sequences which have been modified by the introduction of gaps and similar procedures should be treated with great caution.[104]

### 4.2.5. Histological Components of the Wool Fiber

The obvious histological heterogeneity of the wool fiber led to attempts at a very early stage to separate different cell components and to look for chemical differences. Thus as early as 1925 evidence was obtained to show that the cuticle of wool contained more sulfur than the cortex,[105] a finding that has since been repeatedly confirmed. However, earlier workers also repeatedly claimed that there was little or no tyrosine and histidine in the scales. This finding was based on the reaction of the Pauly reagent with intact scales and seems to have given a misleading result, since more recent

S Y C F S S T V F P G C Y W G S Y G Y P L G Y S V G C G Y G S
T Y S P V G Y G F G Y G Y D G G S A F G C R R F W P F A L Y

Fig. 4-7. Amino acid sequence of high-glycine-tyrosine protein (component 0.62) from Merino wool.

| | | | | | | | | | |
|---|---|---|---|---|---|---|---|---|---|
| S | Y | C | F | S | — | S | T | V | F |
| P | G | C | Y | W | G | S | Y | G | Y |
| P | L | G | Y | S | V | G | C | G | Y |
| G | S | T | Y | S(P) | V | G | Y | G | F |
| G | Y | G | Y | D | G | G | S | A | F |
| G(C) | R | R | F | W | P | F | A | L | Y |

Fig. 4-8. Sequence of component 0.62 rearranged to show a possible decapeptide repeat. A gap (indicated by a dash) has been introduced at position 6, and the proline and cystine residues indicated as (P) and (C) have been treated as insertions.

direct analyses on hydrolysates all confirm the presence of both these amino acids.

The work of Geiger[106] was one of the most convincing of the earlier essays into the field. He reduced and *S*-ethylated wool and subsequently digested it with pepsin at pH 1.1 for 5 days at 35°C. Because of the low pH, it is unlikely that there was any bacterial contamination. After digestion the scales were centrifuged off, the yield being 2.3% of the weight of the original wool. Unfortunately, lack of analytical methods at that time restricted the data obtained but Geiger showed that the scales were greatly enriched in cystine, somewhat enriched in serine, and greatly depleted in arginine compared to the original wool.

More recent workers in this field have tended not to use enzymes, but either to make use of the observation that wool can be slowly disrupted by shaking in formic acid at room temperature[107] or by short treatments with mineral acids e.g., 6 *N* HCl at 60°C for periods of 30 to 120 min.[108] Table 4-8 shows a comparison of amino acid analyses of cuticle by the two methods of independent workers, and it can be seen that the agreement is remarkably close.

Also included in Table 4-8 are analyses of other histological fractions of the wool fiber. There is little worthy of comment in these data except to draw attention to the general similarity of the analysis of the cell membrane fraction to that of the high-glycine-tyrosine fraction. The occurrence of citrulline and ornithine in small amounts in some fractions is interesting, but both these amino acids occur in much larger amounts in the proteins of the medulla and inner root sheath.

Neither the medulla nor the inner root sheath are of any great technological significance in wool chemistry since most commercial wools are nonmedullated and the inner root sheath becomes detached from the fiber in the higher regions of the follicle and rubs off at skin level. Both proteins are interesting, however, in that they are the only proteins definitely known to contain the amino acid citrulline in peptide combination.[109] It has been shown that citrulline is formed from arginine, probably by a desimidase

Table 4-8. Amino Acid Analyses (μmol/g) of Histological Components of Wool[107,108]

| Amino acid | Whole wool | Cuticle | Whole wool | Cuticle | Epicuticle | Cell membranes | Cortical cells |
|---|---|---|---|---|---|---|---|
| Alanine | 490 | 608 | 492 | 566 | 428 | 574 | 517 |
| Arginine | 615 | 467 | 630 | 479 | 397 | 568 | 634 |
| Aspartic acid | 560 | 366 | 587 | 392 | 542 | 671 | 629 |
| Citrulline | — | — | 4 | 31 | 15 | 18 | 0 |
| Cysteic acid | 8 | 23 | 7 | 36 | 1076 | 11 | 13 |
| Cystine | 1035 | 1406 | 967 | 1443 | 31 | 191 | 857 |
| Glutamic acid | 1096 | 914 | 1098 | 894 | 991 | 968 | 1081 |
| Glycine | 761 | 821 | 793 | 961 | 1427 | 1326 | 886 |
| Histidine | 80 | 95 | 86 | 96 | 96 | 160 | 70 |
| Isoleucine | 306 | 256 | 288 | 253 | 234 | 360 | 304 |
| Lanthionine | 3 | 9 | — | — | — | — | — |
| Leucine | 683 | 608 | 708 | 598 | 507 | 757 | 727 |
| Lysine | 260 | 298 | 282 | 288 | 449 | 429 | 265 |
| Methionine | 46 | 54 | 46 | 38 | 3 | 111 | 41 |
| Ornithine | — | — | 3 | 17 | 68 | 17 | 1 |
| Phenylalanine | 239 | 173 | 268 | 186 | 172 | 399 | 290 |
| Proline | 612 | 952 | 546 | 929 | 539 | 371 | 551 |
| Serine | 1038 | 1460 | 945 | 1355 | 1269 | 768 | 957 |
| Threonine | 561 | 441 | 599 | 491 | 334 | 449 | 518 |
| Tyrosine | 342 | 139 | 366 | 292 | 192 | 571 | 389 |
| Valine | 535 | 661 | 509 | 659 | 532 | 504 | 528 |

reaction, since if arginine in which the guanido carbon atom is present as $^{14}C$ is injected into a guinea pig, $^{14}C$-labeled citrulline can subsequently be found in inner root sheath protein.[110] However, no desimidase capable of reacting with free arginine has been discovered in hair roots, and hence it is presumed that the enzyme probably reacts either on the arginine transfer RNA or the protein itself, but no conclusive evidence on this aspect is available.

Another feature of these proteins is the occurrence of covalent crosslinks between the $\gamma$-glutamyl carboxyl groups and the $\epsilon$-amino groups of lysine.[111] In the case of the medullary protein of guinea pig hair, about 36% of the total lysine residues are involved in this crosslink. This corresponds to one crosslink per 60 residues on the average and provides a convincing rationale for the extreme insolubility of this protein in all protein solvents. The inner root sheath protein also contains $\epsilon$-($\gamma$-glutamyl) lysine crosslinks but in much smaller amount. Complete amino acid analyses of guinea pig root sheath and porcupine quill medullary protein are given in Table 4-9. The few peptide sequences that have been determined are given in Table 4-10.[112]

**Table 4-9. Amino Acid Composition (μmol/g) of Medulla and Inner Root Sheath Proteins**[109]

| | Protein source | |
|---|---|---|
| Amino acid | African porcupine quill medulla | Guinea pig inner root sheath |
| Alanine | 370 | 440 |
| Arginine | 260 | 265 |
| Aspartic acid | 460 | 510 |
| Citrulline[a] | 760 | 235 |
| Cystine | Trace | Trace |
| Glutamic acid | 2625 | 1510 |
| Glycine | 310 | 520 |
| Histidine | 95 | 100 |
| Isoleucine | 135 | 250 |
| Leucine | 710 | 680 |
| Lysine | 675 | 630 |
| Ornithine | 90 | 30 |
| Methionine | 45 | 160 |
| Phenylalanine | 215 | 220 |
| Proline | 165 | 245 |
| Serine | 260 | 505 |
| Threonine | 140 | 225 |
| Tyrosine | 175 | 170 |
| Valine | 330 | 345 |

[a] Corrected for citrulline breakdown.

**Table 4-10. Amino Acid Sequences of Tryptic Peptides from Porcupine Quill Medullary Protein**[112]

| |
|---|
| Leu-Cit-Gln |
| Cit[a]-Pro-Pro |
| Asp-Cit-Cit-Phe |
| Asp-Cit-Phe-Cit |
| Phe-Cit-Glu-Glu |
| Leu-Leu-Glu-Cit-Cit |
| Cit-Cit-Val-Cit-Cit-(Glu, Gln)-Val |

[a] Cit, citrulline.

### 4.2.6. Wool Root Proteins

In 1948 Ellis published a method of harvesting wool roots by a wax-sheet technique,[113] and this has been used to investigate proteins in the hair follicle and to compare these proteins with those derived by solubilization procedures on the fully formed and hardened fiber. Rogers[114] showed that SCMK proteins derived from proteins extracted from wool roots with 8 *M* urea could be divided into high- and low-sulfur fractions in precisely the same way as could SCMK from whole wool, and Thompson and O'Donnell[115] showed further, by starch gel electrophoresis, that the wool root fractions were heterogeneous. These observations were of fundamental importance, since the extremely mild conditions used to extract the wool root proteins made it improbable that the heterogeneity is an artifact introduced in the isolation procedure, and hence argued strongly against the protein fractions derived from wool being artifacts of isolation, as had been suggested.

To clinch the matter it would ideally be satisfying to obtain the same proteins from the wool fiber by urea–thioglycollate extraction and from wool roots by urea extraction. This seems possibly to be true of the low-sulfur proteins: the SCMKA fractions from both sources seem to be very similar. Thus Frater[116] attempted the isolation of component 8 from both wool and wool roots and compared them on the basis of amino acid composition and tryptic peptide maps. Some of his data are given in Table 4-11, and his conclusion from peptide mapping studies is that small but significant differences exist between wool and wool root proteins, although the major feature is the overall similarity between the fractions. In view of the known heterogeneity of component 8 (Section 4.2.2), the demonstration of the very great similarities is perhaps more significant than the small differences.

The position with regard to the high-sulfur proteins is much more

**Table 4-11. Amino Acid Composition (μmol/g) of Component 8 Isolated from Wool and Wool Roots**[116]

| Amino acid | Crossbred fleece[a] | | Lincoln fleece[b] | | Border Leicester fleece[b] | |
|---|---|---|---|---|---|---|
| | Wool | Wool root | Wool | Wool root | Wool | Wool root |
| Alanine | 494 | 559 | 544 | 683 | 519 | 636 |
| Arginine | 761 | 651 | 694 | 636 | 689 | 641 |
| Aspartic acid | 836 | 900 | 964 | 922 | 903 | 896 |
| Carboxyethyl cysteine | — | — | 413 | 364 | 442 | 402 |
| Carboxymethyl cysteine | 556 | 490 | — | — | — | — |
| Glutamic acid | 1389 | 1399 | 1564 | 1546 | 1492 | 1542 |
| Glycine | 358 | 317 | 290 | 411 | 337 | 390 |
| Histidine | 52 | 65 | 53 | 58 | 47 | 57 |
| Isoleucine | 286 | 317 | 322 | 369 | 313 | 338 |
| Leucine | 1012 | 1048 | 1080 | 1039 | 991 | 1038 |
| Lysine | 249 | 280 | 281 | 380 | 276 | 368 |
| Methionine | 28 | 30 | 25 | 53 | 29 | 56 |
| Phenylalanine | 187 | 175 | 174 | 185 | 171 | 161 |
| Proline | 336 | 331 | 314 | 263 | 302 | 246 |
| Serine | 665 | 673 | 481 | 418 | 657 | 597 |
| Threonine | 445 | 465 | 382 | 333 | 410 | 371 |
| Tyrosine | 223 | 209 | 203 | 180 | 219 | 209 |
| Valine | 516 | 552 | 553 | 607 | 543 | 582 |

[a] Component 8 isolated as *S*-carboxymethyl derivative.
[b] Component 8 isolated as *S*-cyanoethyl derivative.

obscure. Two high-sulfur fractions can be isolated from wool roots: (a) by 8 *M* urea extraction and (b) by subsequent urea–thioglycollate extraction as for whole wool. The latter fraction is indistinguishable from that prepared from wool, but since wool roots as prepared invariably contain some fully hardened fiber, this is hardly surprising. The urea-soluble high-sulfur fraction shows many differences from the wool high-sulfur fraction: the wool root form gives a much more complex starch gel electrophoresis pattern and contains many proteins not present in the corresponding wool fraction.[117] It has been suggested that some of these may be precursor proteins which are later converted to high-sulfur proteins by some novel biosynthetic mechanism. However, this seems an unnecessary and unlikely hypothesis. The cells of the wool root are actively metabolizing cells and as such must contain many enzymes and other proteins in their cytoplasm. Hence it is hardly surprising that the wool root extract is complex and contains novel proteins. The comparative absence from wool root extract of the sulfur-rich proteins typical of SCMKB need not surprise us, since we know nothing of their properties in the form in which they occur in the wool root. Although they are presumably synthesized initially in the thiol form, there is no guarantee that they remain completely reduced. Because of the frequent occurrence of cysteinyl cysteine sequences, they form extremely active thiol–disulfide interchange systems and may well have only a transitory existence as urea-soluble components in the cell. Alternatively, homogenization of the cells with 8 *M* urea may permit thiol–disulfide interchange to occur. In either case the high-sulfur SCMKB would be converted into a urea-insoluble polymer, and hence would only be found in the urea–thioglycollate fraction.

By injecting $^{35}$S-cystine into the animal it has been possible to study some aspects of wool synthesis. The major finding in this area has been confirmation of the hypothesis that the high- and low-sulfur proteins are synthesized at different levels in the follicle. Thus Ryder,[118] by direct autoradiography, showed that labeled sulfur could be found at the zone of keratinization very soon after an injection of $^{35}$S-cystine, and only after much longer times did it appear at the base of the follicle and begin to be incorporated in this region. It has been shown that for proteins isolated from wool roots within 24 h of an injection of $^{35}$S-cystine, the specific activity of the high-sulfur fraction is greater than that of the low-sulfur fraction, no matter whether the proteins are isolated by an oxidative[119] or reductive procedure.[117]

## 4.3. Physical Studies

### 4.3.1. Physical Studies on Soluble Wool Protein Fractions

A major motivating factor in the search for methods to solubilize keratins is to extend the range of techniques available for study of the proteins. An obvious example of this, of course, is in the protein fractionation work discussed earlier, but two other fields are the study of molecular weights and the optical properties of the solutions.

#### *4.3.1.1. Molecular-Weight Studies*

A variety of techniques have been used to study the molecular weights of derived soluble wool proteins, including the ultracentrifuge (by both sedimentation equilibrium and meniscus depletion methods), osmotic pressure, light scattering, surface balance, and chromatography on calibrated Sephadex columns. Much of the early work, although often technically of very high standard, is now largely only of historical interest, either because it was carried out on material we now recognize as being extremely heterogeneous, or under conditions where interactions between protein molecules make interpretation of the results difficult, if not impossible.

Jeffrey has recently reviewed this field in great detail,[62] and anyone interested in the topic should consult his article. His overall conclusion is that component 8 has a molecular weight of 45,000, while component 7, which is more difficult to isolate, has a molecular weight of 51,000. Two major contributing factors to uncertainty in this field are (1) the difficulty in obtaining preparations of components 7 and 8 that are free of high-glycine-tyrosine proteins, and (2) uncertainty about the values for partial specific volume to use. Recent work by Woods (unpublished) in which particular attention has been paid to these points suggests that Jeffrey's values are too high and that component 8, for example, may have a molecular weight around 40,000.

A very complete study of the molecular weight of SCMKB proteins has been carried out by the group in South Africa.[69,120] They have shown that the molecular weights of SCMKB fractions can range from 10,000 to 60,000 and of course their results have now been confirmed in part by amino-acid sequence determinations.[73–75,81] Interactions between the high-sulfur proteins seem to be less of a problem than in the case of the low-sulfur proteins, and the interpretation of sedimentation data consequently simpler.

#### *4.3.1.2. Optical Studies*

Since keratins may be regarded as the archetype of $\alpha$-proteins it is not surprising that optical rotatory dispersion studies have been made on soluble keratin derivatives. Using the expression

$$\text{helix content} = -\frac{100b_0}{630}\%$$

where $b_0$ is the Moffitt–Yang parameter,[121] it was shown that SCMKA in aqueous solution is approximately 50% helical. Heating to 70°C, or the addition of urea to 8 *M*, reduced the helix content to zero, but this change was completely reversible.[122] It has since been shown that when the high-glycine-tyrosine proteins are completely removed from SCMKA, the measured helical content increases to 62–65%.

Crewther and Harrap[63] made the interesting observation that partial digestion of SCMKA with any of a variety of proteases leads to a decrease in the amount of material that can be precipitated at pH 4 and $I = 0.5$, but this material, when redissolved, is much more helical in solution than the parent SCMKA. Amino acid analysis of this helical material and the parent SCMKA are shown in Table 4-12. As compared with SCMKA, the helical fraction is enriched in glutamic acid, leucine, lysine, aspartic acid, alanine, and isoleucine, while, conversely, it contained less glycine, *S*-carboxymethyl

**Table 4-12. Amino Acid Composition (μmol/g) of Helical, Nonhelical and Soluble Fractions Obtained by Enzyme Digestion of SCMKA[125]**

| Amino acid | SCMKA | Helical fraction (pronase) | Soluble fraction (pronase) | Nonhelical fraction (chymotrypsin) |
|---|---|---|---|---|
| Alanine | 551 | 676 | 526 | 479 |
| Arginine | 648 | 582 | 654 | 617 |
| Aspartic acid | 700 | 915 | 694 | 561 |
| Carboxymethyl cysteine | 621 | 299 | 684 | 911 |
| Glutamic acid | 1234 | 1864 | 1150 | 691 |
| Glycine | 770 | 239 | 883 | 1123 |
| Histidine | 53 | 68 | 50 | 18 |
| Isoleucine | 289 | 342 | 258 | 230 |
| Leucine | 840 | 1180 | 744 | 663 |
| Lysine | 298 | 505 | 258 | 110 |
| Methionine | 44 | 17 | 50 | 9 |
| Phenylalanine | 245 | 163 | 268 | 304 |
| Proline | 350 | 94 | 377 | 773 |
| Serine | 823 | 564 | 912 | 1307 |
| Threonine | 411 | 308 | 456 | 644 |
| Tyrosine | 368 | 291 | 387 | 193 |
| Valine | 551 | 453 | 486 | 571 |

cysteine, serine, proline, threonine, arginine, and tyrosine. In general, these trends fit with the known helix-forming tendencies of the various amino acids.[123,124] As might be anticipated, the nonhelical fraction showed complementary trends in composition.

Later developments in the chemical field leading to the establishment of some sequences for the helical peptides have been mentioned earlier (Section 4.2.2.1). Physicochemical studies have shown that the helix-rich fraction is more than 90% helical compared to the 62–65% helix content of SCMKA. Since SCMKA consists mainly of a mixture of components 7 and 8 in the ratio 2:1, similar studies have been carried out on these components separately and also admixed in various ratios. In this way a picture has been built up of the 41,000-mol-wt helix-rich particle obtained by protease digestion of SCMKA as consisting of three chains about 160 Å long, two of the chains being derived from component 7 material and one from component 8.[126] There seems to be a cooperative effect between the component 7 and 8 chains, leading to increased helicity and stability in the mixture compared to the isolated components, and maximum helicity seems to be reached at a 2:1 ratio of component 7 to 8. It is, of course, difficult to argue from solution studies of this kind as to the conformation in native protein, since the proteins have been considerably modified to make them soluble; moreover, the conformation which they adopt in solution is very much influenced by the solvent in which the protein is studied. Nevertheless, it would seem to be a reasonable assumption that it is the low-sulfur proteins of wool which are responsible for the $\alpha$-type X-ray diffraction photograph given by wool.

The high-sulfur proteins derived from wool—SCMKB or $\gamma$-keratoses—have shown no evidence of $\alpha$ helix in solution. In view of their high content of proline this is hardly surprising, and because of the multiplicity of negatively charged groups, they probably exist as random coils in solution. In the fiber, however, because of the large number of disulfide bonds, they may well have a very definite but nonhelical conformation.

### 4.3.2. X-Ray Studies on the Intact Fiber

The proposal of the $\alpha$-helix structure by Pauling et al.[127] marks a definite watershed in our way of looking at wool and keratin structure. All earlier structural formulations had been inspired guesses aimed at explaining a single aspect of the structure, and in consequence the proposed structures were too imprecisely defined to be capable of critical test. The $\alpha$ helix, on the other hand, was based on data obtained by years of painstaking work on the crystal structure of amino acids and peptides and was very precisely defined. In fact, the structure was so well defined that it was immediately possible to make a specific prediction from it—that structures made up of $\alpha$ helices should show a prominent X-ray reflection corresponding to spacing of 1.5 Å in the direction of the axis of the helix. In fact, such a reflection had been observed by MacArthur[128] in 1943 using porcupine quill tip, but it was left to Perutz[129] to point out the significance of this and extend the observation to other proteins. Further general confirmation of the idea that the structure of keratin was basically $\alpha$-helical came with the working out by Cochran et al.[130] of the diffraction theory for helical molecules. Theory and practice were shown to be in complete harmony in the case of synthetic polypeptides, but discrepancies began to appear when details of the $\alpha$-keratin X-ray photograph were examined closely in the light of the theory. One fact, in particular, obstinately refused to fit into any simple picture of keratin structure made up of simple $\alpha$ helixes. This was the meridional reflection of 5.1 Å, corresponding to one turn of the $\alpha$ helix. In synthetic polypeptides this reflection was about 5.4 Å, agreeing well with theoretical expectation ($1.5 \times 3.6 = 5.4$, i.e., spacing/residue $\times$ residue/turn). For $\alpha$-keratin the observed repeat of 5.1 Å is outside the range considered by Pauling and also outside the range found in model polypeptide studies.

In addition to the problem of the actual repeat distance corresponding to this spacing in $\alpha$-keratin, the character of the reflection was also unexplained. X-ray diffraction patterns from model polypeptides showed quite clearly that the 5.4-Å spacing is split into two off-meridional arcs, whereas for $\alpha$-keratin the 5.1-Å reflection is quite unmistakably a meridional arc and shows no signs whatever of splitting. Crick[131,132] and Pauling and Corey[133] almost simultaneously proposed similar solutions to this dilemma: the $\alpha$ helix itself is coiled into a coiled coil or superhelix. A number of these coiled coils can then twist together to form a ropelike structure: Crick suggested two- or three-strand ropes, but Pauling and Corey envisaged the additional possibility of seven-strand cables. Crick's underlying assumption for the two-strand coiled-coil hypothesis was that the helixes could approach more closely since the side chains of one helix would fit into the inter-side-chain spaces on the other—"knob-hole" packing. Pauling and Corey, on the other hand, suggested that the coiling into cables is due to a regular repetition of amino acids along the chain. It would appear that both views are correct in part, since the knob-hole packing seems to be reinforced by hydrophobic bands, every seventh residue showing a tendency to being an amino acid with a hydrophobic side chain (see Fig. 4-2). The general concept of coiled coils has received considerable support from the work of Cohen and Holmes[134] on muscle proteins. Figure 4-9 shows a diagrammatic representation of a two-strand coiled-coil structure and illustrates how little distortion of the $\alpha$ helix is required when the superhelix has a pitch of about 200 Å. A further point worth noting is that the major and minor helices are of opposite sense (i.e., since the $\alpha$ helix is normally right-handed, the superhelix will be left-handed).[135]

A variation on these ideas, first proposed speculatively in 1955[136] but since developed more critically by Fraser and MacRae,[137,138] is the idea of the segmented rope. In this variant, instead of the $\alpha$ helixes being regularly twisted into a superhelix, peptide chains consisting of shorter lengths of straight $\alpha$ helix separated by short, flexible nonhelical sections are twisted around one another. Using optical diffraction techniques, Fraser and MacRae showed that although the three-strand rope gives satisfactory agreement with the observed pattern in the 1.5- and 5.1-Å regions of the meridian, the agreement with the remainder of the meridional wide-angle pattern is not particularly good. The segmented rope model with segments 27 Å long gives much closer agreement, as can be seen in Fig. 4-10, taken from the original paper. Lundgren and Ward have produced a similar model, but in this case the length of segment (70 Å) would produce a very open structure with a 14-Å separation of the helixes at the end of the segments.[139]

The whole question of the type of coiled-coil structure most compatible with the data has been thrown wide open again by the realization that although the main features of the diffraction patterns can be predicted using only the

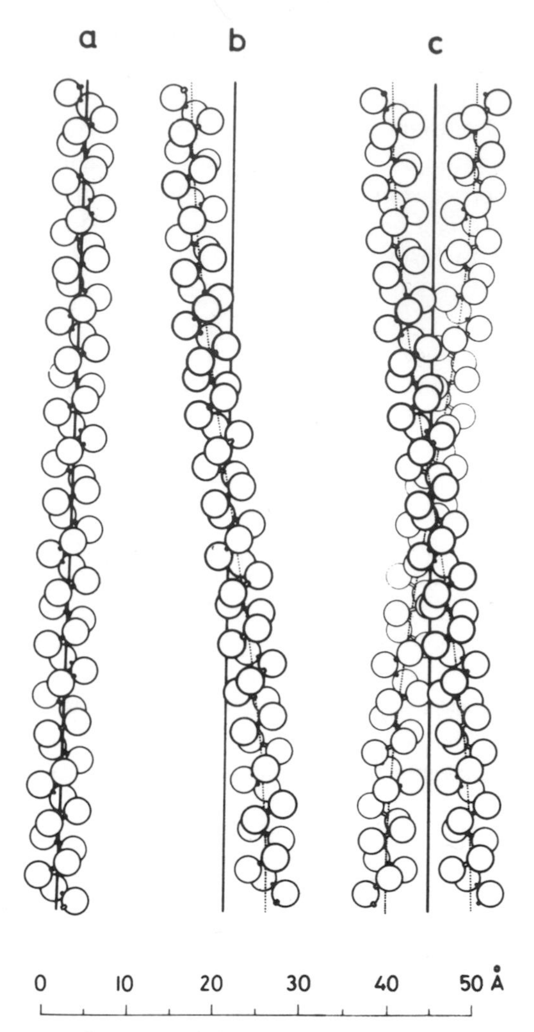


Fig. 4-9. (a) Diagrammatic representation of a straight-chain α-helix. (b) Representation of an α-helix twisted to give a coiled coil of pitch approximately 200 Å. (c) Two such coiled coils fit together to form a two-stranded rope. (Courtesy of *Nature*.)

backbone and β-carbon atoms of the helixes as all earlier work had assumed, background scattering effects of side chains and water molecules also markedly influence some features of the diffraction pattern. A method of introducing a correction for this effect has been developed and used to compare calculated and observed diffraction patterns in some α-proteins.[140] For keratin the conclusion was reached that the data do not permit an unequivocal choice between the various models, but analogy with other more favorable situations in other α-proteins tends to support the two-stranded coiled coil.

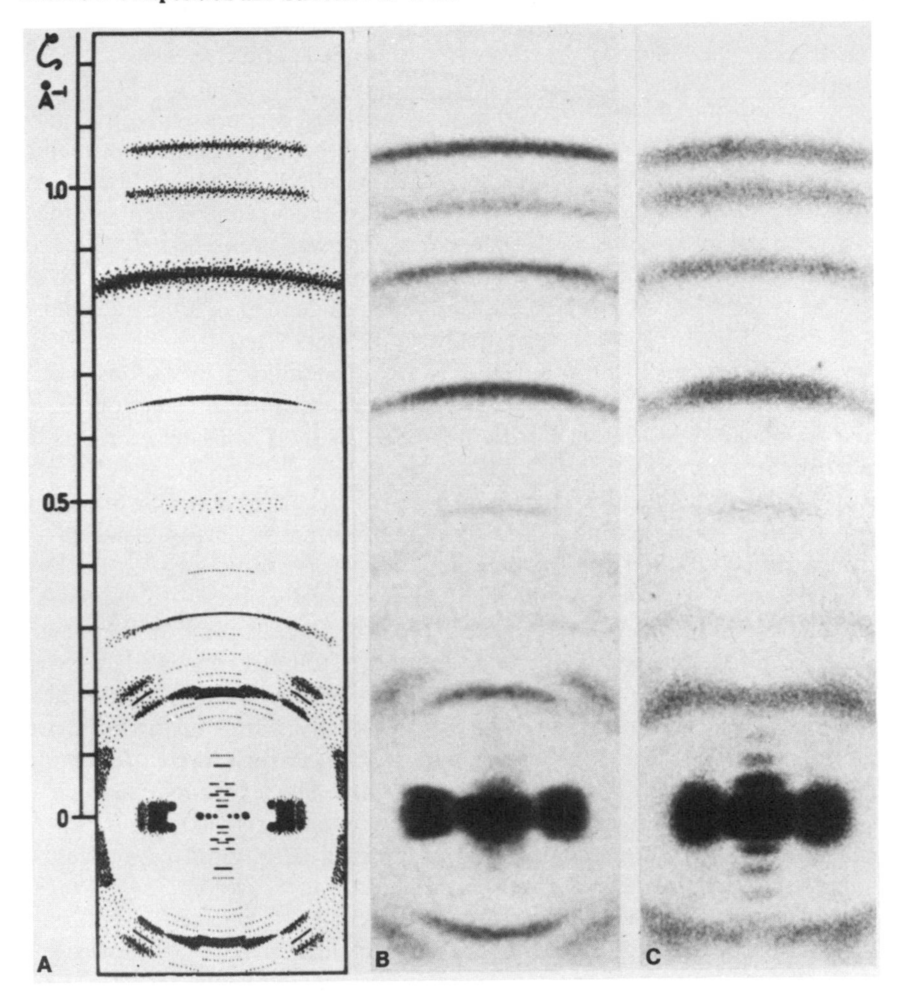


Fig. 4-10. A, Main features of X-ray diffraction pattern of α-keratin; B, optical transform of three-stranded-rope model of α-keratin proposed by Crick; C, optical transform of three-stranded segmented rope model.[137,138] (Courtesy of *Nature*.)

Discussion of the X-ray evidence has so far been concerned almost exclusively with high-angle meridional reflections, since it is these which are diagnostic of variants of the α helix. The X-ray diagrams of keratins also reveals considerable detail in the low-angle meridional region, which requires the postulation of periodicities of at least 198 Å.[128] Direct evidence of this 198-Å period in mohair fibers has been obtained by Spei et al.[141] This reflection and one corresponding to a 66-Å spacing were both intensified by specific mercury staining of the least reactive cystine fraction. Other low-angle meridional periodicities have been intensified after specific chemical treatment.[142–147]

Fraser and MacRae have recently attempted a more ambitious interpretation of the X-ray diffraction photograph of $\alpha$-keratin (i.e., porcupine quill), in which they utilize low-angle near-meridional as well as meridional spacings.[148] The solution which they arrive at is of structural units arranged on a helix of pitch 220 Å and with a unit axial translation of 67 Å. There are thus a nonintegral number of structural units per turn (i.e., $220/67.1 = 3.28$), just as in the $\alpha$-helix itself there are a nonintegral number of amino acids/turn (i.e., $5.4/1.5 = 3.6$). This comparatively simple picture requires one additional complicating factor to allow indexing of the off-meridional reflections. This is that the structure be subjected to a regular axial distortion of period 235 Å, but in practice this factor can be neglected in most considerations of the structure.

At this stage the full implications of the model have not been worked out, but it would seem that additional data are still needed before some of the major structural questions are answered. Thus it is not possible on the basis of the proposed structure and available chemical evidence to establish a definite correspondence between the structural and chemical units, nor is it possible to decide between the two- and three-stranded rope models for the coiled coils. However, the lateral positions of the maxima of the near-meridional reflections give a rough estimate of the radius of the diffracting object and suggest that the low-angle pattern in $\alpha$-keratin is largely determined by projections from the *surface* of the microfibril. The reflections thus indicate strong ordering in the microfibril and the microfibril-matrix regions.

Probably the most satisfactorily interpreted keratin diffraction data at the present time relate to the equatorial reflections. Fraser and MacRae,[149] following on the electron microscope evidence of keratin microfibrillar structure produced by Rogers[150] and Birbeck and Mercer,[151] showed that the equatorial distribution of diffracted intensity can be interpreted in terms of the shape, packing, and contents of the microfibrils. Specifically, the data are consistent with a picture of microfibrils about 75 Å in diameter containing an annulus of protofibrillar $\alpha$ helixes of 20 Å diameter embedded in an osmiophilic matrix. Bailey et al.[152] came to a similar general conclusion but found that still better agreement was obtained by assuming that some of the microfibrils also have protofibrils in the center of the annulus. Confirmation of this view has since been obtained by Fourier synthesis of equatorial X-ray data.[153] In this latter interpretation the sole assumption that has been made is one of radial symmetry of the microfibril, but the data have been shown to support the idea that the microfibril consists of a central core containing one or two protofibrils with an annular ring of protofibrils centered on a radius of about 28 Å. Still other confirmatory evidence using image averaging by optical filtering of electron micrographs has also been obtained,[154] but the theoretical validity of this has been challenged.[155]

Further indirect evidence supporting this overall conclusion comes from a study of the $\beta$ photograph of stretched keratins. If we accept Astbury and Woods'[8,9] assumption that the $\beta$ photograph of stretched wool arises by a direct interconversion of $\alpha$ to $\beta$, then the $\beta$ photograph is a valuable potential source of information relating to the $\alpha$ structure. (In fact, although the $\alpha \rightleftharpoons \beta$ mechanism has been queried,[156] no critical evidence exists to prove or disprove the hypothesis.) The $\beta$ photograph of stretched keratin is compatible with the concept of crystallites containing about 25–30 polypeptide chains (i.e., approximately the same number as are found in a microfibril.[157] Moreover, these chains are made up of antiparallel pleated sheets[158] stacked randomly; Fig. 4-11 shows a diagrammatic representation of an antiparallel $\beta$-pleated sheet. It is also probable that the axial lengths of the $\beta$ crystallites are $< 100$ Å long, suggesting that they are derived from quite short lengths of $\alpha$ helix, which agrees reasonably well with an earlier conclusion from the $\alpha$ photograph that the coherent lengths of coiled coil are only $\sim 60$ Å. This conclusion would tend to favor a segmented-rope type of structure. An additional feature of this type of structure is that since a coherent length of 60 Å would only involve a rope twist of 100° (if the pitch length is 220 Å), it is fairly easy to visualize how an $\alpha$-$\beta$ transformation could occur; the difficulty of converting long lengths of coiled coil (especially in the Pauling–Corey seven-strand cable) has been regarded as evidence against the presence of coiled coils.[136,159] The finding that the $\beta$ chains exist in the antiparallel configuration implied a similar condition for the $\alpha$ protofibrils: this agrees with the finding by Parry and Suzuki[160,161] that coiled coils involving antiparallel chains are favored both on energetic and entropic grounds.

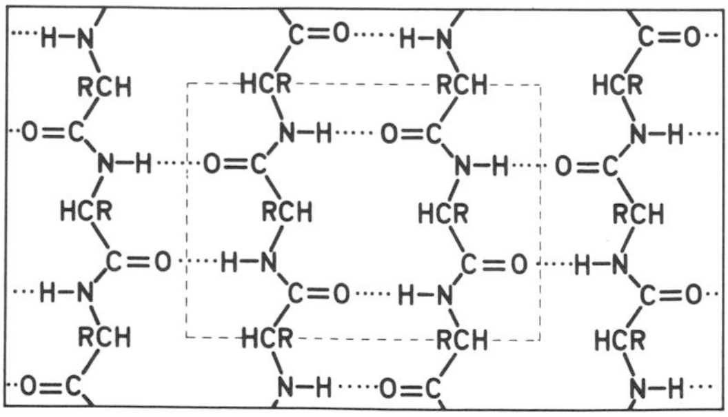

Fig. 4-11. Diagrammatic representation of antiparallel $\beta$-pleated-sheet structure.[146] (Courtesy of *Polymer*.)

## 4.4. An Integrated Picture of the Wool Fiber

Chapter 3 has presented a picture of wool fiber down to the fine histological level, and the problem now remains of relating the chemical and physical studies of the present chapter to this histological picture. Undoubtedly, the great bulk of the fiber is made up of cortical cells, and apart from the specific studies of cuticle and medulla mentioned earlier (Section 4.2.5), most of the chemical and physical studies relate to the cortex of the fiber. Where then do the high- and low-sulfur proteins fit into this picture? The consensus of opinion probably favors the general hypothesis that the low-sulfur proteins represent the protofibrillar component, which is largely responsible for the $\alpha$-X-ray photograph of wool, while the high-sulfur proteins form the matrix surrounding the microfibrils.

Some refinements of this simple view are probably necessary, however. Thus Bendit[162] calculated the percentage volume of the cortex occupied by microfibrils and compared this with the percentage ratio of SCMKA to SCMKA plus SCMKB. On the basis of this calculation, he suggested that the microfibril was not large enough to accommodate the low-sulfur protein and hence that an appreciable fraction of it (perhaps one third) must spill over into the matrix. The only point of weakness in Bendit's conclusions arises from his estimate of the amount of SCMKA. He regards the material obtained by solubilizing wool by reduction and carboxymethylation as being composed entirely of SCMKA and SCMKB. The figure he uses for SCMKA content is obtained by subtracting an experimental SCMKB figure from a total solubility figure. Since other types of protein are present, and since the value found for SCMKB content may well be low, the value he assumes for the amount of SCMKA is undoubtedly high, but it is impossible to say whether it is sufficiently high to invalidate his conclusion entirely.

A reexamination of the problem by Fraser et al.[163] would tend to suggest that the amount of overspill of low-sulfur protein into the interfibrillar matrix is small. These authors were concerned primarily with the location of the high-glycine-tyrosine proteins in the fiber—a problem that had become important with the realization that these proteins may form a very significant fraction of the total protein. Their overall conclusion from data on five different keratins is that the interfibrillar matrix can accommodate both the high-sulfur and high-glycine-tyrosine proteins, but that there would be only a little space left over for any spillover from low-sulfur proteins. These authors therefore favor the view that the matrix is made up largely of the high-sulfur and high-glycine-tyrosine proteins, with the annular outer ring of the microfibril being mainly the helical portion of the low-sulfur proteins. The nonhelical tail material fills up the central core of the microfibril with some small "spillover" into the matrix.

It is worth pointing out, however, that the concept of the matrix as made up entirely of random chains is one that may need revision in the future. It is true that the high-sulfur and high-glycine-tyrosine proteins would be precluded from an $\alpha$-helical conformation by their amino acid composition. Their uniformly high contents of proline, cystine, serine, threonine, and glycine necessarily imply this, and it is also true that, as isolated, the high-sulfur proteins show no evidence of organized structure. However, in the fiber the molecules exist with intact disulfide bonds, and this is a very different situation and, in fact, evidence for some degree of orientation of the matrix in the intact fiber has been found.(148,152)

One further important piece of evidence relates to the occurrence of a decapeptide repeat in some of the high-sulfur proteins. This finding of a regularity of chemical structure suggests very strongly that there may also exist a regularity of conformation. It is interesting in view of this suggestion of interchangeability of high-sulfur proteins and high-glycine-tyrosine proteins in the matrix(163) that it is also possible to fit the one high-glycine-tyrosine protein sequence into a decapeptide repeat, in which hydrophobic bonding between aromatic amino acid side chains may substitute for the covalent disulfide bonds of the high-sulfur proteins (Fig. 4-8).

One structural possibility suggested by analogy with the behavior of synthetic polymers is that the molecules may exist in part as folded-chain polymer crystals. These were originally discovered in polyethylene by a number of workers in 1957, and recently Lindenmeyer(164) has shown that the folded-chain crystal represents the best energy compromise for which a mechanism of crystal formation is normally available. With the added additional energy and entropy contribution from disulfide-bond formation, a structure of this type might well be a favored one for the high-sulfur proteins. Figure 4-12 shows a two-dimensional diagrammatic representation of a highly speculative version of this for the parent protein of SCMK-IIIB2 sequenced by Haylett and Swart.(73)

A feature of this structure is that the disulfide bonds are mainly intrachain. This is in contradistinction to classical views of keratin, in which disulfide bonds have usually been considered as crosslinks between chains. However, this seems an improbable view to take of the structure of the matrix, since if many of the disulfide bonds of the matrix were interchain (and hence intermolecular), the matrix would have physical properties resembling those of Bakelite. In fact, X-ray evidence(165,166) shows that swelling of wool fibers in water is mainly due to swelling in the matrix: the fibrils are further apart at high-regain values, but otherwise there are only comparatively small changes in their parameters. The preoccupation with intermolecular crosslinks which has characterized much of the theorizing around physical models for wool structure seems to have overlooked two important facts: (1) strong

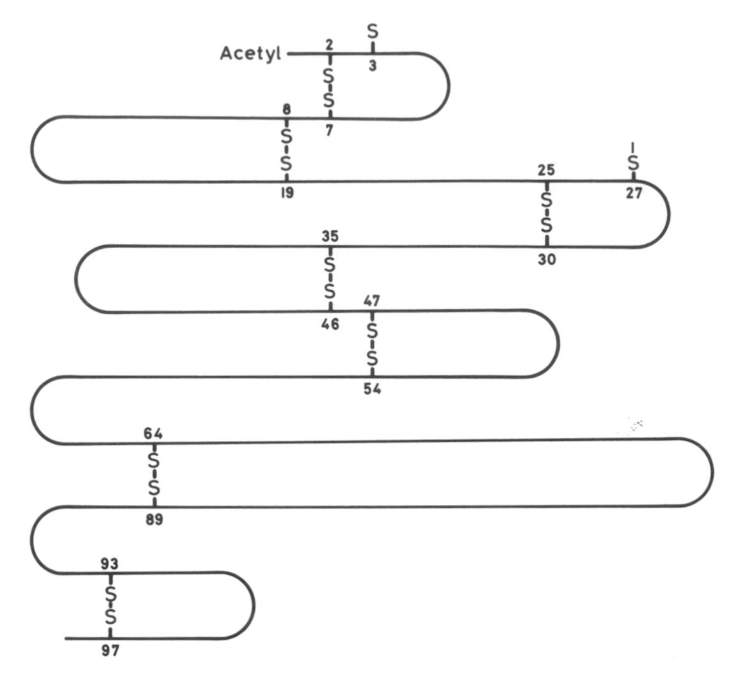


Fig. 4-12. Two-dimensional diagrammatic representation of a speculative structure for SCMKB-IIIB2.[67]

fibers do not necessarily imply covalent crosslinks, as the properties of silk and synthetic fibers demonstrate, and (2) a dramatic effect on fiber strength of breaking disulfide bonds does not necessarily prove the existence of intermolecular crosslinks—obviously breaking any of the disulfide bonds in the structure represented in Fig. 4-12 would have profound effects on its physical properties. It is perhaps worth pointing out that the breaking load of rubber reaches a maximum at a low level of crosslinking[167]: any further crosslinking produces increased hardening and brittleness, with a decrease in breaking load.

A second feature of this structure which is of interest is the way in which cys-cys sequences can be used to stabilize adjacent folds. The fact that cys-cys sequences are a characteristic feature of $\alpha$-keratin structure was pointed out by Lindley and Haylett,[168] and this has been amply confirmed in subsequent sequence studies. Whether incorporated in the way suggested in Fig. 4-12 or not, these cys-cys sequences probably explain the dramatic effect of thiol content of the fiber on creep rate.[169]

It is probably true, of course, that no one simple picture can describe the structure of the matrix, and it may well be that fringed micelles and folded-chain crystallites both play a role. It is possible that the predominating type of packing in the matrix may determine the physical character of the keratin. Thus horny keratins contain a lower percentage of high-sulfur proteins than do the hairs, and furthermore the horn high-sulfur proteins tend to be of the low-molecular-weight, low-sulfur type.[170] However, further examination shows that the hypothesis of a direct relationship between amount of matrix and softness of handle is an oversimplification[171]: obviously, the structure of the matrix is also important. It is even possible that an effect of this kind is responsible for the much debated ortho and para subdivision of keratin fibers.

In contradistinction to the unsatisfactory picture of the matrix material, the outline of the fibrillar structure of the wool fiber is clear. There seems little reason to doubt the conclusion that the microfibril comprises a central core consisting of one or two protofibrils and an annular ring of protofibrils centered on a radius of about 28 Å. Dobb and Sikorski's criticisms of the early evidence based largely on interpretation of electron micrographs undoubtedly have some validity, but the combination of alternative lines of evidence suggests that these criticisms have been overemphasized. The evidence, although circumstantial, would also seem strong that the helical portion of the low-sulfur proteins constitutes the protofibrils, but we have no clear evidence on finer details of protofibrillar structure. The occurrence of an amino acid residue with a hydrophobic side chain as every seventh residue in the helical fraction of the low-sulfur proteins is strong circumstantial evidence for coiled coils of some kind, but the question as to whether a two- or three-strand rope is involved is still unsettled. Fraser and MacRae's interpretation of the X-ray data[148] suggests a pitch of 220 Å for the superhelix with a unit height of 67 Å. The segmented rope structure is another alternative which cannot be ruled out, but neither the coherent length of helix nor the nature of the discontinuities is known. One thing that has been established is that coiled-coil structures are energetically favored with respect to straight $\alpha$ helixes.[149,150]

## 4.5. Unsolved Problems and Future Trends

Most of the chemical problems have been indicated in passing, but some outstanding ones merit special mention. The first is the need for a great improvement in our techniques of protein fractionation. It is no exaggeration to say that the complexity in this area of keratin research is probably only exceeded by that of the immunoglobulins, and unfortunately nothing equivalent to a myeloma protein has so far been discovered among keratins.

The amount of progress that has been made is quite remarkable when judged with a realistic view of the difficulties involved and is a tribute to the tenacity of the workers concerned. The major bottleneck is undoubtedly the difficulty of obtaining a protein fraction that is sufficiently homogeneous to make amino-acid-sequence determination possible. This is especially true of the high-sulfur proteins, where the relatively simple amino acid composition implies that many peptide sequences are very similar. This means that one is constantly faced by the problem of knowing whether two closely related peptides both occur in the same molecule or whether they are alternative sequences from closely related protein molecules. The choice is further complicated by the fact that the simple amino acid composition seriously reduces the possibility of obtaining useful overlap peptides by digestion with a second enzyme.

The second big chemical problem which is so far untouched is the problem of the disulfide-bond arrangement. The techniques that protein chemists have evolved to handle this problem are obviously not capable of coping with a situation of the complexity found in wool, quite aside from the purely practical problem of completely preventing thiol–disulfide interchange reactions in a system that may well have been evolved to facilitate them.[172] It has recently been calculated that there are more than $10^{23}$ possible isomers of the parent disulfide protein corresponding to SCMKB2A, which differ only in disulfide bond arrangement.[173] Moreover, this calculation involves the simplifying assumption that all the disulfide bonds are intramolecular. Three approaches to this problem seem possible.

One is to isolate proteins from wool roots prior to keratinization and, hopefully, to characterize disulfide bonds in these. A second approach might be to concentrate attention on only a relatively small fraction of the disulfide bonds. Thus the finding that reaction of the least reactive cystine bonds with mercury salts produces an enhancement of meridional X-ray spacings corresponding to 66 Å[143] and 198 Å[174] suggests that a study of these might be very rewarding; the possibility that these disulfide bonds may have a special structural significance has been stressed.[175,176] A final approach, which may ultimately be possible, is the theoretical one. The disulfide bonds formed in the final keratinization reaction in the fiber are almost certainly those which are most energetically and entropically favored, and hence the problem may ultimately be amenable to mathematical analysis.[177]

A related problem but in a more biochemical context is the mechanism of keratinization. This is known to involve the oxidation of sulfhydryl groups to disulfide groups, and it is also known that its onset is delayed in copper-deficient animals. This suggests that copper-containing enzymes may be involved, but no such enzymes have yet been characterized in hair follicles. It would also be interesting to know if the disulfide-rearranging enzyme[178] is present in wool roots; if so, this might provide yet another lead to the problem of disulfide-bond arrangement.

In the physical area the major problem is further refining the interpretation of the α photograph, and a special need is to relate it more closely to chemical and physicochemical data. Electron diffraction techniques may help here.

ACKNOWLEDGMENTS

The chapter reflects one aspect of 30 years in the life of the author, and consequently it would be impossible to acknowledge everyone who has played a role in it, but the author could not let the opportunity pass of acknowledging his debt to his one-time mentors Professors W. T. Astbury, J. B. Speakman, and Dr. H. Phillips. More immediately he would like to thank colleagues in Britain, the United States, Germany, South Africa, and New Zealand for help in furnishing reprints and bibliographical lists, and colleagues in Australia for many discussions of various aspects of the topic. He would also like to thank Dr. A. P. Damoglou for his help in converting the tabulated analytical data to a uniform systematic format, and Dr. W. G. Crewther for permitting the use of the data of Fig. 4-2 prior to publication.

## 4.6. References

1. H. Braconnot, *Ann. Chim. Phys.* **13**, 113 (1820).
2. E. Abderhalden and A. Voitinovici, *Hoppe-Seylers Z. Physiol. Chem.* **52**, 348 (1907).
3. J. Barritt and A. T. King, *J. Textile Inst.* **20**, T151 (1929).
4. H. R. Marston, *The Chemical Composition of Wool*, Bulletin 38, Council of Scientific and Industrial Research of Australia (1928).
5. W. T. Astbury, *Fundamentals of Fibre Structure*, p. 135, Oxford University Press, London (1933).
6. R. Brill, *Ann. Chem.* **434**, 204 (1923).
7. K. H. Meyer and H. Mark, *Chem. Ber.* **61**, 1932 (1928).
8. W. T. Astbury and H. J. Woods, *Nature* (*Lond.*) **126**, 913 (1930).
9. W. T. Astbury and H. J. Woods, *Phil. Trans. Roy. Soc.* (*Lond.*) **A232**, 333 (1933).
10. J. B. Speakman and M. C. Hirst, *Trans. Faraday Soc.* **29**, 148 (1933).
11. W. T. Astbury, *Fundamentals of Fibre Structure*, p. 166, Oxford University Press, London (1933).
12. D. M. Wrinch, *Cold Spring Harbor Symp. Quant. Biol.* **6**, 122 (1938).
13. M. Bergmann and C. Niemann, *J. Biol. Chem.* **115**, 77 (1936).
14. A. J. P. Martin and R. L. M. Synge, *Biochem. J.* **35**, 91 (1941).
15. R. Consden, A. H. Gordon, and A. J. P. Martin, *Biochem. J.* **38**, 224 (1944).
16. R. Consden, A. H. Gordon, and A. J. P. Martin, *Biochem. J.* **44**, 548 (1949).
17. R. Consden, A. H. Gordon, and A. J. P. Martin, *Biochem. J.* **42**, 443 (1948).

18. R. Consden, A. H. Gordon, and A. J. P. Martin, *Biochem. J.* **40**, 3 (1946).
19. G. Toennies and R. P. Homiller, *J. Colloid Sci.* **64**, 3054 (1942).
20. F. Sanger, *Nature (Lond.)* **160**, 295 (1947).
21. F. Sanger, *Biochem. J.* **44**, 126 (1949).
22. E. O. P. Thompson and I. J. O'Donnell, *Aust. J. Biol. Sci.* **12**, 282 (1959).
23. M. C. Corfield, A. Robson, and B. Skinner, *Biochem. J.* **68**, 348 (1958).
24. J. M. Gillespie, I. J. O'Donnell, E. O. P. Thompson, and E. F. Woods, *J. Textile Inst.* **51**, T703 (1960).
25. P. Alexander and C. Earland, *Nature (Lond.)* **166**, 396 (1950).
26. P. Alexander, D. Carter, C. Earland, and O. E. Ford, *Biochem. J.* **48**, 636 (1951).
27. I. J. O'Donnell and E. O. P. Thompson, *Aust. J. Biol. Sci.* **14**, 461 (1961).
28. I. J. O'Donnell and E. O. P. Thompson, *Aust. J. Biol. Sci.* **15**, 740 (1962).
29. E. O. P. Thompson and I. J. O'Donnell, *Aust. J. Biol. Sci.* **15**, 552 (1962),
30. M. C. Corfield, *Biochem. J.* **84**, 602 (1962).
31. M. C. Corfield, *Biochem. J.* **86**, 125 (1963).
32. T. Haylett, F. J. Joubert, L. S. Swart, and D. F. Louw, *Textile Res. J.* **33**, 639 (1963).
33. M. Fell, N. H. La France, and K. Ziegler, *J. Textile Inst.* **51**, T797 (1960).
34. M. Fell, *Kolloidzeitschrift* **220**, 107 (1967).
35. M. C. Corfield, J. C. Fletcher, and A. Robson, *Biochem. J.* **102**, 801 (1967).
36. M. C. Corfield and J. C. Fletcher, *Biochem. J.* **115**, 323 (1969).
37. M. C. Corfield, J. C. Fletcher, and A. Robson, in: *Symposium on Fibrous Proteins, Australia, 1967* (W. G. Crewther, ed.), p. 289, Butterworth & Co. (Australia) Ltd., Sydney (1968).
38. D. R. Goddard and L. Michaelis, *J. Biol. Chem.* **106**, 605 (1934).
39. D. R. Goddard and L. Michaelis, *J. Biol. Chem.* **112**, 361 (1935).
40. J. A. MacLaren, D. J. Kilpatrick, and A. Kirkpatrick, *Aust. J. Biol. Sci.* **21**, 805 (1968).
41. A. Kirkpatrick and J. A. MacLaren, *Proc. Aust. Biochem. Soc.* **2**, 77 (1969).
42. A. Schoberl, *Angew. Chem.* **60**, 7 (1948).
43. F. H. White, *Fed. Proc.* **18**, 350 (1959).
44. Discussion E. O. P. Thompson, I. J. O'Donnell, and A. Robson, in: *Symposium on Fibrous Proteins, Australia, 1967* (W. G. Crewther, ed.), p. 310, Butterworth & Co. (Australia) Ltd., Sydney (1968).
45. H. Lindley, J. M. Gillespie, and R. J. Rowlands, *J. Textile Inst.* **61**, 157 (1970).
46. E. O. P. Thompson and I. J. O'Donnell, *Aust. J. Biol. Sci.* **15**, 757 (1962).
47. B. S. Harrap and J. M. Gillespie, *Aust. J. Biol. Sci.* **16**, 542 (1963).
48. J. M. Gillespie, *Aust. J. Biol. Sci.* **15**, 572 (1962).
49. J. M. Gillespie, *Aust. J. Biol. Sci.* **16**, 241 (1963).
50. E. O. P. Thompson and I. J. O'Donnell, *Aust. J. Biol. Sci.* **17**, 277 (1964).
51. B. J. Davis, *Ann. N.Y. Acad. Sci.* **121**, 405 (1964).
52. R. Frater, *Aust. J. Biol. Sci.* **21**, 815 (1968).
53. R. Frater, *J. Chromatog.* **50**, 469 (1970).
54. E. O. P. Thompson and I. J. O'Donnell, *Aust. J. Biol. Sci.* **18**, 1207 (1965).
55. R. Frater, *Aust. J. Biol. Sci.* **19**, 699 (1966).
56. I. J. O'Donnell and E. O. P. Thompson, *Aust. J. Biol. Sci.* **21**, 385 (1968).
57. E. Gross and B. Witkop, *J. Amer. Chem. Soc.* **83**, 1510 (1961).
58. E. O. P. Thompson and I. J. O'Donnell, *Aust. J. Biol. Sci.* **20**, 1001 (1967).
59. I. J. O'Donnell, *Aust. J. Biol. Sci.* **22**, 471 (1969).
60. R. Hosken, B. A. Moss, I. J. O'Donnell, and E. O. P. Thompson, *Aust. J. Biol. Sci.* **21**, 593 (1968).

61. W. G. Crewther and L. M. Dowling, *Appl. Polymer Symp.* **18**, 353 (1971).
62. P. D. Jeffrey, *J. Textile Inst.* **63**, 91 (1972).
63. W. G. Crewther and B. S. Harrap, *Nature (Lond.)* **207**, 295 (1965).
64. D. Hogg, L. M. Dowling, and W. G. Crewther, *Proc. Aust. Biochem. Soc.* **4**, 16 (1971).
65. W. G. Crewther, L. M. Dowling, K. H. Gough, A. S. Inglis, and N. M. McKern, *Proc. Aust. Biochem. Soc.* **6**, 4 (1973).
66. L. M. Dowling and W. G. Crewther, *Proc. Aust. Biochem. Soc.* **5**, 3 (1972).
67. J. M. Gillespie, *Aust. J. Biol. Sci.* **16**, 259 (1963).
68. F. J. Joubert and M. A. C. Burns, *J. South African Chem. Inst.* **20**, 161 (1967).
69. F. J. Joubert, P. J. de Jager, and L. S. Swart, *Symposium on Fibrous Proteins, Australia, 1967* (W. G. Crewther, ed.), p. 343, Butterworth & Co. (Australia) Ltd., Sydney (1968).
70. R. L. Darskus, J. M. Gillespie, and H. Lindley, *J. Aust. Biol. Sci.* **22**, 1197 (1969).
71. R. L. Darskus and J. M. Gillespie, *Proc. Aust. Biochem. Soc.* **2**, 77 (1969).
72. L. S. Swart, T. Haylett, and F. J. Joubert, *Textile Res. J.* **39**, 912 (1969).
73. T. Haylett and L. S. Swart, *Textile Res. J.* **39**, 917 (1969).
74. T. Haylett, L. S. Swart, and D. Parris, *Biochem. J.* **123**, 191 (1971).
75. L. S. Swart and T. Haylett, *Biochem. J.* **123**, 201 (1971).
76. J. M. Gillespie and B. S. Harrap, *Aust. J. Biol. Sci.* **16**, 252 (1963).
77. H. Lindley and T. C. Elleman, *Biochem. J.* **128**, 859 (1972).
78. T. C. Elleman, *Biochem. J.* **128**, 1229 (1972).
79. T. C. Elleman and T. A. Dopheide, *J. Biol. Chem.* **247**, 3900 (1972).
80. T. C. Elleman, *Biochem. J.* **130**, 833 (1972).
81. L. S. Swart and T. Haylett, *Biochem. J.* **133**, 641 (1973).
82. L. S. Swart, *Nature New Biol.* **243**, 27 (1973).
83. T. C. Elleman, H. Lindley, and R. J. Rowlands, *Nature (Lond.)* **246**, 530 (1973).
84. J. M. Gillespie, T. Haylett, and H. Lindley, *Biochem. J.* **110**, 193 (1968).
85. H. Lindley, A. Broad, A. P. Damoglou, R. L. Darskus, T. C. Elleman, J. M. Gillespie, and C. H. Moore, *Appl. Polymer. Symp.* **18**, 21 (1971).
86. P. J. Reis and P. G. Schinckel, *Aust. J. Biol. Sci.* **16**, 218 (1963).
87. P. J. Reis and P. G. Schinckel, *Aust. J. Biol. Sci.* **17**, 532 (1964).
88. J. M. Gillespie, P. J. Reis, and P. G. Schinckel, *Aust. J. Biol. Sci.* **17**, 548 (1964).
89. J. M. Gillespie, A. Broad, and P. J. Reis, *Biochem. J.* **112**, 41 (1969).
90. J. M. Gillespie and A. Broad, *Aust. J. Biol. Sci.* **25**, 139 (1972).
91. D. M. Simmonds and I. G. Stell, *Proc. Intern. Wool Textile Res. Conf., Australia* **C**, 75 (1955).
92. H. Zahn and M. Biela, *Textil-Praxis* **23**, 103 (1968).
93. I. J. O'Donnell and E. O. P. Thompson, *Aust. J. Biol. Sci.* **15**, 740 (1962).
94. H. Zahn and M. Muller, *Z. Ges. Textil Ind.* **65**, 263 (1964).
95. W. G. Crewther, J. M. Gillespie, B. S. Harrap, I. J. O'Donnell, and E. O. P. Thompson, *Proc. 3rd Intern. Wool Textile Res. Conf., Paris* **1**, 475 (1965).
96. R. A. de Deurwaerder, M. G. Dobb, and B. J. Sweetman, *Nature (Lond.)* **203**, 48 (1964).
97. J. H. Bradbury, G. V. Chapman, and N. L. R. King, *Proc. 3rd Intern. Wool Textile Res. Conf., Paris* **1**, 359 (1965).
98. J. M. Gillespie, *Comp. Biochem. Physiol.* **41B**, 723 (1972).
99. J. M. Gillespie and R. L. Darskus, *Aust. J. Biol. Sci.* **24**, 1189 (1971).
100. M. J. Frenkel, J. M. Gillespie, and P. J. Reis, *Aust. J. Biol. Sci.* **27**, 31 (1974).
101. J. M. Gillespie, *J. Chromatog.* **72**, 319 (1972).

102. M. J. Frenkel, J. M. Gillespie, and E. F. Woods, *European J. Biochem.* **34**, 112 (1973).
103. T. A. A. Dopheide, *European J. Biochem.* **34**, 120 (1973).
104. J. E. Haber and D. E. Koshland, Jr., *J. Mol. Biol.* **50**, 617 (1970).
105. S. R. Trotman, E. R. Trotman, and R. W. Sutton, *J. Soc. Chem. Ind.* (*Lond.*) **44**, 1115 (1925).
106. W. B. Geiger, *J. Res. Natl. Bur. Std.* **32**, 127 (1944).
107. J. H. Bradbury, G. V. Chapman, and N. L. R. King, *Symposium on Fibrous Proteins, Australia, 1967* (W. G. Crewther, ed.), p. 368, Butterworth & Co. (Australia) Ltd., Sydney (1968).
108. A. Parisot and J. Derminot, *Appl. Polymer Symp.* **18**, 45 (1971).
109. G. E. Rogers, *Biochim. Biophys. Acta* **29**, 33 (1958).
110. A. K. Allen, H. Lindley, and G. E. Rogers, *Proc. 6th Intern. Congr. Biochem., New York* **5B**, 20 (1964).
111. H. W. J. Harding and G. E. Rogers, *Proc. Aust. Biochem. Soc.* **3**, 56 (1970).
112. P. M. Steinert, H. W. J. Harding, and G. E. Rogers, *Biochim. Biophys. Acta* (*Lond.*) **175**, 1 (1969).
113. W. J. Ellis, *Nature* (*Lond.*) **162**, 957 (1948).
114. G. E. Rogers, *The Epidermis* (W. Montagna, ed.), p. 179, Academic Press, New York (1964).
115. E. O. P. Thompson and I. J. O'Donnell, *Aust. J. Biol. Sci.* **17**, 277 (1964).
116. R. Frater, *Aust. J. Biol. Sci.* **19**, 699 (1966).
117. A. M. Downes, K. A. Ferguson, J. M. Gillespie, and B. S. Harrap, *Aust. J. Biol. Sci.* **19**, 319 (1966).
118. M. L. Ryder, *Proc. Roy. Soc. Edinburgh* **B67**, 65 (1958).
119. A. M. Downes, L. F. Sharry, and G. E. Rogers, *Nature* (*Lond.*) **199**, 1059 (1963).
120. F. J. Joubert, P. J. de Jager, and T. Haylett, *Die Skaap en sy Vag* (J. C. Swart, ed.), p. 235, Nasionale Boekhandel Ltd., Cape Town (1968).
121. W. Moffit and J. T. Yang, *Proc. Natl. Acad. Sci.* **42**, 596 (1956).
122. B. S. Harrap, *Aust. J. Biol. Sci.* **16**, 231 (1963).
123. P. Urnes and P. Doty, *Advan. Protein Chem.* **16**, 401 (1961).
124. R. D. B. Fraser, B. S. Harrap, R. Ledger, T. P. MacRae, F. H. C. Stewart, and E. Suzuki, *Symposium on Fibrous Proteins, Australia, 1967* (W. G. Crewther, ed.), p. 57, Butterworth & Co. (Australia) Ltd., Sydney (1968).
125. W. G. Crewther, M. G. Dobb, L. M. Dowling, and B. S. Harrap, *Symposium on Fibrous Proteins, Australia, 1967* (W. G. Crewther, ed.), p. 329, Butterworth & Co. (Australia) Ltd., Sydney (1968).
126. W. G. Crewther and B. S. Harrap, *J. Biol. Chem.* **242**, 4310 (1967).
127. L. Pauling, R. B. Corey, and H. R. Branson, *Proc. Natl. Acad. Sci.* **37**, 205 (1951).
128. I. MacArthur, *Nature* (*Lond.*), **152**, 38 (1943).
129. M. F. Perutz, *Nature* (*Lond.*) **167**, 1053 (1951).
130. W. Cochran, F. H. C. Crick, and V. Vand, *Acta Crystallog.* **5**, 581 (1952).
131. F. H. C. Crick, *Nature* (*Lond.*) **170**, 882 (1952).
132. F. H. C. Crick, *Acta Crystallog.* **6**, 689 (1953).
133. L. Pauling and R. B. Corey, *Nature* (*Lond.*) **171**, 59 (1953).
134. C. Cohen and K. C. Holmes, *J. Mol. Biol.* **6**, 423 (1963).
135. A. Elliott and B. R. Malcolm, *Proc. Roy. Soc.* (*Lond.*) **A249**, 31 (1959).
136. H. Lindley, *Proc. Intern. Wool Textile Res. Conf., Australia* **B**, 193 (1955).
137. R. D. B. Fraser and T. P. MacRae, *Nature* (*Lond.*) **189**, 572 (1961).
138. R. D. B. Fraser and T. P. MacRae, *J. Mol. Biol.* **3**, 640 (1961).

139. H. P. Lundgren and W. H. Ward, *Arch. Biochem. Biophys., Suppl. 1*, 78 (1962).
140. R. D. B. Fraser, T. P. MacRae, and A. Miller, *J. Mol. Biol.* **14**, 432 (1965).
141. M. Spei, G. Heidemann, and H. Zahn, *Naturwissenschaften* **55**, 346 (1968).
142. R. D. B. Fraser and T. P. MacRae, *Nature (Lond.)* **179**, 732 (1957).
143. R. D. B. Fraser and T. P. MacRae, *J. Mol. Biol.* **3**, 640 (1961).
144. J. Sikorski and H. J. Woods, *J. Textile Inst.* **51**, T506 (1960).
145. W. S. Simpson and H. J. Woods, *Nature (Lond.)* **185**, 157 (1960).
146. G. Heidemann and H. Halboth, *Nature (Lond.)* **213**, 71 (1967).
147. M. Spei, G. Heidemann, and H. Halboth, *Nature (Lond.)* **217**, 247 (1968).
148. R. D. B. Fraser and T. P. MacRae, *Polymer* **14**, 61 (1973).
149. R. D. B. Fraser and T. P. MacRae, *Biochim. Biophys. Acta* **29**, 229 (1958).
150. G. E. Rogers, *Brit. J. Appl. Phys.* **8**, 1 (1957).
151. M. S. C. Birbeck and E. H. Mercer, *J. Biophys. Biochem. Cytol.* **3**, 203 (1957).
152. C. J. Bailey, C. N. Tyson, and H. J. Woods, *Proc. 3rd Intern. Wool Textile Res. Conf., Paris* **1**, 105 (1965).
153. R. D. B. Fraser, T. P. MacRae, and D. A. D. Parry, *Symposium on Fibrous Proteins, Australia, 1967* (W. G. Crewther, ed.), p. 279, Butterworth & Co. (Australia) Ltd., Sydney (1968).
154. R. D. B. Fraser, T. P. MacRae, and G. R. Millward, *J. Textile Inst.* **60**, 343 (1969).
155. M. G. Dobb and J. Sikorski, *J. Textile Inst.* **60**, 497 (1969).
156. E. G. Bendit, *J. Textile Inst.* **51**, T524 (1960).
157. R. D. B. Fraser, T. P. MacRae, D. A. D. Parry, and E. Suzuki, *Polymer* **10**, 810 (1969).
158. L. Pauling and R. B. Corey, *Proc. Natl. Acad. Sci.* **37**, 729 (1951).
159. A. Skertchly and H. J. Woods, *J. Textile Inst.* **51**, T517 (1960).
160. D. A. D. Parry and E. Suzuki, *Biopolymers* **7**, 189 (1969).
161. D. A. D. Parry and E. Suzuki, *Biopolymers* **7**, 199 (1969).
162. E. G. Bendit, *Textile Res. J.* **38**, 15 (1968).
163. R. D. B. Fraser, J. M. Gillespie, and T. P. MacRae, *Comp. Biochem. Physiol.* **44B**, 943 (1973).
164. P. H. Lindenmeyer, *J. Chem. Phys.* **46**, 1902 (1967).
165. R. D. B. Fraser, T. P. MacRae, and A. Miller, *Nature (Lond.)* **203**, 1231 (1964).
166. R. D. B. Fraser, T. P. MacRae, and A. Miller, *J. Mol. Biol.* **14**, 432 (1965).
167. G. Gee, *J. Polymer Sci.* **2**, 451 (1947).
168. H. Lindley and T. Haylett, *J. Mol. Biol.* **30**, 63 (1967).
169. R. W. Burley, *Proc. Intern. Wool Textile Res. Conf., Australia* **D**, 88 (1955).
170. J. M. Gillespie and A. S. Inglis, *Comp. Biochem. Physiol.* **15**, 175 (1965).
171. J. M. Gillespie, *J. Polymer Sci.* **20**, 201 (1967).
172. H. Lindley and T. Haylett, *Biochem. J.* **108**, 701 (1968).
173. H. Lindley and R. W. Cranston, *Biochem. J.* **139**, 515 (1974).
174. M. G. Dobb, R. D. B. Fraser, and T. P. MacRae, *Proc. 3rd Intern. Wool Textile Res. Conf., Paris* **1**, 95 (1965).
175. H. Lindley and H. Phillips, *Biochem. J.* **39**, 17 (1945).
176. H. Lindley, *Textile Res. J.* **27**, 690 (1957).
177. V. Weber and P. Hartter, *Hoppe-Seylers Z. Physiol. Chem.* **355**, 189 (1974).
178. R. F. Goldberger, C. J. Epstein, and C. B. Anfinsen, *J. Biol. Chem.* **238**, 628 (1963).

# 5

# Chemical Reactions of Keratin Fibers

R. S. Asquith and N. H. Leon

## 5.1. Introduction

The chemical and physical properties of proteins can be ascribed to their primary, secondary, and tertiary structure. All these structures are interdependent and can therefore be influenced by chemical reactions modifying the main polypeptide chain and the side chains of the constituent amino acids (i.e., by altering the primary structure). Owing to the presence of a large number of various side chains, keratins possess a wide diversity of chemical reactions.

In general, the majority of protein modifications depend either on the nucleophilicity of the side-chain groups or on their ability to undergo oxidation or reduction. The reactivity of the same functional group in different proteins or even in the same protein is known to vary to a great extent. Depressed or enhanced reactivity of a functional group in the protein can result from (1) the influence of features characteristic of the protein environment, or (2) interaction between the reagent and the protein environment, or both (1) and (2). It is generally recognized that keratin fibers, by their nature and origin, are among the most heterogeneous materials to be subjected to physical and chemical studies. With proteins in solution, many reactions take place rapidly and specifically, but reactions of keratin fibers involve a two-phase system, and hence competing nonspecific reactions may occur.

**R. S. Asquith** • Department of Industrial Chemistry, The Queen's University, Belfast, Northern Ireland **N. H. Leon** • Unilever Research, Isleworth Laboratory, Isleworth, Middlesex, England

Moreover, the amorphous regions of the keratin fiber appear to be more readily accessible by various reagents than the crystalline regions, while the cuticular layers, particularly those in human hair, act as an additional barrier.

Rapid progress in recent years on the modification of the side chains in proteins is now providing new methods for specifically modifying keratin fibers, and considerable effort is being mounted to exploit these methods in research on structure and basic processes of keratin fibers. Much information is being gathered concerning the chemical structure that gives keratin fibers their special characteristics. Possibilities arise for modifying the chemical structure, either temporarily or permanently, so as to change the properties in desirable ways. Sites in the molecules where useful chemical reactions can take place, such as those involved in setting or dyeing, are being identified. Detailed knowledge of the structure and chemistry of keratin is thus providing a sounder basis for explaining the properties of keratin fibers and for devising new chemical modifications of the fibers, as well as for understanding the chemical mechanisms and increasing the efficiency of existing processes.

Most of the reactions to be considered are nonionic and involve the formation of relatively stable covalent bonds. Diverse reactions are grouped together and discussed under common headings, such as hydrolysis, reduction, oxidation, acylation, arylation, alkylation, reaction with electrophilic reagents, and condensation with carbonyl reagents. It is obvious that a sharp distinction between reagents and reactions based on the above classification is not always possible; for instance, reactions with many carbonyl compounds generally result in alkylation of various functional groups in the keratin. Owing to the increasing importance of reactive dyeing in wool industries, a separate heading is allocated for a brief description of reactions with reactive dyes, although modification of keratin with many of the reactive systems used in reactive dyes has been discussed in the appropriate sections.

## 5.2. Hydrolysis

Hydrolysis of the peptide chain involves nucleophilic substitution, in which the —NH— group is replaced by —OH. Under acid conditions hydrolysis involves attack by the water molecule on the protonated amide, while under alkaline conditions it involves attack by the strongly nucleophilic hydroxyl ion on the amide itself. It is generally agreed that protonation of the carbonyl oxygen rather than the amide nitrogen is predominant during acid hydrolysis of amides.[1,2]

$$-\overset{\displaystyle O}{\overset{\|}{C}}-NH- \xrightarrow[\text{acid hydrolysis}]{H^+} -\overset{\displaystyle O-H}{\overset{|}{\underset{+}{C}}}-NH- \xrightarrow[-H^+]{H_2O} -\overset{\displaystyle O-H}{\overset{|}{\underset{\underset{\displaystyle OH}{|}}{C}}}-NH-$$

$$\downarrow OH^- \text{ alkaline hydrolysis}$$

$$-\overset{\displaystyle O^-}{\overset{|}{\underset{\underset{\displaystyle OH}{|}}{C}}}-NH- \xrightarrow{H^+} -\overset{\displaystyle O}{\overset{\|}{C}}-OH + H_2N-$$

## 5.2.1. Acid Hydrolysis

Acid hydrolysis is the most satisfactory general method of converting keratin fibers "quantitatively" into their individual amino acids (see Chapter 1). The hydrolytic process is not a random cleavage of peptide bonds; instead, a degree of specificity is observed.[3,4] Martin and Synge[5] were the first to examine partial acid hydrolysis in a semiqualitative manner, largely because of the development of chromatographic methods for the microdetermination of amino acids and peptides.[5,6] In their studies, wool, edestin, and gelatin were hydrolyzed at 37°C in an excess of 10 *N* HCl for several days. In the course of the hydrolysis, free amino acids were liberated from the outset of hydrolysis, and after about 1 week approximately one third of the total residues were free amino acids. The remaining residues were small peptides, with dipeptides in the greatest proportion. The bonds involving the amino groups of serine and threonine were found to be more labile. Ammonia was liberated rapidly and was almost completely released within 48 h. Thus the estimation of ammonia in acid hydrolysates prepared under conditions that result in only limited hydrolysis of peptide bonds serves as an excellent means for determination of the total amide content of proteins, including wool.[7–9] Similarly, small peptides with serine and threonine as N-terminal groups have been found in significant amounts in the partial hydrolysates of whole wool,[10,11] in $^{35}$S-cystine-labeled wool,[324] in tryptic digests of oxidized and reduced wool proteins,[37] and in keratose fractions.[16,17]

The resistance of dipeptides to acid hydrolysis appears to be the result of inhibitory effects of the terminal protonated $\alpha$-amino group, which tends to repel the hydroxonium ion at the nearest peptide bond.[18,19] Thus, it has also been found that dipeptides represented the overwhelming proportion of the products formed when silk fibroin is hydrolyzed for 42–96 h at 40°C in concentrated HCl.[20] Recently, Asquith and associates have succeeded in a semiquantitative determination of some peptide sequences in wool keratin by controlled hydrolysis of keratose fractions at a stage where dipeptides predominated.[16,17] Partial hydrolysis has also been used in the determination of the amino acid sequence of wool proteins.[21,22,218]

The enhanced lability of peptide bonds formed by the amino groups of serine and threonine residues has been explained by the N → O acyl migration to form the hydrolytically more labile ester bonds. The reaction is indicated by the appearance of a new N-terminal serine or threonine.

$$\begin{array}{c} \text{—CO—NH—CH—CO—NH—} \\ | \\ \text{HO—CH}_2 \end{array} \underset{OH^-}{\overset{H^+}{\rightleftharpoons}} \begin{array}{c} \text{H}_3\overset{+}{\text{N}}\text{—CH—CO—NH—} \\ | \\ \text{—CO—O—CH}_2 \end{array}$$

Apart from a variety of proteins, this rearrangement has been observed in the carrotting of rabbit fur (a method involving treatment of fur with mercuric nitrate or hydrogen peroxide in dilute nitric acid to enhance felting properties)[23] and in the carbonizing of wool (a method involving treatment of wool with sulfuric acid to remove cellulosic impurities).[24–30]

In the carbonizing process the fabric is wetted with dilute (5%) sulfuric acid and subsequently dried and baked (at 120°C) to degrade cellulosic contaminants. This results in a progressive concentration of the acid solution on the fiber. Studies of the reaction of sulfuric acid with wool[31] indicate that some hydrolysis of the fiber may occur during drying, while sulfation of serine and threonine, and N → O acyl shifts, also result, although sulfonation of tyrosine is improbable under normal conditions.[27,32] Relatively stable *O*-sulfate esters of the hydroxyamino acids[41] persist in the fiber after carbonizing and result in dyeing faults[28] that are due to the change in ionic charge on the molecules. Serine-*O*-sulfate has been identified in enzymatic hydrolysates of carbonized wool,[105] indicating that it is probably one of the intermediates involved in the N → O acyl shift. The N → O acyl shift has been employed for the selective cleavage of peptide chains.[33]

The bonds formed by aspartic acid residues were originally shown to be very susceptible to hydrolysis in dilute acids.[34] When wool is treated with boiling dilute acids, aspartic acid is preferentially liberated.[35–38] This specific cleavage is probably due to the fact that in dilute acid solutions of pH 2–3, proton transfer from the un-ionized $\beta$-carboxyl group of aspartic acid residues is more effective than from the hydroxonium ion in the solution, the latter being involved in nonspecific cleavage of all peptide bonds.[38] A number of studies have examined in detail the preferential cleavage of proteins at aspartic acid residues, and a review of this method has been published.[39]

The sensitivity of wool to acid hydrolysis is increased if the cystine is transformed by oxidation to cysteic acid, the peptide bond adjacent to the cysteic acid group being very sensitive.[40] Cysteic acid is liberated from wool oxidized with performic acid at a fairly early stage but at a slower rate than aspartic acid.[157] This is not surprising, since the low pK value of $—SO_3H$

means that at pH between 2 and 4, only a small fraction of the groups would be in the nonionized form.[38] The —$SO_3H$ group of cysteic acid residues is therefore not so effective as the $\beta$-COOH group of aspartic residues in promoting hydrolysis. It has been found that acid dyeing of wool, after an oxidative nonshrink treatment, produces considerable damage. Cys($O_3H$)—Cys($O_3H$) has been shown to be present in partial hydrolysates of oxidized wool,[42] in tryptic digests of oxidized wool proteins,[25] and in keratose fractions.[16,17]

Bonds formed by the carboxyl groups of valine, leucine,[43–46] and isoleucine[47–50] are most stable. Synge[46] attributed the stability to the steric limitation imposed by the isopropyl and isobutyl side chains of valine and leucine on the approach of $H^+$ ions to peptide bonds. Whitfield[51] has applied the well-known "rule of six," which is used to explain steric factors in organic chemistry, to assess the stability of peptide bonds toward hydrolysis. (The rule states: In reactions involving addition to an unsaturated function containing a double bond, the greater the number of atoms in the six position, the greater will be the steric effect.) He has shown that a peptide bond is most stable when atoms occupy position 6 in a dipeptide that is numbered arbitrarily as follows:

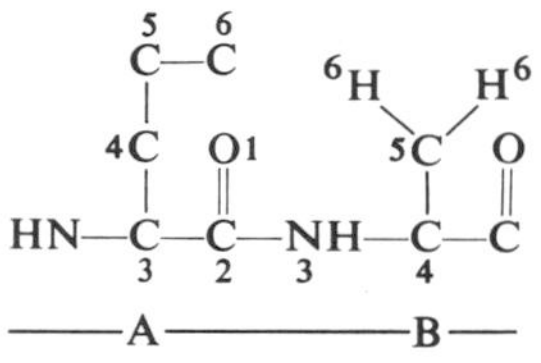

```
      5   6
      C — C      6H     H6
      |            \   /
     4C   O1        5C    O
      |   ||         |    ||
 HN — C — C — NH  —  C  — C
      3   2    3     4
 ———— A ——————————— B ———
```

Degradation of wool by acid hydrolysis of peptide bonds is shown by the loss of tensile strength and may occur to some extent in processes involving the use of acids.[52] The extent of hydrolysis is increased in the presence of high-affinity anions which are attracted to the fiber.[53] It has been suggested that the adsorption of anions will be accompanied by a corresponding adsorption of $H^+$ ions, and hence the $H^+$ ion concentration in the fiber is increased.[54] By boiling wool with aqueous sodium sulfate solutions of different concentrations and pH values, it has been demonstrated that main-chain hydrolysis increases not only as the solution becomes more acidic but also as the sodium sulfate concentration increases.[55]

Although no fission of main polypeptide chains occurs when human hair is exposed to acid solutions of pH less than 2 at room temperature, it has been shown that some structure changes in the fiber have taken place, probably owing to the irreversible dissociation of some disulfide bonds causing permanent weakening of the fiber.[56,57]

### 5.2.2. Alkaline Hydrolysis

Alkali will also hydrolyze keratin fibers but less selectively than with acids; in fact, 0.1 *N* NaOH will rapidly dissolve wool at the boil. The complete destruction of arginine, serine, threonine, cystine, and cysteine preclude the use of this approach for amino acid analysis. On the other hand, tryptophan is not destroyed in alkali, and the analysis of alkaline hydrolysates forms the basis of one method for the quantitative determination of this amino acid in keratin fibers.[58,59]

The extent of the reaction of keratin fibers with alkali depends on the conditions (temperature, concentration, etc.). From the practical point of view, solubility of the fiber in alkali has been used as a parameter for assessing damage that may have occurred during wet processing. Treatment for 1 h with 0.1 *N* NaOH at 65°C has been standardized as suitable.[60] Intact fibers all exhibit fairly low solubilities (4–6% for human hair, 10–12% for wool). The value increases when the fiber has been degraded by hydrolysis of peptide bonds or rupture of disulfide bonds. In addition, the test may be used to determine qualitatively the amount of crosslinking in the fiber, as alkali solubility decreases greatly when the disulfide bond is replaced by a more stable linkage. It is important, however, to realize that alkali itself gives rise to such a linkage (lanthionine) in the fiber, so that alkali solubility cannot be used to assess damage due to peptide-bond hydrolysis occurring in alkaline treatments.

### 5.2.3. Rupture of Disulfide Bonds and Formation of New Amino Acids

From alkali-treated wool, three new amino acids—lanthionine, lysinoalanine, and β-aminoalanine—have been isolated. Although lanthionine was identified from hydrolysates of alkali-treated wool,[61] human hair, and chicken feathers[62] as early as 1941, it was not until 1964 that lysinoalanine was identified in alkali-treated proteins[63,64] and in wool.[65,66] The third new amino acid, β-aminoalanine, was only isolated in 1968 by Asquith and García-Domínguez[67] from alkali-treated wool.

The most probable mechanism for the formation of these amino acids is by alkali-catalyzed bimolecular β elimination of the disulfide group. This is initiated by a proton abstraction from the α-carbon by the attack of a hydroxyl ion, leading to the formation of a dehydroalanine residue and an *S*-thiocysteine residue, which decomposes to give a bound cysteinate ion and sulfur[68,69]:

$$\begin{array}{ccccc} | & & & & | \\ \text{CO} & & & & \text{CO} \\ | & & & & | \\ \text{CH} & -\text{CH}_2- & \text{S}-\text{S} & -\text{CH}_2- & \text{CH} \\ | & & & & | \\ \text{NH} & & & & \text{NH} \\ | & & & & | \end{array} \underset{\text{H}_2\text{O}}{\overset{\text{OH}^-}{\rightleftharpoons}} \begin{array}{ccccc} | & & & & | \\ \text{CO} & & & & \text{CO} \\ | & & & & | \\ {}^-\text{C} & -\text{CH}_2- & \text{S}-\text{S} & -\text{CH}_2- & \text{CH} \\ | & & & & | \\ \text{NH} & & & & \text{NH} \\ | & & & & | \end{array} \rightleftharpoons$$

$$\begin{array}{c} | \\ \text{CO} \\ | \\ \text{C}{=}\text{CH}_2 \\ | \\ \text{NH} \\ | \end{array} + \begin{array}{cccc} & & & | \\ & & & \text{CO} \\ & & & | \\ {}^-\text{S}-\text{S} & -\text{CH}_2- & & \text{CH} \\ & & & | \\ & & & \text{NH} \\ & & & | \end{array} \longrightarrow \begin{array}{c} | \\ \text{CO} \\ | \\ \text{C}{=}\text{CH}_2 \\ | \\ \text{NH} \\ | \end{array} + \begin{array}{ccc} & & | \\ & & \text{CO} \\ & & | \\ {}^-\text{S} & -\text{CH}_2- & \text{CH} \\ & & | \\ & & \text{NH} \\ & & | \end{array} + \text{S}$$

Dehydroalanine
residue

This mechanism is strongly supported by the recent isolation of a pure dehydroalanine containing peptide by alkaline treatment of glutathione.[237] A small amount of dehydroalanine residue may also be provided for by the serine residue.[70,71] The dehydroalanine residue is capable of adding nucleophilic groups across its activated double bond. Thus lanthionine crosslinks could be formed by the inter- or intrachain addition of the bound cysteinate ion,[72,73] lysinoalanine crosslinks by the $\epsilon$-amino groups of lysine residues, and $\beta$-aminoalanine residues by the ammonia derived from the hydrolysis of the amide side chain of asparagine or glutamine[67,357]:

$$\begin{array}{c} | \\ \text{CO} \\ | \\ \text{C}{=}\text{CH}_2 \\ | \\ \text{NH} \\ | \end{array} + \begin{array}{ccc} & & | \\ & & \text{CO} \\ & & | \\ {}^-\text{S} & -\text{CH}_2- & \text{CH} \\ & & | \\ & & \text{NH} \\ & & | \end{array} \longrightarrow \begin{array}{ccccc} | & & & & | \\ \text{CO} & & & & \text{CO} \\ | & & & & | \\ \text{CH} & -\text{CH}_2- & \text{S} & -\text{CH}_2- & \text{CH} \\ | & & & & | \\ \text{NH} & & & & \text{NH} \\ | & & & & | \end{array}$$

Lanthionine crosslink

$$\begin{array}{c} | \\ \text{CO} \\ | \\ \text{C}{=}\text{CH}_2 \\ | \\ \text{NH} \\ | \end{array} + \begin{array}{ccc} & & | \\ & & \text{CO} \\ & & | \\ \text{H}_2\text{N} & -(\text{CH}_2)_4- & \text{CH} \\ & & | \\ & & \text{NH} \\ & & | \end{array} \longrightarrow \begin{array}{ccccc} | & & & & | \\ \text{CO} & & & & \text{CO} \\ | & & & & | \\ \text{CH} & -\text{CH}_2- & \text{NH} & -(\text{CH}_2)_4- & \text{CH} \\ | & & & & | \\ \text{NH} & & & & \text{NH} \\ | & & & & | \end{array}$$

Lysinoalanine crosslink

$$\begin{array}{c} | \\ \text{CO} \\ | \\ \text{C}{=}\text{CH}_2 \\ | \\ \text{NH} \\ | \end{array} + \text{NH}_3 \longrightarrow \begin{array}{c} | \\ \text{CO} \\ | \\ \text{CH}-\text{CH}_2-\text{NH}_2 \\ | \\ \text{NH} \\ | \end{array}$$

$\beta$-Aminoalanine
residue

Yet another two new amino acids, β-aminoalanylalanine and ornithinoalanine, may also be formed. Addition to the dehydroalanine residue of the newly formed β-aminoalanine residue gives β-aminoalanylalanine,[74] while addition to the dehydroalanine residue of the ornithine residue resulting from alkali degradation of arginine residues gives ornithinoalanine[75]:

$$\begin{array}{c} | \\ CO \\ | \\ C{=}CH_2 \\ | \\ NH \\ | \end{array} + H_2N{-}CH_2{-}\begin{array}{c} | \\ CO \\ | \\ CH \\ | \\ NH \\ | \end{array} \longrightarrow \begin{array}{c} | \\ CO \\ | \\ CH \\ | \\ NH \\ | \end{array}{-}CH_2{-}NH{-}CH_2{-}\begin{array}{c} | \\ CO \\ | \\ CH \\ | \\ NH \\ | \end{array}$$

β-Aminoalanylalanine crosslink

$$\begin{array}{c} | \\ CO \\ | \\ C{=}CH_2 \\ | \\ NH \\ | \end{array} + H_2N{-}(CH_2)_3{-}\begin{array}{c} | \\ CO \\ | \\ CH \\ | \\ NH \\ | \end{array} \longrightarrow \begin{array}{c} | \\ CO \\ | \\ CH \\ | \\ NH \\ | \end{array}{-}CH_2{-}NH{-}(CH_2)_3{-}\begin{array}{c} | \\ CO \\ | \\ CH \\ | \\ NH \\ | \end{array}$$

Ornithinoalanine crosslink

These newly formed crosslinks can contribute to the stability of set, especially when they are formed after the necessary disulfide- and hydrogen-bond rearrangements have occurred. It is now generally agreed that the breaking of cystine disulfide bonds and the introduction of new crosslinks is one of the basic methods for attaining permanent set[76,77] and various chemical treatments, particularly those of reduction, oxidation, and crosslinking reactions, have been studied extensively.

The content of some amino acids in wool during treatment at various temperatures for 1 h with sodium hydroxide and sodium carbonate solutions has been investigated. With 0.1 *N* NaOH (pH 12.7), cystine is degraded at ordinary temperature; the change into lanthionine is almost stoichiometric. Lanthionine decomposes at about 65°C and changes, probably into serine. Serine and threonine are destroyed even at low temperatures. Cysteic acid is formed in small quantities at high temperatures. The reaction of lysine as a function of temperature with dehydroalanine residue gives lysinoalanine; the increase in the lysinoalanine content correlates with the decrease in the lysine content. From 60 to 70°C, arginine is changed into ornithine, part of which reacts with dehydroalanine residues to form ornithinoalanine (Fig. 5-1).[78,79] In the case of sodium carbonate solutions, similar trends in the transformation and decomposition of amino acids are observed but take place at higher temperatures, depending on the ionic strength of the solution.[80]

The reactive intermediate dehydroalanine residue, formed during alkali treatment of wool, will also react with added amines or thiols.[356] For

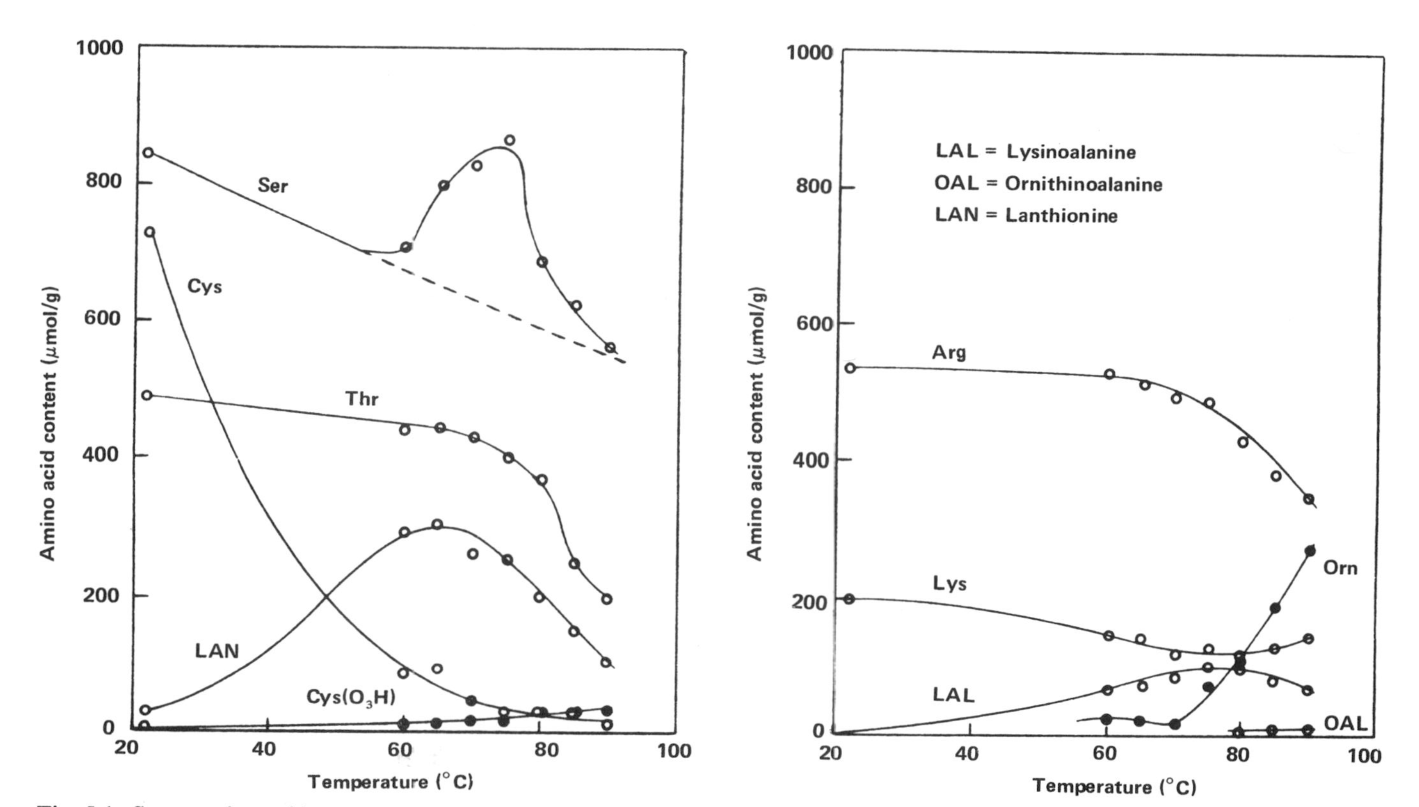


Fig. 5-1. Some amino acid contents of wool after treatment at various temperatures for 1 h with 0.1 *N* sodium hydroxide. Amino acid content (μmol/g) of wool samples before treatment: Arg, 530; Cys, 824; Lys, 227; Ser, 886; Thr, 490; $Cys(O_3H)$, 1.9; lanthionine, 3.5.[78]

example, with alkyl amines, large amounts of new β-*N*-alkyl amino acids can be incorporated into the fiber with only slight solubilization[81,15]:

$$\begin{array}{c} | \\ CO \\ | \\ C{=}CH_2 \\ | \\ NH \\ | \end{array} + H_2N{-}R \longrightarrow \begin{array}{c} | \\ CO \\ | \\ CH{-}CH_2{-}NH{-}R \\ | \\ NH \\ | \end{array} \qquad R = \text{ethyl, propyl, butyl, or pentyl}$$

The introduction of additional basic groups into the fiber by this method has been found useful for increasing affinity of wool toward acid dyes.[82] The combination of aromatic diamines and 4-aminothiophenol with the dehydroalanine residue has been used for covalent dyeing. By diazotizing the introduced aromatic amino group and coupling with β-naphthol, wool dyeings with excellent wet-fastness have been obtained.[83]

Asquith and García-Domínguez[84] have established that lysinoalanine participates in crosslinking in sodium hydroxide–treated wool and demonstrated that at low temperatures crosslinking occurs to a small extent in the fractions of the wool usually isolated as γ-keratose, while at higher temperatures more reactions occur in that fraction of wool usually isolated as α-keratose. The latter reaction is concurrent with the breakdown of the crystalline structure of the fiber.

The formation of cysteic acid in wool treated with sodium hydroxide solution is believed to be the same mechanism.[85] Oxidation of the bound cysteinate ion leads to the cysteic acid residue.

The formation of lanthionine by the action of cyanide and sulfite probably involves the following $S_N2$ type of nucleophilic substitution reactions.[86] A β-elimination reaction to form dehydroalanine residues occurs to only a very small extent, even at a pH as high as 8.6, in the case of treatment with sulfite[87]:

$$\begin{array}{c} | \\ CO \\ | \\ CH{-}CH_2{-}S{-}S{-}CH_2{-}CH \\ | \\ NH \\ | \end{array} \xrightarrow{Nu^-}$$

$$\begin{array}{c} | \\ CO \\ | \\ CH{-}CH_2{-}S{-}Nu \\ | \\ NH \\ | \end{array} + {}^-S{-}CH_2{-}\begin{array}{c} | \\ CO \\ | \\ CH \\ | \\ NH \\ | \end{array} \xrightarrow{-{}^-SNu} \begin{array}{c} | \\ CO \\ | \\ CH{-}CH_2{-}S{-}CH_2{-}CH \\ | \\ NH \\ | \end{array}$$

## 5.3. Reaction with Reducing Agents

A study of the action of reducing agents on keratin fibers has been almost totally confined to the behavior of the disulfide bond not only because of its use for the preparation of soluble derivatives but also because it is the basis of some "setting" methods.

### 5.3.1. Thiols

Reaction of the disulfide bond in keratins with thiols proceeds by an interchange mechanism involving two sequential nucleophilic attacks at sulfur by thiol anions, with mixed disulfides being formed as intermediates:

$$\text{CHCH}_2\text{—S—S—CH}_2\text{CH} + \text{RSH} \rightleftharpoons \text{CHCH}_2\text{—S—S—R} + \text{HS—CH}_2\text{CH}$$

$$\text{CHCH}_2\text{—S—S—R} + \text{RSH} \rightleftharpoons \text{CHCH}_2\text{—SH} + \text{R—S—S—R}$$

The equilibrium constants of the reactions are dependent on the electrode potential of the reducing agent used[88] and on pH. For most thiols, they are near unity for both steps, which means that a large excess of thiol is necessary to approach completion of the reaction. For example, a 400-fold excess of thioglycollic acid at pH 5 gives 85% reduction,[89] and a 10-fold excess of α-toluenethiol in aqueous propanol gives 93% reduction[90] of the disulfide bond in wool fibers.

The diffusion rate of thiol reagents into human hair fibers is pH-dependent and can be related to the electrostatic interactions between the ionizable functional groups of the thiol reagent and the keratin, as well as to the disulfide bond cleavage by thiol anions.[91] Thus thiols containing no ionizable groups other than the sulfhydryl group (e.g., thioglycollamide, 1-thioglycerol, and 2-mercaptoethanol) penetrate the fiber relatively rapidly. Thioglycolhydrazide ($pK_b$ 8.5), which can be protonated, penetrates slowly below the isoelectric point of hair keratin (about pH 4.5). For thioglycollic acid ($pK_a$ 3.5) and thiolactic acid ($pK_a$ 3.7), there is a gradual decrease in penetration rate as the pH increases from 2 to 6. At pH higher than 7, the dissociation of the thiol itself becomes appreciable and rapid, increased disulfide bond cleavage by the thiol anions occurs, which, in turn, promotes fiber swelling. In alkaline conditions, the rate-limiting step in the reduction of keratin disulfide is diffusion control. For example, with ammonium thioglycollate at pH 9.3, wool, which has a small fiber diameter, is reduced much more rapidly than Caucasian or Negro hair (Fig. 5-2).

Deuterium exchange measurements and X-ray diffraction patterns from thioglycollic acid–reduced wool fibers show that reduction does not irreversibly destroy the keratin $\alpha$ helix.[93]

Using $^{35}$S-thioglycollic acid, Human and Springell[94] obtained evidence for the formation of the mixed disulfide, carboxymethyl-3-alanyl disulfide, between wool and thioglycollic acid. The formation of the mixed disulfide in the reaction of thioglycollate with cystine itself and cystine residues in wool has been investigated quantitatively by Asquith and Puri.[95] They have also isolated and identified the mixed disulfide, 2-aminoethyl-3-alanyl disulfide, from wool treated with 2-mercaptoethylamine and indicated that this mixed disulfide may participate in the crosslinking reaction with dehydroalanine residues[96]:

$$\text{>CHCH}_2\text{—S—S—CH}_2\text{CH<} + \text{HSCH}_2\text{CH}_2\text{NH}_2 \rightleftharpoons \text{>CHCH}_2\text{—S—S—CH}_2\text{CH}_2\text{NH}_2 + \text{HS—CH}_2\text{CH<}$$

$$\text{>C=CH}_2 + \text{H}_2\text{NCH}_2\text{CH}_2\text{—S—S—CH}_2\text{CH<} \longrightarrow \text{>CHCH}_2\text{—NHCH}_2\text{CH}_2\text{—S—S—CH}_2\text{CH<}$$

The formation of this new crosslink probably accounts for the fact that, unlike thioglycollic acid, 2-mercaptoethylamine produces setting in wool fibers that is stable to mild after-treatment with soap and anionic detergent solutions.

A novel process of dyeing wool by disulfide interchange has been investigated by Asquith and Puri.[97] This approach consists of reacting wool with a dye containing a disulfide bond in the presence of thioglycollate:

$$\text{>CHCH}_2\text{—S—S—CH}_2\text{CH<} + \text{HSCH}_2\text{COOH} + \text{dye—S—S—dye} \rightleftharpoons \text{>CHCH}_2\text{—S—S—dye} + \text{dye—S—S—CH}_2\text{COOH} + \text{HOOCCH}_2\text{—S—S—CH}_2\text{CH<}$$

A method that has little adverse effect on wool is to reduce the disulfide dye before application to wool. Dyeing human hair with disulfide dyes in the presence of reducing agents was described earlier.[98]

The only thiol used as a reducing agent which is not hampered by the usual equilibrium considerations is 1,4-dithiothreitol (DTT). By virtue of its sterically favorable cyclic disulfide, DTT reduces disulfides nearly quantitatively, according to the following schemes[99]:

$$R{-}S{-}S{-}R + HSCH_2(CHOH)_2CH_2SH \rightleftharpoons R{-}S{-}S{-}CH_2(CHOH)_2CH_2SH + R{-}SH$$

$$\text{(HO, OH-substituted ring-open chain with } S{-}S{-}R \text{ and } SH\text{)} \longrightarrow \text{(cyclic disulfide, HO, OH)} + R{-}SH$$

For the reduction of wool fiber, the reaction can be conveniently followed by spectrophotometric determination of the increase in concentration of the oxidized DTT in the solution surrounding the fiber. The rate-controlling process depends on the pH of the reducing solution and changes from diffusion control to reaction-rate control as the pH is decreased from neutral to 3.5.[100] Cysteine plus cystine in acid hydrolysates of proteins, including wool, can be quantitatively determined as *S*-sulfocysteine by reduction of cystine with DTT followed by oxidative treatment of the reaction solution with sodium tetrathionate.[101]

Reduction with thiols, especially with thioglycollic acid, is the basis of a number of setting treatments for wool and human hair. The mechanism of

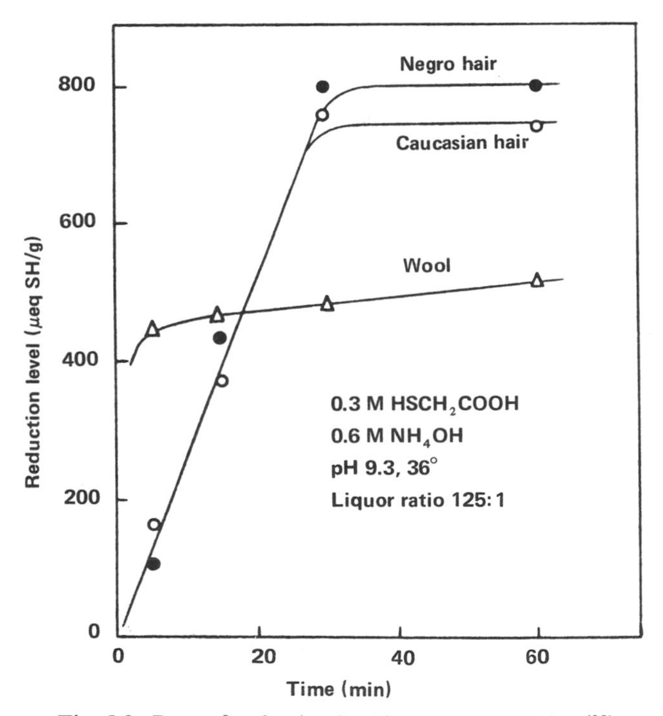


Fig. 5-2. Rate of reduction in thioglycollate solution.[92]

the processes is based on the reduction of the disulfide bond followed by its reformation. In the permanent waving of hair, reformation is usually brought about by a mild oxidative treatment (e.g., with hydrogen peroxide, perborates, or peroxomonosulfate). Alternatively, instead of reoxidizing the sulfhydryl groups, they could be made to react with certain bifunctional compounds, thus forming new covalent bonds between the peptide chains. In durable creasing and pleating of wool fabrics, the chemical spray-on technique has received the widest commercial acceptance. Processes allied to those used in the permanent waving of hair are involved. The Si-Ro-Set process consists of spraying an aqueous solution of a reducing agent (e.g., ammonium thioglycollate, sodium bisulfite, or monoethanolamine sulfite) onto the material, which is then steamed immediately in the press; when employed in garment manufacture, the cloth can also be "sensitized" with reagent and the actual setting delayed until the pressing operation.

### 5.3.2. Phosphines

Reduction of disulfide bonds in wool with tributylphosphine is almost quantitative and specific.[102] The reaction occurs under very mild conditions, little protein is extracted, and coherent fibers are obtained. Extraction of protein fractions from such reduced wool has been studied.[103] The possible mechanism involves an initial nucleophilic attack at a sulfur atom by the tertiary phosphine, followed by a nucleophilic displacement:

$$\mathrm{R{-}S{-}S{-}R} + \mathrm{PR'_3} + \mathrm{H_2O} \longrightarrow \left[\begin{array}{c} \mathrm{R{-}\bar{S}{-}S{-}R} \\ | \\ \mathrm{OH^{-}} \curvearrowright \overset{+}{\mathrm{P}}\mathrm{R'_3} \end{array} + \mathrm{H^+}\right] \longrightarrow$$

$$2\mathrm{RS^-} + 2\mathrm{H^+} + \mathrm{OPR'_3}$$

It can be seen that the reducing system (phosphine → phosphine oxide) has a different chemical basis from that of the keratin system (disulfide → sulfhydryl) and hence the excess reductant will not compete with the sulfhydryl group for the alkylating agent used for subsequent blocking of the cystine residues. Furthermore, partial reduction can be achieved simply by controlling the amount of reagent used.

A new solubility test for wool based on the extraction of the fiber by tributylphosphine in ethanolic sodium iodide solution has been described.[104] Reduction with tributylphosphine (followed by *S*-cyanoethylation or *S*-carboxymethylation) is of particular value in producing a modified wool suitable for enzymic hydrolysis.[105,106] A new method of determining cystine plus cysteine in wool relies on the quantitative reduction of the disulfide bond by tributylphosphine and the selective alkylation of the generated sulfhydryl groups with 2- or 4-vinylpyridine to form, respectively, *S*-β-(2-pyridylethyl)-cysteine or *S*-β-(4-pyridylethyl)cysteine residues, which are stable to acid hydrolysis and can be estimated by ion-exchange chromatography or ultraviolet spectroscopy.[107]

Other tertiary phosphines useful as reducing disulfide bonds in keratin include tetrakis(hydroxymethyl)-phosphonium chloride (THPC).[108] The reducing action of aqueous THPC solution is due to tri(hydroxymethyl)-phosphine (THP), a tertiary phosphine formed by THPC dissociation:

$$\underset{\text{THPC}}{P(CH_2OH)_4Cl} \rightleftharpoons \underset{\text{THP}}{P(CH_2OH)_3} + HCHO + HCl$$

Amino acid analyses of THPC-treated wool show that cystine, tyrosine, lysine, and serine are the only amino acid residues to be significantly affected.[109] The use of THPC in the waving of human hair is said to have the advantage of being odor-free and nondamaging to the fiber.[110]

A valuable characteristic of THP is its selective reactivity toward oxidizing agents to form the corresponding inert phosphine oxide:

$$P(CH_2OH)_3 + \tfrac{1}{2}O_2 \longrightarrow O{=}P(CH_2OH)_3$$

For example, under acid conditions, THP reacts readily with hydrogen peroxide or dissolved oxygen but is inert toward peroxodisulfate. By the use of THPC as an effective oxygen scavenger in peroxodisulfate-initiated vinyl polymerization, methods are now available for the deposition of vinyl polymers in the presence of air in wool[111] and human hair.[112]

## 5.3.3. Sulfites

Interest in the reaction between keratin and sulfite stems from the bleaching of wool by sulfur dioxide or by the application of bisulfite solutions, the use of sulfite solutions as "antichlors" after treatments with halogens to make wool unshrinkable, and the use of sulfite solutions for the setting of wool fibers. If wool is treated with permanganate (either for bleaching or reducing shrinkage), bisulfite is again used to remove the manganese dioxide deposited on the fiber.

Sulfitolysis of wool cystine and residues with sodium sulfite is a reversible one, giving sulfhydryl and *S*-sulfocysteine (Bunte salt) residues[113]:

$$\mathrm{-CHCH_2{-}S{-}S{-}CH_2CH{-}} + NaHSO_3 \rightleftharpoons \mathrm{-CHCH_2{-}SH} + \mathrm{NaO_3S{-}S{-}CH_2CH{-}}$$

Cleavage of the disulfide bond in wool and human hair at equilibrium reaches a maximum at about pH 4.6 in various buffer solutions.[114,115] The apparent equilibrium is displaced to the right by anything that removes the free sulfhydryl group. Thus, in the presence of a small excess of an organomercurial compound, quantitative cleavage of all disulfide bonds can be achieved; these reactions form the basis of convenient analytical methods for combined cystine in proteins.[116–118] In a widely used amperometric method, intact fibers of wool or human hair are treated with methylmercuric iodide in 8 *M* urea in the presence of sulfite, and the excess methylmercuric iodide is determined polarographically[119]:

$$\text{CHCH}_2\text{—SH} + \text{CH}_3\text{HgI} \longrightarrow \text{CHCH}_2\text{—SHgCH}_3 + \text{HI}$$

Owing to its specific reaction with sulfhydryl groups in reduced keratin, methylmercuric iodide is also useful as a selective amino acid label in electron microscope studies.

All the cystine disulfide groups in a protein may be converted into *S*-sulfocysteine residues if the reaction is carried out in the presence of an oxidant such as (a) cupric ions[120] to oxidize the free sulfhydryl groups formed into *S*-sulfocysteine residues:

$$\text{CHCH}_2\text{—S}^- + \text{SO}_3^{2-} + 2\text{Cu}^{2+} \longrightarrow \text{CHCH}_2\text{—S—SO}_3^- + 2\text{Cu}^+$$

(b) 2-iodosylbenzoate[121] to oxidize the free sulfhydryl groups formed back into cystine disulfide:

$$2\text{CHCH}_2\text{—SH} + \text{C}_6\text{H}_4(\text{IO})\text{—COO}^- \longrightarrow \text{CHCH}_2\text{—S—S—CH}_2\text{CH} + \text{C}_6\text{H}_4(\text{I})\text{—COO}^- + \text{H}_2\text{O}$$

(c) tetrathionate[121] to oxidize the free sulfhydryl groups formed, via sulfenylthiosulfate groups, into disulfide and/or *S*-sulfocysteine residues:

$$\text{CHCH}_2\text{—S}^- + {}^-\text{O}_3\text{SS—SSO}_3^- \rightleftharpoons \text{CHCH}_2\text{—S—SSO}_3^- + \text{S}_2\text{O}_3^{2-}$$

$$\text{CHCH}_2\text{—S—SSO}_3^- + \text{CHCH}_2\text{—S}^- \rightleftharpoons \text{CHCH}_2\text{—S—S—CH}_2\text{CH} + \text{S}_2\text{O}_3^{2-}$$

$$\text{CHCH}_2\text{—S—SSO}_3^- + \text{SO}_3^{2-} \rightleftharpoons \text{CHCH}_2\text{—S—SO}_3^- + \text{S}_2\text{O}_3^{2-}$$

and, (d) free *S*-sulfocysteine[122] to oxidize the free sulfhydryl groups formed into mixed disulfides:

$$\overset{|}{\underset{|}{C}}HCH_2{-}S^- + HOOCCH(NH_2)CH_2{-}S{-}SO_3^- \rightleftharpoons$$

$$\overset{|}{\underset{|}{C}}HCH_2{-}S{-}S{-}CH_2CH(NH_2)COOH + SO_3^{2-}$$

Both the cystine disulfide and the mixed disulfides resulting from the above reactions undergo further nucleophilic attack by the excess sulfite, giving an additional *S*-sulfocysteine residue.

Rebuilding of the cystine disulfide bond in Bunte salt human hair by treatment with a thiol solution is rapid and leads to excellent recovery of mechanical properties. This forms a basis of a method for waving and setting of human hair.[123] Alternatively, the sulfite-treated hair may be treated with hexamethylenetetramine, which forms a complex with sulfite and hence displaces the equilibrium, favoring the reformation of the cystine disulfide bond.[124] At pH 6, the rate-determining step in the reduction of keratin fibers with sulfite is diffusion control (Fig. 5-3).

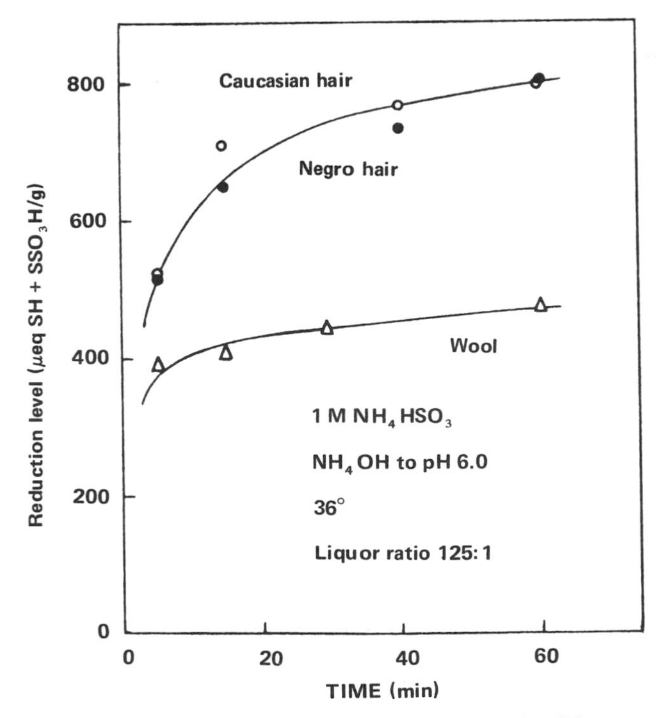


Fig. 5-3. Rate of reduction in sulfite solution.[92]

The Folin–Shinohara colorimetric method for determining cystine in wool hydrolysates[125] consists of reductive fission of the disulfide bond of cystine with sodium sulfite, and colorimetric estimation of the cysteine formed, using phosphotungstic acid. Urea–bisulfite[126] and urea–thioglycollate[127] tests [together with the performic acid–ammonia test (Section 5.4.2)] are useful as qualitative criteria of crosslinking in wool. The bisulfite or thioglycollate component of the reagent breaks acessible reactive crosslinks, while the urea breaks the hydrogen bonds and other secondary interchain forces; the net effect is that part of the wool keratin dissolves or is dispersed in the reagent.

Reaction of wool with sulfite[128,129] with thiols[90] and with alkali[130] is greatly increased if aqueous alcohol solutions rather than aqueous solutions are used. This is probably due to the swelling of the fiber by alcohol solutions. The increase in fiber diameter of mohair in methanol, ethanol, and *n*-propanol (and also dimethylformamide) is around 12%; but in isopropanol, contraction of the fiber occurs.[131]

### 5.3.4. Bunte Salts (*S*-Alkyl Thiosulfates)

Wool and human hair have been modified by reducing some of their disulfide bonds to sulfhydryl groups and then treating with mono-, bi-, or polyfunctional Bunte salts yielding mixed disulfides or new crosslinks:

$$\overset{|}{\underset{|}{C}}HCH_2{-}S^- + RS{-}SO_3^- \rightleftharpoons \overset{|}{\underset{|}{C}}HCH_2{-}S{-}SR + SO_3^{2-}$$

$$2\overset{|}{\underset{|}{C}}HCH_2{-}S^- + {}^-O_3S{-}S(CH_2)_nS{-}SO_3^- \rightleftharpoons$$

$$\overset{|}{\underset{|}{C}}HCH_2{-}S{-}S(CH_2)_nS{-}S{-}CH_2\overset{|}{\underset{|}{C}}H + 2SO_3^{2-}$$

where $n = 2$, 3, or 4. The properties of the modified fibers depend on the organic moiety introduced. The formation of new crosslinks increases the strength of the fiber and reduces the tendency of wool to shrink on washing. Monofunctional compounds containing $C_{12}$ to $C_{18}$ alkyl chains give improved lustre, softness, and feel of reduced hair. In the permanent waving of hair with thioglycollate, the subsequent treatment with a 2% 1,2-ethanebisthiosulfate solution gives similar curl-retention results to the usual treatment with a 3% hydrogen peroxide solution.[132]

The thiosulfate group has been used as the reactive dye-solubilizing group for dyeing of keratin fibers.[133] According to Osterloh,[134] these Bunte salt dyes react initially with the free sulfhydryl groups formed in small quantities by hydrolytic fission of the cysteine disulfide bond:

$$\overset{|}{\underset{|}{C}}HCH_2{-}S^- + dye{-}S{-}SO_3^- \rightleftharpoons \overset{|}{\underset{|}{C}}HCH_2{-}S{-}S{-}dye + SO_3^{2-}$$

$$\Big\updownarrow$$

$$\tfrac{1}{2}\overset{|}{\underset{|}{C}}HCH_2{-}S{-}S{-}CH_2\overset{|}{\underset{|}{C}}H + \tfrac{1}{2}\, dye{-}S{-}S{-}dye$$

The sulfite ion formed then reacts with more cystine disulfide bonds in the fiber to produce further free sulfhydryl groups. It appears probable that part of the unsymmetrical dye–fiber disulfide disproportionates to the symmetrical dye disulfide. Thus good wet-fastness depends not only on the covalent bonding of the dye to the fiber but also on the insolubility of the dye disulfide, which has pigmentlike character. For dyeing human hair, successful results may be obtained if the hair is first treated with a suitable reducing agent to generate sufficient sulfhydryl groups.(135)

### 5.3.5. Metal Hydrides

With sodium borohydride, near-quantitative reduction of the disulfide bond in wool can be achieved.(136) However, peptide bonds are also partly attacked. The use of sodium or potassium borohydrides for permanent waving of human hair has been described.(137) To improve its stability on storage, the borohydride may either be mixed with an aqueous sulfite solution or be dispersed in a paste of inert but water-soluble solvents, such as polyethyleneglycols or sulfates of higher aliphatic alcohols.

Powerful reducing agents such as lithium aluminium hydride and related hydrides reduce carboxyl groups (if methylated) in proteins to primary alcohol groups. Thus C-terminal amino acid is transformed into an amino alcohol, and the glutamic and aspartic acid residues in the chain are transformed into hydroxyamino acids. These reaction products can be isolated and identified after hydrolysis. This procedure constitutes a method for C-terminal group determination in wool.(138) However, this type of reduction of carboxyl groups is limited by excessive reduction of peptide bonds.(139) Recently, use has been made of the fact that diborane ($B_2H_6$) shows a significantly greater preference for the reduction of free carboxyl groups than for amides or even for esters,(140) and thus reduction with diborane may find useful application in keratin chemistry.

## 5.4. Reaction with Oxidizing Agents

A number of protein groups are susceptible to oxidative attack (—SH, —S—S—, thioether, imidazole, phenol, indole), and a number of reagents cause their oxidation under varying conditions. Since oxidation can occur by a variety of mechanisms, it is not unreasonable to expect that significant difference in reactivity of functional groups will be observed and that the orders of reactivity will differ.

Oxidation is an important reaction in the case of wool and hair keratins. Thus hydrogen peroxide is used as a bleaching agent and chlorine, bromine, potassium permanganate, and peroxomonosulfuric acid have all been used to confer shrink resistance on wool, while performic and peracetic acids are used for the preparation of soluble derivatives.

### 5.4.1. Hydrogen Peroxide

Hydrogen peroxide is a potent oxidant of many organic compounds. In proteins, it attacks principally the sulfur-containing amino acids cysteine and methionine. In the presence of certain metal ions or organic acids, it may also attack cystine, tryptophan, and tyrosine residues. The rates of oxidation of methionine and cystine are affected differently by pH. Oxidation of cysteine decreases in rate at low pH, whereas oxidation of methionine increases slightly in going from pH 5 to 1.[141] Cystine is relatively less susceptible to oxidation than either cysteine or methionine.

For bleaching of wool, hydrogen peroxide has, nowadays, very largely superseded sulfur dioxide and sulfites. Improvements in the properties of wool obtained by bleaching with hydrogen peroxide in acid media have been reported.[142] In alkaline media, a higher degree of oxidation of the disulfide bond occurs, although the bleaching effect is better.[143]

The reaction of wool keratin with hydrogen peroxide is relatively slow except under alkaline conditions. When wool is immersed in hydrogen peroxide solutions (pH 2.5–9), some of the reagent is initially adsorbed on the amino and imino groups.[144] The adsorbed peroxide seems to be remarkably stable and may be removed by washing. Oxidation of all the cystine can occur.[145] Simultaneous attack on the peptide bond may also occur, as it was shown that both silk (which contains no cystine) and wool were completely dissolved by 3% hydrogen peroxide solution in 3 days at 60°C; no high-molecular-weight peptides could be isolated.[146] Hydrogen peroxide–treated wool shows a decrease in acid combining capacity but not in amino nitrogen content.[147] This may be accounted for by the formation of oxides of cystine, which are relatively strongly acidic. Cysteic acid is also found in the acid hydrolysate.[148] The effects of peroxide oxidation on 19 of the amino acid residues in wool have been described.[149]

Bleaching of human hair with hydrogen peroxide has been the subject of several recent investigations. Zahn[150] demonstrated that the reaction occurs primarily at cystine residues. Bleached hair shows a marked reduction in acid dye uptake, which may be attributed to the formation of cysteic acid residues. Bleached hair contains less cystine, tyrosine, and methionine residues and cysteic acid is found in the hydrolysate.[151] In practice, about 10–30% of the disulfide bonds are broken during bleaching (3–9% hydrogen peroxide, pH 9–11, 15–60 min). Cysteic acid is the only established end product of the oxidative cleavage of the disulfide bond; neither the monoxide nor the dioxide of cystine is present as a significant end product, probably because in alkaline media hydrolysis of the cystine oxide intermediates is highly competitive with oxidation.[152] Loss of cystine residues is reflected in changes in other properties of the hair, such as alkali solubility, which increases linearly with the cysteic acid content,[153] and wet tensile strength, which decreases with the disulfide content.[152]

The color of hair is due mainly to the inclusion of discrete, darkly colored melanin granules in the keratinized cytoplasmic protein of the fiber-forming cells. During bleaching with alkaline hydrogen peroxide, an initial solubilization of the melanin occurs followed by decolorization.[146]

Acting as a cation-exchange resin, melanin present in the fiber readily binds metals. Thus heavy-metal ions generally catalyze the reaction between keratin and hydrogen peroxide; whether the reaction is identical with that in the absence of these ions is uncertain. Bleaching of keratin fibers involving pretreatment with aqueous ferrous sulfate solutions has been shown to yield the most satisfactory results (i.e., maximum color removal with minimum fiber damage.[154] The catalytic effect of some heavy-metal ions may be due to the promotion in the breakdown of hydrogen peroxide to free radicals. Ferrous ions, for example, bring about the formation of HO· radicals and have been used to mimic the effects of $\gamma$ irradiation on proteins.[155]

Hydrogen peroxide in acid media is the most frequently used oxidizing agent for "neutralizing" the thiol compound and reforming hair disulfide bonds during permanent waving of human hair. Reoxidation of the ruptured disulfide bonds takes place to the extent of only about 85%, and some cysteic acid is found in the hydrolysate of the treated hair.[156] While the cysteic acid can be derived directly from the cysteic acid residue in the fiber, which is the end product of the oxidation of the fiber sulfhydryl and/or disulfide groups, part of it may originate from the hydrolysis of some of the intermediate oxidation products of cystine and cysteine residues (see Section 5.4.8).

### 5.4.2. Organic Peroxyacids*

Organic peroxyacids are the most specific reagents for the oxidation of disulfide bonds in proteins, performic and peracetic acids being the preferred reagents. The advantages of these peroxyacids are their good solubility, ease of removal of excess reagent, their conversion of the cystine disulfides into cysteic acid groups, which may facilitate electrophoric separation of the peptides, and irreversibility.

$$\overset{|}{\underset{|}{C}}HCH_2{-}S{-}S{-}CH_2\overset{|}{\underset{|}{C}}H + 5RCO_3H + H_2O \longrightarrow 2\overset{|}{\underset{|}{C}}HCH_2{-}SO_3H + 5RCO_2H$$

In wool, performic acid[157] and peracetic acid[158] have been shown to react only with tryptophan, methionine, and cystine, cystine being oxidized quantitatively to cysteic acid when excess reagent is used without attack on peptide bonds. Thus the combined cystine may be estimated as cysteic acid, after performic acid oxidation and acid hydrolysis, either by ion-exchange chromatography or by electrophoresis.[159] Higher peroxyacids also react with the disulfide bond in wool.[160] The cysteic acid group in oxidized wool is probably present as the ionized form, forming strong salt links $—SO_3^- \cdots {}^+H_3N—$.

Oxidation of the disulfide bonds of keratins with performic acid or peracetic acid and extraction with ammonia solution leaves an insoluble residue β-keratose. The low-sulfur fraction, α-keratose, is precipitated by acidification of the filtered extract and the high-sulfur fraction, γ-keratose, remains in solution. For the preparation of keratoses, performic acid is often preferred because the conversion of cystine into cysteic acid is complete with little or no peptide-bond hydrolysis.

One of the most popular tests for detecting the introduction of crosslinks qualitatively involves determination of the solubility of the treated wool after oxidation in performic acid–ammonia solution.[84] The presence of crosslinks (other than disulfide) will lower the solubility of wool relative to untreated wool.

Mild oxidation with aqueous peracetic acid followed by treatment with sulfite or bisulfite can confer good shrink resistance on wool with little fiber damage.[172] In this two-stage process, the initial oxidation is thought to be a disulfide-activating step, in which dilute peracetic acid converts some disulfide bond into its monoxide and dioxide, thus facilitating fission by nucleophilic attack by sulfite ions.[161]

* In the IUPAC Nomenclature of Organic Chemistry, Rules C, 1969, acids containing the group —C(=O)—OOH are called peroxy acids. The "peroxy" prefix is used except for the well-established names of performic, peracetic, and perbenzoic acids, which are retained.

### 5.4.3. Inorganic Peroxoacids*

#### *5.4.3.1. Peroxodisulfate*

Degradation and decoloration of wool by peroxodisulfate in aqueous solution at 90°C are greater at pH 1.6 than at pH 4.6. There is a significant attack on cystine, methionine, arginine, histidine, proline, tyrosine, and probably phenylalanine. Carbon dioxide and ammonia are liberated during the reaction. The treatment results in a decrease in total nitrogen but an increase in both amide nitrogen and Van Slyke nitrogen of the fiber. Needles[162] has suggested that the main damage is due to the attack of the peptide chain by hydroxyl or sulfate radicals resulting from the decomposition of peroxodisulfate:

$$S_2O_8^{2-} \longrightarrow 2SO_4^{-}\cdot \xrightarrow{2H_2O} 2HSO_4^{-} + 2OH\cdot$$

$$\underset{\substack{|\\ R}}{-CH}CONH\underset{\substack{|\\ R'}}{CH}CONH- \xrightarrow{OH\cdot \text{ or } SO_4^{-}\cdot} \underset{\substack{|\\ R}}{-CH}CONH\underset{\substack{|\\ R'}}{\dot{C}}CONH- \xrightarrow[-SO_4^{-}\cdot]{S_2O_8^{2-}}$$

$$\underset{\substack{|\\ R}}{-CH}CONH\overset{\substack{O-SO_3^{-}\\ |}}{\underset{\substack{|\\ R'}}{C}}CONH- \xrightarrow{H_2O} \underset{\substack{|\\ R}}{-CH}CONH_2 + R'COCONH- + HSO_4^{-}$$

The first step is the abstraction of an α-hydrogen from the amino acid residues to give a peptide radical, which reacts with a peroxodisulfate ion to give an unstable sulfate peptide, and hence cleavage of the chain takes place between α-carbon and nitrogen to give an amide peptide and an α-oxoacyl peptide that may further decompose liberating carbon dioxide and ammonia. It has been shown by electron spin resonance spectroscopy that the free radicals produced are on cystine, tyrosine, and tryptophan residues.[163] A similar mechanism has been proposed for the formation of glyoxylic and pyruvic acids during heating of wool in air at 110°C.[164]

Kantouch and Bendak[165] demonstrated that the content of sulfur-containing amino acids cystine and methionine, aromatic amino acids tyrosine and tryptophan, and basic amino acids lysine, histidine, and arginine, as well as the acid-combining capacity of wool, decreases on oxidation. The

* In the IUPAC Nomenclature of Inorganic Chemistry, 1970 Rules, the prefix "peroxo," when used in conjunction with the trivial names of acids, indicates substitution of —O— by —O—O—. For example, the compound $K_2S_2O_8$ should be called potassium peroxodisulfate rather than potassium persulfate. It follows that permanganic acid and periodic acid are not peroxoacids; the prefix "per" denotes a fully oxidized state for the central atom.

alkali and the urea–bisulfite solubility of the oxidized wool is higher than that of the untreated wool.

Breuer and Jenkins[166] studied the kinetics of oxidation of human hair with peroxodisulfate and showed that the reaction is highly complex and that the rate of the reaction is proportional to the square root of the peroxodisulfate ion concentration. Disulfate bonds are involved but other groups are also reactive.

Free cystine is oxidized by peroxodisulfate to cysteic acid. The initial step in these reactions probably involves the formation, by a four-center mechanism,[167] of a complex oxosulfur anion $R{-}S{-}O{-}SO_3^-$ [$R = {-}CH_2CH(NH_2)CO_2H$], which is hydrolyzed to cysteinesulfenic acid, followed by immediate deprotonation to give an $R{-}SO^-$ ion:

$$\underset{\substack{\vdots \quad \vdots \\ {}^-O_3S{-}O{-}O{-}SO_3^-}}{R{-}S{-}S{-}R} \longrightarrow 2R{-}S{-}O{-}SO_3^- \xrightarrow{OH^-} 2R{-}S{-}OH + 2SO_4^{2-}$$

$$2R{-}S{-}OH \longrightarrow 2R{-}SO^- + 2H^+$$

Being a strong nucleophile, this ion can attack cystine or $R{-}S{-}O{-}SO_3^-$ ion to give a cystine monoxide:

$$R{-}SO^- + R{-}S{-}S{-}R \longrightarrow R{-}SO{-}S{-}R + R{-}S^-$$

or

$$R{-}SO^- + R{-}S{-}O{-}SO_3^- \longrightarrow R{-}SO{-}S{-}R + SO_4^{2-}$$

The $R{-}S^-$ ion, resulting from the first reaction, is reoxidized to cystine:

$$R{-}S^- + R{-}S{-}O{-}SO_3^- \longrightarrow R{-}S{-}S{-}R + SO_4^{2-}$$

The cystine monoxide is consumed by two processes. The first involves an attack by $OH^-$,

$$R{-}SO{-}S{-}R + OH^- \longrightarrow R{-}SO_2H + R{-}S^-$$

while the second involves a four-center reaction with peroxodisulfate to form an $R{-}SO{-}O{-}SO_3^-$ ion, which also gives cysteinesulfinic acid by the attack of $OH^-$:

$$\underset{\substack{\vdots \quad \vdots \\ {}^-O_3S{-}O{-}O{-}SO_3^-}}{R{-}SO{-}S{-}R} \longrightarrow R{-}SO{-}O{-}SO_3^- + R{-}S{-}O{-}SO_3^-$$

$$R{-}SO{-}O{-}SO_3^- \xrightarrow{OH^-} R{-}SO_2H + SO_4^{2-}$$

The cysteinesulfinic acid formed is then oxidized to cysteic acid by another four-center process:

$$\underset{\substack{\vdots \quad \vdots \\ {}^-O_3S{-}O{-}O{-}SO_3^-}}{R{-}SO{-}O{-}H} \longrightarrow R{-}SO{-}O{-}O{-}SO_3^- + HSO_4^-$$

$$R{-}SO{-}O{-}O{-}SO_3^- \xrightarrow{OH^-} R{-}SO_3H + SO_4^{2-}$$

### 5.4.3.2. *Peroxomonosulfuric Acid* (*Caro's Acid*)

Like the more reactive performic and peracetic acids, peroxomonosulfuric acid oxidizes only cystine, tryptophan, and possibly methionine residues in wool under the conditions normally used to impart shrink resistance, but more vigorous oxidation can produce some destruction of tyrosine residues. A large excess of oxidizing agent can ultimately convert all the disulfide bonds into cysteic acid groups, but when lesser amounts of oxidizing agents are used, products intermediate in oxidation state between disulfide and cysteic acid residues are formed.[168]

Treatment of wool with peroxomonosulfuric acid under normal practical conditions (0.1% solution, pH 1.5, 22°C, 2 h, liquor ratio 35:1) produces a substantial increase in the dyeing rate of both acid and basic dyes but a decrease in acid-combining capacity.[169] This is probably because disulfide bond cleavage promotes fiber swelling and loss of cell membrane material, which outweigh the opposite effect of the introduced cysteic acid groups.

Shrinkproofing with peroxomonosulfuric acid or its alkaline metal salts is followed, in the industrial treatment, with a sulfite reduction.[170] It has also been shown that, in the presence of high concentrations of sodium sulfate, sufficient shrink resistance can be attained from a single treatment with neutral peroxomonosulfuric acid solutions.[171]

### 5.4.3.3. *Perborate*

Perborates are employed in the wool technology in bleaching and washing operations; they are also used as mild oxidizing agents for neutralizing the thiol compound during permanent waving of human hair and reforming hair disulfide bonds. Sodium perborate is a white crystalline powder available in the form of mono- or tetrahydrate. The latter ($NaBO_3 \cdot 4H_2O$) is more commonly used, although its formula should be more correctly written as $NaBO_2 \cdot H_2O_2 \cdot 3H_2O$, because of the rapid hydrolysis of the peroxoborate ion in solution. Sodium perborate, therefore, behaves rather like a hydrogen peroxide addition compound; it dissociates in water, in which it has moderate solubility, and the solution reacts in a manner similar to a solution of hydrogen peroxide, boric acid, and sodium hydroxide.

### 5.4.3.4. *Other Inorganic Peroxoacids*

Peroxomolybdate ($MoO_8^{2-}$) and peroxotungstate ($WO_8^{2-}$) in aqueous phosphoric acid react with wool in a manner similar to peracetic acid. After oxidation, up to 90% of the wool becomes soluble in dilute ammonia solution, leaving a residue of highly swollen gelatinous membranes, and thus they may be used for solubilizing keratins.

Peroxochromic and peroxovanadic acids are not so reactive as the organic peroxyacids in shrinkproofing of wool and appear to behave like a solution of hydrogen peroxide.[158]

### 5.4.4. Permanganate

The extent and course of oxidation by permanganate depends on whether acid or alkaline solution is used. Alkaline permanganate solutions react with tyrosine residues but oxidize very little of the cystine until severe degradation of the fiber has occurred.[173] On the other hand, limited attack on disulfide bonds can occur when wool is treated in acid solutions of permanganate (pH 2) at room temperature. For example, 30% and 18.6% of the total cystine are oxidized when reagent concentrations of 25% and 6.3% (on weight of fiber), respectively, are used. At a concentration as low as 2%, however, very little disulfide bond fission occurs; the treated fiber exhibits a similar rate of absorption of basic dyes to the untreated fiber.[169] The products of oxidation of cystine in wool by acid permanganate solutions were shown to be cysteic acid and sulfate in equal proportions.[160]

Potassium permanganate has been applied as an oxidizing agent to render wool unshrinkable and nonfelting. Treatment with neutral potassium permanganate in concentrated sodium chloride solutions produces shrink-proofed wool at relatively low cost with no adverse effects of practical significance.[174] Inorganic salts added in relatively large amounts increase the degree of initial shrink resistance, probably by preventing swelling and confining the reaction more to the cuticle layers, which results in greater surface modification.

### 5.4.5. Periodic Acid

Kantouch and Bendak[175] observed that the sulfur content of periodate-treated wool decreases gradually with oxidation. The treated sample exhibits infrared absorption bands in the region 1600–1700 $cm^{-1}$, corresponding to methyleneimine groups, shows a considerable decrease in acid-combining capacity but an increase in both amide nitrogen content and alkali-combining capacity. It has been suggested that periodate attacks mostly the carbon $\alpha$ to the peptide linkage. This attack leads to cleavage of the bonds between this carbon and both the —NH— and —CO— groups, with the formation of new carbamic acid derivatives, amide groups, and aldehydes:

$$-\mathrm{NH}-\underset{}{\overset{\mathrm{R'}}{\overset{|}{\mathrm{CH}}}}-\mathrm{CO}-\mathrm{NH}-\overset{\mathrm{R}}{\overset{|}{\mathrm{CH}}}-\mathrm{CO}-\mathrm{NH}-\overset{\mathrm{R''}}{\overset{|}{\mathrm{CH}}}-\xrightarrow{\mathrm{IO_4^-}}$$

$$\left[-\mathrm{NH}-\overset{\mathrm{R'}}{\overset{|}{\mathrm{CH}}}-\mathrm{CO}-\mathrm{NH}\cdot\right]+\left[\cdot\mathrm{CO}-\mathrm{NH}-\overset{\mathrm{R''}}{\overset{|}{\mathrm{CH}}}-\right]+\mathrm{RCHO}\xrightarrow{\mathrm{H_2O}}$$

$$-\mathrm{NH}-\overset{\mathrm{R'}}{\overset{|}{\mathrm{CH}}}-\mathrm{CO}-\mathrm{NH_2}+\mathrm{HOOC}-\mathrm{NH}-\overset{\mathrm{R''}}{\overset{|}{\mathrm{CH}}}-$$

The carbamic acid derivative is probably decarboxylated, giving rise to carbon dioxide and an amino end group. The newly formed aldehydes possibly condense with the free amino and/or amide groups in the fiber to give methyleneimine derivatives (—C=N—). Reactivity of the $\alpha$-carbon is influenced by the adjacent side chain. This is indicated by a high degree of degradation of the two hydroxyamino acids, serine and threonine, as well as tyrosine and tryptophan.

Periodate produces considerably more disulfide bond fission than does peroxomonosulfuric acid under similar conditions. The introduction of cysteic acid residues probably accounts for the fact that periodate-treated wool shows a marked reduction in dyeing rate of anionic dyes. The effectiveness of periodic acid as a setting agent has also been attributed to the formation of cysteic acid groups, which can form strong electrovalent links of the type $-\mathrm{SO_3^-}\cdots{}^+\mathrm{NH_3}-$ with the free amino groups in the fiber.

### 5.4.6. Chromium Compounds

Chromium compounds, especially chromates and dichromates, have long been used for dyeing wool. When wool is dyed by the chrome mordant method, the chromium is generally introduced into the wool by boiling in a solution containing Cr(VI), which is reduced by the wool to Cr(III). According to Hartley,[176–178] the first reaction is the ionic association of chromate or dichromate ions with the positively charged groups of lysine, arginine, and histidine residues; this occurs at room temperature. Above 60°C there is progressive conversion of Cr(VI) to Cr(III); this change takes place rapidly at 100°C. The Cr(VI) is reduced at the expense of cystine, tyrosine, and lysine residues in the wool, depending on the pH conditions. The most important of these residues is cystine, which at low pH values is oxidized directly, whereas at high pH values it is the products of the alkaline hydrolysis of cystine that are oxidized rather than cystine itself. Tyrosine residues are oxidized both under acid and under alkaline conditions; lysine residues are oxidized only under alkaline conditions.

The reduction of Cr(VI) to Cr(III) occurs in four stages: (1) Cr(VI) is reduced to Cr(IV) with simultaneous oxidation of the cystine residues; (2)

Cr(IV) is reduced to Cr(II) with simultaneous oxidation of the cystine and tyrosine residues; (3) Cr(II) then forms a complex with the carboxyl group in the wool; (4) the Cr(II)–carboxyl complex is spontaneously oxidized by air to give the observed Cr(III)–carboxyl complexes.

$$\text{Cr(VI)} \xrightarrow[\text{cystine}]{\text{oxidation of}} \text{Cr(IV)} \xrightarrow[\text{and tyrosine}]{\text{oxidation of cystine}} \text{Cr(II)} \xrightarrow{\text{wool—COOH}}$$

$$\text{wool—COO Cr(II)} \xrightarrow{\text{air oxidation}} \text{wool—COO Cr(III)}$$

When the mordanted wool is treated with a dye (e.g., an *o,o'*-dihydroxyazo dye), the Cr(III) in the fiber reacts with the dye to form a stable complex. There is no direct chemical bond between the chromium and the wool in the final mordanted dyeing, the high wash-fastness of which is due primarily to the size of the chromium–dye complex, which is too large to diffuse out of the fiber. Cr(III) salts (e.g., chromic chloride) cannot be used instead of Cr(VI) salts for introducing chromium into the wool. This is because the direct bonding of Cr(III) to the carboxyl groups in wool to form the Cr(III)–carboxyl complex occurs extremely slowly.

### 5.4.7. Halogens and Derivatives

The reaction of the halogens with wool has been studied extensively. Fluorine, the most active of the halogens, reacts rapidly with wool and is known to render wool unshrinkable.[179] Limited amounts of cysteic acid are formed and there is decomposition of disulfide bonds, as evidenced by the evolution of gaseous sulfur fluoride, $SF_6$.

Although chlorine is capable of oxidizing all the amino acids of wool, it reacts preferentially with tyrosine and cystine.[180] The oxidizing properties of hypochlorite solutions are very pH-dependent, according to the formation of hypochlorite anions, free hypochlorous acid, or free chlorine:

$$H_2O + Cl_2 \xleftarrow{\text{HCl}} \text{HOCl} \underset{\text{HCl}}{\overset{\text{NaOH}}{\rightleftharpoons}} \text{NaOCl} \longrightarrow Na^+ + OCl^-$$

The free chlorine and hypochlorous acid can oxidize all the cystine in wool to cysteic acid, but the hypochlorite anions oxidize only about 25% of the cystine.[173] Valk[181,182] has confirmed the assumption that the reaction mechanism of aqueous chlorine with wool is a function of pH. Between pH 4 and 7, where the active chlorine is mainly present as hypochlorous acid, solubility tests and acid- and alkali-combining capacities indicate most extensive modification of the fiber in this region. Tyrosine and tryptophan residues are degraded to a higher extent in this pH region. Both cystine oxide groups and cysteic acid groups are present in chlorinated wool. There is evidence[183] that the oxidation of wool with chlorine at pH 2 may take place through the formation of chloramines from free amino groups in the fiber. This reaction is followed by an oxidation taking place mainly at the reactive centers of wool, such as cystine and tyrosine. Chlorination is not only the oldest, but it is probably still the most widely used commercial method for rendering wool resistant to felting shrinkage.

Reaction with bromine has not been studied in detail, but the course of reaction would seem to be similar to that of chlorine.[184] Bromic acid ($HBrO_3$) oxidizes mainly cystine and tyrosine residues in wool.[185]

Iodine, being the least reactive halogen, does not oxidize the disulfide bond in wool, although reaction occurs presumably with tyrosine and histidine residues (Section 5.8.1). Treatment of wool with iodic acid ($HIO_3$) probably leads to the formation of stable intermediate oxidation products of disulfides without bond fission. If treatment is sufficiently prolonged, the iodic acid solution itself decomposes to yield iodine and iodination of tyrosine and histidine residues then takes place. The chemical changes brought about by iodic acid and iodine appear to be too small to produce any marked change in the dyeing behavior of the treated wool.[169]

The wide acceptance of alkali-metal bromates and iodates as cold-waving neutralizers for reforming disulfide bonds is due to the fact that these salts can be packed conveniently in dry form. Despite their desirable stability and use characteristics, they are being replaced by the more economical and less toxic sodium perborate.

### 5.4.8. Partial Oxidation

That partial oxidation of wool converts its cystine disulfide bonds to oxidation states intermediate between disulfide and cysteic acid residues was first postulated by Harris and Smith,[186] and subsequent work, using improved analytical methods, has confirmed and amplified this postulate. Some possible oxidation products of cystine disulfide are as follows:

| Cystine oxides (S—S bond intact) | | Sulfur acids (S—S bond fission) | |
|---|---|---|---|
| R—SO—S—R | Monoxide | R—SOH | Cysteinesulfenic acid |
| R—$SO_2$—S—R | Dioxide | R—$SO_2$H | Cysteinesulfinic acid |
| R—$SO_2$—SO—R | Trioxide | R—$SO_3$H | Cysteic acid |
| R—$SO_2$—$SO_2$—R | Tetroxide | | |

Although the analytical methods do not enable each of the intermediates to be estimated quantitatively, there is evidence (e.g., infrared) that cystine monoxide, cystine dioxide, and cysteinesulfinic acid derivatives can be formed in considerable amounts when keratin fibers are oxidized under certain conditions. Eager[187] showed that careful oxidation of wool with performic or peracetic acid followed by hydrolysis gives cysteinesulfinic acid in good yield. The cysteinsulfinic acid obtained by this method is probably derived from cystine monoxide or dioxide residues, since treatment of wool with dilute aqueous peracetic acid results in a large proportion of cystine monoxide and possibly dioxide residues.[168] These oxide residues give some chemical reactions similar to those of thiols. For example, they react with methylmercuric iodide under the conditions used for amperometric determination of thiols, giving *S*-methylmercuric cysteine residue:

$$\text{CHCH}_2\text{—SO—S—CH}_2\text{CH} + \text{CH}_3\text{HgI} + \text{H}_2\text{O} \longrightarrow \text{CHCH}_2\text{—SHgCH}_3 + \text{HO}_2\text{S—CH}_2\text{CH} + \text{HI}$$

This reaction is responsible for the increase in "apparent thiol" content of wool mildly treated with acidic peracetic or peroxomonosulfuric acid and of human hair mildly treated with acidic hydrogen peroxide or acidic peracetic acid.[188]

Other main evidence for the presence of monoxide and probably dioxide in peracetic acid–oxidized wool is based on the fact that both residues and the corresponding free amino acids give similar chemical reactions.[165] For example, treatment of oxidized wool with hydrogen sulfide leads to a decrease in "apparent thiol" and a considerable increase in sulfur content, and 3,3′-dialanyl trisulfide is found in the acid hydrolysate:

$$\text{CHCH}_2\text{—SO—S—CH}_2\text{CH} + \text{H}_2\text{S} \longrightarrow \text{CHCH}_2\text{—S—S—S—CH}_2\text{CH} + \text{H}_2\text{O}$$

Treatment of the oxidized wool with hydriodic acid results not only in a decrease in "apparent thiol" but also in an increase in disulfide content:

$$\text{CHCH}_2\text{—SO—S—CH}_2\text{CH} + 2\text{HI} \longrightarrow \text{CHCH}_2\text{—S—S—CH}_2\text{CH} + \text{I}_2 + \text{H}_2\text{O}$$

In addition, cystine monoxide and dioxide react with thiols, giving mixed disulfides according to the following equations:

$$\text{R—SO—S—R} + 2\text{R}'\text{SH} \longrightarrow 2\text{R—S—S—R}' + \text{H}_2\text{O}$$

$$\text{R—SO}_2\text{—S—R} + \text{R}'\text{SH} \longrightarrow \text{R—S—S—R}' + \text{R—SO}_2\text{H}$$

Peracetic acid–oxidized wool is found to react with thiols and sulfinates, presumably by similar mixed disulfide formation. This reaction might conceivably be employed to link dyes and other compounds covalently with the fiber. Several model dyes containing these functions have been synthesized and shown to be covalently bound to oxidized wool.[189,190]

## 5.5. Acylation

The many reagents employed to acylate proteins differ considerably in structure and reactivity; the common mechanistic feature is the attack of a nucleophile (NuH—) at an $sp^2$ or trigonal carbon atom:

$$\mathrm{R{-}C(=O){-}L} + \mathrm{:NuH{-}} \rightleftharpoons \left[\mathrm{R{-}C(O^-)(L){-}\overset{+}{N}uH{-}}\right] \overset{-\mathrm{H}^+}{\rightleftharpoons}$$

$$\left[\mathrm{R{-}C(O^-)(L){-}Nu{-}}\right] \longrightarrow \mathrm{R{-}C(=O){-}Nu{-}} + \mathrm{L^-}$$

Within a given class of nucleophiles (e.g., primary aliphatic amines) there is a good correlation of nucleophilic reactivity with pH in respect to any particular acylating agent. As expected from the lower p*K*, N-terminal groups react faster than ϵ-amino groups of lysine. Attempts at quantitative correlation between nucleophiles of different classes often fail, as do efforts to relate the reactivities of various acylating agents. Two general principles are, however, useful in formulating conditions for an acylation reaction: (1) Nucleophiles are most effective in their unprotonated forms (e.g., $RNH_2$, $ArO^-$, $RS^-$). High reactivity of a group is usually attributed to a low p*K*. However, increasing the pH of the medium more than one unit above the p*K* of the nucleophile has little effect on the concentration of the attacking species and, consequently, will generally do little to enhance the rate of reaction. (2) Great selectivity toward a variety of nucleophiles is often observed by using a highly reactive acylating agent at a relatively low pH or a weak acylating agent at a high pH.

The above principles can be illustrated by the acylation of various homologous amines. For example, the acylation of alanine (p*K* 9.9) by acetic anhydride proceeds relatively slowly at pH 5–7 as compared with that

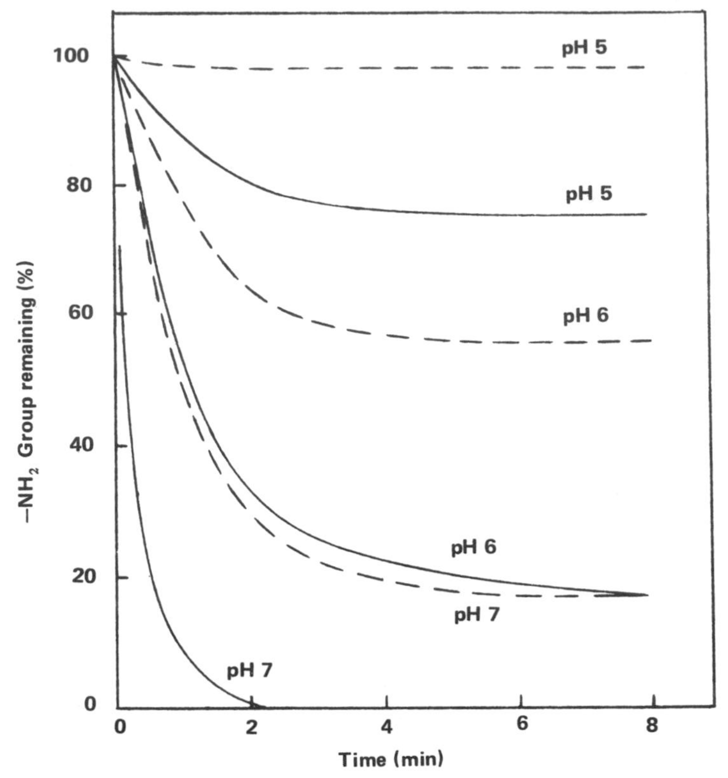


Fig. 5-4. Rate of acylation of alanine (dashed curves) and triglycine (solid curves) by acetic anhydride. Reactions were carried out at 30°C on 0.5 μmol/ml of $NH_2$ group with a twofold molar excess of acetic anhydride.[191]

of triglycine (p*K* 7.9) (Fig. 5-4). At higher pH values, less differentiation is found.

### 5.5.1. Carboxylic Acids and Anhydrides

Mixtures of acetic anhydride and glacial acetic acid in the presence of sulfuric acid are capable of acetylating some of the amino groups in wool with the simultaneous introduction of sulfate groups.[192] Acetic anhydride alone reacts with the amino groups as well as with the guanidino groups of arginine residues.[193] With acetic anhydride–glacial acetic acid–sulfuric acid–dimethylaniline mixtures, acetylation of the amino group to the extent of 90% can be obtained under carefully controlled conditions.[194] Conversion of cationic

—$NH_3^+$ groups into neutral acylamino groups by acetylation is one of the earlier methods for increasing resistance of wool toward anionic dyes.[192]

Dimethylformamide and dimethyl sulfoxide are excellent solvents for the acylation of wool with anhydrides. The incorporation of relatively large numbers (about 1400 $\mu$mol/g wool) of small acyl groups (e.g., acetyl) has little effects on settability, but bulky acyl groups (e.g., benzoyl) cause a large enhancement.[195] These effects have been attributed to electrostatic and hydrophobic interactions. At low levels of acylation ($<200$ $\mu$mol/g wool), where *N*-acylation is predominant, the result is inhibition of set due to a decrease in the internal pH of the fiber by the conversion of cationic —$NH_3^+$ groups into neutral acylamino groups. At high levels of acylation (*N*- and *O*-acylation) with bulkyl anhydrides, these electrostatic interactions are outweighed by hydrophobic interactions, which result in set stabilization.

*O*-Acetylation in keratin fibers can be distinguished from *N*-acetylation by the fact that the former can be reversed readily by the treatment with hydroxylamine at alkaline pH.[196] Acetohydroxamic acid (*N*-hydroxyacetamide) is quantitatively released into the solution and can be determined colorimetrically, in the presence of ferric chloride, as its purple ferric acetohydroxamate complex.[197] By this method, *N*-acetylated hair has been obtained for investigating the mechanisms of some chemical treatments.[198]

$$\text{CH—CH}_2\text{—C}_6\text{H}_4\text{—O—C(=O)—CH}_3 + \text{NH}_2\text{OH} \longrightarrow \text{CH—CH}_2\text{—C}_6\text{H}_4\text{—OH} + \text{CH}_3\text{CONHOH}$$

$$\text{CH}_3\text{CONHOH} \xrightarrow{\text{FeCl}_3} \text{Fe(CH}_3\text{CONHO)}_3$$

Modification of wool by cyclic acid anhydrides such as succinic or citraconic anhydride introduces new carboxyl groups, which are potential sites for further modification. For example, subsequent reaction of acylated wool with zinc acetate gives good shrinkage protection.[199]

### 5.5.2. Active Esters and Amides

Potentially, any of the aliphatic or aromatic carboxylic acids can be activated to provide acylating agents capable of reacting under mild conditions to form covalent bonds with proteins. A variety of acyl groups has been introduced into wool by the use of 4-nitrophenyl esters* (A),[200] *N*-acyloxysuccinimides (B),[201] and *N*-acylimidazoles (C).[202]

* See, however, Chapter 4.

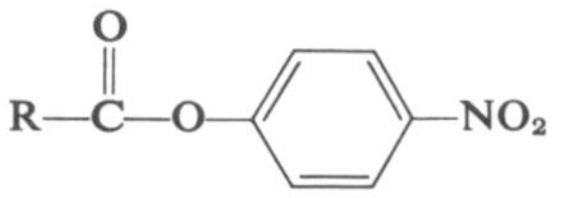

(A) 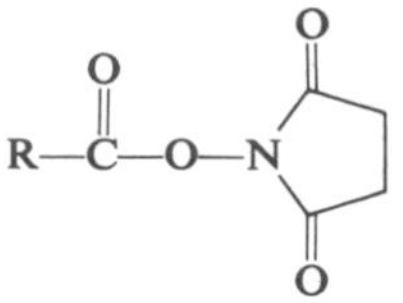 (B) 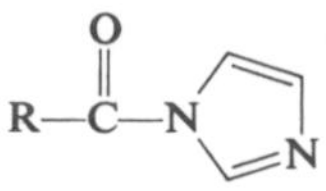 (C)

N-Terminal groups and ε-amino groups are acylated most readily, but on prolonged treatment hydroxyl groups of serine, threonine, and tyrosine residues react. Some of the acyl amino acids have actually been isolated from enzyme digests of wool treated with radioactively labeled reagents.[106] *S*-Acylation of sulfhydryl groups also occurs.[200] *O*-Acyl derivatives are more alkali labile and could be selectively removed by treatment of the acylated wool with hydroxylamine.[196]

The introduction of acyl groups into wool fibers decreases the solubility, inhibits swelling (due presumably to the blocking of water-sorption sites), and lowers the transition temperature for contraction, probably by accentuating electrostatic repulsions resulting from the conversion of $—NH_3^+$ groups into acylamino groups in the helical regions. Bifunctional reagents cause an even greater decrease in solubility and swelling by introducing new crosslinks, preventing the polypeptide chains from moving apart, but raise the transition temperature, indicating that the stabilizing effect of crosslinks between helical chains is more than sufficient to offset the destabilizing effects of acylation.[202] At low levels of acylation ($<200$ μmol/g wool), the decrease in settability has also been attributed mainly to *N*-acylation, thereby lowering the internal pH of the fiber (cf. effect of esterification) and, in the case of bifunctional reagents, to the introduction of new crosslinks, which inhibit conformational changes in the wool proteins.[201]

The possible use of active esters containing chromophoric groups as reactive dyes for wool has been studied.[203] With pentachlorophenyl esters, which are more reactive than 4-nitrophenyl or 2,4,5-trichlorophenyl esters, up to 50% of dye fixation may be achieved.[204]

### 5.5.3. Carboxylic Acid Chlorides

Carboxylic acid chlorides generally react with nucleophilic groups in wool leading to extensive *O*-, *N*-, and *S*-acylation. High levels of acyl groups can be incorporated when dimethylformamide or dimethyl sulfoxide is used as solvent. The increase in settability of wool by treatment with phenylacetyl chloride is probably due to the introduction of the phenyl group.[205]

The introduction of organic chlorine or bromine into wool by reaction with aliphatic halo-organic acid halides in dimethylformamide imparts greatly enhanced flame resistance to wool.[206]

### 5.5.4. Cyanate, Isocyanates, and Isothiocyanates

Cyanate ion or cyanic acid reacts with most protein nucleophiles, for example, amino, carboxyl, sulfhydryl, phenolic hydroxyl, and imidazole to yield carbamoyl derivatives.[207]

$$\text{protein—NuH} + \text{H—N=C=O} \longrightarrow \text{protein—Nu—}\overset{\overset{\displaystyle O}{\|}}{C}\text{—NH}_2$$

The reagent has the advantage of being easily soluble in water, and in most cases reactions occur with maximum rate near pH 7 and are quite insensitive to moderate changes in pH. Selective modification of amino groups can be achieved for most proteins, since all the other reactions can be reversed under slightly alkaline conditions. Potassium cyanate is well suited for N-terminal analysis. In this procedure, it is coupled in mildly alkaline aqueous solution with the amino group to form the carbamoyl derivative. While $N^6$-carbamoyl derivatives slowly revert to free amino groups during acid hydrolysis, $N^2$-carbamoyl derivatives cyclize to hydantoins (rigorously, 2,4-dioxoimidazolidines). After isolation, most of the hydantoins can be hydrolyzed to the free amino acids in good yield. This method has been used to determine N-terminal groups in components from *S*-carboxymethylkeratine.[208]

$$\text{H}_3\overset{+}{\text{N}}\text{CHRCO—Y} + {}^{-}\text{NCO} \xrightarrow{\text{pH 7–8}} \begin{array}{c}\text{RCH—C(=O)—Y} \\ \text{HN} \quad \text{HN—H} \\ \text{C=O}\end{array} \xrightarrow{\text{H}^+} \begin{array}{c}\text{RCH—C=O} \\ \text{HN} \quad \text{NH} \\ \text{C=O}\end{array} + \text{HY}$$

$$\text{Y} = \text{—NHCHR}'\cdots\text{COOH}$$

separation and hydrolysis

$$\downarrow$$

$$\text{H}_2\text{NCHRCOOH} + \text{NH}_3 + \text{CO}_2$$

Phenylisocyanate and phenylisothiocyanate (PTC) add readily to amino groups of peptides to give, respectively, the *N*-phenylcarbamoyl and the *N*-phenylthiocarbamoyl derivatives, which cyclize readily to hydantoins or thiohydantoins under the influence of acid catalysts. With PTC it is possible to split off the N-terminal acid as a 3-phenyl-2-thiohydantoin leaving the rest of the chain intact, and thus PTC is particularly suitable for stepwise degradation of peptides (Edman degradation).[209] Automatic analysis of amino acid sequences in peptides by Edman degradation has been reviewed.[210]

The retardation of photoyellowing of wool by treatment with methyl isothiocyanate has been attributed to the introduction of thioureido groups.[211] Reactive dyes based on the isothiocyanate group have been described.[212,213]

Reaction of wool with mono- and bifunctional isocyanates has been the subject of several investigations. In pyridine, the reaction occurs with amino, carboxyl, and sulfhydryl groups, and probably with phenolic, guanidino, and imidazole groups.[214,215] The treatment reduces fabric felting shrinkage and increases resistance to acids, alkali, and reducing agents. If dimethylformamide or dimethyl sulfoxide is substituted for pyridine, the reaction is much more rapid, resulting in high isocyanate uptake.[216] On the whole, diisocyanates impart more chemical resistance than monoisocyanates, probably because the disocyanates introduce new crosslinks into the fiber structure. The introduction of large amounts of hydrophobic phenyl residues in wool by treatment with aromatic isocyanates leads to a marked increase in settability.[205]

### 5.5.5. Sulfonyl Halides

Aliphatic and aromatic sulfonyl halides are rather similar to acyl halides and anhydrides in their reactions with nucleophiles. Until recently, little use was made of these reagents in protein modification, possibly because they are less reactive than most acylating agents and because they have a greater tendency to denature and insolubilize proteins. The development of 5-dimethylaminonaphthalene-1-sulfonyl chloride (dansyl chloride) as a reagent for the identification of N-terminal residues[217] seems to have stimulated other studies in selective modification of proteins.

Sulfonylation of amino compounds with pipsyl chloride (4-iodophenylsulfonyl chloride) gives *N*-pipsyl derivatives. Pipsyl chloride has been used in the identification of amino acids and of N-terminal residues in wool.[218]

### 5.5.6. Lactones and Thiolactones

$\beta$-Propiolactone has been shown, in organic chemistry, to react with a large number of nucleophilic centers by the $S_N2$ mechanism. Its reaction with proteins was first studied on wool.[219] Treatment with $\beta$-propiolactone in carbon tetrachloride can result in a very high weight increase of the wool. The modified wool is softer and whiter than untreated wool and has greater luster. It is postulated that histidine and lysine side chains are acylated with —$COCH_2CH_2OH$ by acyl-oxygen cleavage of the lactone ring, while methionine, glutamic acid, serine, and threonine, and probably cysteine side chains, are alkylated with —$CH_2CH_2COOH$ by alkyl-oxygen cleavage of the lactone ring. Reactions of these new reactive centers with the reagent will eventually lead to the formation of "anchored" polymers in the wool fiber. In this way, the lysine residue, for instance, might give

$$-NH_2 + \text{(β-propiolactone)} \longrightarrow -NHCOCH_2CH_2O(CH_2CH_2COO)_nCH_2CH_2COOH$$

Alkylation of methionine residues in proteins with β-propiolactone to give the sulfonium salt is specific at low pH values.[220] Treatment of wool with α,α-bis(chloromethyl)-β-propiolactone enhances shrink resistance and lowers flammability.[221]

Thiolation of amino groups may be achieved with *N*-acetylhomocysteine thiolactone, which reacts fairly selectively with amino groups of amino acids and proteins including silk and wool to form the *N*-acetylhomocysteine derivatives.[222] The amount of sulfhydryl group introduced agrees closely with the decrease in the free amino group. The reaction is markedly catalyzed by imidazole and tertiary amines, which function as acyl transfer agents.[223] The use of *N*-acetylhomocystine thiolactone in permanent waving of human hair has been described.[224] During the waving process, hydrolysis of the thiolactone forms *N*-acetylhomocysteine, which acts as a reducing agent for the disulfide bonds, while aminolysis of the thiolactone by the fiber amino group results in thiolation. Many sulfhydryl groups are thus available for the formation of disulfide bonds during the subsequent mild oxidation stage. In this way, hair is strengthened and set in a manner which cannot possibly occur when ammonium thioglycollate is used.

$$\text{(thiolactone)}-NHCOCH_3 \xrightarrow{H_2O} HOOCCH(NHCOCH_3)CH_2CH_2SH$$

$$\text{(thiolactone)}-NHCOCH_3 \xrightarrow{-NH_2} -NHCOCH(NHCOCH_3)CH_2CH_2SH$$

### 5.5.7. Dithiocarboxylic Acid Esters

Carboxymethyl carbodithioates are excellent reagents for the thioacylation of amines[225,226] and thiols.[227,228] They also react readily with amino and sulfhydryl groups of polypeptides and keratin fibers in aqueous solution under mild conditions.[229]

The use of these reactions as a basis for dyeing keratin fibers offers several advantages. The hydrolysis rate of these dyes is slow compared with

$$R{-}\overset{\overset{\displaystyle S}{\|}}{C}{-}SCH_2COOH + \text{keratin}{-}NH_2 \longrightarrow R{-}\overset{\overset{\displaystyle S}{\|}}{C}{-}NH{-}\text{keratin} + HSCH_2COOH$$

$$R{-}\overset{\overset{\displaystyle S}{\|}}{C}{-}SCH_2COOH + \text{keratin}{-}SH \rightleftharpoons R{-}\overset{\overset{\displaystyle S}{\|}}{C}{-}S{-}\text{keratin} + HSCH_2COOH$$

that of the existing reactive dyes. Simultaneous dyeing and reducing of fiber disulfide bonds by treatment with the dye in the presence of a thiol has been shown to be effective in dyeing human hair.[230] For example, in the presence of 2-mercaptoethanol the following equilibrium rapidly sets up in the dye solution by the ester interchange reaction.

$$R{-}\overset{\overset{\displaystyle S}{\|}}{C}{-}SCH_2COOH + HSCH_2CH_2OH \rightleftharpoons R{-}\overset{\overset{\displaystyle S}{\|}}{C}{-}SCH_2CH_2OH + HSCH_2COOH$$

Apart from carboxymethyl esters, other esters (e.g., 2-hydroxyethyl esters) are also reactive. Thus, in wool dyeing, the flexibility of the system enables dyes of varying reactivity to be made easily on the basis of a given chromogen. While carboxymethyl ester dyes are similar to the corresponding Lanasol dyes (Ciba-Geigy) in their high final fixation values, the controllability of fixation by means of pH during the dyeing process to give a satisfactory degree of leveling is more favorable in the case of carboxymethyl ester dyes. Dyes in which the chromophoric group is electronically isolated from the reactive center exhibit little pH sensitivity on the shade and render the oxidative aftertreatment to convert C=S into C=O unnecessary.

Polymer formation within the fiber occurs when the fiber is treated with diamines, dithiols, or 2-mercaptoethylamine followed by bifunctional carboxymethyl carbodithioates. Waving of human hair by this process results in good retention of the curl.[231]

## 5.6. Arylation

Aromatic compounds containing electron-withdrawing substituents such as nitro groups and heterocycles showing aromatic character will undergo nucleophilic substitution reactions with nucleophiles. In the latter case, the activation is due to the electronegativity of the heteroatom(s) in the ring. The order of reactivity in terms of the group being displaced is $F > Cl \approx Br > SO_3H$. In the absence of the special protein factors which limit accessibility or alter reactivity, the general order of nucleophilicity is $SH > NH_2 >$

phenol > imidazole. Arylation is often effective at significantly lower pH values than those needed for acylation.

### 5.6.1. 1-Fluoro-2,4-dinitrobenzene (FDNB)

Arylation has received an enormous impetus through the work of Sanger,[232] who firmly established FDNB as a protein reagent. In addition to the amino group, the phenolic group of tyrosine, the imidazole group of histidine, and the sulfhydryl group of cysteine residues tend to become alkylated. But the products of these reactions are not yellow and thus do not interfere with the prime use of FDNB as an amino reagent. Five water-soluble DNP amino acids are found in the hydrolysate of dinitrophenylated wool, human hair, or mohair: $N^6$-DNP-lysine, *O*-DNP-tyrosine, *O*-DNP-serine, *S*-DNP-cysteine, and $N^{\tau}$-DNP-histidine.[233] It is possible to determine $N^6$-DNP-lysine directly from the hydrolysate of dinitrophenylated fiber by measuring the absorptivity of the solution at 390 nm, where the interference from other water-soluble DNP amino acids is negligible.[234]

In wool keratin 20% of the $\epsilon$-amino groups are unavailable for dinitrophenylation under the normal reaction conditions of pH 8. (Part of the unavailable amino group can be accounted for by natural crosslinks, the total amount of $N^6$-$\gamma$-glutamyllysine crosslink being 15 $\mu$mol/g.[235]) In pH 6 acetate buffer, the reaction of FDNB with the carboxylate anion is faster than that with the protonated $\epsilon$-amino group. Thus, under these conditions, the percentage of $\epsilon$-amino groups unavailable for dinitrophenylation increases to 80. On the other hand, in esterified wool 93% of the lysine residue can be dinitrophenylated.

Treatment of *S*-dinitrophenylated peptides with alkali promotes $\beta$-elimination of *S*-dinitrothiophenoxide and hence the formation of dehydroalanine residues.[236] This method was used for converting all the cystine residues in glutathione and wool into dehydroalanine residues.[81,13]

In the presence of an amine, large amounts of new $\beta$-*N*-alkyl amino acid residues are incorporated into the fiber by the addition of the amine to the double bond of the dehydroalanine residues.[14,15]

Extensive investigations of the crosslinking action of 1,5-difluoro-2,4-dinitrobenzene (FFDNB) and 4,4′-difluoro-3,3′-dinitrophenyl sulfone (FF-sulfone) have been undertaken at the Deutsches Wollforschungsinstitut. Identification of the crosslinked amino acids is possible, since they arc stable to acid hydrolysis conditions. Isolation of these crosslinked amino acids is irrefutable evidence of the spatial arrangement of the amino acids involved, since these can only form bonds if able to approach to a distance of about 0.4 nm (FFDNB) and 1.1 nm (FF-sulfone).

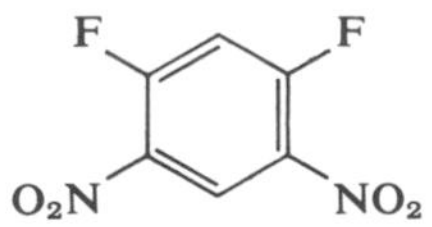

FFDNB

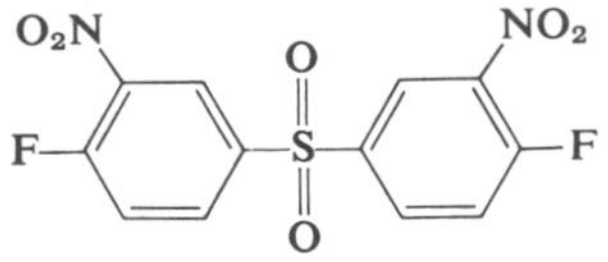

FF-Sulfone

After reaction of wool with FFDNB, the following crosslinked dinitrophenylene (DPE) amino acids can be isolated from the hydrolysate[237–239]:

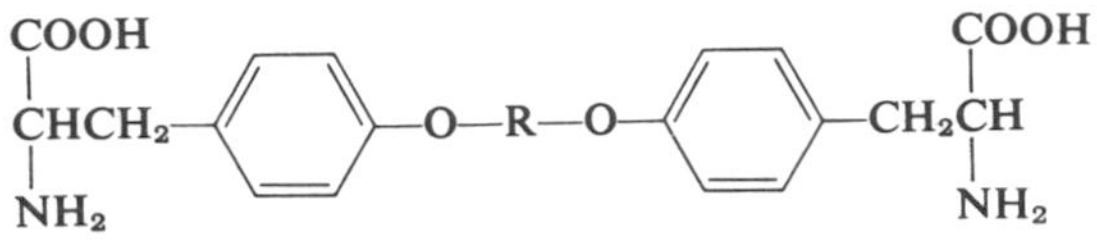

*O*,*O*′-DPE-bis-tyrosine

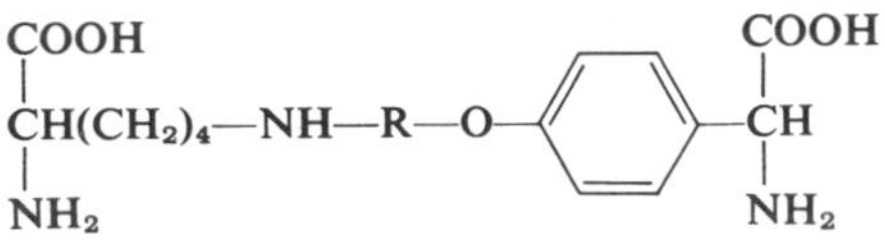

$N^6$,*O*-DPE-lysine-tyrosine

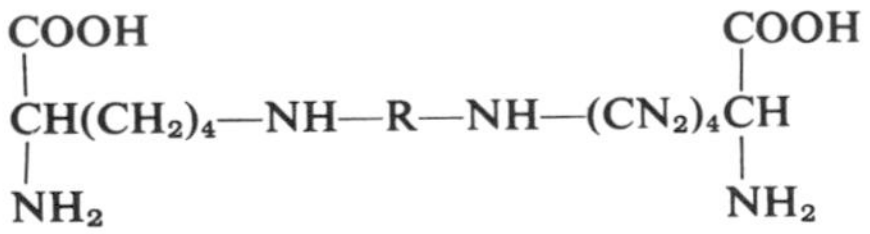

$N^6,N^{6\prime}$-DPE-bis-lysine

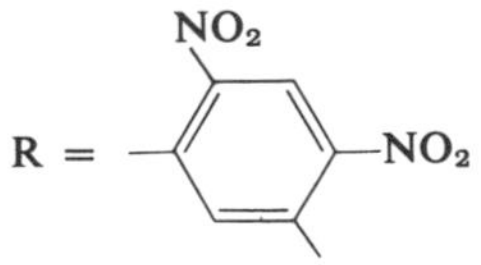

This shows that both lysine and tyrosine residues are involved in the crosslinking reaction.

FF-sulfone will also form crosslinks between tyrosine and lysine residues.[240] The following crosslinked dinitrodiphenyl sulfone (Sulfone) amino acids can be isolated from the hydrolysate: *O*,*O*′-Sulfone-bis-tyrosine, $N^6,N^{6\prime}$-Sulfone-bis-lysine, and $N^6$,*O*-Sulfone-lysine-tyrosine. $N^6$,*O*-Sulfone-lysine-tyrosine is obtained in high concentration, and this means that the phenolic groups of tyrosine residues and the ε-amino groups of lysine residues are close to one another. The crosslinked wool is 80–90% insoluble in ammonia after oxidation and is found, after long periods of ultrasonic radiation, to retain the microfibril network, as shown by electron micrographs.

### 5.6.2. Cyanuric Halides and Derivatives

Cyanuric fluoride (2,4,6-trifluoro-*s*-triazine) is not a particularly useful protein modifying reagent since it reacts with all protein nucleophiles, including tryptophan, with little or no selectivity.[241,242] However, its virtue lies in the fact that the cyanurate ester formed with tyrosine is stable at high pH and that the absorption maximum of the phenolic chromophore is displaced toward the blue in the derivative, owing to its strong electron-withdrawing ability. Thus the unreacted tyrosine residue in a protein may be determined by measuring the absorptivity of its anion at 295 nm (pH 13).

Color derivatives of *s*-triazines and diazines, for example pyrimidines, quinoxalines, and phthalazines, together form an important class of reactive dye. In the dyeing process the halogen atom on the heterocyclic ring is displaced by nucleophiles in the fiber.

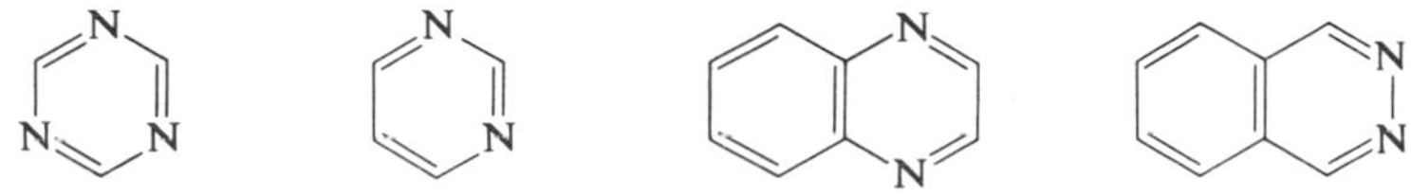

*s*-Triazine Pyrimidine Quinoxaline Phthalazine

The action of 2,4-dichloro-6-methoxy-*s*-triazine and 2,4,6-trichloro-*s*-triazine on wool has been studied in wool stabilization experiments.[243] Tensile strength, elongation, and solubilities in urea–bisulfite, acid, and alkali of the treated wool indicate that chemical and mechanical properties are improved, particularly by the former reagent.

## 5.7. Alkylation

The majority of alkylation reactions proceed by the classical $S_N2$ displacement mechanism. Although much is known regarding the effects of structure, steric hindrance, and environment on rates of nucleophilic displacement for simple organic molecules, the use of such information in protein chemistry is limited by the complexity of the molecule. The order of reactivity of nucleophiles toward the tetrahedral carbon of alkylating agents is somewhat different from that toward the trigonal carbon in acylation reactions or toward aromatic carbon in arylation. In the related $S_N1$ displacement, in which the alkyl halide is, or becomes, partially ionized prior to reaction, differences in nucleophilicities of functional groups are considerably smaller.

Nucleophilic addition to double bonds has stereochemical and mechanistic features which differ from those of displacement. Generally, the double bond must be adjacent to a carbon atom which contains an electron-withdrawing group.

The investigation of keratin fibers often requires the reduction of the disulfide bond and blockage of the resulting sulfhydryl group. Because of the intrinsic reactivity of the sulfhydryl group toward alkylating agents, *S*-alkylation has been a subject of many investigations.

### 5.7.1. Alkyl Halides, α-Haloacids, and Derivatives

The reactivity of these halogen compounds generally depends on the halogens (i.e., $I \geqslant Br > Cl \gg F$) such that iodo and bromoacetates are highly useful protein reagents, chloroacetate is less reactive, and fluoroacetate is quite unreactive and of little value for modifying proteins. A review of haloacetates and their reactions with proteins has been published.[244] Under various conditions, haloacetates react with sulfhydryl, imidazole, thioether, and amino groups of protein, sulfhydryl groups being intrinsically the most reactive. With rather harsh treatment, phenolic side chains may also react.

With proteins in solution, *S*-carboxymethylation with iodoacetic acid takes place very rapidly and specifically, but alkylation of reduced wool fibers involves a two-phase system, and here the alkylation process is far slower and hence competing nonspecific alkylation may occur.[102] While the reaction is almost specific for cysteine residues in reduced wool at pH 8 for 6 h, some modification of histidine, methionine, and tyrosine residues has taken place at pH 8.5 for 24 h. *S*-Carboxymethylcysteine is stable under acid hydrolysis conditions and can be determined by amino acid analysis; it is rapidly eluted, usually before aspartic acid. Iodoacetic acid is capable of reacting rapidly with lanthionine residues in wool to give the carboxymethyllanthioninesulphonium salt, which decomposes, under acid hydrolysis conditions, to β-iodoalanine and carboxymethylcysteine[245]:

$$\begin{array}{c} | \\ CO \\ | \\ CHCH_2{-}S{-}CH_2CH \\ | \\ NH \\ | \end{array} + ICH_2COOH \longrightarrow \begin{array}{c} | \\ CO \\ | \\ CHCH_2{-}\overset{+}{S}{-}CH_2CH\,I^- \\ | \quad\; | \quad\; | \\ NH \;\; CH_2 \;\; NH \\ | \quad\;\; | \quad\;\; | \\ COOH \end{array} \xrightarrow{\text{hydrolysis}}$$

$$\begin{array}{c} H_2N{-}CH{-}COOH \\ | \\ CH_2SCH_2COOH \end{array} + \begin{array}{c} H_2N{-}CH{-}COOH \\ | \\ CH_2I \end{array}$$

A nonhydrolytic method for determining cysteine content in hair is based on the titration with silver nitrate of the iodide ion generated in the reaction between fiber sulfhydryl groups and iodoacetamide.[246]

The amino group of wool can be effectively blocked by the reaction with bromoacetic acid and magnesium oxide in aqueous solution.[247] The treated wool is resistant to anionic dyes, but the cost of bromoacetic acid prohibits the use of the method on industrial scale.

$$2\text{wool—}NH_2 + 2BrCH_2COOH + MgO \longrightarrow 2\text{wool—}NH\text{—}CH_2COOH + MgBr_2 + H_2O$$

With methyl iodide and ethylene dibromide, the alkylation of wool is slower than with iodoacetic acid. In the case of methyl iodide, a considerable proportion of the initially formed $—SCH_3$ groups can be further methylated to give the dimethylsulfonium salt of cysteine, which is unstable in the presence of a base and decomposes by a nucleophilic $\beta$-elimination process to give dimethylsulfide and peptide-bound dehydroalanine[248] (cf. $\beta$ elimination of *S*-dinitrothiophenoxide in Section 5.6.1). After repeated reduction of wool with thioglycollate and alkylation with methyl iodide, *S*-carboxymethylcysteine and lanthionine residues can be formed, respectively, by the addition of cysteine residue and thioglycollate to the double bond of the dehydroalanine residue[249]:

$$\text{CHCH}_2\text{—SH} \xrightarrow{CH_3I} \text{CHCH}_2\text{—S—CH}_3 \xrightarrow{CH_3I} \text{CHCH}_2\text{—}\overset{+}{\text{S}}(\text{CH}_3)\text{—CH}_3 \xrightarrow{OH^-} \text{C=CH}_2 + (CH_3)_2S$$

$$\text{C=CH}_2 \xrightarrow{\text{HS—CH}_2\text{CH}} \text{CHCH}_2\text{—S—CH}_2\text{CH}$$

$$\text{C=CH}_2 \xrightarrow{\text{HSCH}_2\text{COOH}} \text{CHCH}_2\text{—SCH}_2\text{COOH}$$

Under the above treatment conditions, $N^{\tau}$-methylhistidine and $N^{\tau},N^{\pi}$-dimethylhistidine are found in the hydrolysate.

Bifunctional agents such as alkyl dihalides are of special interest because they permit the formation of crosslinks. The disulfide bond of wool can be transformed to the more stable bisthioether linkage, $—S(CH_2)_nS—$, by a two-step process involving reduction of the disulfide bond to sulfhydryl groups followed by treatment of the wool with an aliphatic dihalide.[250] Wool modified by this process is greatly improved in its resistance to degradation by alkali, acids, and oxidizing or reducing agents. Interestingly, the wool with stabilized crosslinks is much less susceptible to attack by moths, carpet beetles, and enzymes. By the use of a reducing agent such as sodium or zinc sulfoxylate formaldehyde which does not react with the alkylating agent, the wool may be reduced and alkylated in a single step.[251] Direct evidence for the formation of new stable crosslinks comes from the isolation of the corresponding polymethylene-*S,S'*-dicysteine from the acid hydrolysate of the reduced wool treated with alkyl dibromides.[252]

Bis(chloromethyl) ether will react with wool, giving a product with increased tensile strength and reduced alkali solubility.[253] Apart from the reaction with fiber sulfhydryl groups, the following reactions involving fiber amino and carboxyl groups may also take place:

$$\text{—NH}_2 + \text{ClCH}_2\text{OCH}_2\text{Cl} \longrightarrow \text{—NHCH}_2\text{OCH}_2\text{Cl} + \text{HCl}$$

$$\text{—COOH} + \text{—NHCH}_2\text{OCH}_2\text{Cl} \longrightarrow \text{—NHCH}_2\text{OCH}_2\text{OOC—} + \text{HCl}$$

### 5.7.2. Aziridines

*S*-Aminoethylation of sulfhydryl groups in reduced wool with 2-bromoethylamine and ethyleneimine (aziridine) is apparently specific.[254] 2-Bromoethylamine is slowly converted, under slightly alkaline conditions, into ethyleneimine, which is the active species:

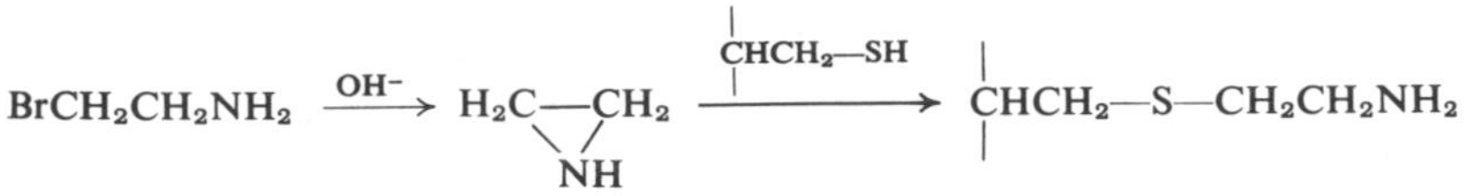

$$\text{BrCH}_2\text{CH}_2\text{NH}_2 \xrightarrow{\text{OH}^-} \underset{\text{NH}}{\text{H}_2\text{C—CH}_2} \xrightarrow{\text{|CHCH}_2\text{—SH}} \text{|CHCH}_2\text{—S—CH}_2\text{CH}_2\text{NH}_2$$

*S*-(2-Aminoethyl)cysteine closely resembles lysine but with a sulfur atom taking the place of a methylene group, and peptide bonds formed by the carboxyl group of aminoethylcysteine in proteins are susceptible to tryptic hydrolysis.[255] The reaction is commonly used to introduce additional trypsin-susceptible bonds into proteins, usually to facilitate degradation in sequence studies. The C-terminal aminoethylcysteine residues are in turn susceptible to cleavage by carboxypeptidase B[256] and a colorimetric procedure for its determination has been described.[257]

Aziridinyl groups have been used as the active group in reactive dyes. In these dyes the reactive group is linked by means of a bridge (usually —$SO_2$—) to the chromophoric group of the molecule:

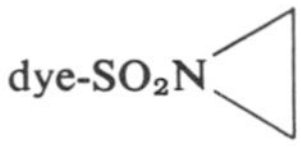

$$\text{dye-SO}_2\text{N}\!\triangleleft$$

However, dyes with reactive groups that are derived from epimines by opening of the ring are described more frequently, in particular, dyes with the $\beta$-chloroethylsulfamoyl group, —$SO_2NHCH_2CH_2Cl$, or the $\beta$-sulfatoethylsulfamoyl group, —$SO_2NHCH_2CH_2OSO_3H$.

Bifunctional epimines have been used in dimensional stabilization of wool (antishrink and antifelt).[258] Most of the successful products of water- and oil-repellent finishing belong to the class of 1-(perfluoroacyl)aziridines, for example, perfluoroglutarylbisaziridine.[259] The electron-withdrawing effect of the fluoroalkyl moiety greatly enhances the reactivity of the carbonyl function and of the aziridine ring.

### 5.7.3. Maleic Acid and Derivatives

*N*-Ethylmaleimide (NEM) enjoys considerable popularity in the chemical modification of proteins and in the quantitative determination of sulfhydryl groups because it is water-soluble and stable, the reaction is rapid at neutral pH and involves no change in pH, the reaction may be followed by a decrease in absorbance at 305 nm, and the adduct is stable to acid hydrolysis. Hydrolysis of NEM-treated protein converts the modified cysteine residues into *S*-(2-succinyl)cysteine and ethylamine. The former can be estimated by amino acid analysis; it is rapidly eluted and usually precedes aspartic acid.[260] NEM is a specific reagent for protein sulfhydryl groups below pH 7. At pH 7, its reaction rate with simple thiols is approximately 100-fold greater than the corresponding simple amino compounds. NEM itself has been used in determining cysteine contents in wool, while other monofunctional N-substituted maleimides [e.g., *N*-phenylmaleimide and *N*-(4-ethoxyphenyl)-maleimide] have been used to modify reduced wool. Extensive crosslinking of reduced wool can be introduced by *o*- or *m*-phenylenebismaleimide[261,262]; acid hydrolysis of the modified wool gives *S*-(2-succinyl)cysteine as the only reagent derivative, thus demonstrating the specificity of these reagents:

$$2\ \overset{|}{\underset{|}{\mathrm{CO}}}\!-\!\mathrm{CHCH_2{-}SH}\!-\!\underset{|}{\mathrm{NH}} + \text{(maleimide)N{-}R{-}N(maleimide)} \longrightarrow$$

$$\mathrm{[CO]CH(NH)CH_2{-}S{-}(succinimide)N{-}R{-}N(succinimide){-}S{-}CH_2CH(CO)(NH)} \xrightarrow{\text{hydrolysis}}$$

$$2\,\mathrm{HOOC\underset{\displaystyle NH_2}{CH}CH_2{-}S{-}\underset{\displaystyle CH_2COOH}{CH}COOH}$$

Reactive dyes containing the maleimide group, which reacts rapidly with sulfhydryl groups in reduced wool, have been described.[189]

According to a patent,[263] the sheen and strength of damaged human hair can be restored by a solution of maleic acid or its esters, which react with the free sulfhydryl and amino groups in the fiber.

### 5.7.4. Acrylonitrile

Nucleophilic addition to acrylonitrile has been used as a means of alkylating sulfhydryl groups in reduced wool. In mildly alkaline media (pH 8), *S*-cyanoethylation appears to be rapid and specific and is much more selective than iodoacetate. In more alkaline media (pH 9.5), reaction with amino, guanidino, or imidazole groups becomes significant, and may even result in *O*-cyanoethylation.[264] Hydrolysis of cyanoethylated wool converts the modified cysteine residues into *S*-carboxyethylcysteine, which is conveniently determined by high-voltage electrophoresis at pH 2.5.[265]

$$\begin{array}{l} | \\ CO \\ | \\ CHCH_2{-}SH \\ | \\ NH \\ | \end{array} + CH_2{=}CHCN \longrightarrow \begin{array}{l} | \\ CO \\ | \\ CHCH_2{-}S{-}CH_2CH_2CN \\ | \\ NH \\ | \end{array} \xrightarrow{\text{hydrolysis}} \begin{array}{l} H_2NCHCOOH \\ \quad | \\ \quad CH_2{-}S{-}CH_2CH_2COOH \end{array}$$

### 5.7.5. Vinyl Sulfones and Phosphonates

Vinyl sulfones are well suited for specific blockage of sulfhydryl groups in reduced wool fibers.[267] If the reaction time is not too long, a simultaneous addition of amino groups to vinyl sulfones can practically be excluded. In dyeing human hair at ambient temperature, the necessity to have free sulfhydryl groups for reaction with vinyl sulfone dyes is indicated by the ratio 10,000:100:1 for the reactivity of $SH:NH_2:OH$ groups toward these dyes.[266] In dyeing of wool keratin, all groups are involved to some extent.[12]

Divinyl sulfone can lead to an extensive crosslink formation in reduced wool.[268] It is apparently specific for sulfhydryl groups under mild reaction conditions (pH 3 and room temperature); neither the $\epsilon$-amino group of lysine nor the phenolic hydroxyl group of tyrosine appears to have reacted.[269] However, if dimethyl sulfoxide is used as the reaction medium, other functional groups in the side chains, especially the amino group of lysine and the imidazole group of histidine, are readily modified by vinyl sulfone or acrylonitrile.[270]

Sulfhydryl groups of reduced wool can be selectively alkylated with bis-($\beta$-chloroethyl) vinyl phosphonate.[271] The expected derivative *S*-bis-($\beta$-chloroethyl)phosphonylethylcysteine has been isolated from the acid hydrolysate of the treated wool. The introduction of chlorine and phosphorus into the fiber by such treatment has been found to be effective for making wool flame resistant.

$$\text{CHCH}_2\text{—SH} + \text{CH}_2\text{=CHPO(OCH}_2\text{CH}_2\text{Cl)}_2 \longrightarrow \text{CHCH}_2\text{—S—CH}_2\text{CH}_2\text{PO(OCH}_2\text{CH}_2\text{Cl)}_2$$

One of the most popular types of reactive dyes for wool contains a double bond activated by an electron-withdrawing group such as vinyl sulfone, acrylamide, or sulfonamide.

### 5.7.6. Vinylpyridines

2-Vinylpyridine[107] and 4-vinylpyridine[272] have been found to be useful reagents for selective modification of sulfhydryl groups in reduced wool and for determining the cystine and/or cysteine content of wool.

### 5.7.7. Unsaturated Aldehydes

The action on wool of acrylaldehyde, $\alpha$-methylacrylaldehyde, crotonaldehyde, and cinnamaldehyde has been studied in some detail by McPhee and Lipson.[273] Of these compounds acrylaldehyde reacts most readily, combining with free sulfhydryl groups and under alkaline conditions with the $\epsilon$-amino group of lysine residues. Reduced wool is markedly more reactive and acylated wool shows a greatly reduced affinity for the aldehydes.

### 5.7.8. 1,4-Benzoquinone

1,4-Addition of thiols to quinones occurs readily at room temperature. The hydroquinone product is oxidized by unreacted quinone, and a 2-alkylthiol-1,4-benzoquinone is normally the product isolated. Multiple addition gives 2,5- and 2,6-disubstitution.[274] 1,4-Addition of sulfhydryl groups in reduced wool and human hair to benzoquinone occurs under mild aqueous

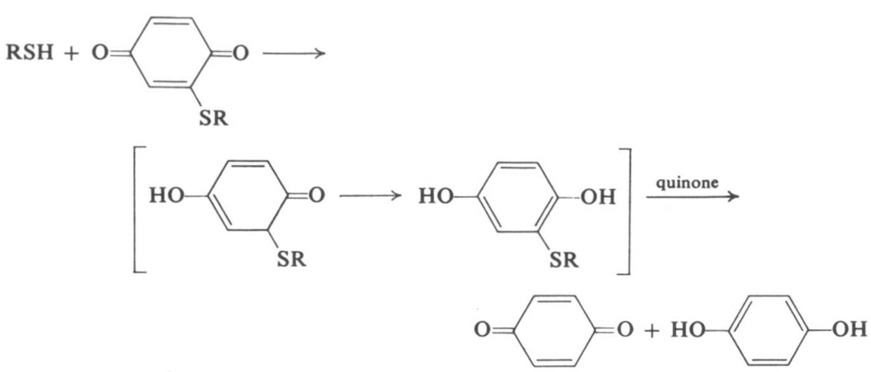

RSH + O=⟨ ⟩=O ⟶
SR

[ HO—⟨ ⟩=O ⟶ HO—⟨ ⟩—OH ] $\xrightarrow{\text{quinone}}$
SR SR

O=⟨ ⟩=O + HO—⟨ ⟩—OH

conditions.[275,276] It is likely that reaction with fiber amino groups and the formation of crosslinks and polymers may also occur. The effect of benzoquinone on the physical properties of wool fibers has been extensively investigated.[277–279]

Benzoquinone and its derivatives, or alternatively such compounds produced during application to wool fabric by oxidation of the corresponding hydroquinone with dichromate, are used to improve the strength and wet abrasion resistance of the woollen filtration felts used on paper-making machines. Such treatments are claimed to prolong the useful life of the felts. As these treatments result in the development of colors (yellow, red, black) on the wool fiber, and these colors are fugitive to light, the process is limited to industrial fabrics. Natural quinones (e.g., henna) have been applied to hair to produce color.

### 5.7.9. *N*-Methylol Compounds

1,3-bis(Hydroxymethyl)-2-imidazolidone, better known as dimethylolethyleneurea, and analogous *N*-methylol derivatives are widely used for the crease-resist finishing of cellulose fabrics and blends. They can also be used to stabilize set in wool fabric.[280] The reaction with wool occurs mainly with lysine and tyrosine residues and also sulfhydryl groups in reduced fibers.[281] The treated fiber shows decreased supercontraction in sodium sulfite solution as well as decreased solubilities in alkali and urea–bisulfite tests, indicating the introduction of crosslinks.[282]

The condensation of an *N*-methylol compound with a nucleophile is subject to acid catalysis. Addition of a proton to the hydroxyl group followed by the elimination of water gives the reactive carbonium ion, which reacts readily with nucleophiles, for example hydroxyl groups in cellulose or amino groups in keratin:

$$-\overset{|}{N}-C(=O)-\overset{|}{N}-CH_2OH + H^+ \rightleftharpoons -\overset{|}{N}-C(=O)-\overset{|}{N}-CH_2\overset{+}{O}H_2 \rightleftharpoons$$

$$\left[ -\overset{|}{N}-C(=O)-\overset{|}{N}-\overset{+}{C}H_2 \longleftrightarrow -\overset{|}{N}-C(=O)-\overset{|}{N}=\overset{+}{C}H_2 \right] + H_2O \xrightarrow{NuH}$$

$$-\overset{|}{N}-C(=O)-\overset{|}{N}-CH_2Nu + H^+$$

Since the electron pair of the nitrogen atoms takes part in the mesomerism of the ureidomethyl carbonium–immonium ions, those *N*-methylol compounds that possess a high electron density at the nitrogen atom, for example dimethylolethyleneurea, are particularly active.

Dimethylolurea and dimethylolthiourea are reported to be able to improve the strength and elasticity of human hair. In acid media, condensation with fiber amino groups and formation of polymers at the fiber surface occur.[(283)]

Reactive dyes containing the *N*-methylol group have been described.[(133)]

### 5.7.10. Esterification of Carboxyl Groups

Esterification of protein carboxyl groups is achieved by nucleophilic attack by $RO^-$ on the carbonyl carbon of the carboxylic acid residues, resulting in a displacement of OH by an OR,

$$\overset{|}{\underset{|}{C}}H(CH_2)_{1,2}COOH + RO^- \longrightarrow \overset{|}{\underset{|}{C}}H(CH_2)_{1,2}COOR + OH^-$$

or in some instances by nucleophilic attack by $—COO^-$ of the protein on suitable reagents, such as epoxides, active halogen-containing compounds, and dimethyl sulfate.

$$\overset{|}{\underset{|}{C}}H(CH_2)_{1,2}COO^- + R—L \longrightarrow \overset{|}{\underset{|}{C}}H(CH_2)_{1,2}COOR + L^-$$

However, the problem of esterification of the carboxyl group of proteins under gentle conditions in aqueous solution has not been solved. None of the reagents currently used produces more than partial carboxyl esterification.

Diazomethane readily esterifies about two thirds of the free carboxyl groups and also alkylates about four fifths of the tyrosine phenolic groups in wool.[284] The reaction with dimethyl sulfate, methyl iodide, or methyl bromide introduces methyl groups into wool, but the carboxyl content is not reduced to the expected degree.[285] With dimethyl sulfate, only about 25% of the free carboxyl groups in wool are esterified and, at the same time, some other amino acid residues are attacked.[286]

#### *5.7.10.1. Alcohols*

Esterification of wool[287,288] with alcohols and a catalytic amount of hydrogen chloride was originally regarded as specific for carboxyl groups. However, later evidence shows that in wool the esterification causes some modification of amide and other groups.[289]

A variation of this technique was proposed by Bello,[290] in which thionyl chloride rather than hydrochloric acid is used as catalyst, and complete esterification of gelatin was attained in a few hours below 0°C. Unfortunately, when applied to wool, this method gives only 14% esterification.[291] Thus, the esterification of wool with methanol-containing hydrogen chloride at 20°C remains the most suitable and dependable method. The number of carboxyl groups in wool accessible to the esterification decreases with increasing size of the alcohol. For example, the extent of the modification of carboxyl groups at 60°C after 7 h in methanol, ethanol, *n*-propanol, and *n*-butanol has been found to be 71, 23, 14, and 7%, respectively.[292]

Like other protein esters, the esterified wool is fairly stable to acid but is very sensitive to even the mildest alkali. Esterification enhances settability by producing a net positive charge on the fiber.[292] As a result, esterified fibers have a higher internal pH than that of the external setting solution, owing to the Donnan effect. Thus the effect of esterification is essentially the same as that produced by increasing the pH of the setting solution, which is well known to promote setting. It is possible to nullify the Donnan effect by increasing the ionic strength of the setting solution. Indeed, in setting solutions containing 1 *M* potassium chloride, esterified fibers set at the same rate as untreated fibers. Esterified wool shows an increase in affinity for acid dyes, and dyed esterified wool shows an increase in dye fastness.[291] The basis for this increase in both cases is the conversion of anionic carboxyl groups into un-ionized ester groups.

#### *5.7.10.2. Epoxides*

Reagents of the epoxide type are not selective even under most favorable conditions; they can undergo nucleophilic substitutions with carboxyl, amino, hydroxyl, or sulfhydryl groups and even with imidazole nitrogens and the sulfur atom of methionine residues:

$$-NuH + H_2C\underset{O}{\diagdown\!\diagup}CHR \longrightarrow -NuCH_2\underset{\displaystyle OH}{\underset{|}{C}}HR$$

With wool, epichlorohydrin (R = $CH_2Cl$) has been found to esterify the carboxyl groups to a maximum of 50%, but other epoxides only react to an extent of 10%.[288] It can also act as a crosslinking agent.[286] The second reactive center in this molecule may be the chlorine atom, which is known to react readily with amino groups or, alternatively, another epoxide group may be formed by the elimination of hydrochloric acid under alkaline conditions. Phenyl glycidyl ether (R = $CH_2OC_6H_5$) has been shown to react rapidly with not only acidic amino acid residues but also lysine, histidine, and tyrosine residues in wool[293]; arginine and serine residues react rather slowly. The reaction between epichlorohydrin or phenyl glycidyl ether and wool in different solvent systems has been shown to take place in the aqueous phase rather than the organic phase.[294]

3,4-Isopropylidene-1,2,5,6-dianhydromannitol,[295] 1,2,3,4-diepoxybutane[296] and bis(2,3-epoxypropyl) ether[194] have been shown to be effective for crosslinking of wool keratin. It is likely that carboxyl and amino groups are both involved, as it is found that esterified fibers can also be crosslinked by 1,2,3,4-diepoxybutane.

In general, polyepoxide or unsaturated epoxide compounds (e.g., glycidyl methacrylate[297]) are used much often than monoepoxide compounds in the field of textile finishing. The main purpose of the application of these compounds to wool is to give shrink resistance. Reactive dyes containing the epoxide ring as the reactive system have been described.[133]

### *5.7.10.3. Carbodiimides*

Perhaps the most general and specific methods for modifying carboxyl groups in proteins involve the use of water-soluble carbodiimides. They react with carboxyl groups at slightly acidic pH to give an *O*-acylisourea, an activated intermediate that can, in turn, either rearrange to a more stable *N*-acylurea or react with a nucleophile. Reactions of carbodiimides have been reviewed.[298]

The introduction of amide and ester crosslinks into wool with water-soluble carbodiimides has been studied.[299] Both solubility and swelling measurements have indicated that diisopropyl carbodiimide in the presence of *N*-hydroxysuccinimide in dimethylformamide is an effective crosslinking agent. Both amino and carboxyl groups are modified, consistent with the

$$-\overset{\overset{O}{\|}}{C}-OH + R-N{=}C{=}N-R' \longrightarrow$$

$$\left[ -\overset{\overset{O}{\|}}{C}-O-C\begin{matrix} {=}N-R \\ \diagdown NH-R' \end{matrix} \right] \begin{matrix} \nearrow -\overset{\overset{O}{\|}}{C}-\underset{\underset{R}{|}}{N}-C\begin{matrix} {=}O \\ \diagdown NH-R' \end{matrix} \\ \underset{NuH}{\searrow} -\overset{\overset{O}{\|}}{C}-Nu + R-NH-\overset{\overset{O}{\|}}{C}-NH-R' \end{matrix}$$

introduction of amide crosslinks. Ester crosslinks are also formed by condensation of hydroxyl and carboxyl groups. The "crosslinked" peptides, $N^6$-$\beta$-aspartyllysine[300] and $N^6$-$\gamma$-glutamyllysine,[301] have been isolated from enzyme digests of wool. Increased amounts are formed in wool treated with diisopropyl carbodiimide.[106] The ratio of $N^6$-$\beta$-aspartyllysine to $N^6$-$\gamma$-glutamyllysine is 1:11, which indicates that in wool the glutamic acid side chains are more favorably situated for condensation reactions with $\epsilon$-amino groups of lysine residues than are the aspartic acid side chains.

Water-soluble carbodiimides also react with phenolic groups and sulfhydryl groups[302] in proteins. The adducts involving the phenolic group are stable at neutral pH, but can be decomposed by treatment with hydroxylamine; the adducts involving the sulfhydryl group appear to be more stable.

Cyanamide can readily undergo a wide variety of addition reactions with nucleophiles. Often these can be carried out in aqueous solution. Although spectroscopic data support the $N$-cyanoamine structure, $NH_2-C{\equiv}N$, the tautomeric carbodiimide structure, $HN{=}C{=}NH$, has at times been involved to explain the high chemical reactivity. The application of cyanamide to wool results in improvement in shrink resistance. It reacts readily above pH 10 with carboxyl, hydroxyl (alcoholic and phenolic), sulfhydryl, and amino groups in the fiber to form, respectively, the derivatives of acid amide, isourea, isothiourea, and guanidine.[303]

## 5.8. Reaction with Electrophilic Reagents

In contrast to the several types of nucleophilic reactions previously described, electrophilic substitution involves, primarily but not exclusively, the interaction of cationic species (or potentially cationic species) with $\pi$-electron systems, such as phenol, imidazole, or indole. Except for the classical methods of iodination and diazotization, and a more recently developed technique for nitration, electrophilic substitution has been a relatively unexplored area.

### 5.8.1. Iodine

Unlike the more reactive halogens, iodine does not oxidize the disulfide bond of wool but under most conditions reacts mainly with tyrosine residues to give the *o*-substituted mono- and diiodo derivatives.[304]

$$\text{CHCH}_2\text{—C}_6\text{H}_4\text{—O}^- \xrightarrow{I_2} \text{CHCH}_2\text{—C}_6\text{H}_3\text{(I)—O}^- \xrightarrow{I_2} \text{CHCH}_2\text{—C}_6\text{H}_2\text{(I)}_2\text{—O}^-$$

The reaction is influenced by the choice of solvent; most of the tyrosine residues in wool react in methanol or ethanol, but only half in *n*-propanol and hardly any in *n*-butanol.[305] These results were explained in terms of the pore size of wool and the ability of the different solvents to penetrate the fiber. It has also been suggested that, because tyrosine residues are almost completely iodinated in ethanol, they are situated in noncrystalline and, hence, accessible regions of the fiber. However, recent study[306] has indicated that the rate and extent of iodination of tyrosine residues with different solvents could arise largely from variations in the degree of ionization of the phenolic group. On this basis, the failure of tyrosine residues in SCMKB high-sulfur fraction or in peracetic acid oxidized wool to react with iodine has been explained in terms of the repression of ionization of the phenolic side-chain groups by the high net negative charge on the protein.

Another classical effect of iodine is the oxidation of sulfhydryl to disulfide groups. The reaction presumably involves the initial transient sulfenyl iodide group (—SI) and its hydrolyzed product, the sulfenic acid group (—SOH). Since both sulfenyl iodides and sulfenic acids are labile groups, products other than disulfide groups (e.g., sulfonic acid groups) can often result under certain conditions.

$$\text{CHCH}_2\text{—SH} \xrightarrow{I_2} \text{CHCH}_2\text{—SI} \xrightarrow{H_2O} \text{CHCH}_2\text{—SOH} \xrightarrow{\text{CHCH}_2\text{—SH}} \text{CHCH}_2\text{—S—S—CH}_2\text{CH}$$

$$\text{CHCH}_2\text{—SOH} \xrightarrow{[O]} \text{CHCH}_2\text{—SO}_3\text{H}$$

It is well established but not generally appreciated that both 5-iodohistidine and 3,5-diiodohistidine are formed under the usual conditions for iodination of proteins,[307] the reaction of histidine with iodine being only slightly slower than that of tyrosine. However, both iodinated derivatives break down during normal acid hydrolysis, making their detection difficult by usual procedures. Under harsh conditions, tryptophan and methionine residues may also be affected by iodination.[308]

### 5.8.2. Tetranitromethane (TNM)

Ortho nitration of tyrosine residues in proteins can be achieved rapidly and most favorably with TNM at pH 8 under mild conditions.[309,310] With longer reaction time, formation of 3,5-dinitrotyrosine occurs. The main side reaction is the oxidation of the sulfhydryl groups to the disulfide or sulfinic acid via the intermediate sulfenyl nitrate. The most probable scheme for the reaction of TNM with tyrosine or cysteine is indicated by the following equations:

$$\text{R}-\text{C}_6\text{H}_4-\text{OH} + (\text{NO}_2)_4\text{C} \longrightarrow \text{R}-\text{C}_6\text{H}_3(\text{NO}_2)-\text{O}^- + (\text{NO}_2)_3\text{C}^- + 2\text{H}^+$$

$$\text{R}-\text{SH} + (\text{NO}_2)_4\text{C} \longrightarrow \text{R}-\text{SNO}_2 + (\text{NO}_2)_3\text{C}^- + \text{H}^+$$

$$\text{R}-\text{SNO}_2 + \text{R}-\text{SH} \longrightarrow \text{R}-\text{S}-\text{S}-\text{R} + \text{NO}_2^- + \text{H}^+$$

$$\text{R}-\text{SNO}_2 + \text{H}_2\text{O} \longrightarrow \text{R}-\text{SOH} + \text{NO}_2^- + \text{H}^+$$

$$\text{R}-\text{SOH} + \tfrac{1}{2}\text{O}_2 \longrightarrow \text{R}-\text{SO}_2\text{H}$$

In the reaction, TNM and a phenoxide anion first form a charge-transfer complex, which undergoes a rate-determining electron transfer step to give $ArO\cdot$ and $NO_2\cdot$, which then rapidly condense to products.[311] The reaction rate is markedly enhanced at high pH and is exceedingly slow below pH 7. The extent of reaction can be followed by the increase in absorbance at 350 nm, owing to the release of nitroformate anion, or at 428 nm, owing to the nitrophenolate anion. Nitrotyrosine is stable in hot acid and may be estimated by amino acid analysis after acid hydrolysis. Tryptophan residues in some proteins may also react with TNM. The product of the reaction of TNM with $N^2$-acetyltryptophan is $N^2$-acetyl-7-nitrotryptophan.[312]

The treatment of wool and mohair with TNM results in only partial conversion of the tyrosine residues into 3-nitro- and 3,5-dinitrotyrosine residues.[313] The cystine content as well as the lysine and arginine content of the nitrated sample is decreased. The treated fiber shows a decreased solubility in the performic acid–ammonia test, presumably through the introduction of new crosslinks of yet-unknown chemical nature involving tyrosine residues. The apparent formation of crosslinks has also been observed in many other proteins.[314]

### 5.8.3. Nitric Acid

The treatment of proteins with aqueous nitric acid under mild conditions results primarily in the nitration of the aromatic amino acid residues. Under more severe conditions, oxidative reactions resulting in partial or complete destruction of various amino acid residues become increasingly more important.

Tyrosine, whether as the free amino acid or incorporated in proteins including silk fibroin and keratin, reacts readily with nitric acid to yield the 3-nitro derivative.[315] The tyrosine residues in human hair can be completely mononitrated by reaction with 1 *N* nitric acid for 24 h at 70°C, with little or no oxidation of the cystine residue.[316] When compared with that of untreated wool, the titration curve of the nitrated wool shows a marked increase in alkali- and acid-combining capacities.[247] The introduction of the nitro group to tyrosine residues decreases the pK value of the phenolic hydroxyl group. This effect, together with the liberation of free carboxyl groups by the hydrolysis of asparagine residues, accounts for the increased alkali-combining capacity. The increased acid-combining capacity is probably due to the titration of carboxyl groups liberated together with α-amino groups by partial hydrolysis of peptide bonds.

### 5.8.4. Nitrous Acid

When treated with nitrous acid, both free tyrosine and tyrosine bound to a protein become diazotized by a multistep reaction.The first step involves the formation of a 3-nitroso derivative, probably by an electrophilic substitution reaction. The 3-nitroso derivative is not very stable, being readily converted to the diazo compound by the reaction with nitric oxide formed by the decomposition of nitrous acid.[317] In the presence of metal ions (e.g., $Cu^{2+}$), the nitroso derivative is trapped as the brightly colored, stable, metal chelate, and no diazo grouping is formed. This complex-forming reaction constitutes a method for dyeing keratin fiber and silk fibroin[318] and has been used for histochemical location of tyrosine-rich protein(s) in keratin fibers.[319]

In wool, the treatment with nitrous acid results in decreased solubility of the wool in alkali and urea–bisulfite, indicating the formation of new crosslinks. In the presence of cupric ions, however, the alkali solubility is increased, indicating that the crosslinking reaction is inhibited by the formation of metal chelates.[320]

Tyrosine content in hydrolysates of wool and silk can be estimated by a method based on the Millon's reaction. In the presence of mercuric sulfate and sodium nitrite in sulfuric acid, tyrosine will give a red-blood coloration, owing to the chelation between the nitrosophenol and mercury.[321]

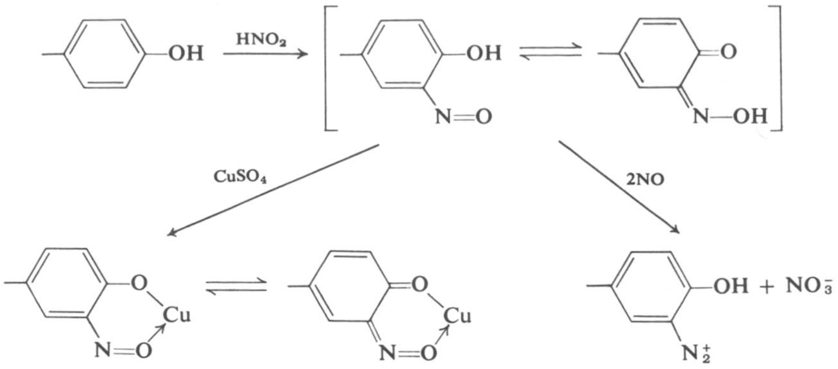

The reaction of primary amino groups in acid solution with nitrous acid leads to the formation of an unstable intermediate diazonium ion, which decomposes to form an equivalent of nitrogen and other products:

$$R{-}NH_2 + HONO \xrightarrow[-2H_2O]{H^+} [R{-}N_2^+] \longrightarrow N_2 + H_2O + \begin{matrix}\text{mixture of alcohols}\\ \text{and alkenes}\end{matrix}$$

This reagent was used by Van Slyke to deaminate and, from the nitrogen liberated, to quantitatively determine $\alpha$-amino acids.

Deamination with nitrous acid is the oldest method of removing the basic groups in wool, and this method was used by Speakman and Stott[322] in their classical research on the "salt link." However, in wool, guanidino groups of arginine residues, which provide the largest number of the free basic groups, react only very slowly. Thus, in order to remove most of the basic groups in wool, a very prolonged treatment (about 6 days) is necessary. Under these conditions, considerable oxidation of the disulfide bond and complete removal of tyrosine residues occurs.[323] Even with a solution of $\frac{1}{8}$ *M* sodium nitrite in $\frac{1}{4}$ *M* sodium acetate at pH 4 and 38°C, contents of all amino acids in wool are more or less altered after 24 h, lysine, phenylalanine, and tyrosine being particularly affected. As a method for the removal of amino groups or basic groups, it cannot, therefore, be regarded as satisfactory.

Deamination with nitrous acid is still widely used as an analytical procedure for determining the free amino content of proteins, including wool (e.g., 324). Asquith and Watson[325] have shown that, in the estimation of amide nitrogen in $\gamma$-keratose, the method of Van Slyke is the most reliable. Amide nitrogen is calculated to be 0.859 g of nitrogen per 100 g of $\gamma$-keratose (i.e., 83.2% of the glutamic or aspartic acids are present as their amides).

### 5.8.5. Diazonium Salts

Diazo coupling reactions involving other diazonium salts have been used extensively in chemical modification of proteins, for example for the affinity labeling of antibodies and for the selective modification of several enzymes.[326] They occur readily at neutral or slightly alkaline pH and low temperature. In addition to phenolic and imidazole groups, the amino, guanidino, and indole moieties can also react with diazonium compounds.

The colored bisazo product obtained when tyrosine and histidine compounds are treated with a diazonium reagent is the basis of several colorimetric procedures for determining these compounds. In wool, tyrosine residues can be estimated by allowing the wool to react with diazotized sulfanilic acid (Pauly reaction), then dissolving in sodium hydroxide solution and taking the absorbance at 325 nm.[327]

$$\text{CHCH}_2\text{—C}_6\text{H}_4\text{—O}^- + \text{Ar—N}_2^+ \longrightarrow \text{CHCH}_2\text{—C}_6\text{H}_2(\text{N=N—Ar})_2\text{—O}^- \qquad \text{Ar} = \text{—C}_6\text{H}_4\text{—SO}_3^-$$

Diazotized sulfanilic acid has been used to render $\beta$-keratose of wool water-soluble.[328]

## 5.9. Condensation with Carbonyl Reagents

### 5.9.1. Formaldehyde

Formaldehyde has been used often and for a longer time than any other protein reagent. Its reaction with wool and other proteins has been widely studied, and many types of side chains, for example amino, amido, guanidino, sulfhydryl, phenolic, imidazole, and indole, have been proposed as sites for mono- or bifunctional reaction.[329]

In neutral or alkaline solutions, the amino groups of proteins can react in a readily reversible manner with formaldehyde. The resulting displacement of the acid–base equilibrium by the reagent is the basis of the well-known formol titration of amino groups.

$$\text{R—NH}_2 + \text{HCHO} \rightleftharpoons \text{R—NHCH}_2\text{OH} \xrightleftharpoons{\text{HCHO}} \text{R—N(CH}_2\text{OH)}_2$$

During the treatment of keratin fibers with formaldehyde, much of the reagent is only loosely bound to the fiber, possibly in the form of labile methylol (hydroxymethyl) groups. Thus, by rinsing the treated fiber with water, there is a gradual loss of the reagent from the sample.

The site of reaction of wool with formaldehyde has been a subject of several investigations. McPhee[327] demonstrated that at least indole, amido, guanidino, and amino groups are all involved under alkaline conditions. In the case of the amino group, there are indications that the aminomethylol group formed can condense with amide groups and with other groups containing active hydrogen to give stable methylene bridges.

$$—NHCH_2OH + NH_2CO— \longrightarrow —NHCH_2NHCO— + H_2O$$

By means of enzymic hydrolysis, lysine and glutamide side chains have been identified as two of the many sites of reaction.[330]

In the case of the indole group of the tryptophan residue, it is likely that the initial product is the 2-methylodindole derivative, which could form a cyclic product by condensation with the amide linkage in the peptide chain or a crosslinked product by condensation with the free amino group[336]:

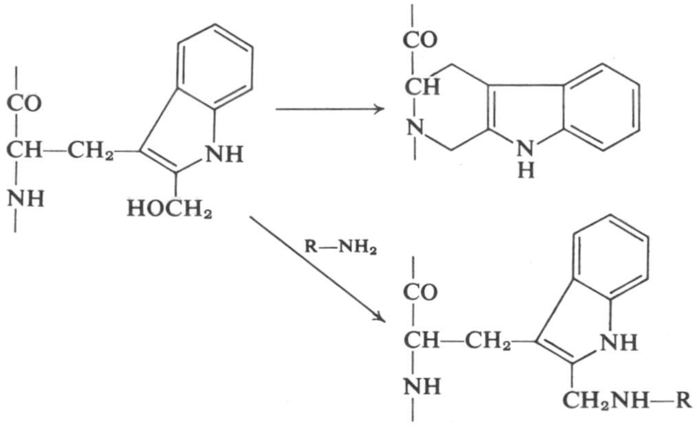


Phenolic groups of tyrosine residues react more extensively at low and high pH values than intermediate pH values to form the ortho mono- or dimethylol derivatives. Under alkaline conditions, electrophilic substitution of the phenoxide ion by formaldehyde occurs; under these conditions the formaldehyde exists mainly as methylene glycol, $CH_2(OH)_2$. At low pH values, it is probably the carbonium ion of formaldehyde, $\overset{+}{C}H_2OH$ (a superior electrophile to methylene glycol), that undergoes the reaction. The methylol derivative can also form crosslinks with amino groups, since the product of the Mannich condensation product has been isolated from acid hydrolysates of formaldehyde-treated proteins[338]:

$$NH_2-\underset{\displaystyle COOH}{CH}-CH_2-C_6H_3(OH)-CH_2-NH-(CH_2)_4-\underset{}{\overset{\displaystyle NH_2}{CH}}-COOH$$

Formaldehyde reacts with both cysteine residues and cystine residues to give thiazolidine-4-carboxylic acid.[337] Presumably, the *S*-methylol derivative is involved in the case of combined cysteines.

$$\overset{|}{CO}-CH(-CH_2-SCH_2OH)-\underset{|}{NH} \xrightarrow[-H_2O]{} \overset{|}{CO}-CH\langle CH_2-S \rangle N- \xrightarrow{\text{hydrolysis}} \text{COOH-thiazolidine (HN, S)}$$

Treatment of reduced wool with formaldehyde also introduces new crosslinks (probably methylenedithiol crosslinks), which stabilize the $\alpha$-helical conformation of keratin.[278,331]

$$\text{CHCH}_2-\text{SCH}_2\text{OH} + \text{HS}-\text{CH}_2\text{CH} \longrightarrow \text{CHCH}_2-\text{S}-\text{CH}_2-\text{S}-\text{CH}_2\text{CH}$$

Although there is no information about the nature of the reaction product concerning the imidazole group of histidine residues in wool, in other proteins it undergoes a similar reaction giving the $N^{\tau}$-methylol derivative.[332]

The treatment of wool with formaldehyde to introduce crosslinks is now well established.[333] A method for stabilizing set uses after-treatment with formaldehyde after setting shrink-resist wool with bisulfite.[76] Clearly, besides crosslinking sulfhydryl and other reactive groups, its effectiveness depends partly on (a) blocking of the sulfhydryl group by forming *S*-methylol derivatives, which could then form combined thiazolidine-4-carboxylic acid; (b) blocking of amino groups by forming *N*-methylol derivatives; and (c) reacting with and removing bisulfite from the system:

$$HSO_3^- + HCHO \rightleftharpoons HOCH_2SO_3^-$$

which would reverse the sulfitolysis reaction of the fiber:

$$\text{CHCH}_2-\text{SSO}_3^- + \text{HS}-\text{CH}_2\text{CH} \rightleftharpoons \text{CHCH}_2-\text{S}-\text{S}-\text{CH}_2\text{CH} + \text{HSO}_3^-$$

and so provide another way of removing sulfhydryl groups from the system, thus inhibiting chemical stress relaxation via the thiol–disulfide interchange.

### 5.9.2. Other Aldehydes

4-Dimethylaminobenzaldehyde condenses with the indole ring of tryptophan under acid conditions to form colored products (Ehrlich test).

Although the detailed chemistry is not certain, reaction of the aldehyde at the 2-position of the indole nucleus is initially involved:

$$(CH_3)_2N-C_6H_4-CHO + \text{3-R-indole} \xrightarrow[-H_2O]{H^+} (CH_3)_2N-C_6H_4-CH{=}\text{(2-indolenine, 3-R)}$$

Thus tryptophan contents of keratin fibers can be determined colorimetrically with 4-dimethylaminobenzaldehyde after hydrolysis with sulfuric acid,[334] hydrochloric acid,[335] or barium hydroxide.[59] In the case of fabrics, the tryptophan content can be conveniently estimated by measuring the reflectance at 480 nm of a piece of fabric that has been treated with the aldehyde in hydrochloric acid solution.[327]

Treatment of wool with bifunctional aldehydes such as glyoxal, glutaraldehyde, and terephthaldehyde results in a diminishing of the total amino content as well as a simultaneous reduction of lysine and histidine contents.[339] The treated fiber shows a decrease in acid, alkali, and urea–bisulfite solubilities, as well as improved mechanical properties, indicating the formation of new stable crosslinks. Glutaraldehyde is superior in the stabilization of wool, but the treated fiber acquires a yellow-brown coloration. It has been shown that this discoloration can be avoided if the aldehyde is used as the sodium bisulfite adduct. During treatment, the aldehyde is gradually released and, at the same time, the sodium bisulfite produced reacts with the cystine disulfide to give *S*-sulfocysteine and sulfhydryl groups, which, under subsequent oxidizing conditions, partly form disulfide linkages again.[340]

$$OHC-(CH_2)_3-CHO + NaHSO_3 \rightleftharpoons NaO_3S-\underset{OH}{\underset{|}{HC}}-(CH_2)_3-\underset{OH}{\underset{|}{CH}}-SO_3Na$$

Owing to the combined effect of the setting action of sodium bisulfite and the stabilizing crosslinking of glutaraldehyde, the glutaraldehyde–sodium bisulfite adduct is particularly suited for the setting of wool. In general, the chemical basis of the reaction between protein and glutaraldehyde remains obscure, but it appears to involve the side chains of lysine, cysteine, histidine and tyrosine residues. The low recovery of lysine from glutaraldehyde-treated proteins suggests that the reaction product involving the amino group is not a Schiff base, since Schiff bases are quite easily hydrolyzed back to amines.

It has been proposed that α,β-unsaturated aldehydes are formed by aldol condensation of glutaraldehyde. The unsaturated aldehydes react with amino groups in a Michael-type condensation to give secondary amines, which are stable to hydrolysis.[341]

$$\mathrm{OHC(CH_2)_3CHO} \longrightarrow \mathrm{OHC(CH_2)_3CH{=}\underset{}{\overset{\displaystyle CHO}{\overset{|}{C}}}CH_2CH_2CHO},$$

$$\mathrm{OHC(CH_2)_3CH{=}\overset{\displaystyle CHO}{\overset{|}{C}}CH_2\overset{\displaystyle CHO}{\overset{|}{C}}{=}CH(CH_2)_3CHO,\ etc.} \xrightarrow{\mathrm{RNH_2}}$$

$$\mathrm{\sim\underset{\underset{\displaystyle RNH}{|}}{CH}{-}\overset{\displaystyle CHO}{\overset{|}{CH}}{-}CH_2{-}\overset{\displaystyle CHO}{\overset{|}{CH}}{-}\underset{\underset{\displaystyle NHR}{|}}{CH}\sim}$$

Recently, using simple alkyl amines as model compounds, Lubig and Zahn[342] have isolated a polyether consisting of 2,6-dihydroxytetrahydropyran (A) and *N*-alkyl-2,6-dihydroxypiperidine (B). They suggest that under aqueous conditions, the cyclic form of glutaraldehyde, 2,6-dihydroxytetrahydropyran (A), reacts with the amine to form the unstable *N*-alkyl-2,6-dihydroxypiperidine (B), which either loses water to give *N*-alkyldihydropyridine (C) or condenses with (A) to yield the cooligomer (D).

$$\text{glutaraldehyde} \underset{}{\overset{\mathrm{H_2O}}{\rightleftharpoons}} \text{(A)} \xrightarrow{\mathrm{RNH_2}} \text{(B)} \xrightarrow{-2\mathrm{H_2O}} \text{(C)} \longrightarrow \text{other products}$$

$$\text{(A)} + \text{(B)} \longrightarrow \text{(D)}$$

The guanidino groups of arginine residues are chemically distinct from the ε-amino groups of lysine residues; they remain protonated within the pH range of protein stability and possess an inherent resonance stabilization. It has been postulated that in wool they may combine with glyoxal bisulfite to give a cyclic derivative.[192] The treated wool acquires good acid dye-resistant properties. The method, however, causes damage to the fiber and hence does not appear to have been used commercially.

$$\text{CH(CH}_2)_3\text{NHC}(=\text{NH})\text{NH}_2 + \text{HO–CH(SO}_3\text{Na)–CHO} \longrightarrow \text{CH(CH}_2)_3\text{NHC}{<}^{\text{N–CH–SO}_3\text{Na}}_{\text{N=CH}}$$

### 5.9.3. 1,2- and 1,3-Dicarbonyl Compounds

Recognition of the enhanced reactivity associated with intramolecular reactions, and of the possible stabilization of condensation products by the formation of heterocyclic rings, has led to studies of the condensation of 1,2- and 1,3-dicarbonyl compounds with arginine residues, for example malonaldehyde,[343] 1,2-cyclohexanedione,[344] and benzil.[345] However, the strong acid or alkaline conditions required for the above condensations may introduce nonspecific peptide and disulfide bond cleavage.

$$\text{CH(CH}_2)_3\text{NHC}(=\overset{+}{\text{N}}\text{H}_2)\text{NH}_2 + \text{O=CH–CH}_2\text{–CH=O} \xrightarrow{10\text{ N HCl}} \text{CH(CH}_2)_3\overset{+}{\text{N}}\text{H}_2\text{–(pyrimidin-2-yl)} + 2\text{H}_2\text{O}$$

$$\text{CH(CH}_2)_3\text{NHC}(=\text{NH})\text{NH}_2 + \text{1,2-cyclohexanedione} \xrightarrow{0.2\text{ N NaOH}} \text{CH(CH}_2)_3\text{N=(bicyclic imidazolidinone spiro-cyclopentane)} + \text{H}_2\text{O}$$

On the other hand, glyoxal[346] and phenylglyoxal[347] have been claimed to react with arginine residues in aqueous solutions at pH 8–10 and 7–8, respectively. Although the nature of the glyoxal condensation product has not been elucidated, phenylglyoxal is believed to react with the guanidino group to give a bicyclic derivative, which is stable to mild acid conditions.

$$\text{CH(CH}_2)_3\text{NHC}(=\text{NH})\text{NH}_2 + 2\text{C}_6\text{H}_5\text{–}\overset{\text{O}}{\overset{\|}{\text{C}}}\text{–}\overset{\text{O}}{\overset{\|}{\text{C}}}\text{–H} \longrightarrow$$

$$\text{CH(CH}_2)_3\text{NH–(bicyclic N,O-ring, C}_6\text{H}_5\text{)–}\overset{\text{O}}{\overset{\|}{\text{C}}}\text{–C}_6\text{H}_5 + \text{H}_2\text{O}$$

Chemical modification of wool with dicarbonyl compounds has been achieved with dimethyl sulfoxide as solvent and triethylamine as base.[348] While 1,2-cyclohexanedione and di(2-pyridyl)glyoxal modify only the arginine side chain, acenaphthenequinone and sodium 1,2-naphthoquinone-4-sulfonate are selective for lysine residues. Other compounds react with both the arginine and the lysine side chain.

## 5.10. Reaction with Reactive Dyes

The introduction of reactive dyes is the outstanding achievement in dyestuff chemistry of recent years. These dyes differ from all other classes in that they contain functional groups capable of forming covalent bonds with the fiber, imparting good wet-fastness properties to the dyeings. The development in the field of reactive dyes since the summary by Zollinger in 1961[349] has been the subject of several valuable reviews (e.g., 350, 351). The main ranges of reactive dyes currently being applied to wool are listed in Table 5-1. In general, they can be allocated either to the nucleophilic-substitution type or to the nucleophilic-addition type, according to the mechanism of their reaction with nucleophilic groups of the fiber.

Table 5-1. Reactive Systems in Reactive Dyes for Wool

| Reactive system | Structure | Dye range |
|---|---|---|
| *s*-Triazinyl (monochloro-, R = H) | 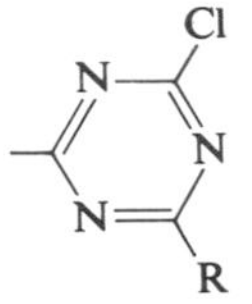 (triazine ring: Cl, N, N, N, R) | Procion H (ICI), Cibacron (CGY), Cibacrolan (CGY) |
| *s*-Triazinyl (dichloro-, R = Cl) | | Procion M (ICI) |
| Pyrimidinyl (trichloro-) | 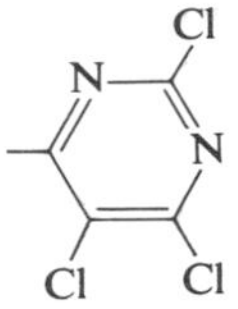 (pyrimidine ring: Cl, N, N, Cl, Cl) | Reactone (CGY), Drimarene (S) |
| Pyrimidinyl (2,4-difluoro-5-chloro-) | 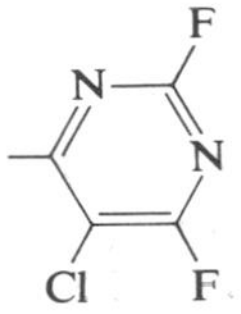 (pyrimidine ring: F, N, N, Cl, F) | Verofix (FBy), Drimalan F (S) |
| Quinoxolinyl (dichloro-) | 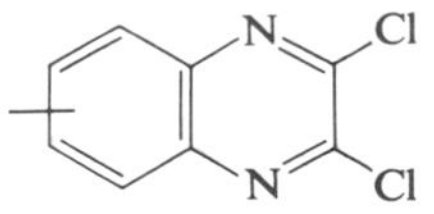 (quinoxaline ring: N, Cl, N, Cl) | Levafix E (FBy) |
| Chloroacetamido | $—NHCOCH_2Cl$ | Cibalan Brilliant (CGY), Drimalan (S) |
| *β*-Sulfatoethylsulfamoyl | $—SO_2NRCH_2CH_2OSO_3H$ | Levafix (FBy) |
| Vinyl sulfonyl | $—SO_2CH{=}CH_2$ | Remazol, Remalon, and Remazolan (FH) |
| Acrylamido | $—NHCOCH{=}CH_2$ | Primazin (BASF) |
| *α*-Bromoacrylamido | $—NHCOCBr{=}CH_2$ | Lanasol (CGY) |

### 5.10.1. Nucleophilic Substitutions

In this type of dye, the reaction involves the attack of fiber nucleophiles on the dye reactive group:

$$\text{fiber—NuH} + \text{dye—X—L} \longrightarrow \text{dye—X—Nu—fiber} + \text{HL}$$

where —NuH = —SH, —$NH_2$, or —OH; L = leaving group; and X = activating group. In Table 5-1, reactive dyes with *s*-triazinyl, pyrimidinyl, quinoxolinyl, chloroacetyl, or $\beta$-sulfatoethylsulfamoyl groups as reactive systems belong to this type of reactive dyes.

The activation of reactive dyes containing the halogenated heterocyclic system is due to the electron-withdrawing properties of the nitrogen atoms. The activity of the reactive halogen is thus determined by the other substituents on the heterocyclic nucleus, including the dye residue, and by the nature of the heterocyclic nucleus itself. The chlorine atoms on *s*-triazines are, in general, more active than those on pyrimidines, and those on dichlorotriazines are the most reactive of all.

Another way in which reactivity is developed is through the use of strained, highly reactive three- or four-membered heterocyclic rings. Ring openings by nucleophilic substitution result in the formation of new bonds with the substrate. In Levafix dyes, the ring is a cyclic aziridinium cation intermediate formed by the participation of a neighboring group[352]:

$$\text{dye—SO}_2\text{N(R)—CH}_2\text{—CH}_2\text{—OSO}_3\text{H} \longrightarrow \left[\text{dye—SO}_2\text{—}\overset{+}{\text{N}}(\text{R})\text{<}\begin{matrix}\text{CH}_2\\ |\\ \text{CH}_2\end{matrix}\right]^+ \xrightarrow{\text{fiber—NuH}} \text{dye—SO}_2\text{—N(R)CH}_2\text{CH}_2\text{Nu—fiber}$$

These dyes differ from the other reactive dyes in that they do not have sulfonic or carboxyl groups, and the only solubilizing groups are the sulfatoethyl groups in the reactive system. Normally, at least two such groups are present on each molecule, and in this case, if the $\beta$-nitrogen carries a hydrogen atom (R = H), intermolecular alkylation of the sulfamide nitrogen may lead to polymeric products, one or both ends of which may or may not be bound to the fiber.

### 5.10.2. Nucleophilic Additions

In this type of dye, covalent-bond formation is by the 1,2-trans addition of a nucleophile in the fiber across a polarized ethylenic bond:

$$\text{dye—X—CH=CH}_2 + \text{fiber—NuH} \rightleftharpoons \text{dye—X—CH}_2\text{CH}_2\text{—Nu—fiber}$$

In Table 5-1, reactive dyes with vinyl sulfonyl, acrylamido, or α-bromoacrylamido groups as reactive systems belong to this type of reactive dyes.

In the commercial forms of Remazol dyes, the vinyl groups are usually stabilized as their sulfuric acid esters. A strong base effects elimination of the protective acid group to give the reactive vinyl form:

$$\text{dye—X—CH}_2\text{CH}_2\text{OSO}_3\text{H} \xrightarrow{\text{OH}^-} \text{dye—X—CH=CH}_2$$

## 5.10.3. Hydrolysis of Reactive Dyes

In the practical application of reactive dyes, the aqueous solvent also provides hydroxyl groups, which compete with the nucleophilic sites in the fiber during the dyeing process.

$$\text{dye—X—L} + \text{H}_2\text{O} \longrightarrow \text{dye—X—OH} + \text{HCl}$$

$$\text{dye—X—CH=CH}_2 + \text{H}_2\text{O} \rightleftharpoons \text{dye—X—CH}_2\text{CH}_2\text{OH}$$

The sensitivity of hydrolysis in a given process is dependent on the chemical structure of the dye, pH, and temperature of the dye bath, and there is little possibility of completely eliminating this side reaction for all the existing commercial reactive dyes. The presence of hydrolyzed dye, which is not covalently bound to the fiber, can give rise to lower wet-fastness properties, and in most cases must be eliminated by thorough washing. In determining the exhaustion-fixation characteristics of reactive dyes on wool, acidic pyridine solution is widely used to remove all the hydrolyzed dye and unreacted dye from the fiber. Although some fiber degradation occurs, it appears that this is less than when constant-boiling pyridine is used.[353]

## 5.10.4. Dyeing Human Hair with Reactive Dyes

The increased permanence of wool dyeings achievable with commercial reactive dyes has attracted the attention of the hair dyers. In general, the application of reactive dyes to hair involves the reducing of some of the cystine bonds and bonding a water-soluble reactive dye to the sulfhydryl groups produced. For example, in the presence of hydrogen bond breakers such as urea or lithium bromide in aqueous solution, reactive dyes can be bonded to reduced hair in practical periods.[354] Processes involving prior treatment of the hair with a peroxy compound have also been described.[355]

To protect hair dyed with reactive dyes from being dyed for a second time during coloring of the newly grown hair, thereby developing a deeper shade, the dyed hair may be treated with a solution of a compound capable of irreversibly blocking the remaining untreated sulfhydryl groups, such as formaldehyde, quinones, epoxy compounds, maleic acid amides, or other *S*-alkylating agents.

## 5.11. References

1. R. B. Martin, *J. Amer. Chem. Soc.* **84**, 4130 (1962).
2. C. O'Connor, *Quart. Rev.* **24**, 553 (1970).
3. R. L. Hill, *Advan. Protein Chem.* **20**, 37 (1965).
4. A. Light, *Methods Enzymol.* **11**, 417 (1967).
5. R. B. Martin and R. L. M. Synge, *Biochem. J.* **35**, 1358 (1941).
6. S. Moore and W. M. Stein, *J. Biol. Chem.* **176**, 367 (1948).
7. S. H. Leach and E. M. J. Parkhill, *Proc. Intern. Wool Textile Res. Conf., Australia* **C**, 92 (1955).
8. A. C. Chibnal, C. Haselback, J. L. Mangen, and M. W. Rees, *Biochem. J.* **68**, 122 (1958).
9. P. E. Wilcox, *Methods Enzymol.* **11**, 63 (1967).
10. R. Consden, A. H. Gordon, and A. J. P. Martin, *Biochem. J.* **44**, 548 (1948).
11. R. Consden, *J. Textile Inst.* **40**, P814 (1949).
12. R. S. Asquith and D. K. Chan, *J. Soc. Dyers Colourists* **87**, 181 (1971).
13. R. S. Asquith and P. Carthew, *Biochim. Biophys. Acta* **278**, 347 (1972).
14. R. S. Asquith and P. Carthew, *Tetrahedron* **28**, 4769 (1972).
15. R. S. Asquith, P. Carthew, D. Hanna, and M. S. Otterburn, *J. Soc. Dyers Colourists* **90**, 357 (1974).
16. R. S. Asquith and D. C. Parkinson, *Proc. Soc. Anal. Chem.* **5**, 206 (1968).
17. R. S. Asquith and T. Shaw, *Makromol. Chem.* **115**, 198 (1968).
18. D. A. Long and J. E. Lilleycrop, *Trans. Faraday Soc.* **59**, 907 (1963).
19. D. A. and T. G. Truscott, *Trans. Faraday Soc.* **59**, 918 (1963).
20. W. H. Stein, S. Moore, and M. Bergmann, *J. Biol. Chem.* **154**, 191 (1944).
21. T. Haylett and L. S. Swart, *Textile Res. J.* **39**, 917 (1969).
22. M. C. Corfield and J. C. Fletcher, *Biochemistry* **115**, 323 (1969)
23. H. G. Fröhlich, *J. Textile Inst.* **51**, T1237 (1960).
24. H. Zahn, *J. Soc. Dyers Colourists* **76**, 226 (1960).
25. E. Hille and H. Zahn, *J. Textile Inst.* **51**, T1162 (1960).
26. N. La France, K. Ziegler, and H. Zahn, *J. Textile Inst.* **51**, T1168 (1960).
27. R. L. Elliott, R. S. Asquith, M. E. P. Hopper, and D. H. Rawson, *J. Soc. Dyers Colourists* **76**, 222 (1960).
28. R. L. Elliott, R. S. Asquith, and B. J. Gordan, *J. Soc. Dyers Colourists* **77**, 345 (1961).
29. J. Knott, *Wool Sci. Rev.* **41**, 2 (1971).
30. J. Knott and H. Zahn, *Appl. Polymer Symp.* **18**, 601 (1971).
31. R. L. Elliott, R. S. Asquith, and D. H. Rawson, *J. Soc. Dyers Colourists* **74**, 173 (1958).
32. R. L. Elliott, R. S. Asquith, and D. H. Rawson, *J. Soc. Dyers Colourists* **74**, 176 (1958).

33. K. Iwai and T. Ando, *Methods Enzymol.* **11**, 263 (1967).
34. S. M. Partridge and H. F. Davis, *Nature* (*Lond.*) **165**, 62 (1950).
35. G. Biserte and P. Pigache, *Bull. Soc. Chim. Biol.* **33**, 1379 (1951).
36. G. Biserte and P. Pigache, *Bull. Soc. Chim. Biol.* **34**, 51 (1952).
37. S. Blackburn and G. R. Lee, *Proc. 3rd Intern. Wool Textile Res. Conf., Paris* **1**, 321 (1965).
38. S. J. Leach, *Proc. Intern. Wool Textile Res. Conf., Australia* **C**, 181 (1955).
39. J. Schultz, *Methods Enzymol.* **11**, 255 (1967).
40. F. Sanger, *Biochem. J.* **45**, 563 (1949).
41. R. L. Elliott, R. S. Asquith, and D. H. Rawson, *J. Soc. Dyers Colourists* **73**, 424 (1957).
42. F. Sanger, A. P. Ryle, L. F. Smith, and R. Kitai, *Proc. Intern. Wool Textile Res. Conf., Australia* **C**, 49 (1955).
43. H. N. Christensen, *J. Biol. Chem.* **151**, 319 (1943).
44. H. N. Christensen, *J. Biol. Chem.* **154**, 427 (1944).
45. R. L. M. Synge, *Biochem. J.* **38**, 285 (1944).
46. R. L. M. Synge, *Biochem. J.* **39**, 351 (1945).
47. E. Harfenist, *J. Amer. Chem. Soc.* **75**, 5528 (1953).
48. R. Hirohata et al., *Z. Physiol. Chem.* **295**, 368 (1955).
49. M. Muramatu et al., *Z. Physiol. Chem.* **332**, 256 (1963).
50. M. Muramatu et al., *Z. Physiol. Chem.* **332**, 271 (1963).
51. R. E. Whitfield, *Science* (*Wash., D.C.*) **142**, 577 (1963).
52. E. Elöd, H. Nowotny, and H. Zahn, *Melliand Textilber.* **23**, 577 (1942).
53. J. Steinhardt and C. H. Fugitt, *J. Res. Natl. Bur. Std.* **29**, 315 (1942).
54. L. Peters and J. B. Speakman, *J. Soc. Dyers Colourists* **65**, 63 (1949).
55. R. V. Peryman, *J. Soc. Dyers Colourists* **73**, 455 (1957).
56. M. M. Breuer, *J. Phys. Chem.* **68**, 2067 (1964).
57. M. M. Breuer and D. M. Pritchard, *J. Soc. Cosmetic Chem.* **18**, 643 (1967).
58. M. C. Corfield and A. Robson, *Biochem. J.* **59**, 62 (1955).
59. K. K. Juneja, A. A. Sule, and V. B. Chipalkalti, *Textile Res. J.* **38**, 461 (1968).
60. I.W.T.O., Specification of Test Method, IWTO-4-60(E).
61. M. J. Horn and D. B. Jones, *J. Biol. Chem.* **139**, 473 (1941).
62. M. J. Horn, D. B. Jones, and S. J. Ringel, *J. Biol. Chem.* **138**, 141 (1941).
63. A. Patchornik and J. Sokolovsky, *J. Amer. Chem. Soc.* **86**, 1860 (1964).
64. Z. Bohak, *J. Biol. Chem.* **239**, 2878 (1964).
65. K. Ziegler, *J. Biol. Chem.* **239**, 2713 (1964).
66. K. Ziegler, *Proc. 3rd Intern. Wool Textile Res. Conf., Paris* **2**, 403 (1965).
67. R. S. Asquith and J. J. García-Domínguez, *J. Soc. Dyers Colourists* **84**, 155 (1968).
68. D. S. Tarbell and D. P. Harnish, *Chem. Rev.* **49**, 1 (1951).
69. O. Gacuron and G. Odstrchel, *J. Amer. Chem. Soc.* **89**, 3263 (1967).
70. M. C. Corfield et al., *Biochem. J.* **103**, 15c (1967).
71. P. Mellet, *Textile Res. J.* **38**, 977 (1968).
72. J. M. Swan, *Proc. Wool Textile Res. Conf., Australia* **C**, 25 (1955).
73. J. M. Swan, *Nature* (*Lond.*) **179**, 965 (1957).
74. J. J. García-Domínguez et al., *Appl. Polymer Symp.* **18**, 269 (1971).
75. K. Ziegler, I. Melchert, and C. Lurken, *Nature* (*Lond.*) **214**, 404 (1967).
76. J. R. Cook and J. Delmenico, *J. Textile Inst.* **62**, 27 (1971).
77. A. J. Farnworth and J. Delmenico, *Permanent Setting of Wool*, Merrow, Watford (1971).
78. A. Parisot and J. Derminot, *Bull. Inst. Textile Fr.* **24**, 603 (1970).

79. A. Parisot et al., *Bull. Sci. Inst. Textile Fr.* **1**, 173 (1972).
80. A. Parisot, J. Deminot, and M. Tasdhomme, *Bull. Inst. Textile Fr.* **25**, 385 (1971).
81. R. S. Asquith and P. Carthew, *Biochim. Biophys. Acta* **278**, 8 (1972).
82. R. S. Asquith and J. D. Skinner, *Textilveredlung* **5**, 406 (1970).
83. A. Robson and R. S. Stringer *J. Soc. Dyers Colourists* **88**, 27 (1972).
84. R. S. Asquith and J. J. García-Domínguez, *J. Soc. Dyers Colourists* **84**, 211 (1968).
85. P. Miro, J. J. García-Domínguez, and J. L. Parra, *J. Soc. Dyers Colourists* **85**, 407 (1969).
86. P. Miro and J. J. García-Domínguez, *J. Soc. Dyers Colourists* **83**, 91 (1967).
87. P. Miro and J. J. García-Domínguez, *J. Soc. Dyers Colourists* **84**, 310 (1968).
88. H. E. Jass and L. S. Fosdick, *Textile Res. J.* **25**, 343 (1955).
89. E. O. P. Thompson and I. J. O'Donnell, *Aust. J. Biol. Sci.* **15**, 757 (1962).
90. J. A. Maclaren, *Aust. J. Chem.* **15**, 824 (1962).
91. K. W. Herrmann, *Trans. Faraday Soc.* **59**, 1663 (1963).
92. J. Menkart, L. J. Wolfram, and I. Mao, *J. Soc. Cosmetic Chem.* **17**, 769 (1966).
93. M. E. Hillburn, P. T. Speakman, and R. E. Yardwood, *Biochem. Biophys. Acta* **214**, 245 (1970).
94. J. P. E. Human and P. H. Springell, *Aust. J. Chem.* **12**, 508 (1959).
95. R. S. Asquith and A. K. Puri, *Textile Res. J.* **40**, 273 (1970).
96. R. S. Asquith and A. K. Puri, *J. Soc. Dyers Colourists* **86**, 449 (1970).
97. R. S. Asquith and A. K. Puri, *J. Soc. Dyers Colourists* **87**, 116 (1971).
98. Oreal, Brit. Pat. 833, 809, 835, 247, 849, 045 (1960).
99. W. W. Cleland, *Biochemistry* **3**, 480 (1964).
100. H. D. Weigmann, *J. Polymer Sci.* **6**, 2237 (1968).
101. A. S. Inglis and T. Y. Liu, *J. Biol. Chem.* **245**, 112 (1970).
102. J. A. Maclaren, D. J. Kilpatrick, and A. Kirkpatrick, *Aust. J. Biol. Sci.* **21**, 805 (1968).
103. J. A. Maclaren and D. J. Kilpatrick, *Aust. J. Biol. Sci.* **22**, 1081 (1969).
104. D. J. Kilpatrick and J. A. Maclaren, *Textile Res. J.* **40**, 25 (1970).
105. M. Cole et al. *Appl. Polymer Symp.* **18**, 145 (1971).
106. B. Milligan, L. A. Holt, and J. B. Caldwell, *Appl. Polymer Symp.* **18**, 113 (1971).
107. M. Friedman and A. T. Noma, *Textile Res. J.* **40**, 1073 (1970).
108. L. J. Wolfram, *Proc. 3rd Intern. Wool Textile Res. Conf., Paris* **2**, 505 (1965).
109. M. J. Williams, *Textile Chem. Colorists* **2**, 41 (1970).
110. A. D. Jenkins and L. J. Wolfram (to Gillette Co.), Brit. Pat. 976, 821 (1964).
111. L. J. Wolfram, *Appl. Polymer Symp.* **18**, 523 (1971).
112. L. J. Wolfram, *J. Soc. Cosmetic Chem.* **20**, 539 (1969).
113. W. R. Middlebrook and H. Phillips, *Biochem. J.* **36**, 428 (1942).
114. G. Valk, *Proc. 3rd Intern. Wool Textile Res. Conf., Paris* **2**, 481 (1965).
115. L. J. Wolfram and D. L. Underwood, *Textile Res. J.* **36**, 947 (1966).
116. S. J. Leach, *Aust. J. Chem.* **13**, 520 (1960).
117. S. J. Leach, *Aust. J. Chem.* **13**, 547 (1960).
118. R. Cecil and R. G. Wake, *Biochem. J.* **82**, 401 (1962).
119. J. A. Maclaren, S. J. Leach, and J. M. Swan, *J. Textile Inst.* **51**, T665 (1960).
120. J. M. Swan, *Aust. J. Chem.* **14**, 69 (1960).
121. J. L. Bailey and R. D. Cole, *J. Biol. Chem.* **234**, 1733 (1959).
122. L. Wolfram and K. Bruggeman, *TGA Cosmetic J.* **2**, 45 (1970).
123. D. Y. Hsiung and L. J. Wolfram (to Gillette Co.), U.S. Pat. 3,644,084 (1972).
124. R. Whitman and M. D. Beste (to Rayette, Inc.), U.S. Pat. 2,840,086 (1958).
125. I.W.T.O., Specification of Test Method, IWTO-15-66(E).

126. I.W.T.O., Specification of test Method, IWTO-11-65(E).
127. J. M. Gillespie, *Aust. J. Biol. Sci.* **17**, 282 (1964).
128. J. B. Speakman, J. L. Stoves, and H. Bradbury, *J. Soc. Dyers Colorists* **57**, 73 (1941).
129. G. Blankenburg, *Melliand Textilber.* **48**, 686 (1967).
130. H. Zahn and F. Osterloh, *Proc. Intern. Wool Res. Conf., Australia* **C**, 18 (1955).
131. G. Heidemann and H. Halboth, *Textile Res. J.* **40**, 861 (1970).
132. A. Schöberl and G. Bauer, *Fette, Seifen, Anstrichmittel* **60**, 1061 (1958).
133. W. F. Beech, *Fibre Reactive Dyes*, Logos, London (1970).
134. F. Osterloh, *Melliard Textilber.* **44**, 57 (1963).
135. R. E. Randebrock, U.S. Pat. 3,415,606 (1968).
136. J. M. Gillespie, *Nature* (*Lond.*) **183**, 322 (1959).
137. H. Bogaty and A. E. Brown (to Gillette Co.), U.S. Pat. 2,766,760 (1956).
138. J. B. Speakman, *Proc. Intern. Wool Textile Res. Conf., Australia* **C**, 474 (1955).
139. A. C. Chilbnall and M. W. Rees, *Biochem. J.* **68**, 105 (1958).
140. S. Takahashi and L. A. Cohen, *Biochemistry* **8**, 864 (1969).
141. N. P. Newmann, *Methods Enzymol.* **11**, 485 (1967).
142. O. Schmidt, *Z. Ges. Textil-Ind.* **66**, 849 (1964).
143. J. Cegarra, J. Gacan, and J. Ribe, *J. Soc. Dyers Colourists* **84**, 457 (1968).
144. P. Alexander, D. Carter, and C. Earland, *Biochem. J.* **47**, 251 (1950).
145. A. L. Smith and M. Harris, *J. Res. Natl. Bur. Std.* **16**, 301 (1936).
146. L. J. Wolfram, K. Hall, and I. Hui, *J. Soc. Cosmetic Chem.* **21**, 875 (1970).
147. H. A. Rutherford and M. Harris, *J. Res. Natl. Bur. Std.* **20**, 559 (1938).
148. R. Consden and A. H. Gordon, *Biochem. J.* **46**, 8 (1950).
149. A. S. Inglis and I. H. Leaver, *Textile Res. J.* **37**, 995 (1967).
150. H. Zahn, *J. Soc. Cosmetic Chem.* **17**, 687 (1966).
151. G. A. Erlemann and H. Beyer, *J. Soc. Cosmetic Chem.* **22**, 795 (1971).
152. C. Robbins, *J. Soc. Cosmetic Chem.* **22**, 339 (1971).
153. G. A. Erlemann and H. Beyer, *J. Soc. Cosmetic Chem.* **23**, 791 (1972).
154. G. Laxer and C. S. Whewell, *Proc. Intern. Wool Textile Res. Conf., Australia* **F**, 186 (1955).
155. E. Slobodian et al., 156th Meeting Amer. Chem. Soc., Biol. Chem. Div., Abstr. 166 (1968).
156. H. Zahn, R. Gerthsen, and M. L. Kehren, *J. Soc. Cosmetic Chem.* **14**, 529 (1963).
157. S. Blackburn and A. G. Lowther, *Biochem. J.* **49**, 554 (1951).
158. P. Alexander, D. Carter, and C. Earland, *J. Soc. Dyers Colourists* **67**, 23 (1951).
159. I.W.T.O., Specification of Test Method, IWTO 23-70(E).
160. P. Alexander, M. Fox, and R. F. Hudson, *Biochem. J.* **49**, 129 (1951).
161. B. J. Sweetman and J. A. Maclaren, *Textile Res. J.* **35**, 322 (1965).
162. H. L. Needles, *Textile Res. J.* **35**, 298 (1965).
163. M. Burke and C. H. Nicholls, *Textile Res. J.* **38**, 891 (1968).
164. A. Meybeck and J. Meybeck, *European Polymer J.* **3**, 223 (1967).
165. A. Kantouch and A. Bendak, *Textile Res. J.* **37**, 483 (1967).
166. M. M. Breuer and A. D. Jenkins, *Proc. 3rd Intern. Wool Textile Res. Conf., Paris* **2**, 447 (1965).
167. F. R. Andrews, Ph.D. thesis, University of Sussex (1969).
168. J. A. Maclaren, W. E. Savige, and B. J. Sweetman, *Aust. J. Chem.* **18**, 1655 (1965).
169. C. S. Whewell, C. B. Stevens, and S. C. Amin, *Appl. Polymer Symp.* **18**, 387 (1971).
170. F. M. Stevenson and M. C. Stevenson, *J. Textile Inst.* **53**, P649 (1962).
171. V. A. Williams, *Textile Res. J.* **34**, 79 (1964).

172. J. R. McPhee, *Textile Res. J.* **33**, 755 (1963).
173. P. Alexander, R. F. Hudson, and M. Fox, *Biochem. J.* **46**, 27 (1950).
174. V. A. Williams, *Textile Res. J.* **33**, 444 (1963).
175. A. Kantouch and A. Bendak, *Textile Res. J.* **39**, 858 (1969).
176. F. R. Hartley, *J. Soc. Dyers Colourists* **85**, 66 (1969).
177. F. R. Hartley, *Wool Sci. Rev.* **37**, 54 (1969).
178. F. R. Hartley, *J. Soc. Dyers Colourists* **86**, 209 (1970).
179. R. F. Hudson and P. Alexander, *Fibrous Proteins Proc. Symp., Univ. Leeds*, 193 (1946).
180. P. Alexander and D. Gough, *Biochem. J.* **48**, 504 (1951).
181. G. Valk, *Proc. 3rd Intern. Wool Textile Res. Conf., Paris* **2**, 371 (1965).
182. G. Valk, *Textilveredlung* **2**, 121 (1967).
183. A. Kantouch and S. H. Abdel Fattah, *Appl. Polymer Symp.* **18**, 317 (1971).
184. R. Consden, A. H. Gordon, and A. T. P. Martin, *Biochem. J.* **40**, 580 (1946).
185. M. Lewin and M. Avrahami, *Proc. 3rd Intern. Wool Textile Res. Conf., Paris* **2**, 351 (1965).
186. M. Harris and A. L. Smith, *J. Res. Natl. Bur. Std.* **18**, 623 (1937).
187. J. E. Eager, *Biochem. J.* **100**, 37c (1966).
188. J. Nachtigal and C. Robbins, *Textile Res. J.* **40**, 454 (1970).
189. I. W. Stapleton, *Textile J. Aust.* **43**, 61 (1968).
190. J. A. Maclaren and A. Kirkpatrick, *J. Soc. Dyers Colourists* **84**, 564 (1968).
191. D. G. Smyth, *J. Biol. Chem.* **242**, 1592 (1967).
192. G. H. Elliott and J. B. Speakman, *J. Soc. Dyers Colourists* **59**, 185 (1943).
193. H. Lindley and H. Phillips, *Biochem. J.* **41**, 34 (1947).
194. P. Alexander et al., *Biochem. J.* **52**, 177 (1952).
195. B. Milligan and L. J. Wolfram, *J. Textile Inst.* **63**, 515 (1972).
196. L. A. Holt, S. J. Leach, and B. Milligan, *Aust. J. Chem.* **21**, 2115 (1968).
197. J. F. Riordan and B. L. Vallec, *Methods Enzymol*, **11**, 570 (1967).
198. N. H. Leon, unpublished data (1968)
199. N. H. Koenig and M. Friedman, *Textile Res. J.* **42**, 646 (1972).
200. J. Gerendas and H. Zahn, *Kolloid-Z. Z. Polymere* **245**, 382 (1971).
201. J. B. Caldwell, S. J. Leach, and B. Milligan, *Textile Res. J.* **39**, 705 (1969).
202. S. J. Leach et al., *Symposium on Fibrous Proteins, Australia, 1967*, p. 373, Butterworth & Co. (Australia) Ltd., Sydney (1968).
203. J. B. Caldwell and B. Milligan, *Aust. J. Chem.* **20**, 793 (1967).
204. R. Krüger, Ph.D. thesis, Technische Hochshüle, Aachen (1968).
205. N. H. Koenig and M. W. Muir, *Appl. Polymer Symp.* **18**, 727 (1971).
206. R. E. Whitfield and M. Friedman, *Textile Res. J.* **42**, 533 (1972).
207. G. R. Stark, *Methods Enzymol.* **11**, 125 (1967).
208. E. O. R. Thompson and I. J. O'Donnell, *Aust. J. Biol. Sci.* **20**, 1001 (1967).
209. P. Edman, *Acta Chem. Scand.* **4**, 283 (1950).
210. M. von Wilm, *Angew. Chem. Intern. Ed.* **9**, 267 (1970).
211. J. B. Caldwell and B. Milligan, *Textile Res. J.* **40**, 194 (1970).
212. R. Price and C. E. Vellino, Brit. Pat. 825,911 (1960).
213. H. H. Bosshard and H. Zollinger, U.S. Pat. 3,045,029 (1962).
214. A. J. Farnworth, *Biochem. J.* **59**, 529 (1955).
215. J. E. Moore and R. A. O'Connell, *Textile Res. J.* **27**, 783 (1957).
216. N. H. Koenig, *Textile Res. J.* **32**, 117 (1962).
217. W. R. Gray and B. S. Hartley, *Biochem. J.* **89**, 59.P (1963).
218. M. C. Corfield, J. C. Fletcher, and A. Robson, *Biochem. J.* **102**, 801 (1967).

219. W. G. Rose and H. P. Lundgren, *Textile Res. J.* **23**, 930 (1953).
220. M. A. Taubman and M. Z. Atassi, *Biochem. J.* **106**, 829 (1968).
221. W. G. Rose, U.S. Pat. 3,617,203 (1971).
222. R. Benesch and R. E. Benesch, *Biochem. Biophys. Acta* **63**, 166 (1962).
223. S. G. Elfbaum, Ph.D. thesis, Northwestern University (1966).
224. J. J. Bartoszewicz, M. M. Breuer, and J. T. Pearson, Brit. Pat. 1,187,568 (1969).
225. G. C. Barrett and A. R. Khokar, *J. Chem. Soc.* **(c)**, 1117 (1969).
226. G. C. Barrett and A. R. Khokhar, *J. Chem. Soc.* **(c)**, 1120 (1969).
227. N. H. Leon and R. S. Asquith, *Tetrahedron* **26**, 1719 (1970).
228. N. H. Leon, *Org. Mass Spectrom.* **6**, 407 (1972).
229. N. H. Leon, Ph.D. thesis, University of Bradford (1970).
230. N. H. Leon and J. A. Swift, Brit. Pat. Appl. 16,383 (1969).
231. N. H. Leon, *J. Soc. Cosmetic Chem.* **23**, 427 (1972); Brit. Pat. Appl. 5,984 (1971).
232. F. Sanger, *Biochem. J.* **39**, 507 (1945).
233. E. Siepman and H. Zahn, *Proc. 3rd Intern. Wool Textile Res. Conf., Paris.* **1**, 303 (1965).
234. L. Hamilton, *A Laboratory Manual of Protein Chemistry*, Vol. 2, p. 62, Pergamon Press, London (1960).
235. R. S. Asquith and M. S. Otterburn, *Appl. Polymer Symp.* **18**, 227 (1971).
236. A. Patchornik and M. Sokolovsky, *J. Amer. Chem. Soc.* **86**, 1206 (1964).
237. H. Beyer and U. Schenck, *J. Chromatog.* **61**, 263 (1971).
238. H. Zahn. and J. Meienhofer, *Melliand Textilber.* **37**, 432 (1956).
239. H. Zahn and E. Siepmann, *Kolloid-Z. Z. Polymere* **250**, 54 (1972).
240. H. Zahn et al., *Melliand Textilber.* **50**, 1319 (1969).
241. M. J. Gorbunoff, *Biochemistry* **6**, 1606 (1967).
242. M. J. Gorbunoff, *Biopolymers* **11**, 2233 (1972).
243. M. Marzona and G. Di Modica, *Textilveredlung* **6**, 806 (1971).
244. F. R. N. Gurd, *Methods Enzymol.* **11**, 532 (1967).
245. P. Miro and J. Jaume, *Appl. Polymer Symp.* **18**, 249 (1971).
246. H. H. Stein and J. Guarnaccio, *Anal. Chim. Acta.* **23**, 89 (1960).
247. M. A. Da Silva, Ph.D. thesis, University of Leeds (1954).
248. G. Ebert and C. Ebert, *Angew. Chem. Intern. Ed.* **8**, 896 (1969).
249. C. Ebert, *Appl. Polymer Symp.* **18**, 221 (1971).
250. W. Kirst, *Melliand Textilber.* **29**, 236 (1948).
251. A. E. Brown and M. Harris, *Ind. Eng. Chem.* **40**, 316 (1948).
252. H. Zahn, *Proc. Intern. Wool. Textile Res. Conf., Australia* **C**, 425 (1955).
253. W. Kirst, *Melliand Textilber.* **28**, 314 (1947).
254. J. A. Maclaren and B. J. Sweetman, *Aust. J. Chem.* **19**, 2355 (1966).
255. S. S. Warg and F. H. Carpenter, *J. Biol. Chem.* **243**, 3702 (1968).
256. B. F. Tietze, J. A. Gladner, and J. E. Folk, *Biochim. Biophys. Acta* **26**, 659 (1957).
257. J. A. Rothfus, *Anal. Biochem.* **30**, 279 (1969).
258. G. C. Tessoro and S. B. Sello, *Textile Res. J.* **34**, 513 (1964).
259. A. G. Pittman, *Textile Res. J.* **33**, 953 (1963).
260. J. F. Riordan and Vallee, B. L. *Methods Enzymol.* **11**, 541 (1967).
261. J. E. Moore and H. P. Lundgren, *Proc. Intern. Wool Textile Res. Conf., Australia* C, 355 (1955).
262. J. E. Moore and W. H. Ward, *J. Amer. Chem. Soc.* **78**, 2414 (1956).
263. H. Berkemer, Ger. Pat. 1,220,969 (1966).
264. N. M. Bikales, J. J. Black, and L. Rapopart, *Textile Res. J.* **27**, 80 (1957).
265. A. Robson, M. J. Williams, and J. M. Woodhouse, *J. Textile Inst.* **60**, 140 (1969).

266. J. F. Corbett, *Proc. 3rd Intern. Wool Textile Conf., Paris* **3**, 321 (1965).
267. A. Schöberl, *Proc. 3rd Intern. Wool Textile Conf., Paris* **2**, 301 (1965).
268. A. Schöberl, *J. Textile Inst.* **51**, T613 (1960).
269. H. D. Weigmann and C. J. Dansizer, *Appl. Polymer Symp.* **18**, 795 (1971).
270. M. Friedman and N. H. Koenig, *Textile Res. J.* **41**, 605 (1971).
271. M. Friedman and S. Tillin, *Textile Res J.* **40**, 1045 (1970).
272. J. A. Maclaren, *Textile Res. J.* **41**, 713 (1971).
273. J. R. McPhee and M. Lipson, *Aust. J. Chem.* **7**, 387 (1954).
274. H. S. Wilgus et al., *J. Org. Chem.* **29**, 594 (1964).
275. J. B. Speakman and P. L. D. Peill, *J. Textile Inst.* **34**, T70 (1943).
276. J. L. Stoves, *Trans. Faraday Soc.* **39**, 301 (1943).
277. I. C. Watt and R. Morris, *J. Textile Inst.* **57**, T425 (1966).
278. R. S. Asquith and D. C. Parkinson, *J. Textile Inst.* **58**, 83 (1967).
279. I. C. Watt and R. Morris, *Textile Res. J.* **40**, 952 (1970).
280. H. D. Feldtman and B. E. Fleischfresser, *J. Textile Inst.* **63**, 596 (1972).
281. E. Kantscher and S. Serafimov, *Melliand Textilber.* **52**, 725 (1971).
282. E. Kantscher and S. Serafimov, *Melliand Textilber.* **51**, 467 (1970).
283. B. Joos, Brit. Pat. 1,057,104 (1967).
284. H. A. Rutherford, W. I. Patterson, and M. Harris, *J. Res. Natl. Std.* **25**, 451 (1940).
285. S. Blackburn, E. G. H. Carter, and H. Phillips, *Biochem. J.* **35**, 627 (1941).
286. P. Alexander, *Melliand Textilber.* **34**, 756 (1953).
287. S. Blackburn and H. Lindley, *J. Soc. Dyers Colourists* **64**, 305 (1948).
288. P. Alexander et al., *Biochem. J.* **48**, 629 (1951).
289. L. A. Holt and B. Milligan, *Aust. J. Biol. Sci.* **23**, 165 (1970).
290. J. Bello, *Biochem. Biophys. Acta* **20**, 426 (1956).
291. D. J. Kilpatrick and J. A. Maclaren, *Textile Res. J.* **39**, 279 (1969).
292. L. S. Holt, S. J. Leach, and B. Milligan, *Textile Res. J.* **39**, 290 (1969).
293. H. Shiozaki and Y. Tanaka, *Makromol. Chem.* **138**, 215 (1970).
294. H. Shiozaki and Y. Tanaka, *Sen-i Gakkaishi.* **28**, 174 (1972).
295. C. W. Capp and J. B. Speakman, *J. Soc. Dyers Colourists* **65**, 402 (1949).
296. C. Fearnley and J. B. Speakman, *Nature* (*Lond.*) **166**, 743 (1950).
297. L. A. Miller, *Textile Res. J.* **31**, 451 (1961).
298. F. Kurzer and K. Douraghi-Zadeh, *Chem. Rev.* **67**, 107 (1967).
299. L. A. Holt and B. Milligan, *J. Textile Inst.* **61**, 597 (1970).
300. R. S. Asquith, K. L. Gardner, and M. S. Otterburn, *Experientia* **27**, 1388 (1971).
301. R. S. Asquith et al., *Biochem. Biophys. Acta* **221**, 342 (1970).
302. K. L. Carraway and R. B. Triplett, *Biochem. Biophys. Acta* **200** 564 (1970).
303. H. Prietzel, *Melliand Textilber.* **52**, 950 (1971).
304. H. R. Richards and J. B. Speakman, *J. Soc. Dyers Colourists* **71**, 537 (1955).
305. D. Harrison and J. B. Speakman, *Textile Res. J.* **28**, 1005 (1958).
306. W. G. Crewther and L. M. Dowling, *Textile Res. J.* **41**, 267 (1971).
307. J. Wolff and I. Covelli, *European J. Biochem.* **9**, 371 (1969).
308. M. E. Koshland et al., *J. Biol. Chem.* **238**, 1343 (1963).
309. M. Sokolovsky, D. Harrel, and J. F. Riordan, *Biochemistry* **8**, 4740 (1969).
310. B. L. Vallee and J. F. Riordan, *Ann. Rev. Biochem.* **38**, 733 (1969).
311. T. C. Bruice, M. J. Gregory, and S. L. Walters, *J. Amer. Chem. Soc.* **90**, 1612 (1968).
312. K. Morihari and K. Nagami, *J. Biochem.* (*Tokyo*) **65**, 321 (1969).
313. H. Beyer and U. Schenk, *Kolloid-Z. Z. Polymere* **233**, 890 (1969).
314. R. W. Boesel and F. H. Carpenter, *Biochem. Biophys. Res. Commun.* **38**, 678 (1970).
315. F. Lieben, *Biochem. Z.* **145**, 535 (1924).

316. H. Zahn and K. Kohler, *Z. Naturforsch.* **56**, 137 (1950).
317. J. St. L. Philpot and P. A. Small, *Biochem. J.* **32**, 542 (1938).
318. B. Nilssen, *Fibrous Proteins Proc. Symp., Univ. Leeds*, 142 (1946).
319. F. R. Andrews, Personal communication (1970).
320. M. Leveau, M. Caillet, and N. Demonmerot, *Bull. Inst. Textile Fr.* **90**, 7 (1960).
321. E. R. Fritze and H. Zahn, *Melliand Textilber.* **36**, 1136 (1955).
322. J. B. Speakman and E. Stott, *Nature* (*Lond.*) **141** 414 (1938).
323. R. Cockburn, B. Drucker, and H. Lindley, *Biochem. J.* **43**, 438 (1948).
324. J. H. Buchanan and M. C. Corfield, *Appl. Polymer Symp.* **18**, 101 (1971).
325. R. S. Asquith and P. A. Watson, *Textile Res. J.* **35**, 381 (1965).
326. H. M. Kagan and B. L. Vallee, *Biochemistry* **8**, 4223 (1969).
327. J. R. McPhee, *Textile Res. J.* **28**, 303 (1958).
328. H. Zahn and R. Krüger, *Kolloid-Z. Z. Polymere* **250**, 576 (1972).
329. J. F. Walker, *Formaldehyde*, 3rd ed., Reinhold Publishing Corp., New York (1967).
330. J. B. Caldwell and B. Milligan, *Textile Res. J.* **42**, 122 (1972).
331. I. C. Watt, *Appl. Polymer Symp.* **18**, 905 (1971).
332. R. G. Kallen and W. P. Jencks, *J. Biol. Chem.* **241**, 5864 (1966).
333. R. N. Reddie and C. H. Nicholls, *Makromol. Chem.* **153**, 153 (1972).
334. J. Cegarra and J. Gacen, *J. Soc. Dyers Colourists* **84**, 216 (1968).
335. P. Miro, *Bull. Inst. Textile Fr.* **17**, 1165 (1963).
336. R. N. Reddie and C. H. Nicholls, *Textile Res. J.* **41**, 841 (1971).
337. P. W. Nicholls and L. C. Gruen, *Aust. J. Chem.* **23**, 533 (1970).
338. J. Blass et al., *Bull. Chim. Soc. Fr.* 3957 (1967).
339. G. Di Modica and M. Marzoni, *Textile Res. J.* **41**, 701 (1971).
340. P. Kusch, R. Lubig, and G. Topert, *Melliand Textilber.* **53**, 1393 (1972).
341. F. M. Richards and J. R. Knowles, *J. Mol. Biol.* **37**, 231 (1968).
342. R. Lubig and H. Zahn, personal communication (1972).
343. T. P. King, *Biochemistry* **5**, 3454 (1966).
344. K. Toi et al., *J. Biol. Chem.* **242**, 1036 (1967).
345. H. A. Itano and A. J. Gottbeb, *Biochem. Biophys. Res. Commun.* **12**, 405 (1963).
346. K. Nakaya et al., *Biochim. Biophys. Acta* **194**, 301 (1969).
347. K. Takahashi, *J. Biol. Chem.* **243**, 6171 (1968).
348. R. E. Whitfield and M. Friedman, *Textile Res. J.* **42**, 344 (1972).
349. H. Zollinger, *Angew. Chem.* **73**, 125 (1961).
350. H. Zollinger, *Textilveredlung* **6**, 57 (1971).
351. J. A. Medley and K. L. Gardner, *Rev. Progr. Coloration* **3**, 67 (1972).
352. K. G. Kleb, E. Siegel, and K. Sasse, *Angew. Chem. Intern. Ed.* **3**, 408 (1964).
353. K. R. F. Cockett, I. D. Rattee, and C. B. Stevens, *J. Soc. Dyers Colourists* **85**, 113 (1969).
354. A. Shansky, *Amer. Perfumer Cosmet.* **81**, 23 (1966).
355. Unilever, French Pat. 1,465,870 (1967).
356. R. S. Asquith and P. Carthew, *J. Textile Inst.* **64**, 10 (1973).
357. R. S. Asquith, A. K. Booth, and J. D. Skinner, *Biochim. Biophys. Acta* **181**, 164 (1969).

# 6

# Crosslinking and Self-Crosslinking in Keratin Fibers

K. Ziegler

## 6.1. Introduction

In recent years much attention has been directed toward the elucidation of a variety of bonds and interactions that exist between the protein molecules in wool fibers. It is argued that these forces are responsible for maintaining the stable structure of the fiber and that many of its mechanical and chemical properties can be explained by them. One of the prominent features of wool is the complete insolubility in polar solvents which results from crosslinking. The most thoroughly characterized crosslinks in wool are the covalent disulfide bonds which stem from the incorporation of cystine residues. As in other proteins, a number of noncovalent bonds also occur in sufficient quantities as to have additional effects. It was observed in the course of wool processing that disulfide bonds tend to become labile, and therefore much research work was aimed at the introduction of additional crosslinks or the replacement of some already present by those more able to resist chemical attack. "Crosslinks" are here defined as those covalent bonds which join two neighboring peptide chains through functional amino acid side-chain groups being sufficiently close to participate in a crosslinkage. When additional crosslinks are to be introduced into the molecule it is now considered useful, for the sake of clarification, to differentiate further with regard to the mode of reaction, that is, to indicate whether crosslinking has been achieved (a) by the

---

K. Ziegler • Deutsches Wollforschungsinstitut Aachen, West Germany.

incorporation of a foreign residue of the reagent applied, or (b) by coupling of protein side-chain groups by a special treatment with agents which themselves do not become part of the molecule. The latter type of reaction is termed a "self-crosslinking" reaction, the former being referred to as "crosslinking."

The largely unknown primary structure of wool makes it difficult to define the terms "intermolecular" and "intramolecular" crosslinkages. As wool is built up of a number of different proteins, these give rise to a complexity of physical and morphological structures. It is therefore only possible to attempt to differentiate roughly between "interchenic" and "intrachenic" crosslinks.

The first section of this chapter is concerned with surveying the various noncovalent and covalent bonds that exist in keratin fibers (exclusively wool), which can function as crosslinks. The subsequent section deals with reactions that give rise to self-crosslinking. A large part of the text discusses methods and reactions by which new crosslinks can be introduced by the reaction of bifunctional reagents, thus altering the chemical and physical resistance of wool fibers. In the final section several methods for the possible assessment of crosslinks are reviewed, and these are discussed in the light of their importance.

## 6.2. Naturally Occurring Crosslinks

### 6.2.1. Noncovalent Bonds

Insufficient knowledge of the sequence of the amino acids in the various polypeptide chains of keratin fibers seems to be one of the main obstacles to further progress in the chemistry of crosslinking. More information is necessary, as the reactivity of the peptide bonds, side-chain groups, and crosslinks is profoundly affected by their environment.

Amino acid analysis of wool shows that the fiber consists of 22 different amino acid residues, 13 of which carry functional reactive groups.[1] The total number of 22 amino acids include those which do not survive acid hydrolysis, such as glutamine, asparagine, cysteine, and tryptophan. Besides the covalent bonds present, it is well established that other noncovalent bonds, or interchenic forces of attraction, must be present to stabilize the conformation of the α-folded helical structure of the polypeptide chains. These bonds can link different side-chain groups together or link side-chain groups with the protein-chain backbone.

Three different types of noncovalent bonds exist in keratin fibers which may act as crosslinks. These are generally classified as (1) hydrogen bonds, (2) hydrophobic bonds (or hydrophobic interactions), and (3) ionic bonds or "salt linkages." When the salt linkages are regarded as hydrogen bonds between two ionized groups,[2] it is not possible to differentiate between those two bonds precisely.

A large number of the hydrogen bonds in wool occur between —CO··· HN— groups within an $\alpha$-helical chain, but these cannot be classified as crosslinks. Some other hydrogen bonding that can exist between suitable donors and acceptors in the amino acid side chains are given here.[3]

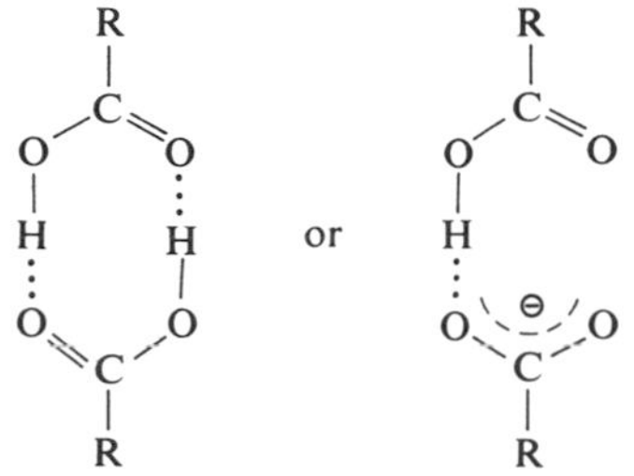

Side-chain hydrogen bond between neutral groups

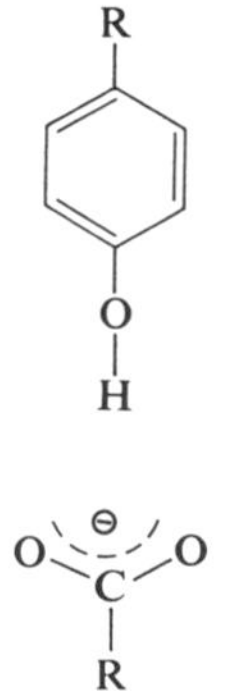

Hydrogen bond between the hydroxyl group of a tyrosyl residue and the carboxylate ion of a glutamyl residue

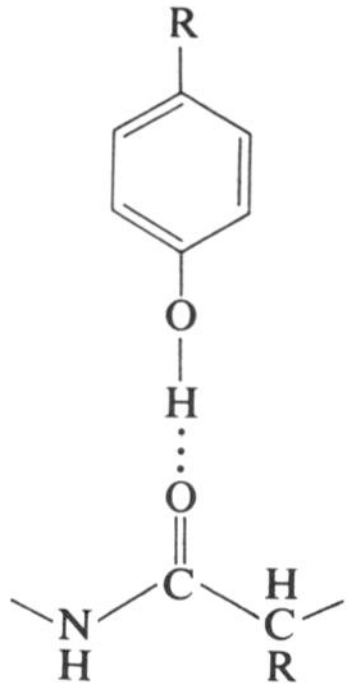

Hydrogen bond between the hydroxyl group of a tyrosyl residue and the carbonyl of a peptide group

*Hydrophobic bonds*[4,5] are entropic interactions between nonpolar side chain residues surrounded by water. These attractive forces are responsible for the tendency of nonpolar residues to avoid contact with the aqueous phase and to adhere to one another in the form of an intramolecular micelle.

The importance of the hydrophobic bonds becomes apparent when one realizes that the content of amino acids with nonpolar side chains accounts for nearly half of the residues present in wool, if glycyl residues are included as hydrophobes. The formation of a hydrophobic bond is seen in Fig. 6-1.

*Ionic bonds* or salt linkages are formed by the mutual attraction of positively and negatively charged groups of the polypeptide chains. They can be formulated as follows:

$$\begin{array}{l} \diagup \\ O{=}C \\ \quad \diagdown \\ \quad H{-}C{-}(CH_2)_4{-}NH_3^+ \\ \quad \diagup \\ H{-}N \\ \quad \diagdown \end{array} \qquad\qquad \begin{array}{r} \diagdown \\ N{-}H \\ \diagup \quad \\ {}^{-}OOC{-}(CH_2)_2{-}C \\ \diagdown \quad \\ C{=}O \\ \diagup \end{array}$$

lysyl glutamyl

The addition of acids or bases causes ruptures of the salt bridges by discharging the carboxylate and amino ions, respectively. In their studies of the elasticity of the fiber Speakman and co-workers[6–8] suggested that these linkages contribute to the fiber strength.

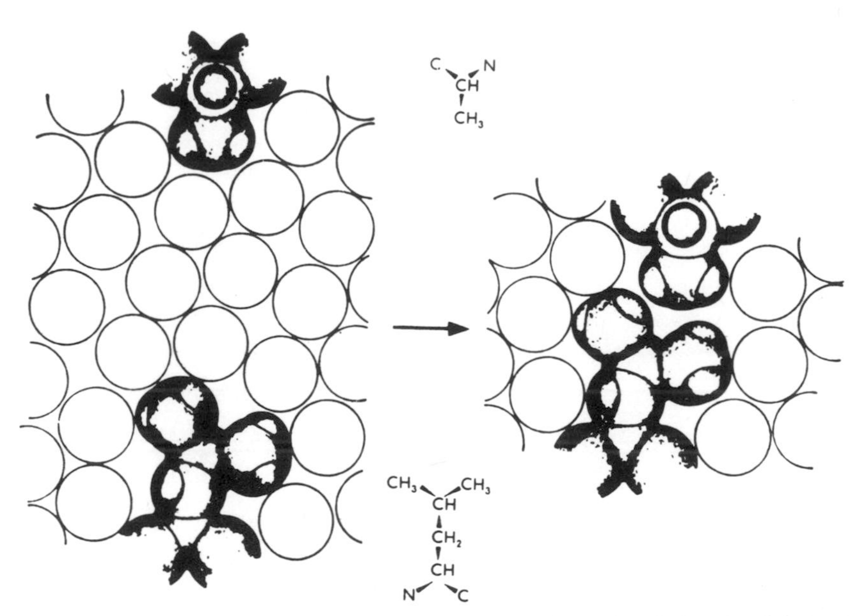


Fig. 6-1. Schematic representation of the formation of a hydrophobic bond between two isolated side chains (alanine and leucine). The bond is formed through an approach of the two side chains until they touch, with a reduction of the number of nearest water neighbors (circles). (Data from Nemethy and Scheraga.[5])

### 6.2.2. Covalent Bonds

One of the most remarkable features of keratin fibers is their high sulfur content. The sulfur is present predominantly in the disulfide group of cystine. Since this amino acid carries four functional groups, it can be incorporated in the protein in two different ways: (1) as a crosslink between two adjacent protein-backbone chains (interchenic); and (2) as a loop within a single peptide chain (intrachenic). These ways of incorporating cystine can be visualized as follows:

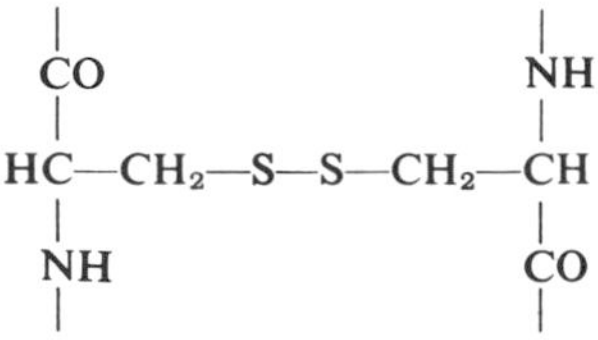

```
 |                        |
 CO                       NH
 |                        |
HC—CH2—S—S—CH2—CH
 |                        |
 NH                       CO
 |                        |

           (1)

        H2C—————————S—S—————————CH2
          |              H                |
 —N—C—C—[HN—C—CO]—N—C—C—
  H  |  O        R          H  H  O
     H

           (2)
```

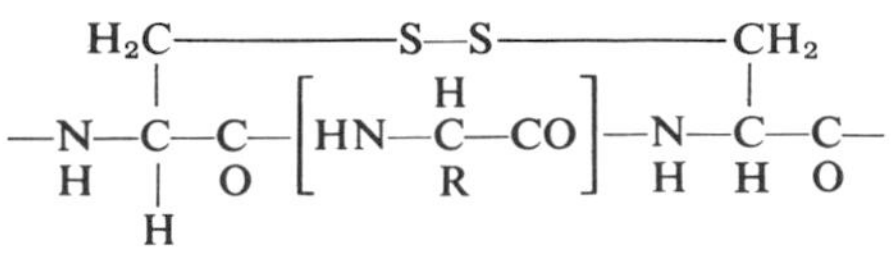

From the early work of Stary,[9] who compared cystine crosslinks with rungs of a ladder, very much attention has been paid to cystine chemistry in wool. This is mainly due to the high reactivity of the disulfide bond, which is reflected in many of the mechanical and chemical properties of wool. Thus treatment of wool fibers with reducing or oxidizing agents or alkalis causes fission of the labile dithio bond, yielding a great variety of sulfur-containing amino acid derivatives and other products. Some of these chemical reactions play an important role in commercial methods such as setting and shrink-resist processes.

Differences in chemical reactivity of the different disulfide bond in the protein molecule have been suggested at several times.[10–12] However, despite the progress in sequence studies, of the many wool fractions isolated so far, only a small number of disulfide-bond arrangements can be assigned. Much speculation was directed to the occurrence of other covalent bonds in native wool fibers. Some possible variations of the well-established $\alpha$-peptide structure would be the presence of isopeptide linkages involving (a) the amino acid side-chain carboxyl and $\alpha$-amino groups; (b) the side-chain amino groups and $\alpha$-carboxyl groups; and (c) the peptide linkages between both carboxyl and amino group side chains. In the first two cases, the $\beta$-aspartic acid or $\gamma$-glutamic acid isopeptide link [or, vice versa, the $\epsilon$-amino(lysine) isopeptide link] would constitute only a branching point rather than a crosslink. The last mentioned of the three possibilities is that the $\epsilon$-amino group of lysine forms crosslinks with $\beta$- or $\gamma$-carboxyl groups of aspartic and glutamic acids, respectively, between different polypeptide chains or between parts of the same chain. Recently, proof of the occurrence of such crosslinks has been obtained, which will be discussed in detail later (see Section 6.3).

## 6.3. Formation of Crosslinks without Incorporation of Artificial Bridge Residues

### 6.3.1. Self-Crosslinking Reactions Induced by the Action of Alkali, Hot Water, Steam, Dry Heat, and Amines

From his studies on setting of wool in steam or alkalis, Speakman[13,14] concluded that disulfide-bond reaction is a fundamental part of the process and that the stabilization of the extended fiber is due either to the rearrangement of disulfide bonds in new positions or to the formation of new covalent linkages. He postulated that alkali-treated wool contains sulfenamide (—S—NH—) crosslinks which were presumed to result from the interaction of basic side chains and sulfenic acid formed by disulfide-bond hydrolysis. The sulfenic acid undergo further breakdown to give aldehyde groups, which in turn reacted with the amino side chains, thus forming methylenimine (—CH=N—) crosslinks.[15] However, despite indirect experimental evidence, both crosslinks have been questioned at various times.[11,16,17] Speakman and Whewell[18] suggested formation of a thioether crosslink as a result of disulfide-bond rupture during alkali treatment of wool. From his experiments on wool fibers with very low disulfide content, Farnworth[19] favored the view that a rearrangement of hydrogen bonds is largely responsible for maintaining the set of the fiber. Thiol–disulfide interchange reactions were first discussed with respect to wool by Burley,[20] who found that the ease of extension of the fibers were markedly dependent on the number of thiol groups, thus playing a certain role during setting. Later, Zahn et al.[21,22] showed that there is a direct correlation between the initial thiol content of wool and the production of lanthionine during treatment with alkali. When thiol groups were destroyed or blocked, both lanthionine and set was inhibited. These authors assumed that the production of lanthionine was responsible for the stabilization of set, whereas Caldwell et al.[23] suggested that the inhibitory effect of thiol blockers is due to their interference with the thiol–disulfide interchange rather than to their inhibition of lanthionine. The rate of lanthionine formation as a crosslink is influenced by the type (i.e., ionic strength, valency, and size) of the ions and the concentration of salt being added during alkali treatment.[24]

The cleavage of disulfide bonds by alkali plays an important role, and for its underlying mechanism, several theories have been proposed so far, which will be briefly mentioned. According to the first-mentioned mechanism, a nucleophilic attack of hydroxyl ion on one of the sulfur atoms of the disulfide bond takes place[13,25]:

$$R—S—S—R + OH^- \longrightarrow R—SOH + R—S^-$$

Subsequent reaction between the thiol and combined α-aminoacrylic acid formed by the decomposition of unstable sulfenic acid lead to the formation of lanthionine. Other mechanisms include the elimination reactions, which initiate a cleavage of carbon–sulfur bond by causing ionization of a hydrogen atom attached to an α-carbon[26] (=α-elimination reaction) or to a β-carbon[27,28] (=β-elimination reaction) to a sulfur atom. The latter mechanism, in which the hydrogen of the β-carbon is ionized, has received much attention. The cleavage of the —C—S— bond was formulated by Tarbell and Harnish[29] as follows:

$$\begin{array}{c} | \\ NH \\ | \\ H{-}C{-}CH_2{-}S{-}S{-}CH_2{-}C{-}H + B^{\ominus} \\ | \\ CO \\ | \end{array} \longrightarrow \begin{array}{c} | \\ NH \\ | \\ {}^{\ominus}C{-}CH_2{-}S{-}S{-}CH_2{-}CH + BH \\ | \\ CO \\ | \end{array}$$

$$\downarrow$$

$$\begin{array}{c} | \\ NH \\ | \\ C{=}CH_2 + {}^{\ominus}S{-}S{-}CH_2{-}CH \\ | \\ CO \\ | \end{array}$$

$$\downarrow$$

$$\begin{array}{c} | \\ NH \\ | \\ S + {}^{\ominus}S{-}CH_2{-}CH \\ | \\ CO \\ | \end{array}$$

In the final step, combination of a cysteine residue with an α-aminoacrylic acid residue will give lanthionine.

Recently, the mechanism of the alkaline hydrolysis of the disulfide bond has been reinvestigated on glutathione,[30] bovine serum albumin,[31] and on ovomucoid including low-molecular-weight disulfides.[32] From these results some evidence is presented that in weakly alkaline media below pH 11, the "classical" —S—S— split mechanism seems to occur, whereas in strongly alkaline media, the β elimination (—C—S split) predominates.

The first detection of self-crosslinks in wool keratin can be dated back to the year 1941, when Horn et al.[33] isolated the amino acid lanthionine from wool, hair, and feathers that have been boiled with a solution of sodium car-

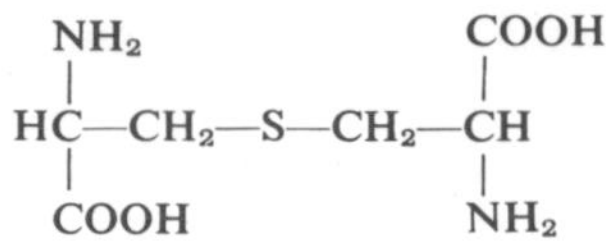

$$
\begin{array}{ccc}
\mathrm{NH_2} & & \mathrm{COOH} \\
| & & | \\
\mathrm{HC{-}CH_2{-}S{-}CH_2{-}} & & \mathrm{CH} \\
| & & | \\
\mathrm{COOH} & & \mathrm{NH_2}
\end{array}
$$

Lanthionine

bonate. This thioether amino acid had been isolated earlier by Küster and Irion,[34] but was incorrectly characterized by these authors. For some time it was suggested that lanthionine is present in intact wool, too, although it could arise from weathering, during storage, and, at a later stage, from scouring processes. So, in the original fiber lanthionine would have to be considered as a naturally occurring covalent thioether bond. However, conclusive evidence has been obtained which shows that it cannot be regarded as a normal constituent of the protein, but rather as a transformation product arising from decomposition of cystine residues. In addition, Dowling and MacLaren[35] have pointed out that analytical results of hydrolysates of reduced and alkylated wool indicate that very small amounts of lanthionine are formed in the early stages of hydrolysis. Any traces of lanthionine detected in hydrolysates of the virgin fiber could thus be formed by the hydrolysis procedure.

Speakman long maintained the hypothesis of the participation of lysine side chains in crosslinking as a result of setting treatments upon wool fibers. He later supported his early mechanochemical evidence[14] with other experimental findings. Thus in set fibers the number of $\epsilon$-amino groups available for reaction with dinitrofluorobenzene was found to be substantially decreased[36,37]; alternatively, dinitrophenylation of fibers greatly decreases the tendency of permanent setting.[38] The evidence apparently offered strong support for Speakman's early postulate of a sulfenamide crosslink formation. However, it ignored the possibility of other reactions of the products of cystine degradation with the lysine side chains. Thus $\alpha$-aminoacrylic acid residues, postulated as an intermediate in the formation of lanthionine, might also react with lysine side chains. Later, Patchornik[39] and Bohak[40] simultaneously identified a new amino acid in hydrolysates of alkali-treated ribonuclease, giving it the trivial name lysinoalanine [$N^{\epsilon}$-(2-amino-2-carboxyethyl)lysine]. Ziegler[41,42] found that the same amino acid is formed when wool is steamed or treated with alkali. It is believed that lysinoalanine is formed by the addition of the $\epsilon$-amino group of a lysyl residue to the double bond of a dehydroalanine residue originating from cystine by a $\beta$-elimination reaction:

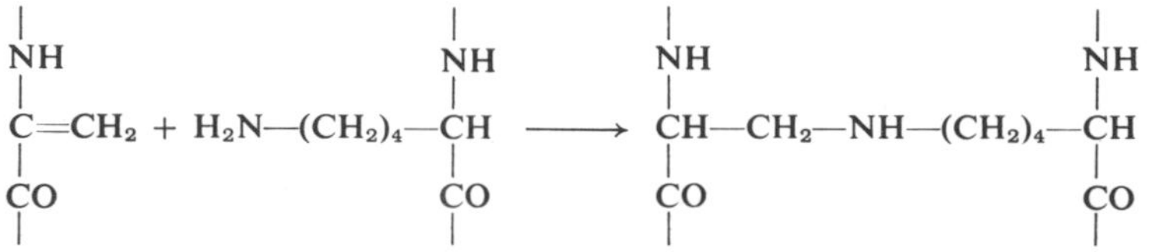

$$
\begin{array}{l}
| \\
\mathrm{NH} \\
| \\
\mathrm{C{=}CH_2} \\
| \\
\mathrm{CO} \\
|
\end{array}
+
\begin{array}{r}
| \\
\mathrm{NH} \\
| \\
\mathrm{H_2N{-}(CH_2)_4{-}CH} \\
| \\
\mathrm{CO} \\
|
\end{array}
\longrightarrow
\begin{array}{lr}
| & | \\
\mathrm{NH} & \mathrm{NH} \\
| & | \\
\mathrm{CH{-}CH_2{-}NH{-}(CH_2)_4{-}} & \mathrm{CH} \\
| & | \\
\mathrm{CO} & \mathrm{CO} \\
| & |
\end{array}
$$

Lysinoalanine residue

Since the detection of lysinoalanine in alkali-treated wool, a considerable amount of research has been carried out with a view toward throwing more light on the underlying reaction mechanism. Thus addition of organic solvents (i.e., propanol, acetone, etc.) to the alkaline reaction bath leads to a remarkable increase in lysinoalanine formation, probably owing to the splitting of hydrophobic bonds or to denaturation.[43] On the other hand, treatment of wool with various *S*-nucleophilic reagents, such as potassium cyanide, causes much lanthionine formation, but the production of lysinoalanine is negligible. It was therefore postulated that in these cases reaction proceeds via a substitution mechanism of the type of reaction $S_N2$, in which lanthionine formation is favored. This mechanism probably involves the following reactions:

$$R{-}CH_2{-}S{-}S{-}CH_2{-}R + CN^- \longrightarrow R{-}CH_2{-}S{-}CN + {}^-S{-}CH_2{-}R$$

$$R{-}CH_2{-}S{-}CN + {}^-S{-}CH_2{-}R \longrightarrow R{-}CH_2{-}S{-}CH_2{-}R + SCN^-$$

Divergent opinions arose as to the source of dehydroalanine residues, when it was proposed that these were derived exclusively from serine residues.[44,45] However, by use of radioactive cystine labeling it was unequivocally demonstrated that half-cystine residues are predominantly involved in the formation of lysinoalanine; about one fourth of lysinoalanine produced stems from other dehydroalanine residues, probably derived from the degradation of serine residues.[46] The rate of the effect of lanthionine and lysinoalanine on contraction is about 1:5.[47]

In addition to these observations, studies on lysine peptides with cystine,

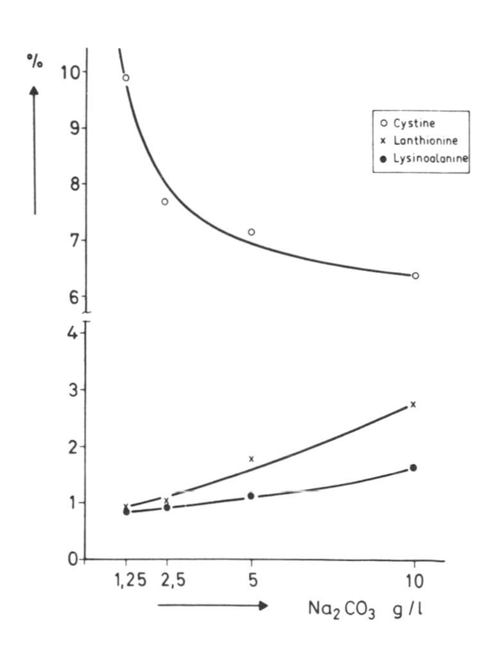


Fig. 6-2. Cystine, lanthionine, and lysinoalanine content in wool after treatment with $Na_2CO_3$ for 1 h at 70°C.[48]

cysteine, and serine in the sequence revealed that all three amino acid residues were able to produce dehydroalanine residues on alkaline treatment which are necessary for the formation of lysinoalanine.[48]

Figure 6-2 demonstrates the degradation of combined cystine and the formation of lanthionine and lysinoalanine in wool which has been given an alkali treatment under mild conditions.

It is well known that free or combined arginine is partially degraded by weak or strong alkali to form ornithine and citrulline as the intermediate product.[49,50]

$$\underset{\text{Arginine}}{H_2N-\overset{\overset{\displaystyle H}{|}}{\underset{\underset{\displaystyle (CH_2)_3-NH-\underset{\displaystyle H_2N}{C}=NH}{|}}{C}}-COOH} \xrightarrow[-\text{urea}]{\text{strong alkali}} \underset{\text{Ornithine}}{H_2N-\overset{\overset{\displaystyle H}{|}}{\underset{\underset{\displaystyle (CH_2)_3-NH_2}{|}}{C}}-COOH}$$

Arginine $\xrightarrow[-NH_3]{\text{weak alkali}}$ $H_2N-CH(COOH)-(CH_2)_3-NH-C(=O)-NH_2$ $\xrightarrow[-NH_3,\ -CO_2]{\text{strong alkali}}$ Ornithine

Such introduced δ-amino groups of combined ornithine react with dehydroalanine residues in a reaction analogous to lysinoalanine formation. This results in a compound called ornithinoalanine:

$$HC(NH_2)(COOH)-CH_2-NH-(CH_2)_3-CH(NH_2)(COOH)$$

After this new amino acid was found in sodium carbonate-treated sericin, very small amounts were also detected in alkali-treated wool.[51] Some results have given supporting evidence that ornithinoalanine does exist in modified wool.[52]

In the course of investigations by treating wool with sodium hydroxide[53] and dry heat,[54] it has been suggested that some other amino acid side chains could also have reacted with dehydroalanine, known as the degradation product of combined cystine. Evidence has been presented that the amido groups of the side chains of asparagine and glutamine residues are added to the double bond of dehydroalanine residues, resulting in the formation of a substituted amide bond which certainly will be split on acid hydrolysis, giving rise to a new compound, 2,3-diaminopropionic acid (= β-amino-alanine). It was shown that the amount of this compound in the hydrolysate is increased when ammonia was present in the alkaline liquor with which the wool had been reacted. However, when other globular proteins were included in these series of investigations, the formation of β-aminoalanine was produced by direct addition of ammonia to dehydroalanine residues and not by initial formation of a crosslink in which asparagine or glutamine residues had participated.[55] The following scheme may illustrate the possible alternative routes to the formation of β-aminoalanine:

$$\begin{array}{c} | \\ NH \\ | \\ C{=}CH_2 \\ | \\ CO \\ | \end{array} + H_2N{-}CO{-}CH_2{-}\begin{array}{c} | \\ NH \\ | \\ CH \\ | \\ CO \\ | \end{array} \longrightarrow \begin{array}{c} | \\ NH \\ | \\ HC \\ | \\ CO \\ | \end{array}{-}CH_2{-}NH{-}CO{-}CH_2{-}\begin{array}{c} | \\ NH \\ | \\ CH \\ | \\ CO \\ | \end{array}$$

+ free ammonia ↘ ↙ acid hydrolysis

$$\begin{array}{c} | \\ NH \\ | \\ HC \\ | \\ CO \end{array}{-}CH_2{-}NH_2 + \text{aspartic acid}$$

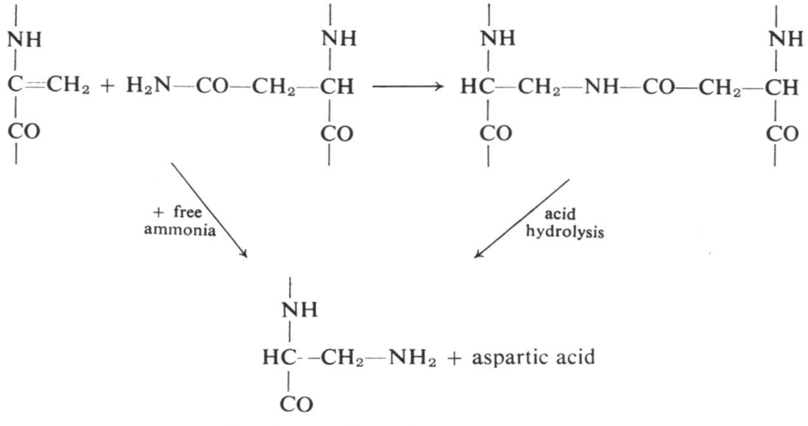

(β-aminoalanine residue)

Further investigations will be required before it can be definitively decided whether such an amide bond might really exist, acting as a crosslink in treated wool. From results regarding the effect of ammonia, ethylamine, and diethylamine[56] solutions upon wool, some indication was obtained that amines in general can compete successfully with the side chains of the protein molecules for the reactive dehydroalanine addition sites formed during the alkaline degradation of cystine.

Very recently, the influence of thioglycollic acid added to different alkaline treatments of wool has been studied in the light of crosslinking formation.[57] An inhibiting effect of this additive at pH 10 toward amino-group addition to dehydroalanine residues has been observed. Closely related to the detection of β-aminoalanine is the identification of a new amino acid, $N^{\beta}$-(2-amino-2-carboxyethyl)-β-aminoalanine (trivial name: β-aminoalanyl-alanine), which should also be mentioned[58]:

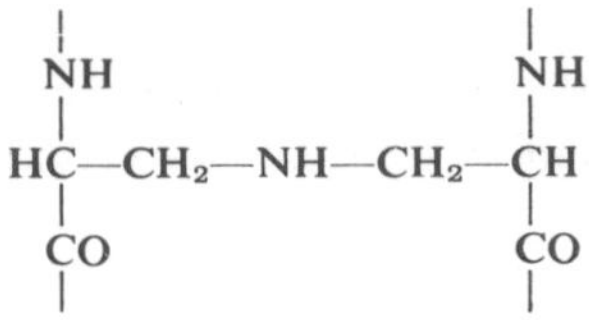

```
 |                       |
 NH                      NH
 |                       |
 HC—CH2—NH—CH2—CH
 |                       |
 CO                      CO
 |                       |
```

β-amino-alanyl-
alanine residue

It is assumed that the β-amino group of combined β-aminoalanine, formed during treatment of wool with ammonia, can be added to the double bond of a second dehydroalanine residue which stems from sodium hydroxide treatment. Thus the simultaneous influence of both ammonia and sodium hydroxide is a prerequisite for the formation of β-aminoalanylalanine in wool. Further confirmation of the suggested double addition mechanism was obtained when synthetic β-aminoalanine was added to the reaction bath containing NaOH. According to the authors, this amino acid probably can be considered as a new crosslink in modified wool.

Numerous investigations revealed that the amount of ε-amino groups of combined lysine in wool which are accessible to the reaction with 1-fluor-2, 4-dinitrobenzene (FDNB) does not coincide with the total lysine content obtained by amino acid analysis. About 20% of the ε-amino groups of lysine in the fiber failed to react with FDNB.[59–61] Other papers [62–65] demonstrated the presence of isopeptide crosslinks between the side chains of lysyl and glutamyl residues in insoluble fibrin. From these results it was tentatively inferred that such unusual isopeptide bonds may also occur in wool. Since acid hydrolysis ruptures all the peptide bonds present, regardless of whether α- or ε-amino groups are involved, acid as a hydrolytic reagent was replaced by successive treatments with a series of proteolytic enzymes such as pepsin, papain, pronase, prolidase, and leucine aminopeptidase. Subsequent amino acid analysis of the enzymatic digests of native wool,[66–67] chemically pretreated wool and fetal wool,[68] and dry-heated wool[68–70] unequivocally showed the presence of two isopeptides, $N^{\epsilon}$-(γ-glutamyl) lysine and $N^{\epsilon}$-(β-aspartyl)lysine.

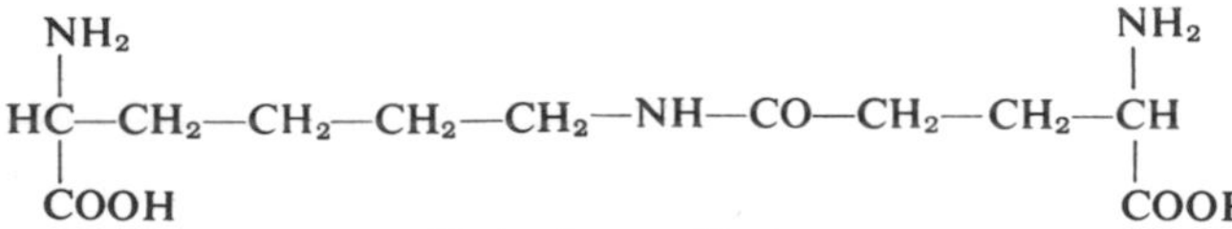

```
NH2                                                  NH2
|                                                    |
HC—CH2—CH2—CH2—CH2—NH—CO—CH2—CH2—CH
|                                                    |
COOH                                                 COOH
```

$N^{\epsilon}$-(γ-glutamyl)lysine

```
NH2                                                  NH2
|                                                    |
HC—CH2—CH2—CH2—CH2—NH—CO—CH2—CH
|                                                    |
COOH                                                 COOH
```

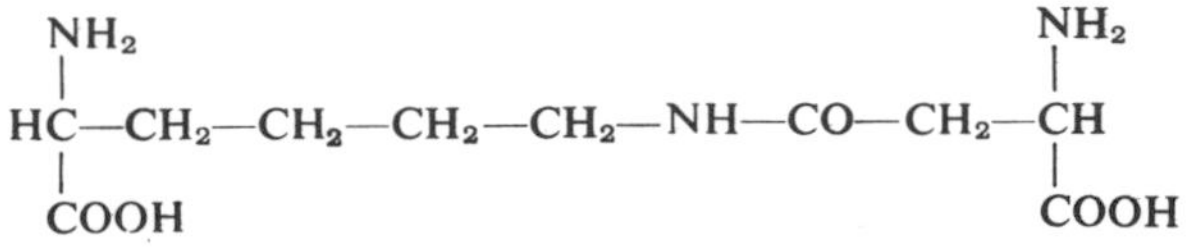

$N^{\epsilon}$-(β-aspartyl)lysine

The identity of both compounds was confirmed by comparison with authentic samples and by hydrolysis to the amino acid constituents. When the ratio of both isopeptides is taken into consideration, $\epsilon$-($\gamma$-glutamyl)lysine is present in far greater quantity. Because of the progressive decrease in solubility of dry-heated wool in urea-bisulfite solutions, and also the remarkable increase of insoluble $\beta$-keratose with heating, which corresponds to the high amount of bound lysine in this fraction, it has been concluded that these peptide bonds act as crosslinks between adjacent polypeptide chains.[70] Another support of the crosslinking hypothesis is evinced by the fact that fetal lamb's wool, and poodle hair,[68] both containing a relatively high amount of $\epsilon$-($\gamma$-glutamyl)-lysine, are less soluble then Merino wool in urea–thioglycollate solutions.

It is apparent that values obtained from solubility tests and changes in soluble/insoluble properties obtained by oxidative or reductive fractionation of wool must be regarded as rather indicative, and final proof will be of great value.

The fact that keratin fibers contain disulfide bonds and thiol groups as well has stimulated Burley[71] to assume that the well-known reaction of thiol–disulfide interchange might be able to explain many effects of chemical changes on physical properties such as setting, supercontraction and recovery of the fiber after release (see also refs. 72–74). The reaction proceeds as follows:

$$\begin{array}{l} \quad | \\ \;NH \\ \quad | \\ HC{-}CH_2{-}S^{\ominus} + R_1{-}CH_2{-}S{-}S{-}CH_2{-}R_2 \longrightarrow \\ \quad | \\ \;CO \\ \quad | \end{array}$$

$$\begin{array}{l} \quad | \\ \;NH \\ \quad | \\ HC{-}CH_2{-}S{-}S{-}CH_2{-}R_2 + R_1{-}CH_2{-}S^{\ominus} \\ \quad | \\ \;CO \\ \quad | \end{array}$$

If the SH–SS interchange mechanism between two adjacent polypeptide chains occurs, initiated by a cysteine thiol group in one of them, the re-arrangement of the disulfide bonds could probable give rise to the formation of new SS crosslinks by a progressive interchange along the chain. However, from a survey of observations, mainly based on their own experiments, and careful collected arguments as to the theories of the thiol–disulfide interchange and the crosslinkage stabilization, Robson et al.[75] concluded that the enhanced stabilization of protein conformation required for the permanent set is brought about by the formation of stable crosslinks and that, if the

SS–SH interchange mechanism operates at all, its effectiveness will probably be restricted to the matrix.

## 6.4. Formation of Crosslinks by Means of Artificial Bridge Residues

### 6.4.1. Bifunctional Reagents

It has been generally accepted that the introduction of additional covalent bridges between amino acid residues of two different peptide chains of the wool proteins can improve the physical and chemical stability of the fibers. To achieve this goal a considerable amount of work was directed toward the search for reagents especially suitable for practical purposes. Prerequisites for this type of crosslinking reaction are that the bifunctional reagents used are sufficiently reactive and that, in addition, at least two reactive sites are available at the required distance for crosslink formation within the substrate.

When the crosslinking reaction is considered primarily from the point of view of the amino acid bridging residues, it is possible to distinguish between the following two reactions: (1) two amino acid side chains of the same type are linked by the reagent giving rise to a symmetrical bridge; and (2) the reagent joins two different amino acid residues, resulting in an unsymmetrical crosslink.

Even when both sites of the crosslinking agent are equal with respect to their reactivity, a monofunctional reaction can result. This occurs when the crosslink, because of its molecular size, does not fit into the space between two residues of adjacent peptide backbones, or when the reaction is restricted or suppressed by steric hindrance.

#### *6.4.1.1. Aryl Dihalides*

Among the various bifunctional reagents capable of forming crosslinks, two are only of academic interest, since they produce bright-yellow colored wool. These two compounds, 1,5-difluoro-2,4-dinitrobenzene (I) (abbreviated FFDNB); and *p,p'*-difluoro-*m,m'*-dinitrodiphenylsulfone (II) (abbreviated FF-sulfone), have been applied as bifunctional reagents to crosslink fibrous proteins, silk fibroin, collagen, and wool.[76–79]

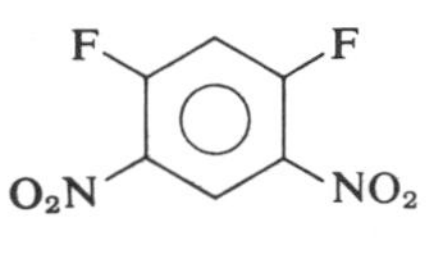

F F $O_2N$ $NO_2$

(I)

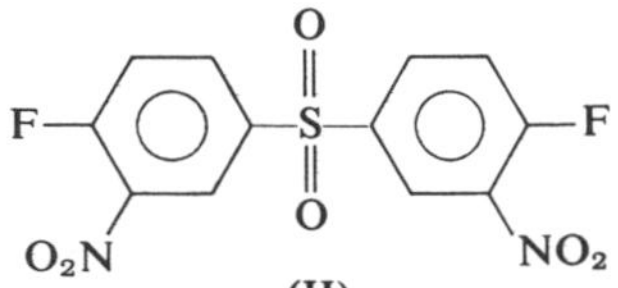

O F S F O $O_2N$ $NO_2$

(II)

Under similar conditions, both reagents react to give rise to monosubstitution and disubstitution. The rate of reaction depends on various parameters, such as temperature, concentration, and time of reaction. Compound (I) is able to link reactive sites within a distance of approximately 5 Å between the two fluorine atoms. The amino acid residues crosslinked by one of these bifunctional reagents have characteristic visible and ultraviolet spectra. Both reagents react easily with N-terminal and side-chain amino groups, tyrosine phenolic groups, and also with thiol and imidazole groups.

When wool is reacted with FFDNB, a predominant bifunctional reaction has been observed. This resulted in one unsymmetrical and two symmetrical pairs of bridged tyrosine and lysine residues, giving rise to the formation of DNPene-bis-$N^{\epsilon}$-Lys, DNPene-bis-*O*-Tyr, and DNPene-$N^{\epsilon}$-Lys-*O*-Tyr. More recently, $N^{\epsilon}$-*S*-dinitrophenylen-lysine-cysteine, an unsymmetrical linked compound has been detected.[80] These crosslinks are fairly resistant to acid hydrolysis and were isolated and subsequently identified by comparison with the respective model compound using paper chromatography[81] and chromatography on nylon-powder columns[82]; see Fig. 6–3.

The reaction of wool with the FF-sulfone compound gave similar results with regard to the types of amino acid residues which were crosslinked by the reagent. To date, sulfone-bis-tyrosine, sulfone-bis-lysine, and sulfone-lysine-tyrosine have been isolated and characterized.[83,84] From the overall yield of the isolated crosslinked amino acid residues, it can be concluded that tyrosyl residues are more closely associated with lysyl residues than are the lysyl residues with themselves.

### 6.4.1.2. *Alkylene Halides*

Wool that has been reduced with thioglycollic acid in neutral or acid solution can be further modified without loss of its fibrous structure by

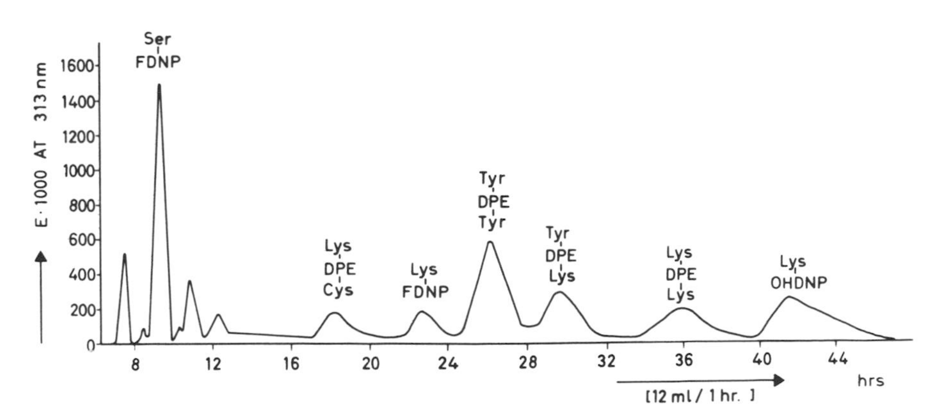


Fig. 6-3. Fractionation of a hydrolysate of FFDNB-treated wool on nylon-powder columns.[80]

treatment with alkyl dihalides. By this procedure the disulfide bonds are split and subsequently reduced to thiol groups which further combine with the reagent, giving rise to bis-thioether groups. The final step in the reaction is given by the following equation, where W represents the portions of the wool connected by the disulfide groups and X represents a halogen atom:

$$W{-}SH + X{-}(CH_2)_n{-}X + HS{-}W \longrightarrow W{-}S{-}(CH_2)_n{-}S{-}W + 2HX$$

Harris and his colleagues[85,86] have investigated the influence of alkylene bromides by varying the number of introduced methylene groups from 1 to 4. From their results, trimethylene dibromide seemed to be most suitable with respect to the increased resistance of wool to the action of alkali. Also, a slight increase in breaking strength and a lower, 30% index of work to stretch was observed. In addition, the modified wool samples exhibited an increased resistance to digestion by enzymes and moth larvae.

More recently the incorporation of those polymethylene-$\alpha\omega$-dihalides between two adjacent polypeptide chains in wool has been confirmed by the isolation of the various *S,S'*-polymethylene-bis-cysteines, with up to six methylene groups within the crosslink and their identification confirmed by comparison with the respective prepared model compounds[87,88] In their studies of various crosslinkages in wool and the effect on the supercontraction properties of the fibers, Crewther et al.[47] have tried to convert all cystine residues in wool to dimethylene-*S,S'*-dicysteine residues by reduction and subsequent alkylation with 1,2-dibromoethane in various solvent systems. However, amino acid analysis revealed that the maximum rate of conversion was only about 40%, reportedly due partly to the formation of *S*-2-hydroxyethylcysteine residues as a result of an additional monofunctional reaction of the agent.

### 6.4.1.3. *Aralkylene Halides*

Among the various reagents employed by Kirst,[89] 1,5-bis(chloromethyl)-2,4-dimethylbenzene might be mentioned, since the treatment of reduced wool with it resulted in an increased resistance to alkali and in low supercontraction data measured in 50% phenol.[90] This compound bridges two cysteine residues, as was shown by the isolation of the respective bis-thioether.[91] Tetrafunctional 1,2,4,5-tetrakis (chloromethyl)benzene also reacted with reduced wool, as was proved by isolation of the crosslink and comparison with a prepared model compound, having the formula:

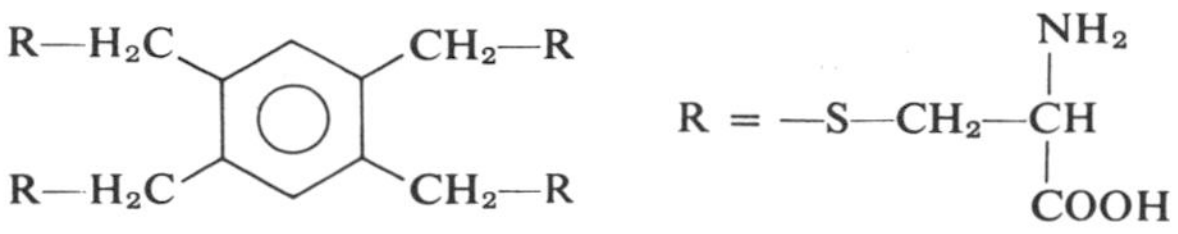

$$\text{C}_6\text{H}_2(\text{CH}_2\text{—R})_4 \qquad \text{R} = \text{—S—CH}_2\text{—CH(NH}_2\text{)—COOH}$$

The treated wool exhibited an improvement in physical properties, including increased tensile strength and reduced solubility in urea–bisulfite solution.

### 6.4.1.4. *Acylating Reagents*

The group of acylating reagents includes a vast variety of compounds of different dimensional shapes, sizes, and reactivities. Among the activated esters and amides, the bifunctional active esters of alkylene-$\alpha\omega$-dicarboxylic acids have acquired theoretical interest in relation to wool chemistry. It was first expected that these esters would react preferentially with primary amino groups of lysine side chains, the latter being accumulated in the microfibrillar regions of the wool fiber.

One of the acylating reagents extensively used in peptide synthesis is the reaction of activated esters such as *p*-nitrophenyl esters.[92] Under mild conditions, activated dicarboxylic acid derivatives were used as crosslinking reagents for proteins, including wool.[93–95] The following scheme represents bifunctional acylation of primary amino groups in proteins with di-*p*-nitrophenylsebacate.

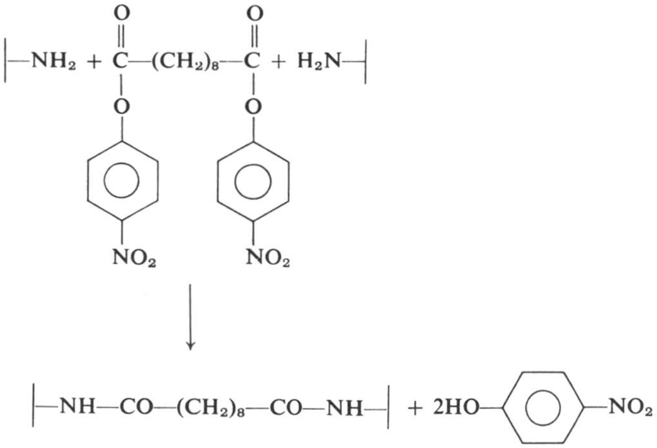

$$|\text{—NH}_2 + \text{O}_2\text{N—C}_6\text{H}_4\text{—O—CO—(CH}_2)_8\text{—CO—O—C}_6\text{H}_4\text{—NO}_2 + \text{H}_2\text{N—}|$$

$$\downarrow$$

$$|\text{—NH—CO—(CH}_2)_8\text{—CO—NH—}| + 2\text{HO—C}_6\text{H}_4\text{—NO}_2$$

Activation of the dicarboxylic acid was also achieved via *N*-hydroxysuccinimide and imidazole derivatives.[96] From the analytical and solubility data of all these bifunctional reagents, including di-*p*-nitrophenyl carbonate, it has been concluded that they are incorporated between polypeptide chains. This suggestion has been strongly supported by measurements (after reduction) of the thermal contraction temperature, which was considerably raised (by at

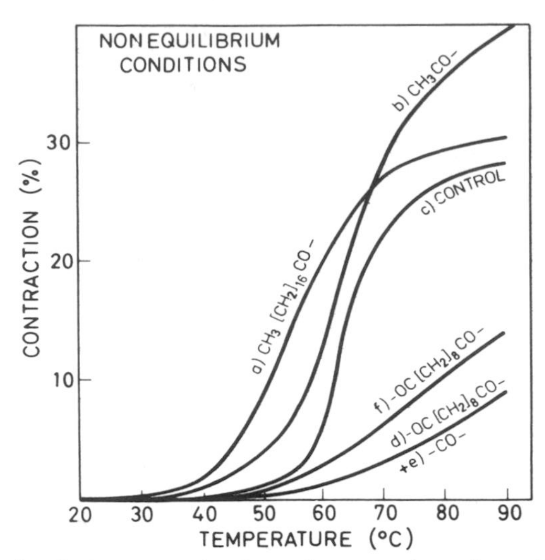


Fig. 6-4. Effect of substitutes on the thermal contraction of wool[97]: (a) *N*-stearylimidazole, (b) *N*-acetylimidazole, (d) *N,N'*-sebacoxydisuccinimide, (e) di-*p*-nitrophenylcarbonate, (f) *N,N'*-sebacyldiimidazole.

least 30°C) due to prior treatment.[96,97] The curve in Fig. 6–4 clearly demonstrates this effect, in contrast to monofunctional substituents. The aliphatic chain length of the activated molecules and their concentration can influence the ratio of the monobifunctional reaction. Such active esters are, however, not specific for the ε-amino groups of lysyl residues, as was suggested earlier. They also react to a certain extent under more severe conditions with hydroxyl groups of wool. These results have been obtained by direct radioactivity measurements, including the removal of *O*-acyl-bound groups by treatment with hydroxylamine[98] and by determining the content of covalently bound sebacic acid as DNP hydrazide in the hydrolysate.[99,100] The effect of acylation on the setting of wool fabrics has been extensively studied.[101] It has been shown that fabrics which have been acylated with bifunctional reagents such as *N,N'*-sebacoxydisuccinimide exhibit a decreased settability which, at least in part, is due to the introduction of crosslinks, mainly involving amino groups.

Since active esters are best applied to wool from nonaqueous solvents such as dimethyl sulfoxide or dimethylformamide, it cannot be predicted at present whether they would be suitable for commercial application.

### *6.4.1.5. Aldehydes*

Numerous investigations have been undertaken with regard to the reactions of aldehydes on wool. The very first studies in this direction can be dated back to the beginning of this century, when formaldehyde was used to improve the properties of the wool fibers.

Contradictory theories concerning the mechanism of reaction were proposed until Fraenkel-Conrat and Olcott[102,103] threw more light on the reactions of proteins with formaldehyde. The primary reaction is the addition of a reactive hydrogen atom to the carbonyl double bond, resulting in a methylol derivative. Crosslinking by methylene groups is then believed to proceed via a secondary reaction involving primary amide, guanidyl, and indole groups:

$$R{-}NH_2 + H_2CO \rightleftharpoons R{-}NH{-}CH_2OH$$

$$R{-}NH{-}CH_2OH + H_2N{-}C(=O){-}R \longrightarrow R{-}NH{-}CH_2{-}NH{-}CO{-}R'$$

Recently, evidence was presented that newly formed thiol groups are capable of reacting with formaldehyde, giving rise to methylene dithiol crosslinks indicated by enhanced thermal stability.[104] On the other hand, it has been found that formaldehyde cannot form a stable link between two amino or two amide groups.

It has been suggested that the observed decrease in solubility of formaldehyde-treated wool in alkali– and urea–bisulfite solutions presumably stems from a few newly introduced methylene crosslinks.[105] The extent of supercontraction in bisulfite is remarkably diminished, and, under certain conditions, some improvement in the tensile properties of the treated fibers has been obtained.[106] Better mechanical properties are also claimed for dyed wool if formaldehyde is added to the dye bath instead of crosslinking before dyeing.[107] Other experiments clearly indicated that formaldehyde, after treatment of steam-set wool fabrics, improves the degree of set and also decreases the rate of release of set.[108] With respect to the sites of crosslinking in keratin due to formaldehyde, the reagent presumably forms crosslinks in the matrix as well as in the microfibrils, and here in particular within the low-sulfur protein fraction (SCMKA or $\alpha$-keratose), generally assumed to stem from the microfibrils—in other words, from the oriented crystalline components of the fiber.[109–111] The observed increase of thermal stability additionally supports the hypothesis that crosslinking reaction has taken place in intimate associations with the helical regions of the fiber.[112] By the use of digestion with a mixture of proteolytic enzymes followed by isolation of the respective components, it was unequivocally demonstrated that in formaldehyde-treated wool, lysine and glutamine side chains represent two of the sites of reaction.[113]

In a series of studies for producing improved hospital bed pads, glutaraldehyde-treated wool appeared to enhance chemical stability, especially in the presence of alkali. It has been postulated that glutaraldehyde combines chemically with wool keratin to produce strong crosslinks.[114] According to the amino acid composition of glutaraldehyde-stabilized wool, only lysine and probably histidine[115] residues seem to be involved in the reaction. It is further believed that other residues may have reacted to form bonds unstable to acid hydrolysis.[116]

From the reaction of reduced wool with a number of unsaturated aldehydes, some evidence has been obtained for a crosslinking reaction.[117] Of the $\alpha$, $\beta$-unsaturated aldehydes, acrylaldehyde (acrolein) reacts most readily with reduced wool, combining with most of the free thiol groups. Also, since no lysine residues were found, following this reaction, either or both of the following crosslinking reactions may have occurred:

$$\vdash CH_2—SH + H_2C{=}CH—CHO + HS—CH_2 \dashv$$

$$\downarrow$$

$$\vdash CH_2—SH—CH_2—CH_2—CH(OH)—S—CH_2 \dashv$$

and/or

$$\vdash (CH_2)_4—NH_2 + H_2C{=}CH—CH(OH) + HS—CH_2 \dashv$$

$$\downarrow$$

$$\vdash CH_2—NH—CH_2—CH_2—CH(OH)—S—CH_2 \dashv$$

A very pronounced decrease in the supercontraction in bisulfite, together with other physical data, support this view.

### 6.4.1.6. *Diisocyanates*

In reactions with wool,[118,119] several diisocyanates, which imparted increased chemical resistance, as shown by the reduced alkali solubility and solubility in peracetic acid followed by ammonia, have been investigated. Dimethylformamide is a suitable swelling agent which easily enables the reagent to penetrate the keratin structures. The modified fabrics were more resistant to shrinkage during washing. Further, a remarkable decrease in supercontraction was observed. It has been postulated that diisocyanates introduce new crosslinks into the wool fiber. The major reaction probably takes place with amino groups, resulting in urea derivatives which are cleaved by acid hydrolysis. Some of the diisocyanates applied for wool modification are as follows:

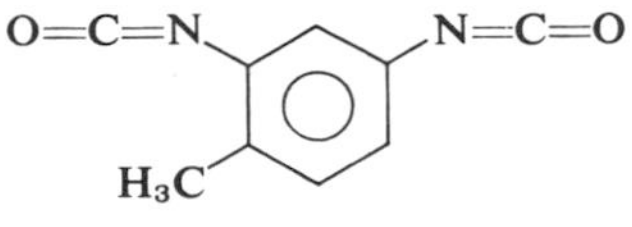

$O{=}C{=}N$ $N{=}C{=}O$

$H_3C$

Tolylene-2,4-diisocyanate

$$O{=}C{=}N{-}(CH_2)_6{-}N{=}C{=}O$$

Hexamethylenediisocyanate

From thermal-contraction measurements on wool treated with hexamethylene diisocyanate, the existence of crosslinks in helical zones was excluded.[112] When monofunctional isocyanates (e.g., phenylisocyanate, octadecyliso-cyanate) were applied, comparatively low solubility data were obtained only after greatly increasing the uptake of the reagent.[120] Crease retention of wool can be improved by treatment with monoisocyanates but not with the corresponding bifunctional compounds, presumably because crosslinking by diisocyanates resists new crease formation.[121]

### 6.4.1.7. *Dimaleimides*

It is well established that N-substituted maleimides react with thiol groups under mild conditions and with a minimum of side reactions. This is the case also for bifunctional maleimides.[122] Reduced wool, which was subsequently treated either with *o*-phenylenedimaleimide (see formula) or its meta isomer, showed decreases in solubility in three different solubility tests. In addition, the decrease in supercontraction of the treated fibers in boiling bisulfite solution was striking.

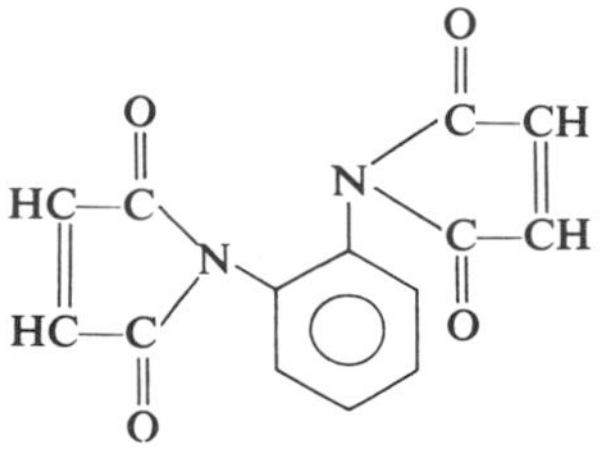

O O
C—CH
HC—C N
N C—CH
HC—C O
O

*N*,*N'*-(1,2-phenylene)bismaleimide

The effects of both bifunctional reagents on wool were quite significant when the data given in Table 6–1 were compared with those obtained by reaction with the monofunctional *N*-phenylmaleimide.

Since the imide ring is cleaved by acid hydrolysis, all three modified

**Table 6-1. Results of Selected Tests on Reduced Wool Treated with Maleimides** [123]

| | Treatment | | | |
|---|---|---|---|---|
| Tests | *N*-Phenyl-maleimide | *m*-Phenylene dimaleimide | *o*-Phenylene dimaleimide | Buffer alone |
| Uptake (%) | 8.36 | 5.52 | 4.32 | 0 |
| 30% index[a] | 0.81 | 0.88 | 0.88 | 0.77 |
| Cystine (%) | 5.71 | 6.52 | 7.50 | 11.4 |
| Supercontraction (%) | 22 | 0.3 | 0.8 | 28 |
| Alkali solubility (%) | 24.4 | 6.9 | 7.6 | 25.3 |
| Acid solubility (%) | 12.8 | 6.4 | 7.0 | 13 |
| Peracetic acid–$NH_3$ solubility (%) | 73 | 44 | 56 | 89 |

[a] The 30% index of work to stretch was initially 0.67 in all cases.

wools yielded the same monofunctional amino acid derivative, 2-amino-2-carboxy-ethyl mercaptosuccinic acid. Hence no direct proof is available that a bifunctional reaction takes place. However, from all tests applied to reduced wools treated with bis-maleimides, it was concluded that a considerable number of disulfide bonds were replaced by stable crosslinks between adjacent polypeptide chains of the wool proteins.

#### *6.4.1.8. Benzoquinone*

In one of the earliest treatments of improving the strength of intact fibers, an aqueous solution of benzoquinone has been applied to wool.[124] A similar strengthening effect by boiling in aqueous alcoholic quinone has been observed.[125] The assumption that the main reaction of quinone is with amino groups is probably correct, but, since the uptake of benzoquinone of the wool exceeded by far the number of amino groups that could have reacted, Speakman and co-workers[126] considered that the benzoquinone formed grafted polymer and loosely deposited polymer within the fiber.

From the change in properties such as the increase in torsional rigidity,[127] thermal contraction, and reduced swelling in formic acid[128] it was assumed that the sites of additional crosslinks induced by benzoquinone treatment are found in the matrix.[129,130] This interpretation is in contradiction to that from authors[109] who suggested that crosslinking takes place in the $\alpha$-helical regions. This was evinced by the estimation of the relative properties of keratose fraction, derived from the microfibrillar regions, showing an increased insolubility in the benzoquinone-treated wool. Despite the fact that such treated wool is highly resistant to fungus and bacterial attack,[131] it has little practical importance in textiles because benzoquinone discolors the wool and probably gives rise to dermatitis if unreacted reagent is not removed completely. In the special field of production of woven woollen felts for paper-making machines, treatment with benzoquinone and its derivatives is widespread. The increased resistance of the treated wool to degradation in wet conditions is more important than coloration effect of the reagent. Such treatment is believed to increase the usable life of the felts considerably.

During the oxidation of wool and silk fibers with potassium nitrosyldisulfonate ON $(SO_3K)_2$, Fremy's salt, Earland and co-workers[132] found this reagent being extremely effective for rendering the fibers less soluble in solution of alkali and of urea–bisulfite.[133] In addition to that, an increase in resistance of the fibers to contractions in sodium bisulfite might indicate that new crosslinkages are formed by this treatment. No precise structure for these newly introduced crosslinks are given. However, since the tyrosine content was remarkably diminished, the hypothesis has been put forward that modified tyrosine residues are involved in the formation of these linkages. It was shown[170] that such crosslinking did not arise through ether links between tyrosine or serine residues. Presumably, tyrosine is involved in an amino-substituted hydroquinone crosslink.

### 6.4.1.9. *bis-Diazohexane*

The introduction of highly reactive diazo groups into wool is based on experiments by Kirst,[134] who applied bis-diazohexane to untreated and reduced wool. The observed effects of stabilization did not preclude the formation of crosslinks, although no direct proof was available. Subsequently, Zahn and Waschka[135] succeeded in the isolation of the bridged amino acid pairs of combined tyrosine and cysteine from hydrolysates of treated wool.

```
H—Tyr—OH                    H—Tyr—OH
  |                            |
  O———————(CH2)6———————O
```

*O,O'*-Hexamethylene-bis-tyrosine

```
H—Cy—OH                     H—Cy—OH
  |                            |
  S———————(CH2)6———————S
```

*S,S'*-Hexamethylene-bis-cysteine

From the tyrosine content of wool (every twenty-second residue is tyrosine) it can be concluded that two tyrosine residues are located within a distance of 7.5 Å, when the dimension of the incorporated hexamethylene bridge is considered. Similar distances have been found with other crosslinking agents.[78]

### 6.4.1.10. *Divinylsulfone*

Divinylsulfone-treated wool was found to have a low solubility, which suggested that crosslinks have been introduced.[136] Subsequent work with divinylsulfone clearly demonstrated that lysine and histidine side chains had reacted to a remarkable extent.[137] A corresponding weight gain and an increase in sulfur support the assumption of the crosslinking reaction of sulfone.

Table 6-2 summarizes a number of reagents which, from solubility and other physical data, apparently crosslink wool protein. Such a list is, however, by no means complete. Methods of detecting crosslinks have recently been seen in a more critical light, and from this point of view it is rather doubtful whether all the listed compounds are actually crosslinking agents. Revision of earlier results is required by applying modern techniques of analysis to demonstrate unequivocally that crosslinking reactions are involved.

## 6.4.2. Reaction of Reactive Dyes with Wool

One of the several classes of modern wool dyes that has increased in interest is that of reactive dyes. One reason for this lies in the fact that reactive dyes can be covalently bonded to the substrate. If the dye molecule carries two activated groups, both are capable of reacting with different positions of the peptide chains to produce crosslinks.

When dichlorotriazine dyes of the Procion type were applied to wool

**Table 6-2. Reagents That Crosslink Wool Protein**

| Compound | Formula | Reference |
|---|---|---|
| Chloromethyl ether | $ClH_2C—O—(CH_2)_n—O—CH_2Cl$; $n = 4, 6$ | 138 |
| Epoxy compounds | R—CH(—O—)CH$_2$ (epoxide ring: $R—\overset{H}{C}—CH_2$ bridged by O) | 139,141 |
| Aziridines | R—N(CH$_2$—CH$_2$) (aziridine ring) | 139,140 |
| Mesyloxy derivatives | $R—(CH_2)_4—R$<br>$R = CH_2SO_2O—$ | 139 |
| Dicarbonic azides | $N_3—CO—(CH_2)_n—CO—N_3$; $n = 1–14$ | 142 |
| Heavy-metal salts | E.g., mercury acetate, chloride | 14,18,143 |
| Aminothioethanol | $HS—CH_2—CH_2—NH_2$ | 144 |

fibers, an appreciable fall in the urea–bisulfite solubility of the dyed yarn was observed.[145] From these results it was tentatively suggested that crosslinking reactions had taken place. The reaction of cyanuric chloride with wool as a model compound of reactive dyestuff has been investigated.[146] Although no indication is given with regard to crosslinking formation, the usefulness of throwing more light on the elucidation of the dyeing mechanism is apparent.

Studies of the reaction of a vinylsulfone dye (Remazol type) revealed unequivocal evidence that the primary amino groups of the combined lysine residues were the main sites in wool that were involved in the reaction. This has been proved by the isolation of a lysine adduct of the dye from the hydrolysate of dyed wool.[147] By comparison of the data of solubility tests and supercontraction measurements obtained after reaction of both monofunctional and bifunctional Remazol dyes, a significant decrease in solubility was observed in the case of the bifunctional dye Remazol black B, shown in Fig. 6-5.[148] The main component of Remazol black R is said to contain a derivative with the following formula[149]:

$$CH_2{=}CH{-}O_2S{-}C_6H_4{-}N{=}N{-}C_{10}H_2(OH)(NH_2)(SO_3K)(KO_3S){-}N{=}N{-}C_6H_4{-}SO_2{-}CH{=}CH_2$$

Owing to the instability of the chromophoric system during acid hydrolysis, attempts to isolate derivatives of amino acids crosslinked by the reactive dye molecule will fail. Even when the bisazonaphthol chromophore of the above-mentioned dye was replaced by an anthraquinone compound resistant

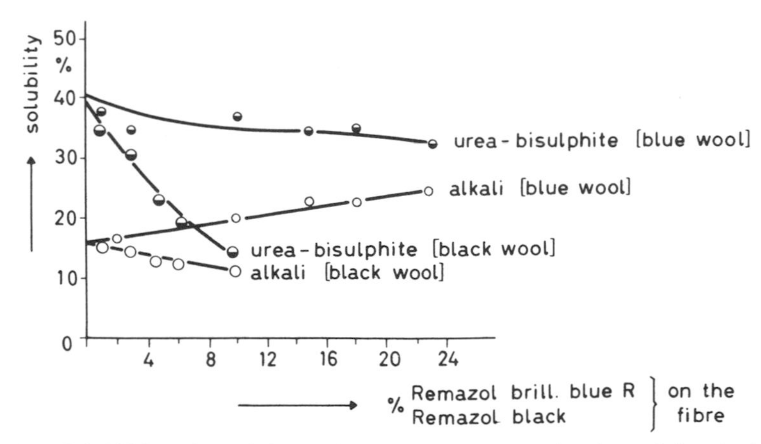


Fig. 6-5. Solubilities of wool dyed with two different reactive dyes of the vinylsulfone type, according to Reinert et al.[148]

**Table 6-3. Solubility Data of Dyed Wool (% of weight)**

| Dyed at: | | Dyed with: | | |
|---|---|---|---|---|
| °C | pH | Monofunctional dye | Bifunctional dye | Blank-dyed |
| | | | I: Urea–bisulfite | |
| 90 | 2 | 49.8 | 16.4 | 59.7 |
| | 3 | 42.3 | 16.6 | 51.5 |
| 100 | 3 | 40.2 | 2.3 | 43.0 |
| | | | II: Urea–thioglycollate | |
| 100 | 3 | 65.3 | 1.3 | 60.5 |
| | | | III: Sodium hydroxide | |
| 90 | 3 | 28.2 | 8.2 | 26.0 |

to acid hydrolysis, no bifunctional adduct of the dye with two amino acids was obtained.[150]

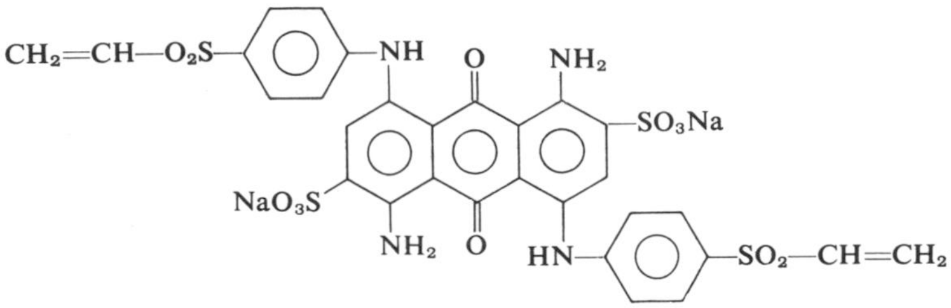

$CH_2{=}CH{-}O_2S$ NH O $NH_2$ $SO_3Na$ $NaO_3S$ $NH_2$ O HN $SO_2{-}CH{=}CH_2$

On the other hand, decreases in solubility in several media presented some evidence of a crosslinking reaction, as shown by Table 6-3.[150] Despite these values, the question arises whether the observed decrease in solubilities is caused by crosslinkages or merely by the introduction of bulky groups of the large bifunctional dye molecule into the wool fiber.

## 6.5. Analysis and Methods for Studying Crosslinking Reactions

### 6.5.1. Solubility Tests

The determination of the number of new crosslinks which have been introduced in wool is by no means simple. This is due mainly to the fact that most types of crosslinking agents produce amino acid derivatives which are unstable to hydrolysis, so in many cases their formation cannot be demonstrated directly. An indirect approach to the determination of the relative proportions of mono- and bifunctional reactions involves analysis of the number and types of functional groups which have reacted and of the amount of reagent bound to the wool. Sometimes it is desirable to detect small amounts of crosslinks in the presence of much monofunctionally bound reagent. Furthermore, it is expected that the incorporation of any additional crosslinks will not alter extensively the physical and chemical properties of the fiber, because of the large number of disulfide crosslinks already present in the wool.

A great variety of physical measurements is employed for assessing crosslinking in wool. These methods are based on changes in physical properties such as diametral swelling,[(96,151)] supercontraction,[(152,47)] settability,[(153)] load extension,[(154,155)] lateral compression,[(156)] torsional moduli,[(157)] thermal contraction,[(97)] swelling in formic acid,[(158)] and several solubility measurements. Special attention has been given to those solubility tests by which disulfide and noncovalent bonds are disrupted. Two such frequently employed tests for crosslinking involve measurement of the decrease in solubility in mixtures containing a cystine reducing agent such as bisulfite[(159)] or thioglycollate[(160)] and urea as a reagent causing denaturation and probably splitting of hydrogen bonds. Performic acid oxidation of cystine followed by extraction with ammonia[(161)] represents another solubility test. After the introduction of any new crosslinks, which are resistant to reduction or oxidation, a large decrease in solubility will be observed. The alkali solubility test,[(162)] which is sometimes used both for detecting the damage done to wool and for the introduction of new alkali-stable crosslinks, is nonspecific and hence inapplicable in this respect.

Because of the fact that monofunctional reagents which introduce bulky groups into wool also reduce its solubility, especially when the urea–bisulfite solubility test is used, great care has to be taken in the interpretation of any results. Therefore, it is considered advisable to react wool with a monofunctional compound of appropriate size as a control. Furthermore, one has to consider that denaturation as a consequence of heat treatment or of the influence of certain organic solvents may also decrease the solubility. In spite of such precautions, it must be emphasized that any of the solubility tests can be used as a qualitative criterion of crosslinking only in connection with other tests.[(163)]

### 6.5.2. Separation Procedures

Closely related to the solubility tests are the methods of separation, which are also applied for determining the degree of new crosslinks that might have occurred in wool keratin. These are based on measuring the changes in the proportions of three fractions comparable to the $\alpha$-, $\beta$-, and $\gamma$-keratose fractions, obtained by oxidation with performic acid followed by extraction with ammonia.[139] From increases in the insoluble $\beta$-keratose fraction, some coherent suggestions can be made regarding the increase in the degree of crosslinking within the fiber structure.[164,94] An alternative procedure of fractionation is also used in which disulfide bonds are reduced with potassium thioglycollate, subsequently alkylating the formed thiol groups to produce *S*-carboxymethyl keratein. This can be fractionated into soluble kerateine fractions.[165]

### 6.5.3. Estimation of Reacted and Unreacted Side-Chain Groups

The most reliable although experimentally inconvenient methods, are based on chemical analyses, from which, however, it can only be assumed that crosslinking reactions have occurred. One of these methods determines the difference of the reactive side-chain groups in the substrate before and after the reaction with the crosslinking compound. Reactive amino acid residues which were not blocked by the agent can easily be estimated by treating the wool with a solution of 1-fluoro-2,4-dinitrobenzene,[166] which, under mild conditions, will react with amino, hydroxy, and thiol groups, leading to dinitrophenyl amino acids. These, on hydrolysis of the protein, can be readily separated and determined.[167,59,61] However, it must be emphasized that such a procedure is not very conclusive for defining crosslinking, since it does not differentiate between mono- and bifunctional reactions but indicates accurately the quantity of fiber-reactive groups blocked by the crosslinking agent.

### 6.5.4. Isolation of Crosslinked Amino Acid Residues

Direct proof that incorporation of a reagent as a crosslink in the protein has taken place can only be established by the isolation of the crosslink itself attached to two amino acid residues that have participated in the reaction. From the few examples known so far, some have been already mentioned in this chapter. Thus the various amino acid residues linked by FFDNB[80] and by FF-sulfone[83,84] are excellent proofs for a crosslinking reaction. Further, *S,S'*-polymethylene-bis-cysteines were isolated from reduced wool that was treated with polymethylene-$\alpha\omega$-dihalides [-Br$(CH_2)_n$-Br-], where $n$ is 2–6.[87] However, in cases where the introduced bridges turned out to be unstable to acid hydrolysis, their isolation was impossible.

The recently propagated complete hydrolysis of wool using a mixture of enzymes which attack α-peptide bond exclusively has already opened up a new field in the isolation of newly introduced crosslinks.

## 6.6. Relationship between Morphological Fractions and Crosslinking Activity

In several investigations[109,127–130,164] it was affirmed that selective crosslinking or self-crosslinking reactions occur in specific morphological parts of the wool fiber. The sites of action within the fiber mainly depend on the type of reagent used and on the fiber-reactive groups participating herein. In order to identify these components of the wool complex, fractionation into the three keratoses after subsequent oxidation and change in their relative proportions of insoluble fractions is sometimes advantageous, giving some indication where the crosslinking might have occurred.[164] The three keratose fractions in virgin wool are assumed to arise from the microfibrillary material (α), which has less sulfur than the original wool; the intercell membranes (β); and the high-sulfur matrix protein (γ), respectively. For instance, in steam-set wool, the decrease in γ-keratose corresponding with an increase in α-keratose was attributed to lanthionine formation, which stems probably from the high-sulfur-matrix proteins. It was later shown[168] that such crosslinking must be due to lysinoalanine rather than lanthionine.

When crosslinks are formed by treatment with formaldehyde, controversial opinions have been proposed; however, the favorite sites of the crosslinking reactions are probably located in the matrix surrounding the microfibrils, which render the α- and γ-keratose fractions less soluble. In alkali treated wool the proportion of the undissolvable β-keratose fraction increases with the increasing temperature of treatment up to 50°C,[168] the α-keratose fraction decreasing proportionately.

When dealing with the α-, β-, and γ-keratose fractions, obtained after various modifications of wool, it must be pointed out that the term "keratose" should be abandoned because any connection with the original morpological components has been invalidated by the previous chemical treatments.

## 6.7. Concluding Remarks

Surveying the literature that covers the topics of this chapter up to 1973 indicates that research work on crosslinking and self-crosslinking reactions in wool has steadily increased within the last 10 years. A detailed article that surveys the literature on crosslinking agents to wool keratin has been published.[169] Many results are directly related to the progress that was achieved through a more detailed understanding of the internal structure of the keratin. On the other hand, the inhomogeneity of the wool fiber has proved to be an obstacle hindering the expected success of its application on an industrial scale. Size, shape, and the nonspecificity of many of the introduced reagents and the distance between reactive side-chain groups, as well as the reactivities of the substrate and the reagents, play a most important role in crosslinking. However, it has been shown that certain reactions which occur readily with simple model compounds do not occur in wool owing partly to the inaccessibility to the reagent of the reactive groups within the fiber.

A further aspect of this work is that a great number of reagents or reactions have thrown light on some new features of fiber structure. The effects of stabilization achieved by introducing new links have led to improvements in fiber properties. Technical processes such as permanent pleating and flat setting have been successfully developed from these findings.

## 6.8. References

1. W. G. Crewther, R. D. B. Fraser, F. G. Lennox, and H. Lindley, *Advan. Protein Chem.* **20**, 191 (1965)
2. M. Laskowski and H. A. Scheraga, *J. Amer. Chem. Soc.* **76**, 6305 (1954).
3. D. Poland and H. A. Scheraga, in: *Poly α-Amino Acids* (G. D. Fasman, ed.), p. 398, M. Dekker, Inc., New York (1967).
4. W. Kauzmann, *Advan. Protein Chem.* **14**, 1 (1959).
5. G. Nemethy and H. A. Scheraga, *J. Phys. Chem.* **66**, 1773 (1962).
6. J. B. Speakman and M. C. Hirst, *Trans. Faraday Soc.* **29**, 148 (1933).
7. J. B. Speakman and E. Stott, *Trans. Faraday Soc.* **30**, 539 (1934).
8. J. B. Speakman, *J. Textile Inst.* **32**, T83 (1941).
9. Z. Stary, *Hoppe-Seylers Z. Physiol. Chem.* **175**, 178 (1928).
10. W. R. Middlebrook and H. Phillips, *Biochem. J.* **36**, 428 (1942).
11. H. Lindley and H. Phillips, *Biochem. J.* **39**, 17 (1945).
12. H. Lindley and R. W. Cranston, *Biochem J.* **139**, 515 (1974).
13. J. B. Speakman, *Nature* (*Lond.*) **132**, 930 (1930).
14. J. B. Speakman, *J. Soc. Dyers Colourists* **52**, 335 (1936).
15. H. Phillips, *Nature* (*Lond.*) **138**, 121 (1936).
16. W. R. Cuthbertson and H. Phillips, *Biochem. J.* **39**, 7 (1945).
17. R. Cockburn, B. Drucker, and H. Lindley, *Biochem. J.* **43**, 438 (1948).
18. J. B. Speakman and C. S. Whewell, *J. Soc. Dyers Colourists* **52**, 380 (1936).
19. A. J. Farnworth, *Textile Res. J.* **27**, 632 (1957).
20. R. W. Burley, *Proc. Intern. Wool Textile Res. Conf., Australia* **D**, 88 (1955).
21. H. Zahn, F. W. Kunitz, and D. Hildebrand, *J. Textile Inst.* **51**, T740 (1960).
22. H. Zahn, F. W. Kunitz, and H. Meichelbeck, *Colloque structure de la laine*, Institut Textile de France, p. 227 (1961).

23. J. B. Caldwell, L. M. Dowling, S. J. Leach, and B. Milligan, *Textile Res. J.* **34**, 933 (1964).
24. L. M. Dowling and W. G. Crewther, *Textile Res. J.* **34**, 1109 (1964).
25. A. Schöberl and M. Wiesner, *Biochem. Z.* **306**, 268 (1940).
26. N. A. Rosenthal and G. Oster, *J. Soc. Cosmetic Chem.* **5**, 286 (1954).
27. B. H. Nicolet and L. A. Shinn, Abstr. 103rd Meeting Amer. Chem. Soc. (April 1942).
28. L. R. Mizell and M. Harris, *J. Res. Natl. Bur. Std.* **30**, 47 (1943).
29. D. S. Tarbell and D. P. Harnish, *Chem. Rev.* **49**, 1 (1951).
30. L. O. Andersson and G. Berg. *Biochim. Biophys. Acta* **192**, 534 (1969).
31. L. O. Andersson, *Biochim. Biophys. Acta* **200**, 363 (1970).
32. J. W. Donovan and T. M. White, *Biochemistry* **10**, 32 (1971).
33. M. J. Horn, D. B. Jones, and S. J. Ringel, *J. Biol. Chem.* **138**, 141 (1941).
34. W. Küster and W. Irion, *Hoppe-Seylers Z. Physiol. Chem.* **184**, 225 (1929).
35. L. M. Dowling and J. A. Maclaren, *Biochim. Biophys. Acta* **100**, 293 (1965).
36. R. S. Asquith and J. B. Speakman, *Nature* (*Lond.*) **170**, 798 (1952).
37. R. S. Asquith and J. B. Speakman, *Proc. Intern. Wool Textile Res. Conf., Australia* **C**, 302 (1955).
38. A. J. Farnworth and J. B. Speakman, *Nature* (*Lond.*) **161**, 890 (1948).
39. A. Patchornik and M. Sokolovsky, *J. Amer. Chem. Soc.* **86**, 1860 (1964).
40. Z. Bohak, *J. Biol. Chem.* **239**, 2878 (1964).
41. K. Ziegler, *J. Biol Chem.* **239**, PC2713 (1964).
42. K. Ziegler, *Proc. 3rd Intern. Wool Textile Res. Conf., Paris* **2**, 403 (1965).
43. P. Miro and J. J. García-Domínguez, *J. Soc. Dyers Colourists* **83**, 91 (1967).
44. P. Mellet, *Textile Res. J.* **38**, 977 (1968).
45. P. Mellet and O. A. Swanepoel, *Naturwissenschaften* **52**, 495 (1965).
46. M. C. Corfield, C. Wood, A. Robson, M. J. Williams, and J. M. Woodhouse, *Biochem. J.* **103**, 15c (1967).
47. W. G. Crewther, L. M. Dowling, A. S. Inglis, and J. A. Maclaren, *Textile Res. J.* **37**, 736 (1967).
48. Kl. Ziegler, *Appl. Polymer Symp.* **18**. Pt. I, 257 (1971); and unpublished results.
49. D. E. Rivard and H. E. Carter, *J. Amer. Chem. Soc.* **77**, 1260 (1955).
50. P. B. Hamilton and R. A. Anderson, *Biochem. Preparations* **3**, 96 (1953).
51. Kl. Ziegler, I. Melchert, and C. Lürken, *Nature* (*Lond.*) **214**, 404 (1967).
52. J. Derminot, M. Tasdhomme, and N. Allard, *Bull. Inst. Textile Fr.* **24**, 71 (1970).
53. R. S. Asquith and J. J. García-Domínguez, *J. Soc. Dyers Colourists* 84, 155 (1968).
54. R. S. Asquith and M. S. Otterburn, *J. Textile Inst.* **60**, 208 (1969).
55. R. S. Asquith, A. K. Booth, and J. D. Skinner, *Biochim. Biophys. Acta* **181**, 164 (1969).
56. R. S. Asquith and J. D. Skinner, *Textilveredlung* **5**, 406 (1970).
57. R. S. Asquith and P. Carthew, *J. Textile Inst.* **64**, 10 (1973).
58. J. J. García-Domínguez, P. Miro, F. Reig, and S. Anguera, *Appl. Polymer Symp.* 18, Pt. I, 269 (1971).
59. E. Siepmann and H. Zahn, *Proc. 3rd Intern. Wool Textile Res. Conf., Paris* **1**, 300 (1965).
60. H. Beyer and U. Schenk, *J. Chromatog.* **39**, 482 (1969).
61. R. S. Asquith, D. Chan, and M. S. Otterburn, *J. Chromatog.* **43**, 382 (1969).
62. J. J. Pisano, J. S. Finlayson, and M. P. Peyton, *Science* (*Wash., D.C.*) **160**, 892 (1968).
63. J. J. Pisano, J. S. Finlayson, and M. P. Peyton, *Biochemistry* **8**, 871 (1969).

64. S. Matacic and A. G. Loewy, *Biochem. Biophys. Res. Commun.* **30**, 356 (1968).
65. L. Lorand, J. Downes, T. Gotoh, A. Jacobsen, and T. Tokura, *Biochem. Biophys. Res. Commun.* **31**, 222 (1968).
66. M. Cole, J. C. Fletcher, K. L. Gardner, and M. C. Corfield, *Appl. Polymer Symp.* **18**, Pt. I, 147 (1971).
67. R. S. Asquith, M. S. Otterburn, J. H. Buchanan, M. Cole, J. C. Fletcher, and K. L. Gardner, *Biochim. Biophys. Acta* **221**, 342 (1970).
68. B. Milligan, L. A. Holt, and J. B. Caldwell, *Appl. Polymer Symp.* **18**, Pt. I, 113 (1971).
69. R. S. Asquith and M. S. Otterburn, *J. Textile Inst.* **61**, 569 (1970).
70. R. S. Asquith and M. Otterburn, *Appl. Polymer Symp.* **18**, Pt. I, 277 (1971).
71. E. W. Burley, *Nature (Lond.)* **175**, 510 (1955).
72. P. T. Speakman, *J. Soc. Dyers Colourists* **75** 252 (1959).
73. B. Caldwell, S. J. Leach, A. Meschers, and B. Milligan, *Textile Res. J.* **34**, 627 (1964).
74. H. D. Weigmann, L. Rebenfeld, and C. Dansizer, *Proc. 3rd Intern. Wool Textile Res. Conf., Paris* **2**, 319 (1965).
75. A. Robson, M. J. Williams, and J. W. Woodhouse, *J. Textile Inst.* **60**, 140 (1969).
76. H. Zahn, *Melliand Textilber.* **31**, 762 (1950).
77. H. Zahn, *Kolloid-Z.* **121**, 39 (1951).
78. H. Zahn and H. Zuber, *Chem. Ber.* **86**, 172 (1953).
79. H. Zahn, A. Würz, and R. A. Räuchle, *Melliand Textilber.* **36**, 125 (1955).
80. H. Beyer, Ph.D. thesis, Technische Hochschüle, Aachen (1967).
81. H. Zahn and J. Meienhofer, *Melliand Textilber.* **37**, 432 (1956).
82. H. Zahn, *Leder* **19**, 144 (1968).
83. H. Zahn and M. M. Hammoudeh, *Kolloid-Z. Z. Polymere* **251**, 289 (1973).
84. H. Zahn, H. Beyer, M. M. Hammoudeh, and A. Schallah, *Melliand Textilber.* **50**, 1319 (1969).
85. W. I. Patterson, W. B. Geiger, L. R. Mizell, and M. Harris, *Amer. Dyestuff Reporter.* **30**, 425 (1941).
86. W. B. Geiger, F. F. Kobayashi, and M. Harris, *J. Res. Natl. Bur. Std.* **29**, 381 (1942).
87. H. Zahn and B. Wollemann, *Makromol. Chem.* **10** 122 (1953); (see also J. A. Schikanowa, *J. Angw. Chem. (USSR)* **23**, 667 (1950).
88. H. Zahn and K. Traumann, *Ann. Chem.* **581**, 168 (1953).
89. W. Kirst, *Melliand Textilber.* **28**, 394 (1947).
90. H. Zahn and W. Gerstner, *Forschungsber. Land Nordrhein Westfalen* **498**, 12 (1957).
91. H. Zahn and H. Steuerle, *Ann. Chem.* **622**, 175 (1959).
92. M. Bodansky, *Nature (Lond.)* **175**, 685 (1955).
93. H. Zahn, F. Schade, and E. Siepmann, *Leder* **14**, 299 (1963).
94. H. Zahn, H. K. Rouette, and F. Schade, *Proc. 3rd Intern. Wool Textile Res. Conf., Paris* **2**, 495 1965.
95. H. K. Rouette, *Melliand Textilber.* **47**, 903 (1966)
96. S. J. Leach, B. Milligan, A. S. Inglis, L. A. Frazer, and L. A. Holt, in: *Symposium on Fibrous Proteins, Australia, 1967*, Butterworth & Co. (Australia) Ltd., Sydney p. 373 (1968).
97. L. A. Frazer, S. J. Leach, and B. Milligan, *J. Appl. Polymer Sci.* **12**, 1992 (1968).
98. L. A. Holt, S. J. Leach, and B. Milligan, *Aust. J. Chem.* **21**, 2115 (1968).
99. J. Gerendas, Ph.D. thesis, Technische Hochschüle, Aachen (1968).
100. J. Gerendas and H. Zahn, *Kolloid-Z. Z. Polymere* **245**, 382 (1971).

101. J. B. Caldwell, S. J. Leach, and B. Milligan, *Textile Res. J.* **39**, 705 (1969).
102. H. Fraenkel-Conrat and H. S. Olcott, *J. Biol. Chem.* **174**, 827 (1948).
103. H. Fraenkel-Conrat and H. S. Olcott, *J. Amer. Chem. Soc.* **70**, 2673 (1948).
104. I. C. Watt and R. Morris, *Textile Res. J.* **40**, 952 (1970).
105. J. C. Griffith *Textile Res. J.* **35**, 1046 (1965).
106. P. Alexander, D. Carter, and K. G. Johnson, *Biochem. J.* **48**, 435 (1951).
107. P. Ponchel, G. Mazingue, and M. van Overbeke, *Bull. Inst. Textile Fr.* **103**, 1155 (1962).
108. J. B. Caldwell, S. J. Leach, B. Milligan, and J. Delmenico, *Textile Res. J.* **38**, 877 (1968).
109. R. S. Asquith and D. C. Parkinson, *J. Textile Inst.* **58**, 83 (1967).
110. P. Mason and J. C. Griffith, *Nature (Lond.)* **203**, 484 (1964).
111. I. C. Watt and R. Morris, *J. Textile Inst.* **57**, T425 (1966).
112. I. C. Watt and R. Morris, *Textile Res. J.* **38**, 674 (1968); and *Kolloid-Z. Z. Polymere* **229** 34 (1969).
113. J. B. Caldwell and B. Milligan, *Textile Res. J.* **42**, 122 (1972).
114. W. F. Happich, W. Windus, and J. Naghski, *Textile Res. J.* **35**, 850 (1965).
115. G. di Modica and M. Marzona, *Textile Res. J.* **41**, 701 (1971).
116. W. F. Happich, M. M. Taylor, and S. H. Feairheller, *Textile Res. J.* **40**, 768 (1970).
117. J. R. McPhee and M. Lipson, *Aust. J. Chem.* **7**, 387 (1954).
118. J. E. Moore and R. A. O'Connell, *Textile Res. J.* **27**, 783 (1957)
119. N. H. Koenig, *Textile Res. J.* **32**, 117 (1962).
120. J. E. Moore, *Textile Res. J.* **26**, 936 (1956).
121. N. H. Koenig and M. W. Muir, *Appl. Polymer Symp.* **18**, Pt. I, 727 (1971).
122. J. E. Moore and H. P. Lundgren, *Proc. Intern. Wool Textile Res. Conf., Australia* C, 355 (1955).
123. J. E. Moore and W. H. Ward, *J. Amer. Chem. Soc.* **78**, 2414 (1956).
124. L. Meunier, *J. Soc. Dyers Colourists* **27**, 214 (1911).
125. J. B. Speakman, *Amer. Dyestuff Reporter* **27**, 168 (1938).
126. R. C. Ghosh, J. R. Holker, and J. B. Speakman, *Textile Res. J.* **28**, 112 (1958).
127. I. C. Watt, *Proc. 3rd Intern. Wool Textile Res. Conf., Paris* **2**, 259 (1965).
128. I. C. Watt and R. Morris, *J. Textile Inst.* **57**, T425 (1966).
129. I. C. Watt and R. Morris, *J. Textile Inst.* **58**, 84 (1967).
130. I. C. Watt, *Appl. Polymer Symp.* **18**, Pt. II, 905 (1971).
131. H. Zahn, A. Würz, and A. Räuchle, *Textile Res. J.* **25**, 120 (1955).
132. C. Earland, J. G. P. Stell, and A. J. Wiseman, *J. Textile Inst.* **51**, T817 (1960).
133. C. Earland and J. G. P. Stell, *Polymer* **7**, 549 (1966).
134. W. Kirst, *Melliand Textilber.* **28**, 169 (1947).
135. H. Zahn and O. Waschka, *Makromol Chem.* **18**, 201 (1956).
136. M. Friedman and N. H. Koenig, *Textile Res. J.* **41**, 605 (1971).
137. N. H. Koenig, M. W. Muir, and M. Friedman, *Textile Res. J.* **43**, 682 (1973).
138. W. Kirst, *Melliand Textilber.* **28**, 314 (1947).
139. P. Alexander, M. Fox, K. A. Stacey, and L. F. Smith, *Biochem. J.* **52**, 177 (1952).
140. G. C. Tesoro and S. B. Sello, *Textile Res. J.* **34**, 523 (1964).
141. C. Fearnley and J. B. Speakman, *Nature (Lond.)* **166**, 743 (1950).
142. W. G. Rose and H. P. Lundgren, U.S. Pat. 2,881,046 (1959).
143. J. D. Leeder, *Textile Chem. Colorist* **3**, 193/35 (1971).
144. R. S. Asquith and A. K. Puri, *J. Soc. Dyers Colourists* **86**, 449 (1970).
145. F. Manchester, *J. Soc. Dyers Colourists* **74**, 421 (1958).
146. P. E. Meier and H. D. Weigmann, *Textile Res. J.* **43**, 74 (1973).

147. H. Zahn and P. F. Rouette, *Textilveredlung* **3**, 241 (1968).
148. G. Reinert, K. Mella, P. F. Rouette, and H. Zahn, *Melliand Textilber*, **49** 1313 (1968).
149. J. Panchartek, Z. J. Allan, and J. Poskocil, *Collection Czech. Chem. Commun.* **27**, 268 (1962).
150. C. Neumann, Ph.D. thesis, Technische Hochschüle, Aachen (1970).
151. J. H. Bradbury and G. V. Chapman, *Textile Res. J.* **33**, 666 (1963).
152. W. G. Crewther, *J. Polymer Sci.* **A2**, 131 (1964).
153. K. Lees, R. V. Peryman, and F. F. Elsworth, *J. Textile Inst.* **51**, T717 (1960).
154. P. Alexander, R. F. Hudson, and C. Earland, *Wool: Its Chemistry and Physics*, Chapman & Hall Ltd., London (1963).
155. J. C. Griffith, *Textile Res. J.* **35**, 1046 (1965).
156. P. Mason, *Textile Res. J.* **35**, 736 (1965).
157. I. C. Watt, *Proc. 3rd Intern. Wool. Textile Res. Conf., Paris* **2**, 200 (1965).
158. J. B. Caldwell and B. Milligan, *J. Textile Inst.* **61**, 588 (1970).
159. K. Lees and F. F. Elsworth, *Proc. Intern. Wool Textile Res. Conf., Australia* **C**, 363 (1955).
160. J. M. Gillespie, *Aust. J. Biol. Sci.* **17**, 282 (1964).
161. E. O. P. Thompson and I. J. O'Donnell, *Aust. J. Biol. Sci.* **12**, 294 (1959).
162. M. Harris and A. I. Smith, *J. Res. Natl. Bur. Std.* **17**, 577 (1936).
163. J. B. Caldwell, S. J. Leach, and B. Milligan, *Textile Res. J.* **36**, 1091 (1966).
164. R. S. Asquith and D. C. Parkinson, *Textile Res. J.* **36**, 1064 (1966).
165. B. S. Harrap and J. M. Gillespie, *Aust. J. Biol. Sci.* **16**, 542 (1963).
166. H. Zahn, *Proc. Intern. Wool Textile Res. Conf., Australia* **C**, 127 1955.
167. H. Steuerle and E. Hille, *Biochem. Z.* **331**, 220 (1959).
168. R. S. Asquith and J. J. García-Domínguez, *J. Soc. Dyers Colourists* **84**, 211 (1967).
169. E. H. Hinton, *Textile Res. J.* **44**, 233 (1974).
170. R. S. Asquith, I. Bridgeman, and A. I. Smith, *Proc. 3rd Intern. Wool Textile Res. Conf., Paris* **2**, 385 (1965).

# 7

# The Dyeing of Wool

C. L. Bird

## 7.1. Introduction

In 1856, when W. H. Perkin inaugurated the era of synthetic dyes, there were four principal fibers used for clothing. Two of these were protein fibers, wool and silk; the other two (cotton and linen) consisted of cellulose. Proteins, unlike cellulose, contain salt-forming acidic and basic groups, so they readily adsorb a variety of coloring matters. Before the advent of synthetic dyes, the problem was not so much how to dye wool but how to obtain, from the comparatively few coloring matters available, a range of colors that would withstand the action of sunlight and be resistant to wet treatments.

With a few exceptions, notably indigo, the necessary fastness was achieved by mordanting. In this process the wool is treated in a boiling acidified solution of a metallic salt—usually a salt of aluminium, tin, chromium, iron, or copper—leading to the deposition of metallic hydroxide within the fiber. It is then possible to apply a substance which, as described later, will form a sparingly soluble colored "lake" with the metal, thus providing fast dyeing.

The color obtained depends on the metal employed, and the natural coloring matter is valueless in the absence of a mordant. This "polygenetic" character is illustrated by camwood, which gives the following colors on wool with different mordants: red (Al), bluish red (Sn), red-violet (Cr), violet (Fe), and brown (Cu). Natural dyes that do not require a mordant include indigo, cudbear (orchil), and turmeric.

---

C. L. Bird • Department of Colour Chemistry, The University, Leeds, England

Apart from logwood, which is still used to a small extent for blacks, the natural coloring matters have now been replaced by the much superior synthetic dyes.

## 7.2. Synthetic Dyes

The first synthetic coloring matters were basic dyes. These were welcomed by wool dyers, as they made possible the production of hues that hitherto had not been obtainable (e.g., bright violets and brilliant greens). Unfortunately, these colors had very poor fastness properties and the new "aniline" dyes soon acquired a bad reputation with the public, in spite of the fact that some of the natural coloring matters also exhibited very inferior fastness. This prejudice still persists in some quarters. However, it was not long before natural coloring matters were replaced by synthetic dyes that were not only brighter but had better fastness, as well as being, in general, easier to apply.

Synthetic dyes belong to many different chemical classes, most of which have no representatives among the natural coloring matters—for example, the very important azo class is not represented. There are about 100 times as many synthetic dyes as there are natural coloring matters, so the dyer now has a much wider choice.

From the point of view of method of application, the natural dyes can be divided into (1) those requiring a mordant, (2) those requiring no mordant, and (3) vat dyes. The synthetic dyes include many additional classes, not all of which are suitable for application to wool. Dyes for application to wool may be subdivided as follows: (1) acid dyes, (2) metal-complex dyes, (3) mordant dyes, (4) reactive dyes, and (5) vat dyes.

These five classes will be considered in turn, but it is first desirable to discuss the substrate, which may be dyed in any of the following forms: loose wool, slubbing, yarn, knitted or woven cloth, pile fabric, and garments.

## 7.3. Wool Fiber as a Substrate for Dyes

Because of the large number of basic side chains in keratin, wool readily adsorbs large amounts of dyes, since most wool dyes are essentially complex organic molecules containing acidic groups. There is consequently no problem in dyeing in full as well as pale and medium depths, and in this respect wool is superior to any other fiber, whether natural or man-made.

On the other hand, wool, like other natural fibers, is inherently variable. At the present time, man-made fibers also vary from batch to batch, but the variations are not of the same order of magnitude as those found with wool, and they are not inherent, since it is possible to visualize control of manufacture of such a high order that all variations have been eliminated. With wool we are dealing with the hair fibers from widely different breeds of sheep, ranging from fine Merino fibers to coarse carpet wools. The existence of this wide range enables a variety of different types of cloth to be produced, which is one of the virtues of the fiber. Clearly, however, problems will arise if the dyer is asked to dye material containing widely different qualities of wool. For example, in a mixture containing fine and coarse wool, the fine fibers will appear lighter than the coarse fibers, owing to an optical effect, although all the fibers contain the same amount of dye. Blending of different qualities of wool, which is an essential operation, is one requiring considerable skill.

Wool is also subject to another source of variation. Like other natural fibers, but to a greater extent, it is subject to the effects of sunlight and weathering before it reaches the textile manufacturer. This takes the form of attack on the exposed tips of the fibers on the back and shoulders of the sheep. Some of the cystine linkages undergo hydrolytic attack, with the formation of aldehyde and thiol groups, which exert a reducing action. Damaged tips also show a deficiency or total loss of scales, and they are much more readily wetted and swell to a greater extent in the dyebath than the undamaged portions of the fiber. Especially when a mixture of dyes has been used, the color of the tips may be quite different from that of the remainder of the fibers.

The extent to which "tippy" dyeing is observed in practice depends on the migrating power of the dye. Dyes may be divided into two groups[1]: (1) those that give "negative" dyeings, the tips remaining lighter than the rest of the fibers, and (2) those that give "positive" dyeings, the tips dyeing darker. Negative dyeings result from the reduced affinity of the tips; although the dye penetrates the tips first, it readily migrates to the undamaged parts of higher affinity. This transfer does not take place with dyes that have little migrating power, so the tips remain heavily dyed and a positive dyeing is obtained.

A related effect is "skittery" dyeing, which results from damage to the epicuticle, the fine membrane covering intact wool fibers. If the epicuticle has been removed by chemical treatment or worn away by abrasion, dyes readily penetrate the fiber, even at low temperatures. Even with virgin wool the condition of the epicuticle varies, not only from fiber to fiber but along individual fibers and from one side to the other. If this fiber selectivity occurs to a marked degree, a skittery dyeing is obtained, particularly when a mixture of dyes is used, and the material has a speckled appearance. The extent to which skittery dyeing is experienced in practice is greatly influenced by the hydrophilic character of the dyes employed. The epicuticle is hydrophobic and resists penetration by hydrophilic dyes (i.e., dyes containing several sulfonic acid groups[2]). Consequently, the least skittery dyeings are obtained with monosulfonated dyes or dyes containing no sulfonic acid groups (e.g., 1:2 metal-complex dyes).

Other causes of modified dyeing behavior are (1) severe abrasion during wear, as in recovered wool fibers; and (2) chemical modification resulting from scouring, bleaching, carbonizing, chlorinating, etc. The most important of these are the various treatments given to wool to render it nonfelting. Many of these treatments attack the cuticle and surface scales. In consequence, the fiber becomes more permeable and dyeing takes place more rapidly. Desorption of dye also occurs more readily, so dyes usually show lower fastness to washing on shrink-resisted wool. Since this type of wool is used for knitted garments that should be washable, the dyes chosen must be those that exhibit good fastness to wet treatments.

If the chemical treatment is uneven, it may be impossible to obtain level dyeing. Thus, when wool is carbonized, strongly acidic groups are introduced into the keratin, leading to reduced affinity for acid dyes. Normally, the effect is slight, but severe local overcarbonizing leads to light spots or streaks in the dyed cloth, and it is almost impossible to eliminate this fault.

It will be clear from what has been written that wool is a variable commodity, but every type of fiber presents problems to the dyer. In fact, wool is one of the easiest fibers to dye, provided that suitable dyes and suitable methods of application are chosen. Good progress is being made toward the ultimate goal of bright colors in any hue that are more durable than the wool itself and can be obtained without damage to the fiber at any stage in the production of knitted or woven cloth.

## 7.4. Acid Dyes

These dyes are almost always the sodium salts of aromatic sulfonic acids, typical examples being

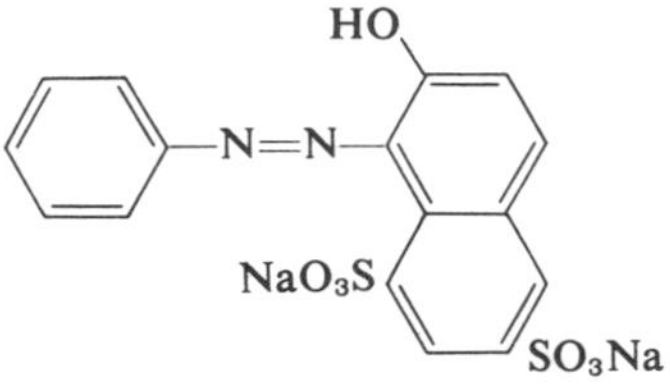

HO, N=N, $NaO_3S$, $SO_3Na$

C.I. Acid Orange 10

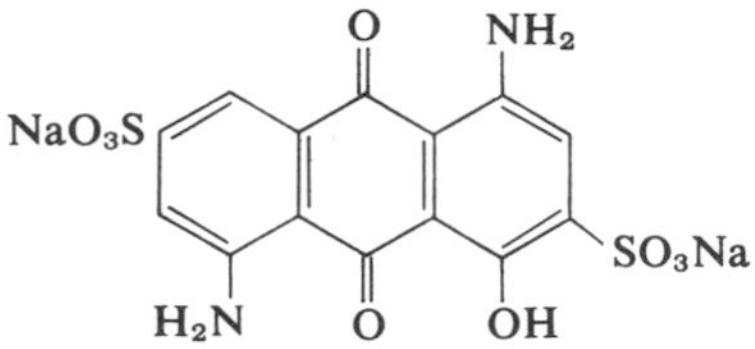

O, $NH_2$, $NaO_3S$, $SO_3Na$, $H_2N$, O, OH

C.I. Acid Blue 45

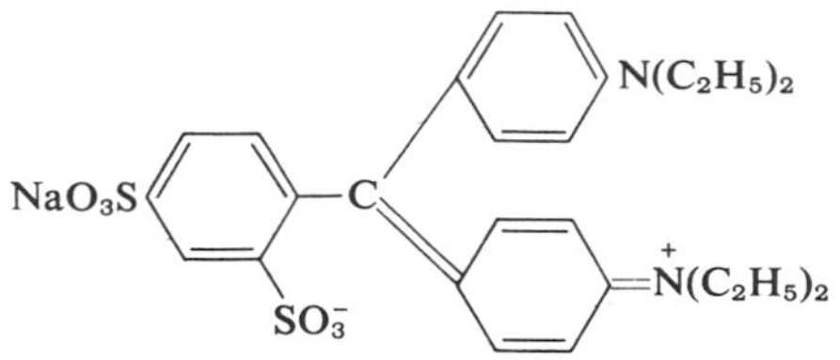

C.I. Acid Blue 1

which represent the azo, anthraquinonoid, and triarylmethane classes, respectively. The azo class is predominant and accounts for about two thirds of all the acid dyes. Anthraquinonoid dyes provide the fast to light blues, in particular, while triarylmethane dyes provide brilliant blues, greens, and violets of only moderate fastness to light. A few important acid dyes belong to the xanthene and phthalocyanine chemical classes.

The molecular weights of most acid dyes lie in the range 300 to 800. The dye molecules consist of a large anion containing one to four sulfonic acid groups associated with a corresponding number of small sodium cations. The free sulfonic acids, of which the dyes are the sodium salts, are, in spite of their high molecular weights, almost as strong acids as sulfuric acid. Their p*Ka* values lie between 1 and 2, and solutions of their sodium salts are neutral.

Owing to their large size and hydrophobic character, dye anions exhibit considerable attraction for one another, leading to the formation of aggregates.[3] In general, the greater the molecular weight and the fewer the number of sulfonic acid groups, the greater the tendency to aggregation. Thus Acid Orange 2G (C. I. Acid Orange 10), shows little sign of aggregation in solution. On the other hand, Polar Yellow R (C. I. Acid Orange 63), which has the formula

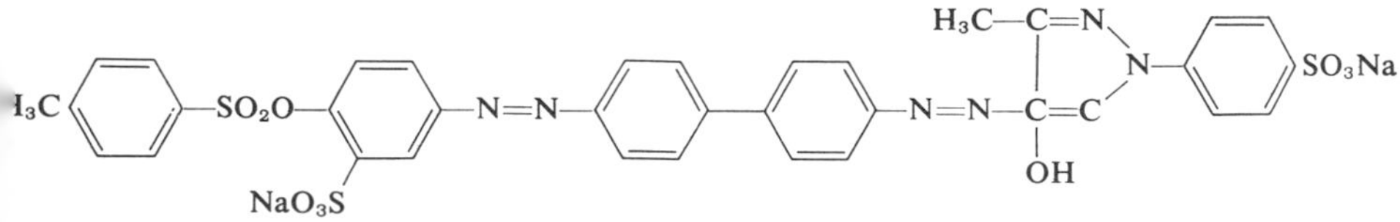

readily aggregates in solution, as is seen in the following table[4]:

**Aggregation of Polar Yellow R in 0.05 *M* NaCl**

| Temperature (°C) | Aggregation number |
|---|---|
| 25 | 500 |
| 60 | 50–60 |
| 90 | 2 |

Aggregation is promoted by the presence in the dyebath of acids or salts (e.g., sulfuric acid or sodium sulfate) and is reduced by raising the temperature. With large dye molecules in particular (e.g., that of Polar Yellow R), any appreciable aggregation produces a particle that is too large to penetrate the water-swollen wool fiber, so dyes of this type must be applied at the boil; otherwise, dyeing will be largely superficial.

The aqueous solubility of anionic wool dyes is generally adequate in both cold and hot water, but certain dyes are known to precipitate under adverse conditions. Aqueous solubility decreases with fall in temperature and in the presence of increasing amounts of acids or neutral salts, and some dyes precipitate in hard water (i.e., water containing traces of calcium and magnesium). The free dye acids, which are present under acidic conditions, are much less soluble than their sodium salts.

The affinity of wool for acid dyes arises mainly from the presence in the fiber of primary amino groups, most of which are found in combination with carboxyl groups (i.e., as salt linkages between the main peptide chains). Similarly, basic dyes are taken up through combination with the carboxyl groups. If wool is deaminated, its total combining power for acid dyes is greatly diminished, while its attraction for basic dyes is increased, owing to the presence of free carboxyl groups.[5] Further, if the basic side chains are eliminated by treating the wool with a mixture of acetic anhydride and acetic acid, the acetylated wool

$$R{-}NH_2 + (CH_3CO)_2O \longrightarrow R{-}NHCOCH_3 + CH_3COOH \qquad (7\text{-}1)$$

has little affinity for acid dyes.

When wool is immersed in solutions of hydrochloric acid, the acid combines with the wool by back-titration of the salt links:

$$R{-}\overset{+}{N}H_3\overset{-}{O}OC{-}R + \overset{+}{H}Cl^- \rightleftharpoons R{-}\overset{+}{N}H_3Cl^- + HOOC{-}R \qquad (7\text{-}2)$$

The amount of acid adsorbed at room temperature, when combination with all the salt links has taken place, is 80–90 ml of N acid per 100 g of dry wool. This figure, which is reached at about pH 1, is in good aggreement with the known amino-group content of keratin. Since the majority of acid dyes are the sodium salts of strong acids, it is to be expected that the corresponding free dye acids will combine with wool in the same way as hydrochloric acid and, in terms of ml of N acid, to the same extent. Many research workers[6] have immersed wool in solutions containing an excess of dye acid (or, more commonly, the dye plus mineral acid) and have obtained results in accordance with expectation (i.e., adsorption of 80-90 ml of N-dye acid per 100 g of dry wool).

Other workers,[6] however, have obtained appreciably higher values, but these high values could be due to hydrolysis of peptide groups and consequent liberation of additional amino groups. In fact, wool is badly damaged at pH 1 in the presence of large amounts of dye, especially with dyes of high affinity. One difficulty is that, with polysulfonated dyes, not all the sulfonic acid groups can be located close to positively charged amino groups in the fiber. It is probable, however, that close approach is not necessary, provided that an equivalent number of dye anions are adsorbed on the surface of the crystallites in order to neutralize the positive charge.

A further factor is the existence of nonionic as well as electrostatic attraction between dye acids and wool. This will lead to a greater maximum combining power, compared with that for hydrochloric acid. It is probable that the maximum combining power is not much greater with the simpler dye acids, but that with dyes of higher molecular weight, and consequently higher affinity, much higher values are to be expected.

From the point of view of the dyer, acid dyes are divided into three overlapping groups: (1) equalizing acid dyes, (2) milling acid dyes, and (3) neutral-dyeing acid dyes. Equalizing acid dyes have relatively low affinity. They are rapidly adsorbed and desorbed, and in a boiling dyebath a level, well-penetrated dyeing is readily obtained. They require addition of sulfuric acid and sodium sulfate to the dyebath in order to obtain adequate exhaustion and level dyeings, and the dyed material shows rather poor fastness to wet treatments.

Milling acid dyes, unlike equalizing acid dyes, show moderate fastness to milling and other wet treatments. Because of their higher affinity for wool they require a less strongly acidic dyebath, so acetic acid is used in place of sulfuric acid. Initial unlevelness is not readily corrected by desorption.

Neutral-dyeing acid dyes show high affinity, and little or no acid is required in the dyebath, which normally contains an ammonium salt. They migrate very slowly and the dyed material shows good fastness to wet treatments.

In the classification adopted by the Society of Dyers and Colourists,[7] there is a small fourth group which comes between the equalizing and milling acid dyes; formic acid is used to give a moderately acid bath showing a final pH of about 4.0.

To obtain level well-penetrated dyeings, careful control is necessary, particularly with milling and neutral-dyeing acid dyes. Apart from mechanical factors such as rate and uniformity of liquor flow, dyeing is mainly controlled by three variables: (1) pH, (2) addition of sodium sulfate, and (3) temperature (and time). A further means of control is provided by certain nonionic auxiliary products which are described later.

The pH of the dyebath is a matter of vital importance. It is seen from Eq. 7-2 that, when wool combines with acid, ionization of the carboxyl groups is repressed and the fiber acquires a positive charge which is neutralized by adsorption of the inorganic anion. The positive charge increases with the amount of acid present and is at a maximum at about pH 1. With the simpler acid dyes (i.e., those of low molecular weight and low affinity), dyeing is carried out at pH 2.5–3.0, obtained by adding 3–5% of sulfuric acid. The sulfate ions adsorbed to restore electrical neutrality have very low affinity (like chloride ions) and are displaced by dye anions which have much higher affinity:

$$\begin{matrix} R-\overset{+}{N}H_3 \\ R-\overset{+}{N}H_3 \end{matrix} SO_4^{2-} + 2\text{dye}^- \rightleftharpoons \begin{matrix} R-\overset{+}{N}H_3-\text{dye}^- \\ R-\overset{+}{N}H_3-\text{dye}^- \end{matrix} + SO_4^{2-} \qquad (7\text{-}3)$$

With these dyes a low pH must be used to obtain adequate exhaustion.

With dyes of higher molecular weight and much higher affinity, the reaction shown in Eq. 7-3 would lead to very rapid superficial adsorption of dye which, because of its high affinity, would not transfer to other basic sites, the dye remaining largely on the surface of the fibers and being unevenly distributed throughout the mass of material. Consequently, it may be necessary to dispense with the use of acid, in which case the reaction with wool from a substantially neutral dyebath (pH 6.0–8.5) may be represented by the equation

$$R-\overset{+}{N}H_3\overset{-}{O}OC-R + \overset{+}{N}aD^- = R-\overset{+}{N}H_3D^- + \overset{+}{N}a\overset{-}{O}OC-R \qquad (7\text{-}4)$$

Equation 7-4 implies that the amino groups in the wool are still involved, although, in fact, nonpolar forces are mainly responsible for the affinity of the dye. More commonly, an ammonium salt is used (e.g., ammonium sulfate or acetate). These salts decompose in a boiling solution, with volatilization of ammonia (provided that the system is not enclosed) and gradual liberation of a small amount of acid:

$$NH_4OOC-CH_3 \longrightarrow NH_3\uparrow + CH_3COOH \qquad (7\text{-}5)$$

With an enclosed system, which, nowadays, is more usual, they serve to neutralize any residual alkali in the wool and, at the boil, they provide either an approximately neutral dyebath (ammonium acetate) or one that is slightly acidic (ammonium sulfate).

For dyes of medium affinity, a week acid such as formic or acetic acid is used, giving a dyebath pH of 4.0–5.5. The effect of pH on the exhaustion of dyes of low, medium, and high affinity is illustrated in Fig. 7-1. It is seen that, with dyes of low affinity, in contrast to the other two groups, exhaustion is not complete at any pH and is insensitive to changes in pH in the region covered when applying dyes of this type. The sensitivity to pH shown by milling and neutral-dyeing acid dyes indicates the importance of establishing a

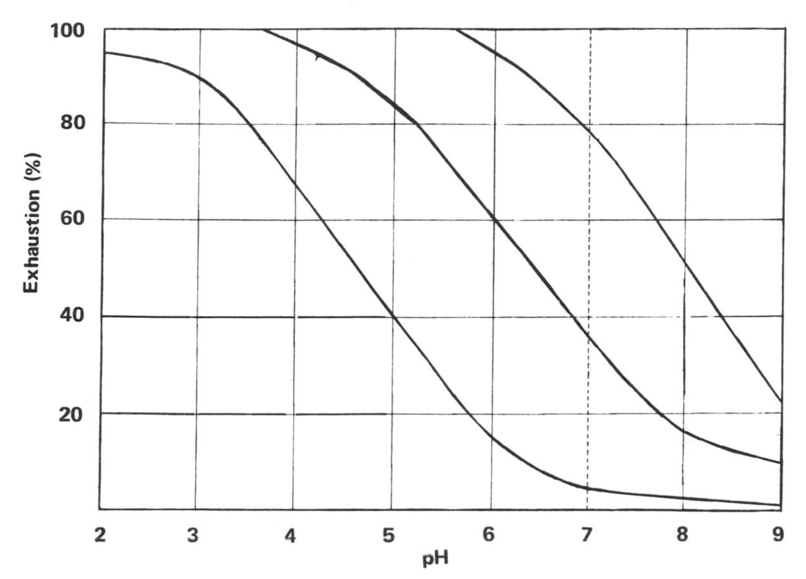


Fig. 7-1. Effect of pH on exhaustion of acid dyes.

uniform pH throughout the material before dyeing is commenced. This is achieved by running the goods in the machine for 15–20 min at the starting temperature with only the assistants present. If the assistants include acid, this will neutralize any residual alkali from scouring that may be present in the wool.

When dyeing is carried out in the presence of sulfuric acid, it is usual to add sodium sulfate (Glauber's salt) as well, to the extent of 10–20% of the weight of wool. By this means the reversible reaction indicated in Eq. 7-3 is shifted to the left, so desorption and transfer of dye to other sites is facilitated, thus promoting leveling. Glauber's salt is also added to dyebaths containing acetic acid, but its leveling effect is less with milling acid dyes, owing to their higher affinity. In a neutral dyebath, combination with the fibre takes place mainly, if not entirely, through nonpolar forces, so Glauber's salt exerts no leveling action. In fact, it promotes transfer of the hydrophobic dye to the essentially hydrophobic fiber phase and may be used to increase the exhaustion of some neutral-dyeing acid dyes, the effect being the reverse of that observed with equalizing acid dyes in an acid dyebath.

When dyeing wool, the dyebath is heated (1) to prevent aggregation of the dye, (2) to increase the swelling of the fiber and so render it more permeable to dye, and (3) to accelerate the rate of diffusion of dye within the fiber. Provided that the correct pH has been chosen, the dye will usually be taken up

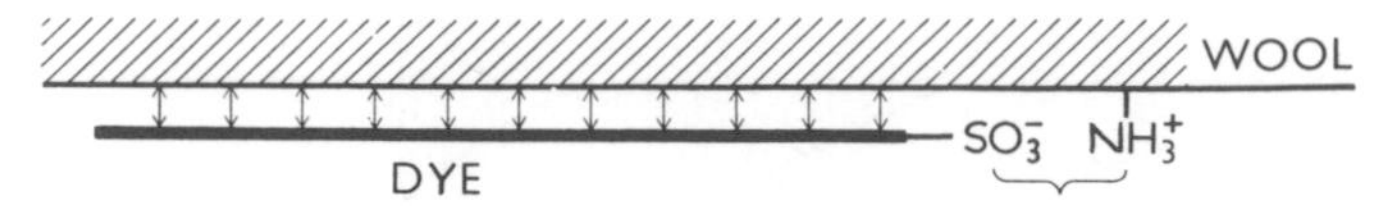


Fig. 7-2. Dye adsorption by (a) ionic and (b) nonpolar van der Waals' forces (Derbyshire) ↔ nonpolar van der Waals' forces.

slowly and uniformly if dyeing is commenced at about 60°C and the temperature is then raised steadily to 100°C over 30–60 min. Dyeing is completed at or near 100°C, as this facilitates leveling.

With acid dyes of high affinity, any initial unlevelness in the dyed material cannot be corrected by subsequent migration of dye from heavily dyed to lightly dyed parts of the material. Consequently, the dye must be taken up uniformly throughout the dyeing, and this can only be achieved if the dye is taken up slowly. Some dyes show a critical temperature range located between 60 and 95°C when adsorption of dye is very rapid. It is, therefore, necessary to raise the temperature very slowly when the critical temperature range is reached, or even, for a short period, to stop raising the temperature at this point. For this purpose it is virtually essential that the dyeing machine should be fitted with an automatic temperature control.

In general, acid dyes are adsorbed by wool as the result of two quite different forces of attraction: (1) ionic attraction between sulfonic acid groups in the dye and positively charged amino groups in the wool, and (2) nonpolar van der Waals' forces exerted between the hydrophobic dye anion and hydrophobic parts of the wool adjacent to positively charged amino groups. The two types of linkage are illustrated[8] in Fig. 7-2.

With equalizing acid dyes, both types of linkage are important, so acid must be present to ensure the presence of a positive charge on the fiber, while the presence of additional sulfate ions promotes competition for the basic sites, leading to desorption and level dyeing. On the other hand, with neutral-dyeing acid dyes, which show high affinity, ionic attraction must be largely eliminated, conditions being adjusted so that transfer of the hydrophobic dye to the fiber phase takes place gradually, dyeing being unassisted by migration of adsorbed dye. Milling acid dyes occupy an intermediate position; ionic attraction is no longer predominant, but affinity is not restricted to nonpolar forces, as is largely the case with neutral-dyeing acid dyes.

## 7.5. Metal-Complex Dyes

Chemically, metal-complex (or premetallized) dyes are closely related to the metal complexes produced in the fiber with mordant dyes, but from the dyer's point of view they are acid dyes, and they are treated as such in the *Colour Index*. It is, therefore, appropriate to consider them at this stage.

From the earliest times, mordanting has been associated with dyeings of good all-round fastness. As mentioned previously, it consists of treating the textile material with a solution of an appropriate metallic salt which will deposit the corresponding metallic hydroxide on the fiber. Subsequent dyeing then produces a colored, insoluble metal-dye complex. Clearly, the operation of mordanting prolongs the process of dyeing, so it was natural for the dyemakers to seek a way in which the metal (normally chromium) and the dye could be combined before dyeing, thus eliminating the mordanting stage. This aim was achieved and led to the production of two ranges: (1) acid-dyeing premetallized dyes, and (2) neutral-dyeing premetallized dyes, introduced in 1919 and 1951, respectively, the second range being considerably the more important.

### 7.5.1. Structure of Metal-Dye Complexes

The pioneers in this field were Morgan and co-workers,[9] who showed that, for complexing with chromium, the dye must contain (1) a salt-forming group such as hydroxyl or carboxyl, and (2) a coordinatively unsaturated atom such as a keto oxygen or an azo nitrogen atom. Normally, two salt-forming groups in appropriate positions are required for stability.

Drew and Fairbairn[10] elucidated the structure of the chromium complexes of azo dyes, which they obtained by heating a solution of the dye with chromium chloride:

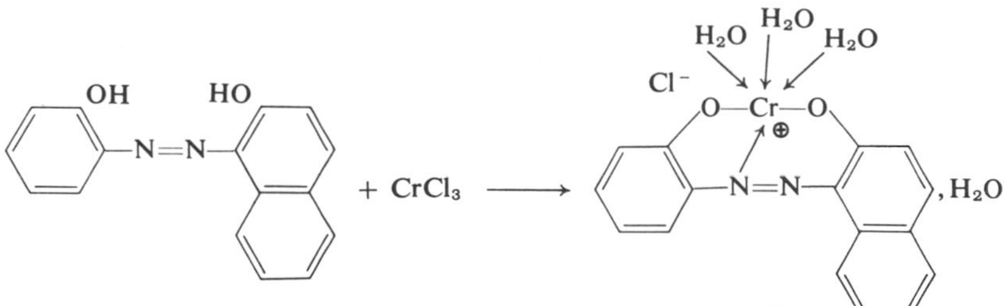

The $o,o'$-hydroxyl groups and one of the azo nitrogen atoms combine with the chromium atom to form two additional rings, one with five members and the other with six members, the whole structure having high stability.

It is seen from formula (I) that one atom of chromium has combined with one molecule of dye (i.e., we have a 1:1 complex). Combination has taken place through the two hydroxyl groups, the third valency being accounted for by the chloride ion.

Under suitable conditions, complexes of type (I) condense with a further molecule of dye, by virtue of the third valency, yielding a 1:2 acidic complex (II).

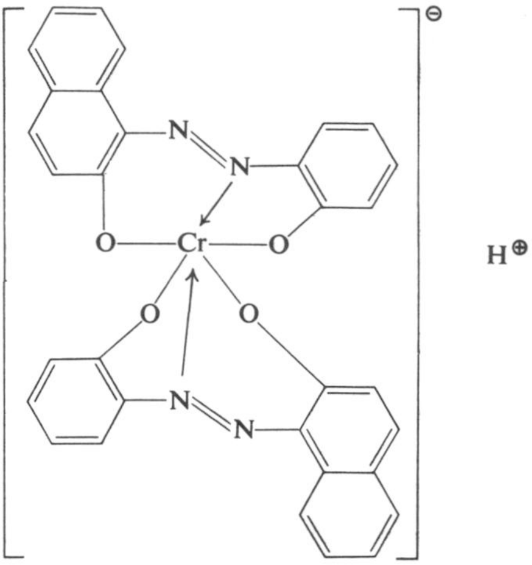

N N O Cr O O O N N ⊖ $H^{\oplus}$

(II)

### 7.5.2. Acid-Dyeing Premetallized Dyes

The 1:1 metal-complex dyes are derived from monoazo dyes, giving structures of type (I). Usually, one or two sulfonic acid groups are also present to impart good aqueous solubility, as illustrated by formula (III) for C.I. Acid Blue 158.

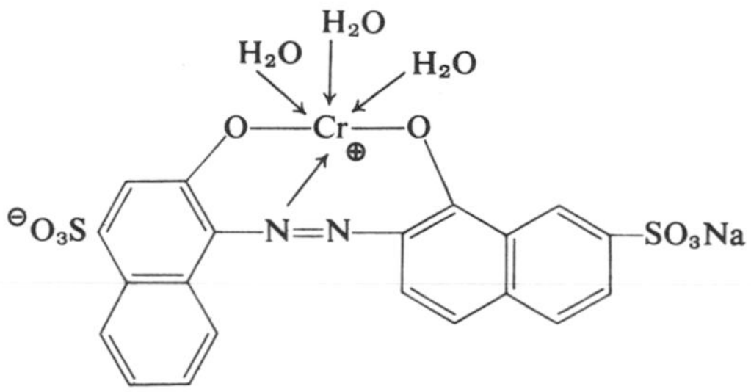

$H_2O$ $H_2O$ $H_2O$ O Cr ⊕ O $^{\ominus}O_3S$ N=N $SO_3Na$

(III)

This dye has two sulfonic acid groups and so has a net negative charge, but the majority of these dyes are apparently monosulfonates.

Dyes of this type [e.g., the Neolan (CGY) dyes] are applied to wool from a strongly acidic dyebath, and a prolonged treatment at the boil is necessary to obtain complete development of the shade.

The affinity of 1:1 metal-complex dyes for wool is partly accounted for by electrostatic linking to positively charged basic side chains and by van der Waals' forces, as with acid dyes. In addition, further linkages are possible through coordination between the chromium of the dye and suitable groups in the keratin[11] (e.g., the imino groups in some of the side chains), giving

$$\text{dye—Cr} \leftarrow \text{NH}\langle$$

Imino groups are only very weakly basic, however, and at pH 2, as used in practice, they become charged—

$$\rangle\text{NH} + \text{H}^{+} \longrightarrow \rangle\overset{+}{\text{N}}\text{H}_2$$

and will no longer coordinate with chromium. Hence, the addition of a relatively large amount of sulfuric acid (8% on the weight of wool) reduces the amount of dye adsorbed and produces a leveling action.

Dyes of the 1:1 metal-complex type are used mainly for dyeing wool cloth. They combine ease of application, good leveling, and good fastness properties, but they are not as bright as acid dyes and not as fast to wet treatments as 1:2 metal-complex or chrome dyes.

### 7.5.3. Neutral-Dyeing Premetallized Dyes

These dyes [e.g., the Irgalan (CGY) dyes] are soluble 1:2 chromium (or cobalt) complexes of type (II), as illustrated by formula (IV) for C.I. Acid Violet 78.

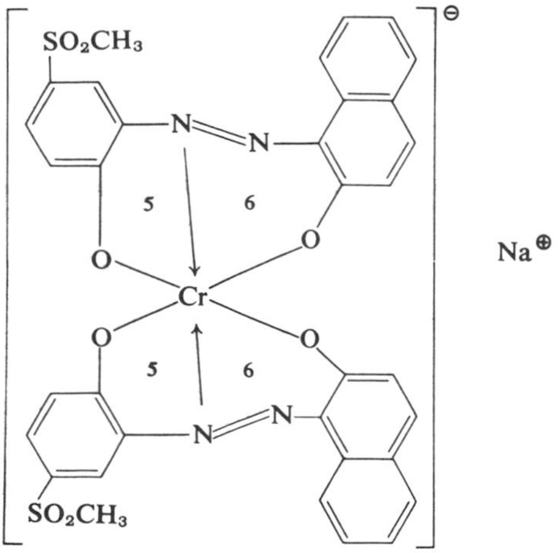

$SO_2CH_3$ $SO_2CH_3$ N N O Cr O 5 6 $Na^{\oplus}$

(IV)

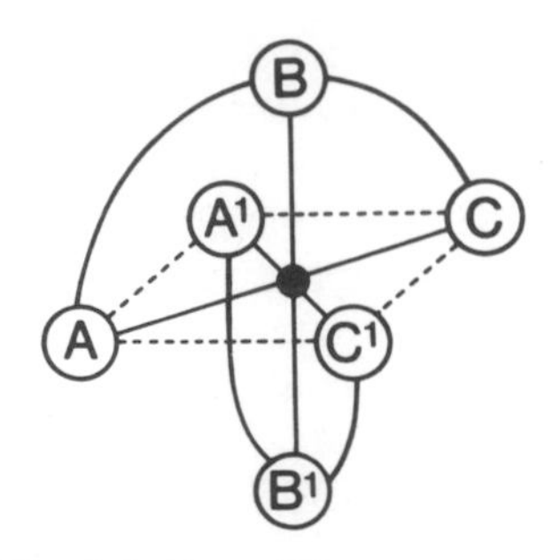


Fig. 7-3. Drew–Pfitzner structure.

The parent dye molecules may be the same or different, symmetrical complexes being the more stable. It is seen that the complex is negatively charged, and, in fact, these dyes are salts of strong acids. Sulfonic acid groups are absent, since the presence of these groups leads to poorer leveling and less satisfactory fastness properties. Instead hydrophilic nonionic groups (e.g., —$SO_2CH_3$ or —$SO_2NH_2$ groups) are incorporated to improve the aqueous solubility and decrease the rate of dyeing.

In the Lanacron (CGY) dyes, however, sulfonic acid groups are present, but these dyes, which are cheaper than the usual 1:2 metal-complex dyes, are intended only for the production of dark colors, where fiber selectivity is of minor importance.

Complexes of type (II) have been studied extensively by Schetty.[12] If the benzene and naphthalene rings are ignored, the central structure can be drawn as shown in Fig. 7-3. The four oxygen atoms (A, $A^1$, C, $C^1$) and the central chromium atom (●) are in the same horizontal plane. The two azo nitrogen atoms (B, $B^1$) coordinated to the chromium atom are vertically above and below it, respectively. (If the circles representing the oxygen and nitrogen atoms are joined by straight lines, a regular octahedron is obtained.) The planes of the two azo dye residues are indicated by the curved lines ABC and $A^1B^1C^1$,

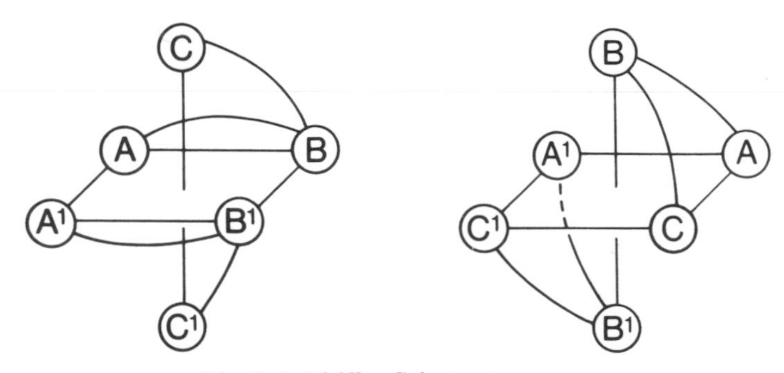


Fig. 7-4. Pfeiffer–Schetty structures.

so the residues are perpendicular (at right angles) to each other. This is the Drew–Pfitzner structure.

An alternative structure was postulated by Pfeiffer, as illustrated in Fig. 7-4. Here, the three bonding atoms (O, N, O) of the azo dye (i.e., ABC and $A^1B^1C^1$) are situated at the corners of two of the parallel equilateral triangles of the valency octahedron. This "sandwich" arrangement—the chromium atom (not shown) is in the middle—is known as the Pfeiffer–Schetty structure. Figure 7.4 illustrates two of the five possible sandwich arrangements. In the case of the 1:2 chromium complex of anthranilic acid → β-naphthol,

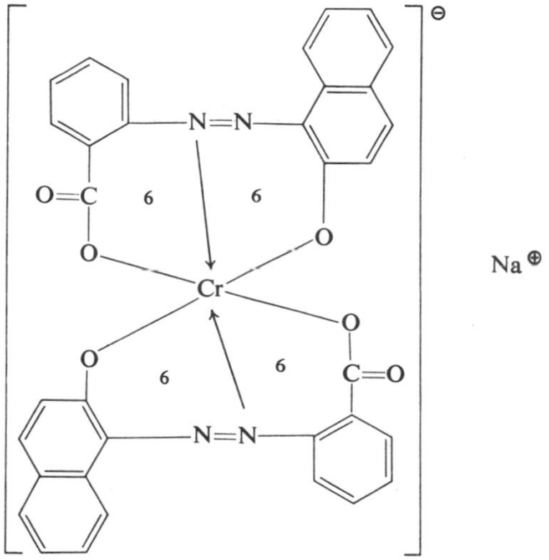

(V)

Schetty was able to separate four sandwich isomers, which showed different physical properties.

From his work on 1:2 complexes, Schetty has concluded that, in general, dyes which form two six-membered rings containing chromium, as in (V), give complexes of the sandwich type, whereas dyes which form five- and six-membered rings, as in (IV), give complexes having the Drew–Pfitzner structure. He has also shown that the type of complex affects fastness as well as physical properties. Compared with Drew–Pfitzner structures, chromium complexes of the sandwich type are more sparingly soluble, brighter in hue, poorer in wet-fastness, and faster to light.

Unlike 1:1 complexes, which can coordinate with suitable groups in the keratin, 1:2 complexes combine with wool only through electrostatic attraction and, in particular, van der Waals' adsorption. Owing to the negative charge on the complex, adsorption is sensitive to pH, although to a smaller extent than with milling acid dyes, probably because the charge is not localized.

Since the molecule is large, powerful van der Waals' forces will operate, leading to high fastness to wet treatments.

The neutral-dyeing premetallized dyes comprise several closely related ranges, each of some 30 members. Like the 1:1 complexes, they are rather dull in hue compared with the acid dyes, but they have very good all-round fastness properties. They are adsorbed gradually and evenly from a neutral or slightly acidic (pH 5.5–7.0) bath, and some of them are very level-dyeing. They give good penetration of tightly woven material and are not fiber-selective. Comparatively slow dyeing compensates for relatively low migration, which varies from fair to very poor; it is probable that migration properties are governed by the nature of the nonionic solubilizing groups, which vary from one range to another.

The usual method of application is to treat the wool first at 40°C for 10 min in a bath set with 2–4% ammonium sulfate or acetate. Dissolved dye is then added and the temperature raised to the boil in 45 min. After 30–60 min at the boil the bath should have exhausted to the extent of about 95%.

## 7.6. Chrome Dyes

Chemically, chrome dyes are closely related to acid dyes, but additional groups are present which enable the dye to form a stable coordination compound with chromium within the fiber. Coordination is accompanied by a bathochromic shift, for example a change in hue from red to blue, as well as a great increase in fastness to light and to wet treatments. With few exceptions, chrome dyes are unsuitable for use as acid dyes, because they are not fast to alkali in the unchromed state. As a range, chrome dyes are not as bright as acid dyes; from the dyer's point of view, 1:2 metal-complex dyes are particularly suitable for pale to medium depths, while chrome dyes are best for full depths. With dyes of both classes the dyeings can often be brightened by adding to the recipe a small amount of a suitable milling acid dye.

The great majority of chrome dyes belong to the azo class, but there are a few important members in the anthraquinonoid and triarylmethane classes. The examples shown,

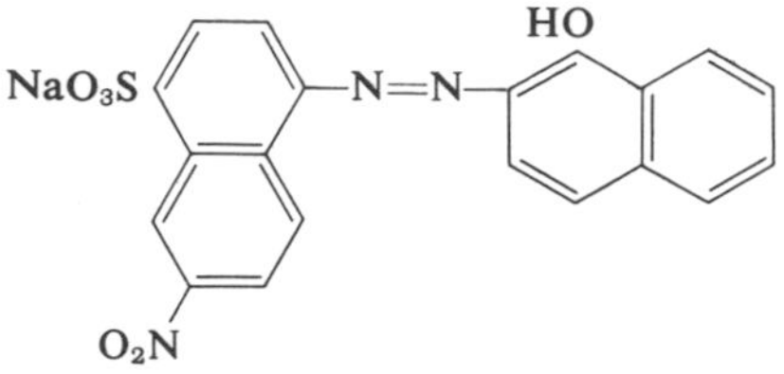

$NaO_3S$, HO, –N=N–, $O_2N$

C.I. Mordant Black 11

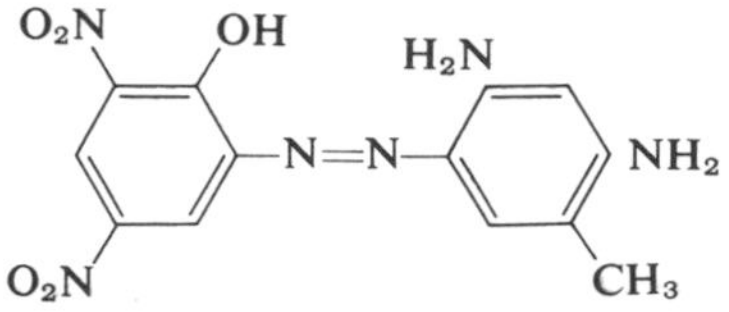

$O_2N$, OH, $H_2N$, –N=N–, $NH_2$, $O_2N$, $CH_3$

C.I. Mordant Brown 4

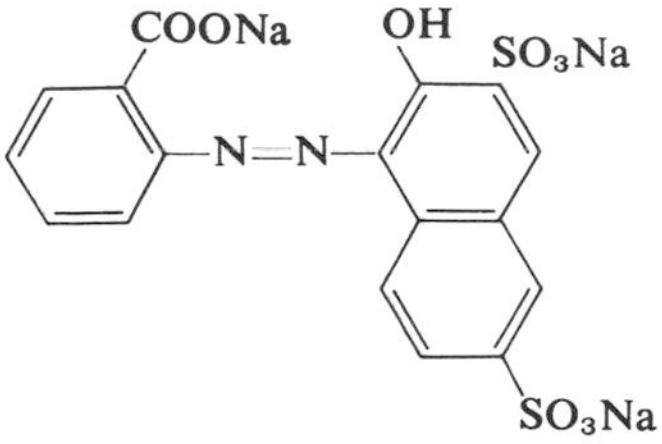

C.I. Mordant Red 9

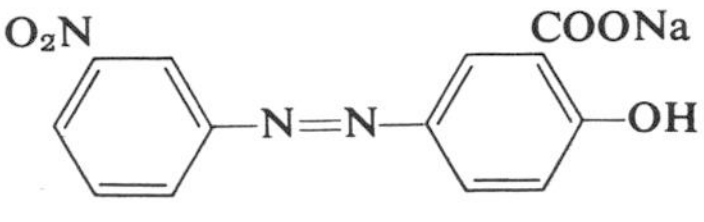

C.I. Mordant Yellow 1

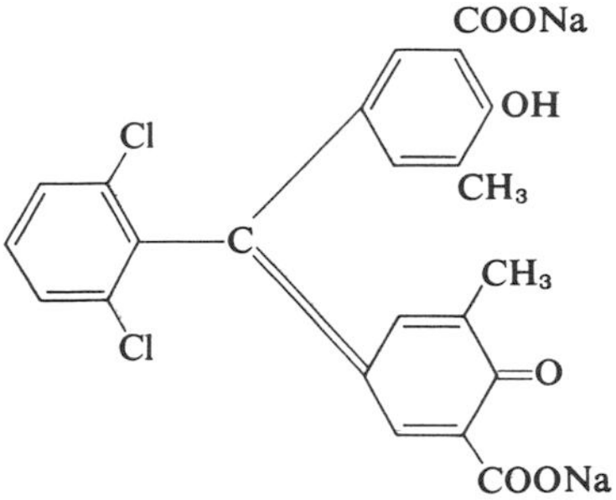

C.I. Mordant Blue 1

illustrate the structures commonly used to enable the dye to form a coordination compound with chromium: *o,o'*-dihydroxyazo, *o*-hydroxy-*o'*-carboxyazo, *o*-amino-*o'*-hydroxyazo, and the salicylic acid residue. With a few azo chrome dyes, coordination is accompanied by oxidation to the quinone form.

Some chrome dyes (e.g., C. I. Mordant Black 11 and Brown 4) have rather poor aqueous solubility, especially in hard water or in the presence of acid, and this can lead to solid dye being present on the surface of the dyed material. To overcome this difficulty, ranges of afterchrome dyes [e.g., Francolane (CFMC) dyes] have been introduced, the dye being present in the form of the very finely dispersed free acid. In this form, the dye is not filtered off by the wool material and does not combine with it, but, as the temperature of the dyebath is raised between 70 and 85°C, the dye gradually dissolves and combines with the wool to give a level, well-penetrated dyeing.

It will be noticed that the *Colour Index* designates these dyes as *mordant* dyes, and a number of different metals could be employed to form the metal-dye complex (i.e., to act as the mordant). In practice, chromium is the only metal to be used, owing to the stability of the complexes produced and its ease of application in the form of the dichromate.

It is generally accepted that, when chrome dyes are applied to wool, the complexes produced are of the 1:2 type, but a small amount of 1:1 complex may be formed if dyeing is carried out at pH values below 3, leading to an alteration in hue and lower wet fastness.[13] In general, chrome dyes show higher wet-fastness than corresponding neutral-dyeing premetallized dyes, in spite of the fact that they usually contain sulfonic acid or carboxyl groups. This is attributed to penetration of the unchromed dye into regions of the keratin that are inaccessible to the large 1:2 complex molecules; subsequent chroming then produces a large molecule which is unable to diffuse out of the fiber.

When applying chrome dyes to wool, it is necessary not only to obtain fixation of the dye on the wool, as with acid dyes, but to apply the chromium mordant as well, under conditions which will ensure that combination of dye and mordant takes place inside the fiber. For this purpose, the chromium is almost always in the form of sodium or potassium dichromate, in which form it reacts with the fiber but not, in general, with the dye.

The first stage of the mordanting process consists of reaction of dichromate with basic side chains in the wool, as with inorganic acids and dye acids:

$$2R{-}NH_3^- OOC{-}R + Cr_2O_7^{2-} + 2H^+ \longrightarrow (R{-}\overset{+}{N}H_3)_2Cr_2O_7^{2-} + 2R{-}COOH \quad (7\text{-}6)$$

As dichromate ions are adsorbed, hydrogen ions are taken up simultaneously to restore electrical neutrality, thus leading to a slight rise in pH.

The next stage is a complicated one.[14] The adsorbed dichromate ion is gradually reduced, from Cr(VI) to Cr(III), at the expense of certain groups in the wool, in particular the cystine groups:

$$\underset{\text{Anionic}}{Cr(VI)} \xrightarrow{100\,^\circ C} \underset{\text{Cationic}}{Cr(III)} \quad (7\text{-}7)$$

The trivalent chromium then combines with the carboxyl groups in the wool:

$$R{-}COO^- + Cr^{3+} \longrightarrow R{-}C\genfrac{}{}{0pt}{}{\diagup O \diagdown}{\diagdown O \diagup}Cr^{2+} \quad (7\text{-}8)$$

The final stage probably consists of transfer of trivalent chromium from the wool to form a 1:2 complex with the dye in which there is no direct linkage with the wool, bonding of the complex to the wool being mainly through van der Waals' forces.

Instead of applying the dichromate before the dye, it may be applied either together with the dye or after dyeing, so there are three methods of applying chrome dyes: (1) chrome mordant, (2) metachrome, and (3) afterchrome. Some chrome dyes can be applied by all three methods, but others are restricted to one or two of the three alternatives.

The chrome mordant method, as outlined above, is the oldest but now by far the least important method. It is the traditional method for applying mordant dyes and was the only suitable method when dyeing was restricted to natural coloring matters, most of which have little or no affinity for unmordanted wool. As well as the dichromate (1–3% on the weight of wool), a small amount (1–3%) of a weak acid or an acid salt is added. Chroming is commenced at 60°C, the temperature then being raised to the boil in 45 min and kept at the boil for 1–1½ h. The material is then rinsed well and dyed in the presence of 1–5% acetic acid (30%), commencing at 50°C, raising to the boil in 45 min, and continuing at the boil for 1–1½ h.

For application by the metachrome method[15] the dye must exhaust well at pH 6.0–8.5 and must not react with Cr(VI). Three reactions take place simultaneously: chroming of the wool, adsorption of neutral-dyeing chrome dye, and formation of a 1:2 chromium-dye complex in the fiber. The chromium is applied in the form of "metachrome mordant," which consists of one part of sodium chromate and two parts of ammoniun sulfate. The ammonium sulfate reacts with the alkali resulting from adsorption of chromic acid by the wool:

$$(NH_4)_2SO_4 + 2NaOH = Na_2SO_4 + 2NH_3\uparrow + 2H_2O \tag{7-9}$$

thereby maintaining the pH of the dyebath between 6.0 and 8.5. In an enclosed machine, however, the liberated ammonia cannot escape and the dyebath becomes increasingly alkaline, which is undesirable.

The dyebath is prepared by adding the dissolved dye and raising the temperature to 50°C. Well-diluted metachrome mordant (2–8%) is added next. The wool is then introduced and the temperature is raised to the boil in 45 min. Dyeing is continued at the boil for 1 h. Even adsorption of dye is essential, since the chromium complex cannot migrate. The metachrome method is suitable for pale and medium depths, particularly on loose wool and slubbing, and it is the shortest of the three methods.

When dyeing is carried out by the afterchrome method, the process consists of an acid dyeing followed by addition of dichromate and further boiling. It is essentially a two-bath process, but by exhausting the dyebath it is possible to use the same liquor for the dichromate treatment. During this afterchroming the wool takes up chromium as described above and the Cr(III) produced reacts with the dye on the fiber to produce a 1:2 chromium–dye complex, both reactions taking place simultaneously.

The usual method of dyeing is as follows. The dyebath is prepared with the dye, 2–5% of acetic acid (30%), and 10% of Glauber's salt. After the temperature has been raised to 50°C, the wool material is introduced. The temperature is then raised to the boil in 45 min and kept at the boil for 30 min. If the dyebath is not completely exhausted, a little more acid may be added and dyeing continued. When the dyebath has been exhausted completely, it is cooled slightly, 0.25–3.0% of dichromate is added, the temperature raised to the boil, and boiling continued for a further 30–60 min. This prolonged boil is necessary to ensure that complete reduction of the chromium to Cr(III) has taken place. Cooling of the bath before adding the dichromate is essential. Usually, chroming is commenced at 70°C and, in raising the temperature once more to the boil, 20 min are allowed to ensure even adsorption of dichromate. Uneven adsorption leads to uneven production of alkali in the wool (cf. Eq. 7–6) and this in turn leads to uneven desorption of dye and so to unlevel dyeing.

The amount of dichromate used for afterchroming is always greatly in excess of that required by theory for complex formation, since part of the chromium combines solely with the wool. It is important, however, not to use too much dichromate, especially with dyes that are sensitive to oxidation. Usually, the amount should be one fourth to one third of the weight of dye.

One difficulty that may arise with chrome dyes, especially when they are applied by the afterchrome method, is local or general dulling of the shade caused by the presence of traces of iron or copper.[16] These metals readily form complexes with some chrome dyes, in particular, the widely used C. I. Mordant Blacks 11 and 17, and the color of these complexes is different from that of the chromium complexes. Iron is more likely than copper to be present and as little as 1 part of iron per million in the water supply will cause trouble. At concentrations of this order the iron may be rendered harmless by adding a sequestering agent such as sodium hexametaphosphate or ethylenediamine tetraacetic acid (EDTA).

The afterchrome method is the most important of the three. It is not much longer than the metachrome method and is particularly suitable for dyeings of very high fastness in full depths (e.g., dark browns, navy blues, and blacks), where the change in hue during afterchroming does not present a serious obstacle to matching. Some afterchrome dyes are very level-dyeing, owing to good migration in the unchromed state, so they are suitable for coloring yarn and cloth.

## 7.7. Reactive Dyes

The introduction of the reactive dyes is the outstanding development in dyestuff chemistry of recent years. These dyes differ from all other classes in that they form a permanent covalent linkage with the fiber, which is completely fast to normal wet treatments. Their chemical structure consists essentially of a simple, highly sulfonated acid dye of the azo, anthraquinonoid, or phthalocyanine type, to which is attached a grouping through which reaction with appropriate groups in the fiber can take place. In contrast to metal-complex and chrome dyes, high wet-fastness is associated with bright hues, so reactive dyes offer the possibility of producing shrink-resist wool garments in almost any color, the dye as well as the fabric being resistant to the hot detergent solutions used in domestic washing machines.[17]

Reactive dyes were originally developed for application to cellulosic fibers and the first range were the Procion (ICI) dyes, introduced in 1956 and represented by

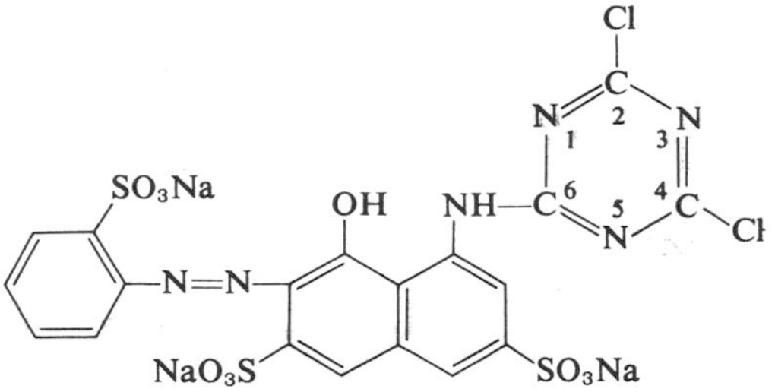

Procion Brilliant Red MX-2B

In the Procion H (ICI) and Cibacron (CGY) dyes, only one chlorine atom is present.

It is seen that, in Red MX-2B, an *s*-triazinyl ring with two chlorine atoms has been attached to a trisulfonated acid dye molecule. The chlorine atoms react under alkaline conditions with the hydroxyl groups of cellulose and, to a smaller extent, with water:

$$\text{dye—Cl} + \text{cell—OH} \longrightarrow \text{dye—O—cell} + \text{HCl} \qquad (7\text{-}10)$$

$$\text{dye—Cl} + \text{HOH} \longrightarrow \text{dye—OH} + \text{HCl} \qquad (7\text{-}11)$$

At the completion of dyeing the fiber contains (1) dye that has reacted with the fiber, (2) dye that has reacted with water, and (3) unreacted dye. The conditions are adjusted so that most of the dye reacts with the cellulosic fiber, the remaining dye (2 and 3) being readily removed by washing.

Many other ranges of reactive dyes, containing a variety of reactive groupings, are now available. Thus, in the Reactone (CGY) dyes, the *s*-triazinyl is replaced by the trichloropyrimidyl ring:

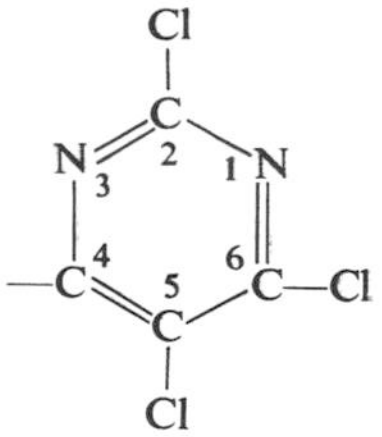

In the Drimalan F (S) and Verofix (FBy) dyes, the chlorine in the 2 and the 6 positions of the pyrimidyl ring is replaced by fluorine, which is highly reactive.[18]

From the chemical point of view, reactive dyes have been divided into two classes,[19] according to whether they react (1) by substitution of a labile halogen atom, or (2) by addition at a double bond. Examples of reactive groups containing a double bond are

| | |
|---|---|
| $—SO_2CH{=}CH_2$ | vinylsulfone |
| $—NHCOCH{=}CH_2$ | acryloylamino |
| $—NHCOC(Br){=}CH_2$ | α-bromoacryloylamino |

Keratin contains a variety of groups that are capable of reacting with reactive dyes.[20]

$$\boxed{H_2N}—CH_2—CO—NH—\underset{\underset{\boxed{NH_2}}{(CH_2)_4}}{CH}—CO—NH—\overset{CH_2\boxed{OH}}{CH}—CO—NH—\underset{\underset{\boxed{SH}}{CH_2}}{CH}—CO—NH—CH_2—CO—NH—\underset{\underset{C_6H_4\boxed{OH}}{CH_2}}{CH}—CO—$$

Thus reaction can occur with terminal and side-chain amino groups, with the —SH groups of cysteine and with the hydroxyl groups of the tyrosine residue. The cysteine groups are highly reactive, but, unless the wool has undergone a reducing treatment, the number of these groups present is small. The most abundant groups are the side-chain hydroxyl groups (e.g., those of serine), although it is believed that the basic groups of the lysine residues are those with which reaction mainly takes place under normal dyeing conditions.[21] On the basis of analysis, this assumption has recently been questioned, and it has been suggested that the significance of lysine residues in reaction with dyes is not greater than that of the hydroxyl groups.[36,37]

The application of reactive dyes to wool is much less straightforward than their application to cellulosic fibers, the following difficulties being encountered: (1) uneven dyeing, (2) inadequate reaction with the fiber in the weakly acidic range suitable for dyeing wool, and (3) the difficulty of removing hydrolyzed and unreacted dye at the completion of dyeing.

Reactive dyes are highly sulfonated to impart good aqueous solubility, but this renders them fiber-selective. They dye the tips of the fibers more strongly than the roots, leading to tippy or skittery dyeing which cannot level up, since most of the dye has formed a permanent linkage with the fiber. Fortunately, this difficulty can be overcome by adding a suitable auxiliary product to the dyebath [e.g., Albegal B (CGY)]. General unlevelness, which cannot be corrected by migration, is prevented by very careful temperature control.

The problem of obtaining almost complete reaction between dye and fiber under weakly acidic conditions has largely been solved by the appropriate choice of reactive groups. Normally, only that portion of the dye that has reacted with the fiber shows high fastness to wet treatments. The presence of a small amount of unreacted or hydrolyzed dye is indicated in a washing test not so much by loss of depth as by staining of adjacent white wool and cotton. This dye has sufficient affinity for wool to prevent it being removed in a final rinse, whereas it could readily have been removed in this way from a cellulosic fiber. To obtain maximum possible fixation, particularly with full shades, dyemakers commonly recommend a final, slightly alkaline treatment which is achieved, for example, by cooling to 80°C, adding a little ammonia, and dyeing for a further 15 min. Under these alkaline conditions (pH 8.0–8.5) the residual unfixed dye present either reacts with the fiber or is desorbed from it, depending on the type of reactive dye.

There are now several ranges of reactive dyes designed for application to wool. At present, the ranges are somewhat restricted, but the essential primary hues—yellow, red, and blue—are included. The dyebath conditions are so arranged that the dye is gradually adsorbed by the wool as the temperature is first raised to the boil and then held at the boil. Simultaneously, but at a slower rate, the adsorbed dye reacts with the wool to form a permanent linkage.

In the Hostalan (FH) dyes,[22] the vinylsulfone group is combined with *N*-methyltaurine, which gradually splits off in a boiling dyebath at pH 4.5–5.5:

$$\text{dye—}SO_2CH_2CH_2\underset{}{\overset{CH_3}{\overset{|}{N}}}CH_2CH_2SO_3H \longrightarrow \text{dye—}SO_2CH{=}CH_2 + H\overset{CH_3}{\overset{|}{N}}CHCH_2SO_3H \qquad (7\text{-}12)$$

Reaction then takes place with the amino groups in the fiber:

$$\text{dye—}SO_2CH{=}CH_2 + \text{wool—}NH_2 \longrightarrow \text{dye—}SO_2CH_2CH_2NH\text{—wool} \qquad (7\text{-}13)$$

The gradual liberation of the reactive vinylsulfone retards the rate of adsorption and thus produces more level dyeing. Dyeing is carried out with sufficient acetic acid to give pH 5, together with a weakly cationic leveling agent [e.g., Remol GE (FH)]. The temperature of the dyebath is raised to the boil in 1 h and kept at the boil for 1 h.

Lanasol (CGY) dyes contain the $\alpha$-bromoacryloylamino reactive group. The bromine atom activates the double bond, bromine being superior to other halogens in this respect, and it is claimed that reaction with wool proceeds almost quantitatively, in the following manner.[20]

$$\text{dye—NHCOC}(\text{Br})\text{=CH}_2 + \text{wool—NH}_2 \longrightarrow \text{dye—NHCOCH}(\text{Br})\text{CH}_2\text{NH—wool} \xrightarrow[-\text{HBr}]{+\text{H}_2\text{O}} \text{dye—NHCOCH}(\text{OH})\text{CH}_2\text{HN—wool} \quad (7\text{-}14)$$

Dyeing is carried out at pH 4.0–5.5 in the presence of 2% of acetic acid, 5% of anhydrous sodium sulfate, and 1% of Albegal B to prevent skittery dyeing. Dyeing is commenced at 50°C, the temperature is then raised to the boil in 45 min, and dyeing is continued at the boil for 30–90 min, according to the depth of shade.

The Lanasol range includes three dyes—Yellow 4G, Red 6G, and Blue 3G—which have almost identical dyeing properties (including coverage of tippy wool), so they are suitable for trichromatic dyeings (i.e., for the production of a wide range of shades obtained by mixing three primary colors in different proportions).

Drimalan F (S) and Verofix (FBy) dyes at present comprise two yellows, an orange, three reds, and a blue, all possessing bright hues. In standard depths, their fastness to light varies between 5 and 7 and they show very good fastness to washing. Dyeing is carried out between pH 4 and 6 in the presence of a leveling agent [e.g., Avolan RE (FBy)]. They show high reactivity with wool[18] and, since reacted dye cannot migrate, very careful control of temperature is necessary. For example, 20 min is given to raising the temperature of the dyebath from 40 to 65°C, followed by 30 min at 65°C, 30 min to raise to the boil, and between 10 and 60 min for fixation at the boil.

The Procilan (ICI) dyes, which contain the acryloylamino reactive group,[23] differ from the other ranges of reactive dyes in that they are based on sulfonated 1:2 metal-complex dyes instead of on acid dyes. They, therefore, possess the advantage that any unreacted dye has high affinity for the wool. On the other hand, they lack the brightness of hue of other ranges of reactive dyes and are, perhaps, best regarded as a range of 1:2 metal-complex dyes possessing exceptionally high wet-fastness.

Procilan dyes are applied to wool at pH 6.0–6.5 in the presence of 3% of Atexal LC-L (ICI), which is a cationic leveling agent. Dyeing is commenced at 70°C, the temperature is raised to the boil in 30–45 min, and dyeing is continued at the boil for 1 h. At this point, reaction of the dye with the fiber will have taken place to the extent of 80%, with very little hydrolysis (in contrast to the Procion dyes), and the dyed material will possess wet-fastness which is adequate for nearly all purposes. Even higher wet-fastness can be obtained (1) by prolonging the period at the boil to 2 h, (2) by dyeing at 105°C, or (3) by adding ammonia at the end of the dyeing and treating the material at pH 8.0–8.5 for 15–20 min at the boil.

In addition to the standard range of Procilan dyes, Procilan Brilliant Yellow 4GS is available. This dye, which is not based on a 1:2 metal-complex structure, is much brighter than the other dyes but is compatible with them.

In concluding this section, mention must be made of padding processes that have been developed for applying reactive dyes to wool. The first of these processes[23c,24] is for applying to shrink-resisted wool slubbing the highly reactive Procion MX dyes, which normally give very skittery dyeings on wool. The dyes are applied at room temperature at pH 10.5, using a pad-batch process. Under these conditions very little hydrolyzed dye is adsorbed by the wool, so high wet-fastness is obtained. The process depends for its success on complete and uniform removal of the epicuticle during the Dylan X shrink-resist treatment, thus enabling even penetration of the fiber by the dye to take place at low temperatures.

An alternative method[25] for the application of Procion MX dyes, as well as reactive dyes of other types, entails the use of a concentrated solution of urea and does not require a preliminary shrink-resist treatment. Wool cloth is padded in a cold liquor containing dye, urea (300 g/liter), wetting agent, acetic acid to give pH 5, a suitable thickener, and an antifoaming agent. Under these conditions, Procion MX dyes undergo negligible hydrolysis. The cloth is then passed through a mangle to a batching roller, where it remains at room temperature, with slow rotation, for 24–48 h. This time can be reduced (e.g., to less than 8 h for Procion MX dyes) by incorporating sodium bisulfite (10–20 g/liter) in the pad liquor, provided that the dye does not react by addition at a double bond. The increase in rate and extent of fixation obtained extends the range of suitable dyes and is probably due to the reducing action of the bisulfite, leading to the formation of highly reactive thiol groups in the wool. The process is completed by washing off in dilute ammonia, which removes unfixed dye, followed by water, and, finally, cold dilute acetic acid to bring to pH 5. A wide range of bright colors of very good wet fastness can be obtained in this way, without any yellowing of the wool.

## 7.8. Vat Dyes

Vat dyes possess outstanding fastness, largely owing to the fact that they are present in the dyed fiber as insoluble particles. They are very important for application to cellulosic fibers, and some of them (e.g., Caledon Jade Green) provide bright, very attractive hues. In general, however, they are unsuitable for application to wool because they require a strongly alkaline dyebath. Some vat dyes require less alkali than others, and these have been used on the Continent for dyeing loose wool for uniform cloths, dyeing being carried out at 50–65°C.

The main exception to this general rule is indigo. Originally derived from plants, it was the standard coloring matter for obtaining different depths of blue on wool until the advent of synthetic dyes. Since the beginning of this century, synthetic indigo has been used, but its importance has steadily declined, although it is still specified for some uniform cloths.

Indigo, like other vat dyes, is insoluble in water, and for use in dyeing it must first be reduced to the alkali-soluble leuco form by means of sodium dithionite:

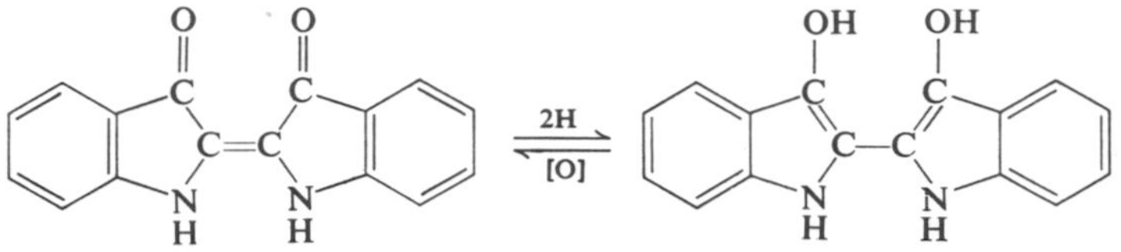

$$\text{Indigo (blue)} \quad \underset{[O]}{\overset{2H}{\rightleftharpoons}} \quad \text{Leuco-indigo (yellow)}$$

Indigo (blue) Leuco-indigo (yellow)

Indigo is sold in the reduced form as a 20% solution which is used, together with ammonia and sodium dithionite, to prepare the "vat," at 50°C. Glue may also be present to stabilize the vat and protect the wool from the alkali. The color of the vat should be greenish yellow.

Wool in loose form or as cloth is dyed in the vat for 10–30 min, keeping the material below the surface of the liquor. It is then squeezed and allowed to oxidize in the air, when the color changes from yellow to light blue. If a dark blue is required, the dyeing operation is repeated. Usually, only one "point" is given and the wool is subsequently "topped" to give a navy blue by dyeing with a chrome dye.

A navy blue based on indigo has high fastness to light and a more attractive appearance, especially in artificial light, than the same shade obtained entirely with chrome dye. One drawback, however, is that any indigo on the surface of the wool tends to rub off on to adjacent white material, but this is minimized by limiting the depth with indigo to light blue and by giving a scour after dyeing.[26]

Vat dyes are also available in a solubilized form. The first example was Indigosol O, the disodium salt of

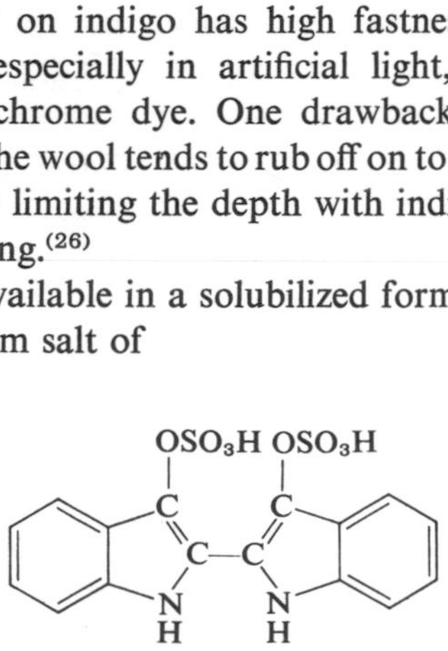

Ranges of these soluble, sulfuric esters of leuco vat dyes are available and they are used, in particular, for printing cellulosic materials.

Solubilized vat dyes may be applied to wool, like milling acid dyes, from a weakly acidic dyebath. It is then necessary to regenerate the parent vat dye by vigorous oxidation at 85°C, using, for example, 5–7½% of ammonium persulfate and 10–20% of sulfuric acid, the percentages being based on the weight of wool. Some damage to the wool occurs during this treatment and the dye may not be oxidized completely (i.e., there may be incomplete development of the shade).

Owing to the problems associated with the application of vat dyes to wool, the aim of high all-round fastness and brightness of hue is more readily achieved on the commercial scale with reactive dyes, but it is possible that reactive dyes will never equal some of the vat dyes in respect to fastness to light.

## 7.9. Dyeing with Minimum Damage to the Fiber

All wet treatments damage wool to some extent. With most wet processes the damage is slight, but it can be serious, especially with fine wools. It is shown by loss of tensile strength, loss in resistance to abrasion, and by felting. Under the microscope, damage to the scale structure of the fibers may be observed.

Keratin undergoes hydrolytic attack of the main-chain peptide linkages if subjected to prolonged treatment in hot water. This attack is increased in strongly acidic solutions. In alkaline solutions the principal attack is on the cystine linkages:

$$\text{CH—CH}_2\text{—S—S—CH}_2\text{—CH} + H_2O = \text{CH—CH}_2\text{—SOH} \quad \text{HS—CH}_2\text{—CH} \qquad (7\text{-}15)$$

Minimum attack occurs in the region between pH 3 and pH 5, as illustrated[27] in Fig. 7-5. Wool that has received a chemical treatment (e.g., bleached or shrink-resisted wool) is even more susceptible to damage.[28] Damage to the fiber increases with increasing temperature and by prolonging the time of treatment, so the most important factors are pH, temperature, and time.

It is usually necessary to dye wool at the boil to obtain satisfactory penetration and even distribution of the dye throughout the fiber. If a pressurized machine is available, the same result without any more damage to the fiber can be achieved by dyeing for a shorter time at 105°C within the pH range 3.5–6.5. A normal period for dyeing at the boil is 1–2 h, but much longer times are often found in practice when the dyer makes a large number

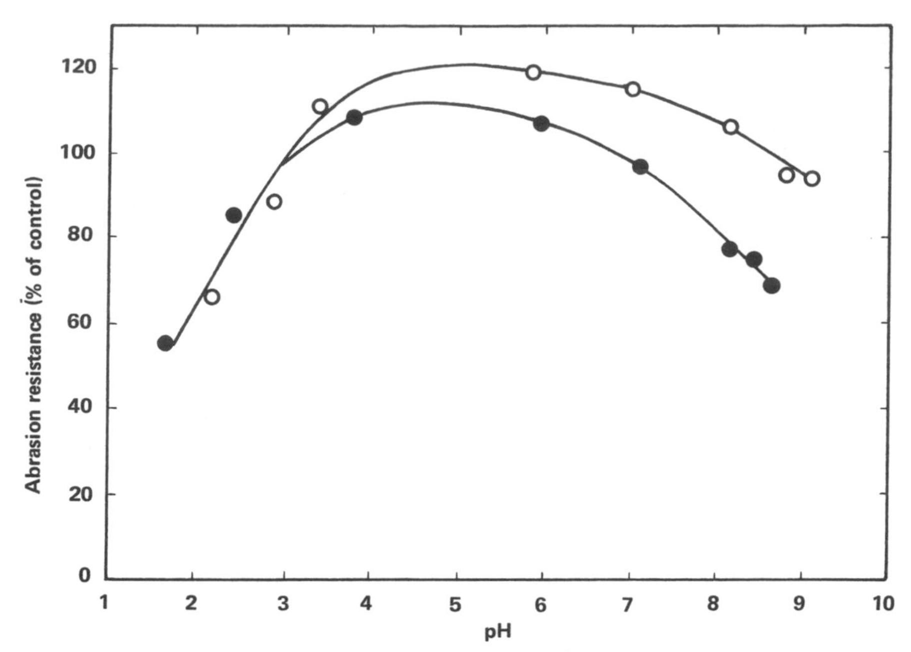


Fig. 7-5. Effect of pH of boiling solution on abrasion resistance of yarn. ○ = No salt; ● = 9.2 g $Na_2SO_4$ per liter (Peryman, 1954[27]).

of small additions of dye in attempting to obtain a very accurate match to pattern. Under such conditions, the wool may suffer considerable damage. Unduly prolonged dyeings are largely avoided by carrying out preliminary small-scale matching and by using machinery fitted with automatic temperature control.

Wool is normally dyed in the pH range 2–8, depending on the class of dyes employed. At one end of the range are 1:1 metal-complex dyes, which are applied at pH 2. Examination of Fig. 7-5 shows that appreciable damage occurs at this pH value, and the need to use a large amount of sulfuric acid is recognized as a deterrent to the use of this class of dyes.

At the other end of the range are the milling acid and metachrome dyes, which are usually applied in the presence of an ammonium salt at pH 6–8, an undesirably high value. In this case, however, the difficulty may be overcome by the use of an appropriate auxiliary product. Products of this type [e.g., Atexal LC-CWL (ICI)] have weakly cationic as well as nonionic properties. They combine loosely with milling acid and 1:2 metal-complex dyes, which can then be applied from a weakly acidic bath, the dye being liberated gradually for combination with the wool as the temperature is raised. A further advantage of dyeing under weakly acidic conditions is that the pH is more readily controlled than when dyeing from an approximately neutral bath, but care must be taken to avoid using excess of auxiliary product, since this leads to poor exhaustion.

Another way to preserve the wool, as well as to prevent yellowing, is to dye at temperatures below 100°C without prolonging the time of dyeing. However, satisfactory dyeings can only be achieved if steps are taken either to prevent aggregation of dye in solution or to produce swelling of the wool, or both. Addition to the dyebath of 2% (by volume) of benzyl alcohol or 0.4% of Irgasolvent CLW (CGY) enables satisfactory dyeings to be obtained in 30 min at 80–90°C with selected milling acid and 1:2 metal-complex dyes, applied in the presence of acetic acid.[29] The extra cost of the solvent is well worthwhile when dyeing an expensive and easily damaged fiber such as cashmere.

Alternatively, a suitable nonionic auxiliary product may be used. Thus, with the condensate from nonylphenol and ethylene oxide (6–10 residues), 0.5–2.0% on the weight of material is claimed[30] to enable satisfactory dyeings to be obtained with 1:2 metal-complex dyes at pH 3.6–4.2 in 20–30 min at 90°C, after raising the temperature very gradually from 40°C. The washing fastness, however, is slightly lower than that obtained by normal dyeing (e.g., dyeing for 1 h at the boil at pH 5–6). Suitable commercial auxiliary products are available [e.g., Lissapol N (ICI), Albegal B (CGY) and Irgasol SW (CGY)], and dyeing at 90°C is also possible with other types of wool dyes.

A method that depends on powerful swelling of the wool, caused by breakage of hydrogen bonds, employs 75% formic acid in place of water in the dyebath. Subsequent rinsing in water restores the wool to its original condition. Totally enclosed machinery is necessary and the formic acid must be recovered by distillation. A continuous process using this method has been described.[31] It is suitable for milling acid, 1:1 metal-complex, and reactive dyes. The sliver passes through the formic acid dyeliquor at 60°C, a 5:1 liquor ratio being used, and only 2 min in the dyeliquor is required before washing in water.

Ideally, wool should be dyed at room temperature, which would also have the advantage of eliminating the cost of heating the dyebath. One method of obtaining rapid dyeing at room temperature is to use a concentrated solution of urea, which permits dyeing with acid dyes and metal-complex dyes to be completed in only 10 min,[32] but the method is mainly of interest for applying reactive dyes,[25] as described on p. 325. The urea acts by disaggregating dyes at low temperatures. Whether it also swells the wool is still debated.[33,34] Swelling appears to be dependent on the treatments given to the fiber before dyeing; if the wool is undamaged, suppression of swelling occurs.[35]

From what has been written, it will be clear that the wool dyer now has a fairly clear picture of the chemical behavior of wool in relation to dyeing processes. In applying this knowledge he has been greatly aided in recent years by the introduction of new auxiliary products and much better dyeing machinery. Automatic temperature control has largely eliminated one important source of error. The quality of wool dyeing is likely to be increased still further by recent developments in instrumental match prediction and in methods of continuous dyeing.

## 7.10. References


1. W. von Bergen, *Melliand Textilber.* **7**, 451 (1926).
2. F. Townend, *J. Soc. Dyers Colourists* **61**, 144 (1945); H. R. Hadfield and D. R. Lemin, *J. Soc. Dyers Colourists* **77**, 97 (1961).
3. E. Coates, *J. Soc. Dyers Colourists* **85**, 355 (1969).
4. F. L. Goodall and C. Hobday, *J. Soc. Dyers Colourists* **55**, 542 (1935).
5. J. B. Speakman and E. Stott, *J. Soc. Dyers Colourists* **50**, 341 (1934).
6. T. Vickerstaff, *The Physical Chemistry of Dyeing*, 2nd ed., Oliver & Boyd Ltd., London (1964).
7. S. D. C. Committee on Dyeing Properties of Wool Dyes, *J. Soc. Dyers Colourists* **66**, 213 (1950).
8. A. N. Derbyshire, *Hexagon Dig.* **21**, p. 12 (1955).
9. G. T. Morgan and J. D. Main-Smith, *J. Soc. Dyers Colourists* **41**, 233 (1925).
10. H. D. K. Drew and R. E. Fairbairn, *J. Chem. Soc.* 823 (1939).
11. W. Ender and A. Müller, *Melliand Textilber.* **19**, 65, 181, 272 (1938); I. D. Rattee, *J. Soc. Dyers Colourists* **69**, 288 (1953).
12. G. Schetty, *Amer. Dyestuff Reporter* **54**, 589 (1965); *Textilveredlung*, **3**, 3 (1968).
13. W. Beal, *Proc. 3rd Intern. Wool Textile Res. Conf., Paris* **3**, 223 (1965).
14. F. R. Hartley, *J. Soc. Dyers Colourists* **85**, 66 (1969); *Wool Sci. Rev.* **37**, 54 (1969).
15. C. B. Stevens, F. M. Rowe, and J. B. Speakman, *J. Soc. Dyers Colourists* **59**, 165 (1943).
16. C. L. Bird and E. Molloy, *J. Soc. Dyers Colourists* **55**, 560 (1939); **57**, 224 (1941).
17. D. M. Lewis and I. Seltzer, *J. Soc. Dyers Colourists* **88**, 93 (1972).
18. H. Klapper, *Bayer Farben Rev.*, **20**, 2 (1971); D. Hildebrand and G. Meier, *Bayer Farben Rev.*, **20**, 12 (1971); *Textil-Praxis* **26**, 499, 557 (1971).
19. O. A. Stamm, *J. Soc. Dyers Colourists* **80**, 416 (1964).
20. W. Mosimann, *Textile Chem. Colorist* **1**, 182 (1969).
21. J. Shore, *J. Soc. Dyers Colourists* **84**, 408, 413, 545 (1968); **85**, 14 (1969).
22. F. Osterloh, *Textil-Praxis* **26**, 164 (1971).
23. (a) A. N. Derbyshire and J. G. Graham, *Intern. Dyer* **131**, 31 (1964); (b) J. G. Graham, *Intern. Dyer* **138**, 367 (1967); (c) A. N. Derbyshire and G. R. Tristram, *J. Soc. Dyers Colourists* **81**, 584 (1965).
24. F. M. Stevenson and M. C. Stevenson, *J. Textile Inst.* **53**, 649 (1962).
25. D. M. Lewis and I. Seltzer, *J. Soc. Dyers Colourists* **84**, 501 (1968); J. D. M. Gibson, D. M. Lewis, and I. Seltzer, *J. Soc. Dyers Colourists* **86**, 298 (1970); I.C.I. Ltd., Organics Division, *Tech. Inform. D1191* (1970).

26. B. Campbell, S. E. Inkumsah, and W. C. Tandoh, *J. Soc. Dyers Colourists* **80**, 583 (1964).
27. R. V. Peryman. *J. Soc. Dyers Colourists* **70**, 83 (1954); **71**, 165 (1955); **73**, 455 (1957).
28. E. A. Cookson, R. J. Hine, and J. R. McPhee, *J. Soc. Dyers Colourists* **80**, 196 (1964); H. Zahn, *Wool Sci. Rev.* **32**, 1 (1967).
29. L. Peters, C. B. Stevens, J. Budding, B. C. Burdett, and J. A. W. Sykes, *J. Soc. Dyers Colourists* **76**, 543 (1960); W. Beal, K. Dickinson, and E. Bellhouse, *J. Soc. Dyers Colourists* **76**, 333 (1960).
30. R. J. Hine and J. R. McPhee, *Intern. Dyer* **132**, 523 (1964); *Proc. 3rd Intern. Wool Textile Res. Conf., Paris* **3**, 261 (1965); *J. Soc. Dyers Colourists* **81**, 268 (1965).
31. B. W. Wilson, *J. Soc. Dyers Colourists* **86**, 122 (1970).
32. I. B. Angliss and J. Delmenico, *J. Soc. Dyers Colourists* **80**, 543 (1964).
33. K. R. F. Cockett, I. D. Rattee, and C. B. Stevens, *J. Soc. Dyers Colourists* **85**, 461 (1969).
34. R. S. Asquith and A. K. Booth, *Textile Res. J.* **40**, 410 (1970).
35. R. S. Asquith, I. D. Rattee, A. K. Booth, K. R. F. Cockett, and C. B. Stevens, *J. Soc. Dyers Colourists* **88**, 62 (1972).
36. R. S. Asquith and D. K. Chan, *J. Soc. Dyers Colourists* **87**, 181 (1971).
37. T. J. Abbott, R. S. Asquith, D. K. Chan, and M. S. Otterburn, *J. Soc. Dyers Colourists* **91**, 133 (1975).

# 8

# The Chemistry of Wool Finishing

## C. S. Whewell

### 8.1. Introduction

Although wool finishing is as old as clothmaking itself, it is only comparatively recently that the subject has been treated in a scientific manner. Cloth as it is received from the loom or the knitting machine is generally unacceptable to the consumer, as it is often greasy and dirty and is aesthetically unattractive. The object of finishing is to convert the "gray" cloth into a more desirable state. Until comparatively recently this was done by a sequence of operations which involved such processes as inspection of gray material, setting (crabbing or blowing), scouring, milling, hydroextracting, drying, raising, brushing, cutting, dry blowing (semidecating), and pressing. The precise details of these operations vary according to the type of cloth and differ from factory to factory. Examples of the routine adopted in finishing several of the more common wool fabrics are described in the literature[1] and form the basis of all wool finishing.

Finishing is frequently regarded as a craft, and indeed the most modern organization still owes much to the skilled operative. In the past 50 years, however, great progress has been made in understanding the fundamentals of the basic finishing operations, and in addition there has been developed a whole range of new chemically based finishing processes, such as mothproofing, waterproofing, rotproofing, shrink-resist finishing, and permanent finishing. This change has been made possible because of advances in knowledge

---

**C. S. Whewell** • Professor and Head of Department of Textile Industries, University of Leeds, Yorkshire, England

of the structure and reactivity of the wool fiber, and in the development of suitable equipment. It must be accepted that the theory of textile finishing is only in its infancy, partly because of the complexity of the phenomena involved and partly because of the lack of precise information on the structure of wool.

The theories put forward in the succeeding pages must often be regarded as working hypotheses and may perhaps be regarded as oversimplifications. Nevertheless, they will form a sound basis of advances in a subtle and essential branch of textile processing.

## 8.2. Scouring

The object of scouring is twofold: (1) to remove size, oils, and dirt from loom-state cloth; and (2) to develop cover, handle, and other essential textile characteristics in the cloth.

In deciding on a scouring technique, two limitations must be borne in mind: (1) damage or chemical modification to the cloth must be kept to a minimum, and (2) the appearance and dimensions of the scoured cloth must be those required by the cloth designer. The method selected depends on many factors, but of special importance are the type of cloth, the nature of the impurities present, the machines and facilities available, and the characteristics required in the finished cloth.

The impurities are the most important and include size, wool lubricants, and dirt. Comparatively few wool cloths contain size, but some warps are sized and the size can be removed by soaking the cloth in water or enzymes (e.g., Gelatase for gelatine size, or Nervanase for starch-based sizes) before the conventional scouring. Particulate dirt is removed by emulsification using detergents.

### 8.2.1. Fiber Lubricants

The lubricants applied initially to facilitate yarn manufacture vary considerably in chemical composition. Some of the more important are as follows.

#### *8.2.1.1. Water-Soluble Lubricants*

These are not used to a great extent, but they are increasing in significance. The following types of compound have been suggested:

a. Polyglycols. The main disadvantage of these reagents is their tendency to form sticky substances.

b. Blends of oils and detergents. One of the disadvantages of these blends is their property of extracting the wool grease from wool being carded. This results in the formation of sticky complexes which interfere with the carding operation. Some of the new water-soluble "oils" are selected because of their inability to dissolve wool grease.

c. Compounds based on fatty acid/amino carboxylic acid derivatives and *o*-phosphoric acid esters (e.g., Lanigan Lubricant SG).

d. Alkyl aryl polyglycol ethers.

All these compounds can be removed by water alone, but since most have no detergent power, particulate dirt present on the fabric is not removed. Treatment in a detergent solution is, therefore, necessary to obtain a clean piece.

#### *8.2.1.2. Alkali-Soluble Lubricants*

These are the reagents that can be removed by chemical reaction with simple cheap reagents, such as sodium carbonate at moderate temperatures to give water-soluble products. In this class are the fatty acids (oleines) and modified oleines which are used largely in woollen processing.

#### *8.2.1.3. Solvent-Soluble Lubricants*

These products do not react with mild alkalis. They form the largest group and include fatty oils and modified mineral oils. They can be removed by solvent extraction or by emulsification.

### 8.2.2. Principles of Oil Removal

The general principles of the various methods of oil removal are as follows:

1. Water extraction. This is limited to the removal of a small number of substances.

2. Solvent extraction. Superficially this is a most attractive method, because it avoids the problem of water softening and effluent treatment, and the wool oils can be recovered. The high cost of solvent has, until recently, been the prohibiting factor, but this is becoming less important as the price of chlorinated hydrocarbon solvents falls. Moreover, by adopting some of the modern "dry" cleaning procedures, some water can be incorporated in the solvent together with a dry cleaning soap, and the mixture will then be

suitable for promoting shrinkage during scouring to an appropriate extent. One continuous technique consists in passing the cloth through hot trichloroethylene and then flashing off the solvent with jets of hot water. The process has been found to be successful for scouring blanket fabric (70 in wide and containing 11.5% oleine) in rope form. The residual oil content was 0.07–0.10% and the solvent losses are 2%.

3. Emulsion scouring. This is the most important, not only because it is a method for removing fatty oils and modified mineral oils, but also because success, even in saponification scouring, relies on the soap produced to emulsify the dirt and any unsaponificable components. The reagents normally employed are water and steam, alkalis (usually sodium carbonate), and detergents, either soap or anionic or nonionic synthetic detergents. The mechanism of oil removal by emulsification is complex, but the following stages are important.[2]

The fibers are wetted by the detergent solution, which must penetrate between the fibers, a process that is aided by mechanical action. The film of oil then rolls up to form globules, which are subsequently detached from the fibers as a result of the mechanical action. The globules of oil must remain suspended in the liquid, and coalescence must be prevented. Absorption of detergent is important in creating this state of affairs, because as a result of absorption, the globules become charged and this hinders coalescence. The suspension or emulsion is subsequently diluted by the addition of water, and the dilute emulsion is removed from the system. The mechanism of the removal of oil and particulate dirt is probably more complex than this and has been discussed in some detail by Durham.[3] For example, after "rolling up," the globules are believed to become surrounded by a voluminous mass formed by interaction between globules and detergent. These "complexes" absorb considerable amounts of detergent, and this is of economic importance.

### 8.2.3. Factors Affecting the Results Obtained in Scouring

Many factors affect the results of scouring, and these can be considered as follows. The type of fiber has become more important with the increasing use of blends. Thus oil is more difficult to remove from polyesters than from wool, and hence difficulties have arisen in the removal of oil from wool/polyester blended fabrics. This may be overcome by increasing the severity of the scour or by incorporating polar compounds in the wool oil.

The various man-made fibers differ in the ease with which oil can be removed from them.[4] In the same way, the ease of removal of oil from chemically modified wools varies according to the chemical treatment. (See Tables 8-1 and 8-2.)

**Table 8-1. Removal of Oil from Various Fibers[a]**

| Fiber | % Residual oil[b] |
|---|---|
| Dynel | 71 |
| Courlene | 69 |
| Thermovyl | 59 |
| Terylene | 42 |
| Terylene (treated with NaOH) | 29 |
| Orlon | 18 |
| Tricel | 14 |
| Celafibre | 12 |
| Nylon | 12 |
| Silk | 11 |
| Viscose (bright) | 10 |
| Viscose (matt) | 9 |
| Fibrolane BX | 9 |
| Cotton (purified) | 8 |

[a] Identical scouring condition used in each case.
[b] 10% mineral oil applied initially.

The nature of the wool oil has a profound effect on its ease of removal. Mineral oils are more difficult to remove than fatty oils, but by blending suitable oil-soluble polar compounds with fatty (or even mineral) oils, products can be obtained which are very easily scoured out of wool. Some of the more effective polar compounds are oleyl alcohol,[(5)] mono- and diglycerides, petroleum sulfonates, wool grease derivatives, and ethylene oxide derivatives

**Table 8-2. Removal of Oil from Treated Wool**

| Treatment | % Residual oil[a] |
|---|---|
| Silicone | 73 |
| Untreated | 57 |
| Hydrogen peroxide | 55 |
| Hydrochloric acid | 42 |
| Sulfuric acid | 41 |
| Bleaching powder | 32 |
| Permonosulfuric acid | 12 |
| Caustic potash in alcohol | 4 |
| Mixture of potassium permanganate and sodium hypochlorite | 2 |
| Acidified sodium hypochlorite | 1 |
| Sodium hypochlorite | 1 |
| Sulfuryl chloride | 1 |

[a] 10% mineral oil applied initially.

**Table 8-3. Removal of Blends of Mineral Oil and Various Polar Compounds**

| Polar compound | % of polar compound in blend | | | | | | | | |
|---|---|---|---|---|---|---|---|---|---|
| | 0 | 5 | 10 | 20 | 30 | 40 | 50 | 75 | 100 |
| Glycerol monooleate | 4 | 40 | 43 | 41 | 35 | 30 | 37 | 31 | 76 |
| Glycerol dioleate | 4 | 25 | 30 | 34 | 37 | 41 | 44 | 49 | 52 |
| Glycerol diricinoleate | 4 | 30 | 42 | 55 | 62 | 68 | 71 | 77 | 87 |
| Glycol monooleate | 4 | 42 | 48 | 51 | 53 | 54 | 56 | 60 | 66 |
| Diethylene glycol monooleate | 4 | 45 | 56 | 67 | 73 | 78 | 81 | 85 | 80 |
| Diethylene glycol dioleate | 4 | 50 | 59 | 69 | 76 | 79 | 81 | 84 | 86 |
| Monoglyceride of arachis oil fatty acids | 4 | 56 | 58 | 56 | 52 | 49 | 47 | 43 | 90 |
| Diglyceride of arachis oil fatty acids | 4 | 62 | 68 | 70 | 65 | 63 | 62 | 63 | 64 |
| Pentaerythritol ester of arachis oil fatty acids | 4 | 50 | 54 | 51 | 47 | 44 | 41 | 44 | 82 |
| Emulphor A | 4 | 31 | 59 | 87 | 95 | 97 | 99 | 99 | 99 |
| Methyl oleate | 4 | 8 | 12 | 20 | 29 | 36 | 45 | 65 | 80 |
| Ethyl oleate | 4 | 10 | 16 | 27 | 37 | 46 | 53 | 66 | 71 |
| Butyl oleate | 4 | 16 | 28 | 40 | 47 | 53 | 48 | 67 | 74 |

such as Emulphor A and Lissapol NX.[6] The effectiveness of some of these additives is shown in Table 8-3.

The type of detergent and conditions of scouring are of great significance. Both soap and synthetic detergents are used, but in the U.K. the soap–soda ash scour is still most common. The function of the soda ash is fourfold[7]: (1) to neutralize any fatty acid in the piece, (2) to neutralize any mineral acid present, (3) to act as a soap builder, (4) to combine with the wool.

Thus the amount of soda ash required can be calculated. The amount of soap should be sufficient to make the total soap present equal to half the weight of nonsaponifiable matter.

## 8.3. Setting Processes

Some cloths, especially those made in plain or hopsack weaves or those made from crossbred wools, must be set if undue distortion during scouring, and particularly piece dyeing, is to be avoided.

Several techniques, for example those known as crabbing, wet blowing, and dry blowing, are used to set greasy or scoured wool fabrics. Each has its own merits, but in all cases accurate control is essential if the best results are to be obtained.

Although setting of cloths and yarns is an old process, it has recently assumed added significance because of the emphasis on permanent finishing, pleating, and creasing of wool fabrics in later stages of finishing and in garment making. In commercial practice, therefore, setting is associated with such operations as crabbing, dry blowing (semidecating), wet blowing, torpedoing, or pressure setting (full-decating), but setting is of importance in many other processes (e.g., tentering, pressing) where the effects may or may not be desirable.

It is essential to define precisely what is meant by set, for much of the controversy which has arisen in this field has been due to various authors associating different meanings to the same term. Set is the retention of deformations intentionally or unintentionally introduced into textile materials. It should be distinguished from the stabilization of dimensions of unstrained materials and from relaxation, as, for example, relaxation during cheese dyeing of yarns, which is often referred to as "setting." *Cohesive set* is set retained in air but which is lost when the material is immersed in water; *temporary set* is retained in warm water but lost in boiling water; *permanent set* is set retained even after boiling freely in water for a prolonged period, usually 1 or 2 h.

### 8.3.1. General Theory of Setting

Although many different views have been expressed about the nature of permanent setting, it is now generally agreed that the process involves bond fission and bond rebuilding. The main difference between the views of the various investigators in this field are concerned with the nature of the bonds broken and rebuilt. Speakman and his colleagues[8] laid great emphasis on the fission of disulfide bonds and the rebuilding of disulfide and other covalent bonds, while others have stressed the importance of fission and rebuilding of hydrogen bonds.[9] It is almost certain that both types of linking are involved, but there is little doubt about the key role of covalent bonds in the production of permanent set. It would appear that some covalent bonds must be broken as a preliminary to setting and that the formation of new covalent bonds facilitates setting. The covalent bonds formed by this rebuilding operation include (1) re-formed —S—S— linkings,[10] (2) lanthionine (—C—S—C) linkings,[11] (3) —S—NH— links, and (4) lysinoalanine linkings[11] see Eqs. 8-1–8-5.

Fission and rebuilding of hydrogen bonds will, however, accompany these reactions involving covalent linkings, and the hydrogen bonds in the set fibers are of considerable significance, for set that is permanent to boiling water can be removed if the set fibers are immersed in hydrogen-bond-breaking agents such as lithium bromide and cuprammonium hydroxide.

$$\mathrm{R{-}CH_2{-}S{-}S{-}CH_2{-}R \xrightarrow{H_2O} R{-}CH_2{-}SOH + HS{-}CH_2{-}R \quad (R = {-}NH{-}CH{-}CO{-})} \tag{8-1}$$

$$\mathrm{R{-}CH_2{-}SOH \longrightarrow \begin{array}{c} | \\ NH \\ | \\ C{=}CH_2 \\ | \\ CO \\ | \end{array} + H_2O + S} \tag{8-2}$$

$$\mathrm{R{-}CH_2{-}S{-}OH + NH_2{-}(CH_2)_4{-}R \longrightarrow R{-}CH_2{-}S{-}NH(CH_2)_4{-}R{-}} \tag{8-3}$$

$$\mathrm{\begin{array}{c} | \\ NH \\ | \\ C{=}CH_2 \\ | \\ CO \\ | \end{array} + HS{-}R{-}CH_2{-}SH{-}R \longrightarrow \underset{\text{(lanthionine)}}{R{-}CH_2{-}S{-}CH_2{-}R}} \tag{8-4}$$

$$\mathrm{\begin{array}{c} NH \\ | \\ C{=}CH_2 \\ | \\ CO \\ | \end{array} + H_2N{-}(CH_2)_4{-}R \longrightarrow \underset{\text{(lysinoalanine)}}{R{-}CH_2{-}NH{-}(CH_2)_4{-}R}} \tag{8-5}$$

Until comparatively recently the main phenomena of setting were interpreted in terms of a relatively simple model of the wool molecule. In the light of new ideas on the fine structure of the wool fiber and on the distribution and composition of the $\alpha$ helixes and of other proteins containing different amounts of sulfur, it will be necessary to extend and perhaps to modify some of the current explanations of setting.

Two aspects of setting deserve special consideration: (1) the setting of chemically modified wool, and (2) the influence of the setting medium on the amount of set induced.

### 8.3.2. Setting of Chemically Modified Fibers

Values have been measured for the amount of permanent set obtained when stretched (40%), chemically treated wool is boiled in water for 2 h and then released in boiling water for 1 h. The following conclusions can be drawn.

1. Hydrolysis of the peptide linkages has little effect on the setting (Table 8-4), but stabilization of the —S—S— bonds by conversion to —C—S—C— reduces the ability of the fiber to take a set.[10]

2. Fibers in which the amino groups have been rendered inactive do not set in water,[12] although it is realized that loss of amino groups may not

**Table 8-4. Setting of Fibers Treated with Hydrolyzing Agents**

| Treatment | % Set |
|---|---|
| None | 14.5 |
| Boiled in water for 48 h | 6.3 |
| Boiled in 0.1 *N* HCl | 14.7 |

be the only reaction which occurs when wool is treated with the reagents indicated in Table 8-5. Some oxidation of the —S—S— links almost certainly takes place when hair is treated with nitrous acid, and nitroso or nitro compounds may also be formed.

3. Esterified keratins set more readily than untreated materials.[13] Few details of the effect of various factors on the reaction are available, but the following data indicate some of the more obvious changes: (a) treatments for a comparatively short time bring about considerable change in setting characteristics; (b) the temperature of treatment is important, the effectiveness of the esterification increasing with temperature; (c) if methanol is replaced by higher alcohols, the effectiveness of the treatment is reduced; and (d) the presence of water reduces the effectiveness of the methanol/HCl treatment (Table 8-6).

It is not possible to explain satisfactorily the setting of esterified fibers, although some workers[14] attribute it to changes in internal pH, but it would appear possible that esterification of the hydroxyl groups in the fiber enhances the reactivity of such groups as the amino and disulfide groups. It is interesting to note that esterification of oxidized or deaminated hair improves the setting of these materials.

4. Tyrosine residues are of importance in determining set (Table 8-7), substitution reducing the amount of set obtained.[13] This may be due to steric hindrance or to stabilization of the —S—S— linkages, either by

**Table 8-5. Setting of Fibers Treated with Reagents that Attack Amino Groups**

| Treatment | % Set |
|---|---|
| None | 14.5 |
| Nitrous acid | −1.1 |
| Difluoronitrobenzene | −1.3 |

**Table 8-6a. Setting of Fibers Treated with Hydrogen Chloride in Methanol at 50°C for Various Times**

| Treatment time (min) | % Set |
|---|---|
| 0 | 14.9 |
| 5 | 27.6 |
| 10 | 25.0 |
| 15 | 24.9 |
| 30 | 27.9 |
| 60 | 28.8 |
| 120 | 28.5 |
| 180 | 28.2 |

**Table 8-6b. Setting of Fibers Treated with Hydrogen Chloride in Methanol for 1 h at Various Temperatures**

| Treatment temperature (°C) | % Set |
|---|---|
| Untreated | 14.9 |
| 5 | 18.5 |
| 10 | 22.2 |
| 15 | 18.9 |
| 30 | 22.5 |
| 40 | 24.9 |
| 50 | 28.8 |

**Table 8-6c. Setting of Hair Treated with Hydrogen Chloride in the Presence of Various Alcohols at 50°C for 1 h**

| Alcohol | % Set |
|---|---|
| None | 14.5 |
| Amyl | 12.0 |
| Butyl | 17.2 |
| Propyl | 16.3 |
| Ethyl | 18.2 |
| Methyl | 26.1 |

**Table 8-6d. Setting of Hair Treated with Hydrogen Chloride in Aqueous Solutions of Methanol**

| Concentration of methanol | % Set |
|---|---|
| — | 14.5 |
| 100 | 26.1 |
| 60 | 24.5 |
| 40 | 23.3 |
| 20 | 23.0 |
| 10 | 21.3 |
| 5 | 16.6 |

**Table 8-7. Setting of Fibers Treated with Reagents That Attack the Tyrosine Residues**

| Treatment | % Set |
|---|---|
| None | 14.5 |
| Concentrated $H_2SO_4$ for 30 s | 10.1 |
| Concentrated $HNO_3$ for 60 s | 11.1 |
| 40% $HNO_3$ for 18 h. | 7.2 |
| Iodination | −3.0 |

oxidation or by the substitution affecting the reactivity of the linkage. Some support for this view is provided by the observation that the nitrated or iodinated fibers can be set in solutions of such reducing agents as THPC, which are much more effective as setting agents than water.[15]

5. Reduced fibers set more readily than untreated fibers.[13] This would suggest that the —S—S— bonds split by the reducing agent are not seriously modified by the subsequent atmospheric oxidation. The fiber therefore retains its increased susceptibility to setting. Such pretreatments have been exploited for textile purposes, such as the presensitizing of fabrics to give them enhanced permanent-pleating characteristics.

6. The action of oxidizing agents on hair is complex,[13] for while in general oxidized fibers do not take a set in water, those oxidized by such reagents as peracetic acid, permonosulfuric acid, and periodic acid under well-defined conditions take a set more readily than untreated fibers (see Table 8-8). This would suggest that the action of the reagents is different from that of hydrogen peroxide, leading perhaps to the formation of residues more reactive than —$SO_3H$.

**Table 8-8. Action of Oxidizing Agents on Hair**

| Treatment | % Set |
|---|---|
| None | 14.5 |
| 2% Peracetic acid (5 min) | 6.1 |
| 2% Peracetic acid (60 min) | 22.5 |
| 2% Peracetic acid (120 min) | 30.1 |
| 10-vol hydrogen peroxide | 1.6 |
| Iodic acid (4 h at 40°C) | 23.5[a] |
| Periodic acid (0.5% at room temperature for 7 h) | 24.0 |

[a] Untreated control, 5.9%.

### 8.3.3. Setting in Various Media

Since the setting of hair can be regarded as a series of chemical reactions, the nature of the setting medium is of great importance, for different reagents promote or retard bond fission and bond rebuilding. Studies on the effectiveness of various reagents as setting media are of considerable commercial significance, for they are the basis on which many commercial processes are based. Some of the more important findings from a theoretical point of view are as follows:

1. Untreated wool cannot be set in solutions of strong acids.[16]
2. Boiling alkaline solutions are better setting agents than boiling water, the maximum setting taking place at pH 9.2[16] (see Fig. 8-1).
3. Wool can be set in solutions of reducing agents even at comparatively low temperatures.[17] Particularly effective reagents include ammonium thioglycollate, sodium and monoethanolamine bisulfites,[17] THPC $P^+(CH_2OH)_4Cl^-$,[15] mercaptans, phosphorous acid, and thiourea dioxide.[18]

$$C(NH_2)_2{=}SO_2$$

The effectiveness of THPC is illustrated by the following data, from which it is evident that hair sets much more readily in THPC than in water. For example, stretched human hair treated for 2 min in a 2% solution of

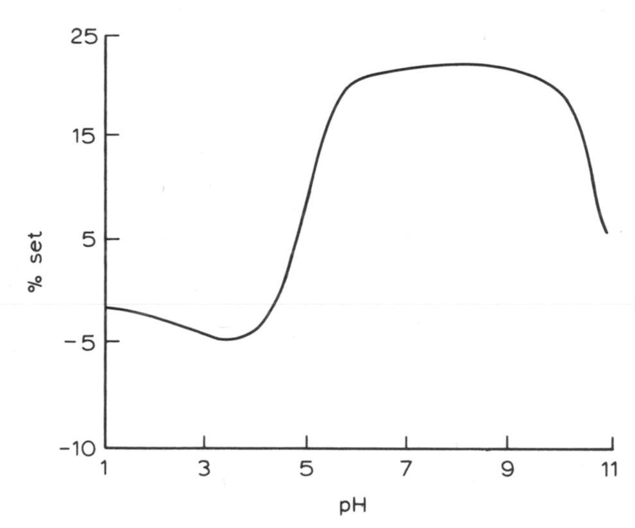


Fig. 8-1. Influence of the pH value of the setting solution on the extent of permanent set obtained at 100°C.

THPC at pH 6.1 acquires a set of 18.5%, whereas in water a supercontraction of 7.9% is obtained. After 2 h of treatment, 33.2% set is recorded on fibers initially extended 40%. The efficiency of the setting solution is clearly dependent on the temperature, concentration, and pH of the solution, but marked improvement is obtained with 0.1% solutions at the boil. Of particular interest is the observation that THPC in presence of 0.1 *N* hydrochloric acid sets hair. In absence of THPC, no set is obtained. Furthermore, although deaminated, iodinated or oxidized fibers cannot be set in water, they do take a set in 2% THPC solution. On the other hand, fibers treated with potassium cyanide cannot be set in either water or THPC solution. This suggests that the effectiveness of THPC is connected with its ability to split the —S—S— links, presumably by reduction.

4. Oxidizing agents are not in general good setting agents, but under particular circumstances certain oxidizing agents such as periodic acid, peracetic acid, and permonosulfuric acid do facilitate the process.(13)

5. While it is not possible in general to set fibers in solutions of hydrogen-bond-breaking agents, mixtures of reducing agents and hydrogen-bond breakers (e.g., urea) are better setting agents than reducing agents alone.(9)

### 8.3.4. Setting Processes

It is not appropriate to dicuss in detail all aspects of setting processes used in industry, but some of the newer findings deserve special mention.

1. Semidecating is an important process which is of growing significance. The set obtained under normal conditions is, however, often only temporary, and if permanent set is required, the time of semidecating must be considerably prolonged (see Fig. 8-2).

To increase the efficiency of the process and to obtain a higher degree of permanent set, the fabric should first be treated with a reagent which promotes setting. The reagent may be applied either by spraying or by padding, but uniform treatment is essential. The fabric is then blown with steam, usually for periods of about 5 min. Several reagents have been suggested, but the most popular is monoethanolamine bisulfite. It is advisable to test the fastness of a dyed fabric to treatments with bisulfite before commencing large-scale treatment, for some dyes are altered in shade by the reagent. The effect produced is good and the fabric has an attractive handle.

A mixture of urea and diethanolamine carbonate can be used instead of bisulfites.

2. Crabbing is perhaps the most common method of obtaining a high degree of set in the early stages of finishing. The pH should be carefully

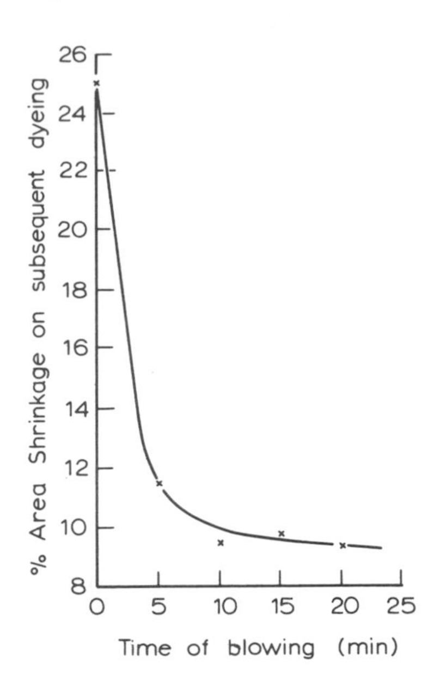


Fig. 8-2. Influence of time of blowing on the subsequent area shrinkage during dyeing. (In these experiments the efficiency of setting was assessed by determining the shrinkage that occurs when the fabrics, set under different conditions, are subsequently scoured and dyed; the smaller the shrinkage the more effective the setting process has been.)

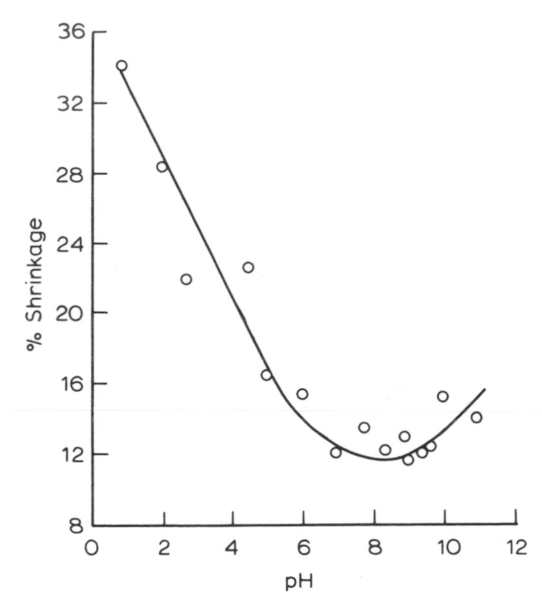


Fig. 8-3. Influence of pH value on the shrinkage of fabric. (Set assessed as in Fig. 8-2.)

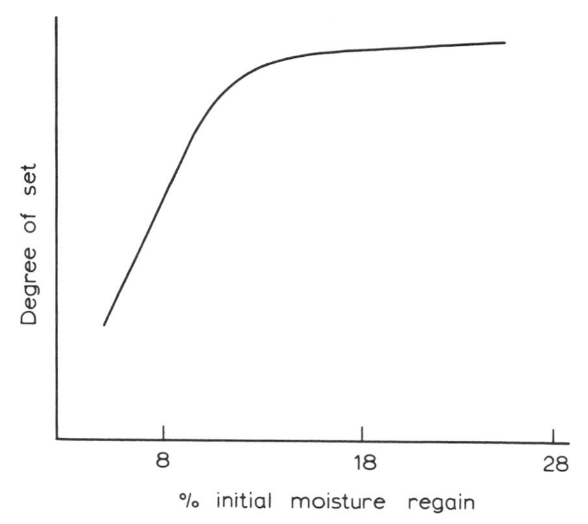


Fig. 8-4a. Influence of the initial moisture regain of fabric on the degree of set obtained during pressure setting.

controlled, preferably at about pH 6[6] (see Fig. 8-3). Solutions of higher pH values may occasionally give a higher degree of set, but the possibility of color bleeding is high.

3. Pressure setting or full decating is increasing in popularity as a final finishing process for giving a fabric dimensional stability and permanent finish. Several machines are available, notably those of Gessner, Gladstone-

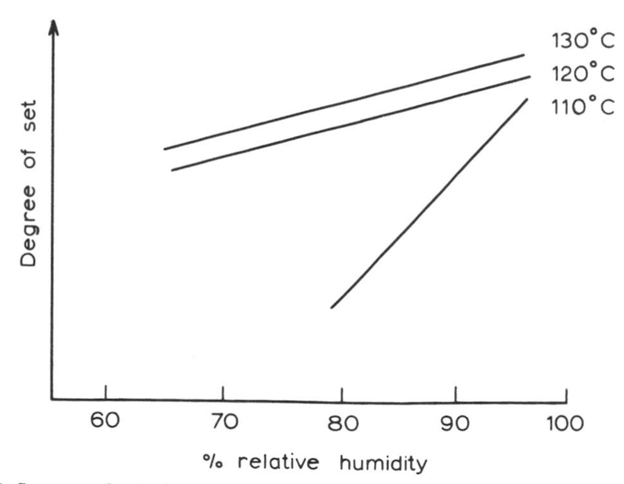


Fig. 8-4b. Influence of relative humidity of steam on the degree of set obtained during pressure setting.

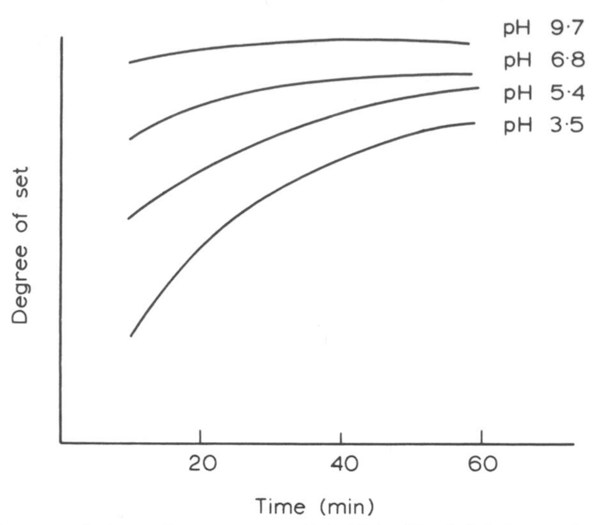


**Fig. 8-4c.** Influence of time of treatment and initial pH of fabric on the degree of set obtained during pressure setting.

Sellers, Kettling-Braun, and the K. D. Machine. The fabric is wound on a roller and then placed in the machine, which is essentially an autoclave (i.e., a cylinder into which high-pressure steam can be injected and maintained at an appropriate pressure). The time and pressure of treatment vary according to the type of cloth, but steaming at 30 psi for 5 min gives a considerable degree of permanent setting. It has recently been shown that the principles governing setting in high-pressure steam are similar to those governing set in boiling water.[19]

The effectiveness of the process has been shown to be dependent on such factors as the initial moisture regain of the cloth (see Fig. 8-4a), the relative humidity of the steam (see Fig. 8-4b), and the pH of the cloth (see Fig. 8-4c).

## 8.4. Milling and Felting

The ability of wool to felt is in many ways the basis of the wool cloth trade but was, until recently, the chief cause of complaint in so far as finished washable garments are concerned.

Felting was originally carried out by vigorous working of cloth, either by the hands or the feet in various aqueous solutions. Now all such "fulling" is carried out in machines, the oldest of which imitates the vigorous pounding associated with working the cloth with the feet. These machines, known as fulling stocks in the felt trade and "beaters" in finishing knitted goods, are still in use, but they have been largely replaced by rotary machines.

The essential features of a modern milling machine are as follows. The wetted cloth, in the form of an endless belt, passes through a throat and then between a pair of squeeze rollers before entering a restricting channel of decreasing width in which the cloth bunches up. This channel, known as the spout, is fitted with a hinged lid which is raised when the pressure of the advancing cloth becomes sufficiently great. The fabric then falls into the base of the machine and remains there until it again passes through rollers via a throat. During milling the cloth undergoes repeated intermittent compression in the various parts of the machine, but after each compression, fabric is allowed to recover. Some of the features of modern machines include plastic milling rollers, chain drives, hydraulic control of pressure on milling rollers and spout, stainless steel and plastic parts, and large capacity units.

The shrinkage that occurs in fulling and the amount of cover on the fulled cloth depend on many factors. Machine factors include the loads on the spout and the milling rollers, the size of the throat, the nature of the surfaces with which the fabric is in contact, and the dimensions and design of the machine. Factors other than those concerned with the machine include the raw materials from which the fabric is made, the construction of the fabric, the chemical history of the fabric, and the conditions under which the fulling is carried out.

### 8.4.1. Causes of Felting

The primary cause of felting is the scaliness of the wool which gives rise to the directional frictional effect (DFE) (i.e., the difference in friction when the scales oppose or do not oppose the motion of the fibers). Because of this DFE, wool fibers are able to travel in the direction of their root ends when intermittent forces are applied. The progression has been likened to that of an earthworm, and it can be seen that fiber travel is a consequence of localized fiber extension under the action of appropriate forces and recovery when the forces are withdrawn. The felting of a mass of fibers can, therefore, be interpreted[20] as shown in Fig. 8-5. In (a) is represented a mass of fibers, three of which are regarded as operative fibers moving under the influence of the fulling forces. On application of a compressive force the operative fibers travel, and by interlacing and interlocking with other fibers lock the mass in position (c). When all the forces are removed (d), full recovery is not possible, owing to interlocking and the mass of fibers is regarded as felted.

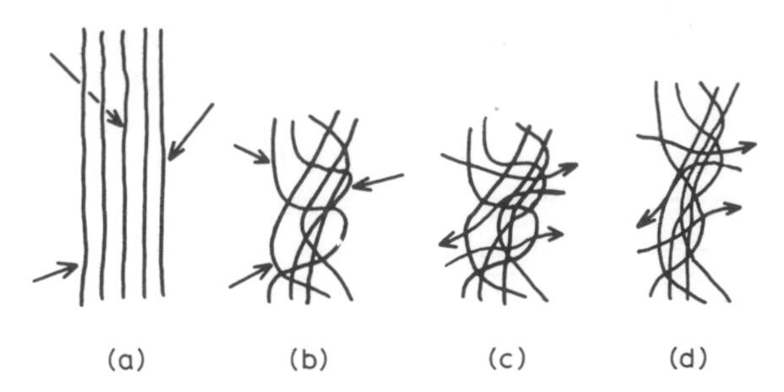


Fig. 8-5. Diagrammatic interpretation of the felting of a mass of fibers. (The fibers indicated by arrows are moving under the felting force.)

It is clear, therefore, that the following fiber properties are important in felting:

1. Scaliness (i.e., $\mu_2 - \mu_1$, where $\mu_2$ is the coefficient of friction when the scales oppose the motion and $\mu_1$ is the coefficient of friction when the fibers move in the opposite direction).
2. The ease of deformation of the fibers.
3. The ease with which the fibers recover from deformation.
4. The freedom of movement of the fibers.

The relative significance of these properties is evident from the following discussion of the influence of various factors on the shrinkage which occurs when cloths are fulled under different conditions. Only wool felts, although certain viscose rayon fabrics consolidate when washed under certain conditions. Fabrics made from the finer wools felt faster and to a greater extent than those made from coarser wools. In general, the lower the quality of the wool, the smaller is the scaliness, although scale profile is important. If the scales are removed by abrasives the fabric does not felt.[21] If the scales are coated with a thin layer of metal so that the profile remains distinct, the material still felts, indicating that the chemical nature of the surface is not the determining factor.[22]

Treatment of wool by processes normally used in conventional cloth finishing have little effect on the felting power of wool, although excessive treatments with hydrogen peroxide, sodium hydrosulfite, or alkalis impair felting. In one series of experiments it was found that wool fabric boiled for 3 h in 0.1 *N* sulfuric acid shrank more readily than untreated wool when the two were milled under the same conditions.

Wool dyed with 30% Solway Green G does not felt[23] because the fibers are more difficult to stretch and have a reduced power of recovery from deformation. Similarly, treatment of wool with crosslinking agents such as benzoquinone, mercuric acetate, or ammonium thioglycollate followed by

**Table 8-9. Fiber Treatment Related to Resistance to Compression of Felt**

| Fiber treatment | Resistance to compression of felt |
|---|---|
| 1. None | 0.72 |
| 2. Roots stiffened with mercuric acetate | 0.76 |
| 3. Tips softened with sodium bisulfite | 0.79 |
| 4. Combination of 1 and 2 | 0.83 |

1,3-dibromopropane prevents the fibers from felting when milled in soap or acid solutions.

The effect of treating different parts of a wool staple is of interest, for the result of felting a mass of fibers treated as shown in Table 8-9 emphasizes the significance of the root ends of fibers in promoting fiber travel during milling.[24]

### 8.4.2. Conditions of Fulling

Cloth felts more readily in the presence of water than when dry. In fact, fabrics do not mill in the absence of liquid water. Acids and soap solutions facilitate milling.

The effectiveness of the various reagents can be explained in terms of their effect on scaliness or elasticity. Thus acids increase the scaliness slightly but make the fibers more easily deformed without adversely affecting their power of recovery.[25]

Alkalis are not so effective as water or acids[25] because although they facilitate deformation, they cause some loss of recovery power at the same time. Simple alkali milling is seldom used. The process normally known as soap milling is more usual, and in this case the effect of the alkali is supplemented by the action of the soap. In a peculiar way the soap acts as a lubricant, and it can be shown that lubricants facilitate milling. For example, ether-extracted wool cloth mills slowly as compared with unextracted cloth.[26] The amount of soap used in commercial practice is about 30 lb of 10% soap solution per 100 lb. of hydroextracted cloth. The type of soap does not seem to be very important except for very heavy milling.[27] It is, therefore, advisable to use a soap that is readily removed in subsequent washing-off wherever possible.

The effectiveness of grease milling was formerly attributed to the formation of soap by the interaction of the wool oil (oleine, a fatty acid) and sodium carbonate, but grease milling in a modern sense is much more complicated because it is sometimes carried out by impregnating the loom-state cloth with a dilute solution of a synthetic detergent. The complex of fatty or mineral oil and the detergent must function as a lubricant. It is of interest that emulsions of liquid paraffin can be used as milling agents,[28] as can solutions of monoethanolamine bisulfite[29] or cationic soaps. Blends of an anionic detergent and a nonionic detergent are more efficient milling agents than either reagent alone.[30]

Whatever the reagent used, the effect of temperature is of considerable importance.[25] Thus in acid milling, the higher the temperature, the greater is the felting, but with alkali or soap milling there appears to be an optimum temperature of about 40°C. This can be explained largely in terms of the effect of temperature on the elastic properties of the wool, for in acid media, ease of deformation increases with temperature while the recovery from deformation is unaffected. In alkaline or neutral media, the ease of deformation increases with temperature, but this is offset by a fall in power of recovery when the temperature exceeds about 40°C.

### 8.4.3. Methods of Preventing Felting

Since the felting of wool arises from the scaliness of the fiber surface and is greatly affected by the ease of extension and power of recovery of the wool fibers, any method of reducing the scaliness or the power of recovery of the fibers or of increasing the resistance of the fibers to extension will be effective in reducing the felting power of the wool. Most commercial methods, however, rely upon reducing the scaliness of the wool either by degrading the scales or the layer of fiber immediately underneath them, or by coating the fiber with some high-molecular-weight substance which forms a resistant film.

The oldest nonshrink process is wet chlorination. Originally, this was carried out by running the cloth in a dilute solution of an acid and then treating it with a dilute solution of sodium hypochlorite containing 3% of available chlorine on the weight of the material. This process certainly rendered the cloth unshrinkable, but it was difficult to control and the results were often uneven. For example, because of the high speed of the reaction between wool and the chlorine, the outside of the yarns may be overchlorinated while the inside is untreated. It is desirable, of course, that the whole yarn should be uniformly treated throughout. The process, therefore, fell into disrepute, although if there is rigid chemical control, it is possible to obtain results which are satisfactory commercially. The limitations of the process did, however, stimulate investigation into the search for alternative methods of chlorination.[31]

One of the most successful is the dry chlorination technique developed by the Wool Industries Research Association.[32] The wool is first conditioned to have a regain of about 8% and is then placed in an autoclave which is evacuated. About 1% of chlorine (on the weight of the goods) is admitted into the autoclave, and this reacts with the wool over a period of 30–60 min. Any residual gas is removed by sweeping out the autoclave with air and the goods are then removed.

It is clearly desirable to confine the chemical reaction selected to bring about the nonshrink effect to the surface of the fibers, since the sole object is to destroy the scaliness of the fibers. This can be done by replacing the aqueous solution of the active component by a solution in an organic solvent. Good results can, for example, be obtained by using solutions of chlorine in carbon tetrachloride.[33] Chlorination is also the fundamental reaction involved in the Dri-Sol process, in which the air-dry goods are immersed for 1 h at room temperature in a 2% solution of sulfuryl chloride in white spirit. Organic solvents are, however, expensive, and other methods of controlling the chlorination have found favor with many finishers. In the modern wet chlorination methods, the conditions are chosen so that the rate of reaction between the wool and the hypochlorite is slow. This is done by careful control of the pH of the solution. In the Negafel process,[34] the chlorination liquor is a mixture of formic acid and sodium hypochlorite and the pH is maintained at about pH 4.5 during the process. The goods are treated with about 3% of chlorine (on the weight of the goods) for about 30 min and are then dechlorinated by immersion in sodium bisulfite solution and thorough rinsing. The Dylan Z[35] process utilizes a mixture of potassium permanganate and sodium hypochlorite at a pH of about 8. Presumably, the permanganate brings about a preliminary fission of the disulfide bonds which makes the fibers susceptible to the action of alkaline hypochlorite. The reagent can be prepared by dissolving 1.5% potassium permanganate (on the weight of the material to be treated) in water and adding to it 7% of calcium chloride (the addition of this compound improves the color of the final product) and sufficient sodium hypochlorite to provide 2.5% of active chlorine. The pH of the mixture is then adjusted to 8.5, usually by adding zinc sulfate. An alternative method of controlling chlorination is to carry out the process in the presence of a resin. This is the basis of the Melafix processes,[36] in which the essential reagents are hypochlorite and an amino resin.

More recently, a great deal of attention has been given to processes based on the reagent dichloroisocyanuric acid (DCCA)[37] marketed under the name of Fi-Chlor[38] (Fison's Ltd.) and Basolan (BASF).[39] DCCA can be used in many different ways, both alone or along with other reagents, and successful procedures have been developed for treating tops, yarns, and fabrics. The pH of the solution must be controlled, and in many ways DCCA behaves as if it were a source of sodium hypochlorite. It has, for example, been suggested that the following reaction takes place:

```
            O                              O
            ‖                              ‖
            C                              C
         /     \                        /     \
Cl—N            N—Cl    H2O        HN           NH
     |          |      ——————>      |           |      + 2NaOCl
   O=C          C=O     NaOH      O=C           C=O
         \     /                        \     /
            N                              N
            Na                             Na
```

Although chlorine is particularly effective in making wool unshrinkable, it is not surprising, in view of the difficulty in carrying out some of the older chlorination techniques on the large scale, that attempts have been made to use other reagents. The number of possible compounds that can render wool fibers nonfelting is considerable, for almost any compound capable of breaking the disulfide linkings in the fiber will affect the felting properties. Not all such compounds are, however, equally effective, and the most successful nonchlorine finishes seems to be based on the use of alkalis or oxidizing agents. In the Tootal process[(40)] and that of Freney and Lipson,[(41)] the active compound is caustic soda. The Tootal method is based on treating the wool with a solution of caustic soda in a mixture of butyl alcohol and white spirit and then rinsing first in dilute acid and finally in water. In the Freney-Lipson process, the alkali is dissolved in industrial alcohol, and it is recommended that the subsequent acidification is carried out with sulfuric acid dissolved in industrial alcohol. Oxidizing agents are used in the process developed by the British Cotton and Wool Dyers' Association.[(42)] The material to be processed is treated with a dilute solution of a copper salt and then with a solution of hydrogen peroxide. The copper catalyzes the decomposition of the peroxide and accelerates the attack on the surface of the fibers.

In another class are the processes based on the treatment of wool with enzymes. The most commonly used enzyme is papain[(43)] and in the method developed by the Wool Industries Research Association, the wool is treated with a solution of this enzyme in the presence of sodium bisulfite, which activates the system. Attack is confined to the surface of the fibers, presumably because the large enzyme molecules cannot penetrate the fiber. More recent reagents include permonosulfuric acid[(44)] used in the Dylan process, which can be applied to tops as well as yarns and fabrics. Another effective treatment is to use potassium permanganate dissolved in a concentrated salt solution.[(45)] Manganese dioxide is subsequently removed from the fabric by means of sodium bisulfite.

An alternative method of rendering the scales ineffective is to coat the fiber with a compound of high molecular weight, or monomers used to form such compounds on the fiber. This not only renders the wool nonfelting but also increases the resistance to wear. One of the earliest methods consisted of treating the fabrics for 6 h at 50°C with a 5% solution of anhydrocarboxy-glycine[46] dissolved in ethyl acetate to which 2% of water has been added. Polyglycine was thus deposited on, and possibly reacted with, the scale surfaces. Fabrics treated in this way are unshrinkable and have improved resistance to wear. In the Lanaset[47] process, methylated methylol melamine–formaldehyde resin is formed on the fiber, while in the Lipson–Jackson[48] process, the fabric is impregnated with an alcoholic solution of nylon which had been treated with formaldehyde in presence of methyl alcohol. Subsequent treatment with acid causes a film of nylon to be deposited on the scales.

Other methods of polymer deposition involve pretreatment with a metal salt (usually $Fe^{2+}$) followed by immersion in hydrogen peroxide and an acrylic monomer.[49] Free-radical polymerization follows and substantial amounts of polymer can be deposited mainly in, but possibly also on, the fibers. An alternative technique is to pretreat the fabric with persulfates[50] or to use cerium catalysts.[51] The methods have been used to deposit polymers such as polyacrylonitrile (PAN) and other related vinyl polymers. To make a fabric unshrinkable, however, deposition of some 25% of PAN is necessary, and this makes the process unattractive.

Another method, known as interfacial polymerization (IFP),[52] found greater acceptance. The finish is known by the trademarks Wurlan or Bancora. A typical procedure consists of impregnating the fabric, first with an aqueous solution of hexamethylene diamine and then with a solution of the acid chloride of sebacic acid in an organic solvent. In later developments 1,3-diaminopropane is preferred to hexamethylene diamine, but in all processes the water content of the bath containing the organic solvent must be controlled. Interaction takes place at the interface formed by the two immiscible solvents on the cloth and a polyamide is formed on the surface of the fibers. Several other polymers can be deposited in this way, but polyamides and polyureas seem to be preferred.

In extensions of this IFP technique advantage is taken of the high chemical reactivity at interfaces. These processes are known as phase-boundary crosslinking (PBLC).[52] They involve crosslinking on the fiber surface of a reactive preformed polymer. Polymer and crosslinking agent are applied successively in immiscible solvents, and crosslinking occurs at room temperature. Suitable polymers are reactive polyethylene derivatives, polyurethanes, polyacrylates, amino polyamides, and polyethers while di- or polyamines are used as crosslinking agents. As an example of the process, 1.5% of a polyurethane containing isocyanate groups, crosslinked by means of a diamine, gives a permanent nonshrink effect.

Preformed polymers can be applied from solvents to alter the properties of wool.[53] Usually soft acrylates with active groups are used, and these may be subsequently crosslinked when heated in the presence of a catalyst. Although the deposition of preformed polymers has been studied over a period of some 40 years, it was not until recently that commercially acceptable results were obtained.

Of particular commercial importance are the processes known as Zeset TP (Du Pont), based on a reactive polyolefin containing acid chloride groups, and Synthappret LKF (BASF), which is a polyurethane. Both reagents can be applied from dry cleaning solvents in conventional dry cleaning equipment. Polymer treatments are much more efficient if the wool is first treated with a reagent such as acidified sodium hypochlorite, which alters the surface of the fiber so that the fiber–polymer union is facilitated. It has been claimed that effective pretreatments alter the critical surface tension of the wool and that almost any polymer can be used to make wool nonfelting provided that the wool has been suitably pretreated. One commercially successful process, the Hercosett process, involves prechlorination followed by treatment with a polyamide–epichlorohydrin reagent known as Hercosett 57.

The polyamide part of this resin is made by condensing adipic acid with diethylenetriamine to form a linear polyamide with secondary amine groups spaced regularly along the polymer backbone.

$$H_2N—CH_2—CH_2—NH—CH_2—CH_2—NH_2 + HOOC(CH_2)_4COOH \longrightarrow$$
$$-[NH—CH_2—CH_2—NH—CH_2—CH_2—NH—CO—(CH_2)_4—CO]_n-$$

This polyamide is treated with epichlorohydrin and this forms crosslinks and reactive "azetidinium cations" as in the following equation:

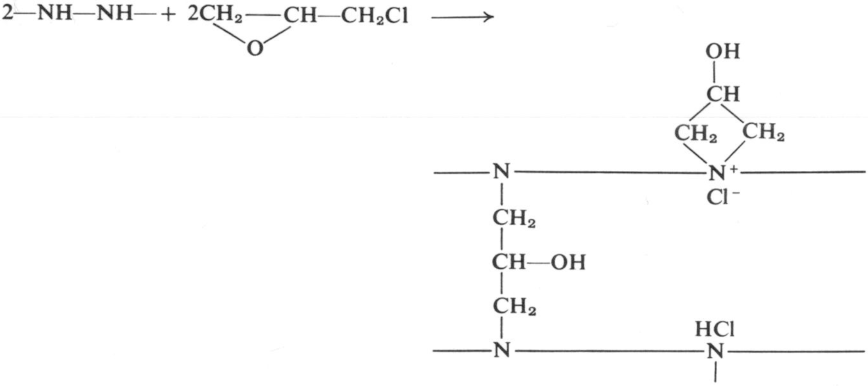

$$2—NH—NH— + 2\,\underset{\diagdown O \diagup}{CH_2—CH}—CH_2Cl \longrightarrow$$

```
                                    OH
                                    |
                                    CH
                                   /  \
                                CH2    CH2
                                   \  /
   ——N——————————————————————————————N+——————
     |                              Cl-
     CH2
     |
     CH—OH
     |
     CH2
     |                              HCl
   ——N——————————————————————————————N———————
                                    |
```

## 8.5. Mothproofing

### 8.5.1. Mothproofing Agents

Many of the earliest were simple fluorides, but these had limited fastness. Halogens were, however, proved to be essential components of several mothproofing agents, and later pentachlorophenol laurate applied along with a mixture of degraded proteins (Catophrene) was found to give adequate protection against attack. In view of the potency of DDT as an insecticide, this compound has found great use in the formulation of mothidicides, both as a solution in an appropriate organic solvent or as a component of a dispersion. The highest degree of fastness is obtained by applying compounds which are attached like wool dyes (e.g., Martius Yellow—2,4-dinitro-α-naphthol). This is colored, but colorless compounds have been developed, such as Eulan N (IG) and Mitin FF (Geigy). Structures of the important compounds are shown in Fig. 8-6. They should be applied in the same manner as acid dyes and can be included, along with the dye, in the normal wool dyeing process.

The main limitation of these processes was the comparatively high price. This was overcome by the observation that fabrics treated with

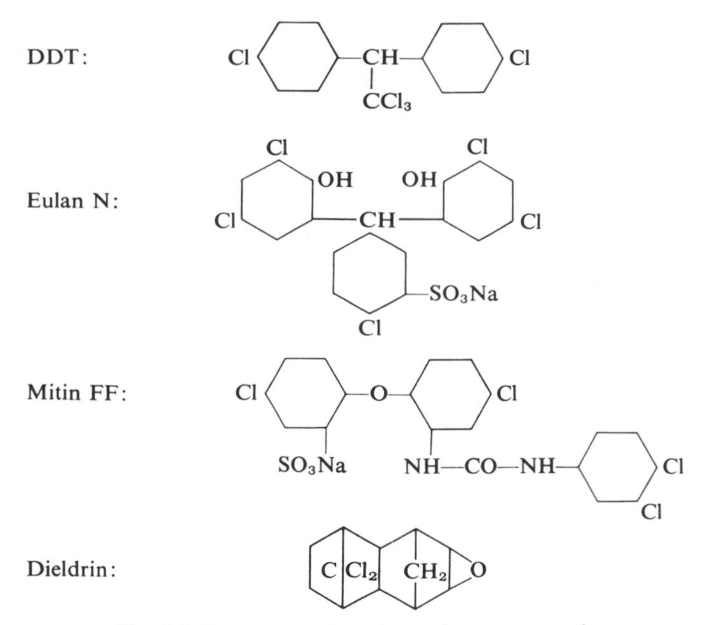


Fig. 8-6. Structures of mothproofing compounds.

Dieldrin were resistant to attack by clothes-moth grubs, and so this reagent attracted a great deal of attention. It is cheap and efficient and was regarded by many as being a complete answer to the mothproofing problem. Unfortunately, however, some doubts have been expressed about its toxicity to humans, especially when applied carelessly or under conditions not recommended by the manufacturers. This has had considerable repercussion, and the finishes based on this reagent are now regarded with some reservation. Research to produce an equally effective reagent without undesirable features is being actively undertaken. Some of the more recent substances examined include *O,O*-diethyl phosphorothionate *O*-ester with phenylglyoxylonitrile oxime (Bay 77488); *O,O*-diethyl-*O*-3,5-6-trichloro-2-pyridylphosphorothioate (Durstran); Allethrin; a mixture of di- and tribromo salicylanilides (Diaphene); and various quaternary ammonium compounds.

## 8.6. Wool Bleaching

This process is necessary for the production of "white" cloths and for preparation of the fabric when pale shades or very bright shades are necessary. The fabric may be impregnated with an aqueous solution (pH 9.2) of hydrogen peroxide (4-vol strength) containing Stabiliser C, then allowed to stand until the bleaching effect is complete. Alternatively, the fabric or yarn is immersed in 2-vol solution for a shorter period.

Bleaching may also be carried out in acid solutions at about pH 4, by adding Stabiliser MP (a fluoride) to the peroxide solution, instead of an alkali. The fabric is then batched and subsequently dried in a tenter, where the bleaching action is completed.

There has been much controversy about the advisability of adding formaldehyde to bleach baths or pretreatments or after treatments involving this reagent. Little evidence is available to support the view that the reagent is effective in commercial processes, although there is no doubt about the ability of formaldehyde to strengthen and protect wool fibers when the treatment is carried out under the appropriate conditions.

The bleaching of pigmented wool and other keratins, such as horse hair and pig bristles, presents special problems. The coloring matter mainly responsible for the many shades of black and brown hairs is termed "melanin" and is found as a rule in the form of granules. The pigment granules in black or brown wools are best isolated by boiling the hair in a mixture of phenol hydrate and thioglycollic acid. Electron microscopic examination of granules isolated by this method reveals that the granules are elongated egg-shaped bodies,[54] the ratio of major and minor axes varying with the

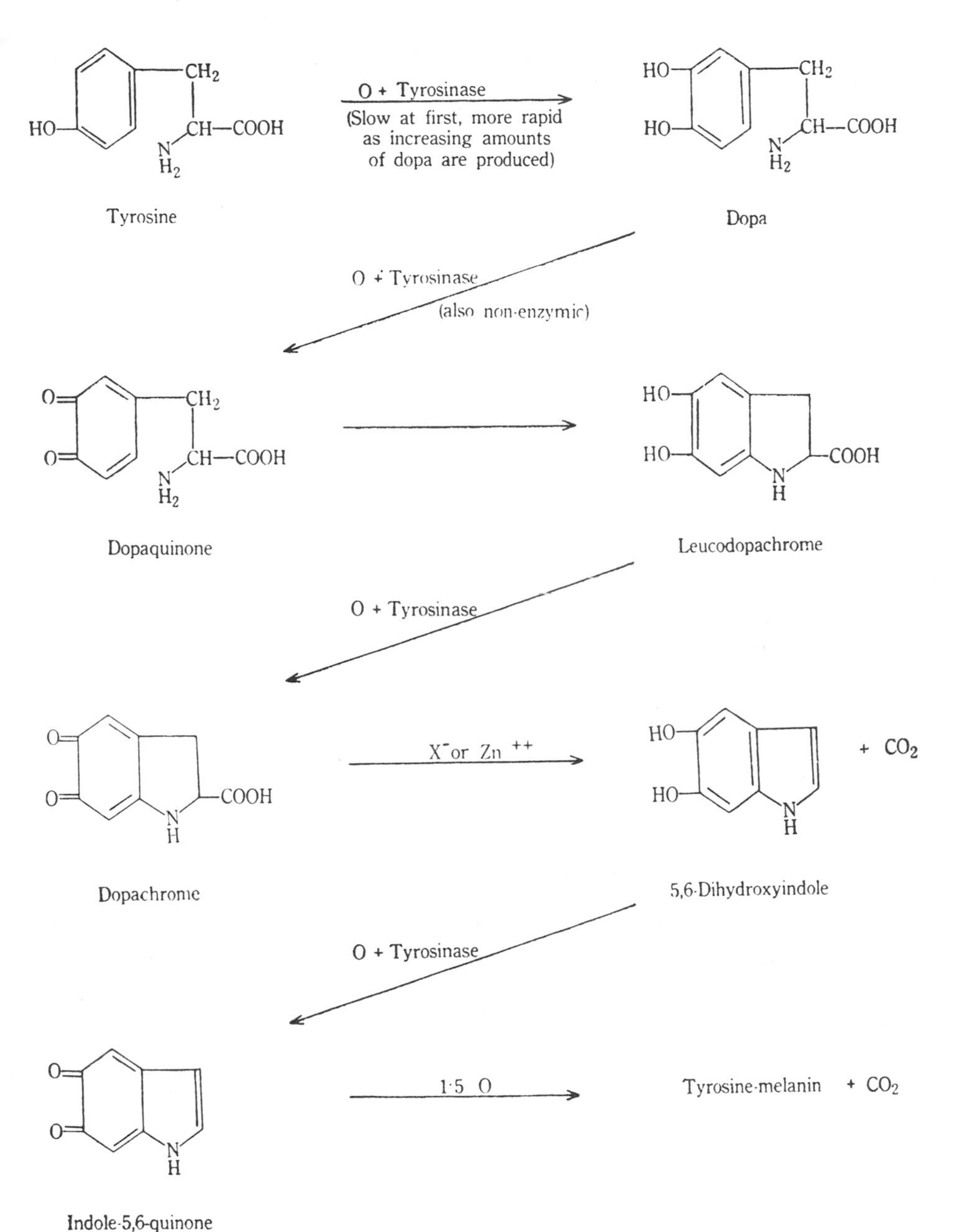


Fig. 8-7. Postulated mechanism of formation of melanin.

origin of the granules. The granules are probably complex in structure, being built up from layers of protein on which are deposited layers of pigment.

The chemical inertness of the natural melanins suggests that they are polymers of high molecular weight, and although the precise structure of the material is not known, it is generally accepted that they are formed by the action of tyrosinase or its derivative to yield indole-5-6-quinone which then polymerizes (see Fig. 8-7). There is also much evidence which suggests that the natural melanins are closely associated with heavy metals, especially iron.

Pigmented fibers often need to be bleached, and this is usually done by mordanting the material with an iron salt and then treating it with a solution of hydrogen peroxide. Few data are available on this process. It was found,[54] however, that when pigmented fibers are treated with iron salts, they become darker. More iron is taken up by pigmented fibers than by unpigmented material, which suggests that the iron is adsorbed by the melanin. More important, however, is the finding that the iron adsorbed by pigmented fibers is more firmly held than that adsorbed by white fibers. This has an important practical consequence, for it indicates the desirability of rinsing mordanted fibers before they are bleached. If this is done, most of the remaining iron will be where it is most required (i.e., absorbed on the pigment granules). When the material is subsequently immersed in hydrogen peroxide, the granules will be preferentially attacked because of the presence of the ferrous iron catalyst and the reaction will be concentrated on the granules. This will minimize the degradation which occurs when keratin is treated with hydrogen peroxide in presence of iron salts.

## 8.7. Durable-Press Finish for Wool

This type of finish has been particularly successful in the finishing of cotton–polyester garments and consists in impregnating fabric with a reactive resin (usually dimethylol dihydroxycyclic ethylene urea) and a suitable catalyst and softener, then making up the garment, after which the pressed garment is baked at temperatures around 160°C to fix the creases, pleats, etc. The garment retains its fresh appearance throughout its useful life and can be washed and dried without being seriously affected. As would be expected, efforts have been made to develop a similar finish for wool fabrics. A first attempt was made several years ago, the process involving presensitizing the cloth with monoethanolamine sulfite. The standard of durable press attained was not particularly high; nevertheless, the principle was firmly established. More modern processes meet higher standards. The effect produced by setting shrink-resisted wool by steaming in the presence of a reducing agent is not sufficiently durable, as creases and the like, previously inserted during making up, are removed or become significantly less sharp when the garment is subsequently machine-washed and tumble-dried. This is because the hydrogen bonds and disulfide rearrangements responsible for set are reversed during the laundering cycle.

The set can, however, be established by (1) inhibiting thiol–disulfide interchange by removing thiol groups or converting disulfide bonds into —S— or —S—$CH_2$—S— bonds, (2) introducing new cross links, or (3) depositing reactive polymers.

Commercial finishes may include one or all of these, and a typical procedure involves setting nonshrink wool with a reducing agent such as sodium bisulfite and then crosslinking with formaldehyde or a compound which yields formaldehyde. In another process the garment is treated with a reactive polymer, and then the steam-pressed garments are set in a steam oven using slightly superheated steam. Shrinkproofing and stabilization of set can be attained by pressing the garment in the usual way, then treating with a solution of a reactive polyurethane in perchlorethylene, and finally hanging in a steam oven. A similar result is obtained by using an aziridine-terminated polyetherpolyurethane. The most suitable polymers have a poly(butylene glycol) backbone terminated with tolylene diisocyanate and finally end-capped with ethylene imine. The polymer is dissolved in a hydrocarbon solvent, emulsified with a nonionic surfactant, and 1–2% sodium bisulfite is added. The garment is padded with this mixture, dried gently, pressed, and baked. Creases inserted in the garment are retained after five launderings. A process developed by Deering Milliken Limited consists in applying to the garment a solution of a polyurethane containing reactive isocyanate groups, then reducing the —S—S— linkages in the wool to facilitate —S—S—/—SH interchange, and finally stabilizing the rearranged internal fiber structure in the pressed garment by means of formaldehyde or a carbamate.

## 8.8. Wrinkle Resistance

Although the wrinkle resistance of most woven and knitted wool fabrics is good, some lightweight fabrics do not have the high wrinkle resistance expected in high-quality goods. The problem is a difficult one to solve, and the well-established methods for improving the crease resistance of cellulosic fabrics are not successful when applied to wool fabrics. In recent years increasing attention has been given to methods for increasing the crease resistance of lightweight wool fabrics. Storage of the cloth for long periods at room temperature or heating the fabric at a constant regain improves the resistance to creasing. The process has been termed "annealing" and is attributed to rearrangement of hydrogen bonds. Chemical crosslinking concurrent with or subsequent to annealing greatly improves resistance to wrinkling, multifunctional crosslinking compounds capable of forming a crosslinked polymer network medium and attached to the wool being preferred.

## 8.9. Oil- and Water-Repellency

Although many wool fabrics exhibit a limited degree of water-repellency, this is not usually sufficient to enable them to be classified as water-repellent or even showerproof. Wool fabrics can be made truly waterproof only by being coated with a continuous layer of water-repellent material (e.g., rubber or a synthetic resin). Water-repellent or showerproof wool fabrics are, however, in demand, and to obtain the best results, a suitably constructed fabric must be selected. These are frequently gabardines or felted fabrics.

Several reagents are used to obtain the appropriate hydrophobic surface on the fabric. They include aluminum acetate, aluminum soaps, and waxes applied from either organic solvents or aqueous dispersions. Modern commercial wax emulsions usually contain aluminum acetate. Other dispersions, such as Mystolene KP (Catomance Ltd.), are applied along with aluminum formate, and a protein–aluminum complex is formed on the surface of the fabric. None of the above reagents is fast to dry cleaning, even with pure solvents, but Velan PF—stearamidopyridinium chloride—gives results which are acceptably fast and without adversely affecting the handle. The reagent is applied in presence of sodium acetate by a pad–dry–bake sequence.

Excellent water-repellency is realized by treating the fabric with silicones [e.g., Silicone Finish MS 2207 (Midland Silicones) and reagents such as I (8-6) have a strong affinity for the fiber and render it water-repellent].

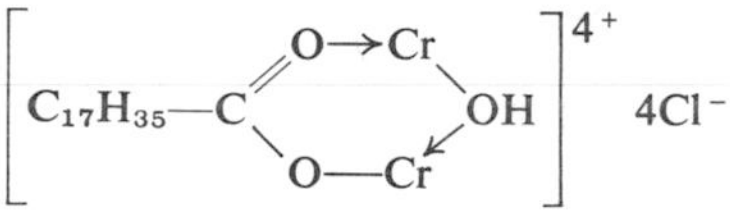

$$\left[ C_{17}H_{35}\text{—}C \begin{matrix} \overset{\diagup\!\!\diagup}{} O \rightarrow Cr \diagdown \\ \qquad\qquad\quad OH \\ \diagdown O\text{—}Cr \swarrow \end{matrix} \right]^{4+} 4Cl^- \qquad (8\text{-}6)$$

Fluorochemicals endow wool with excellent oil-repellency, and some make the fiber both oil- and water-repellent. The effects are associated with the presence on the fiber of low-energy fluorinated groups, and large numbers of fluorinated compounds have been suggested, including PFOH [poly-(1,1-dihydroxyperfluorooctylacrylate] and PFS [poly(3-heptafluoroisopropoxy)propyl silane]. They are, however, expensive but can also be effective when mixed with the cheaper polyurethane in the proportion 20 parts fluorochemical and 80 parts polyurethane.

Closely related to oil- and water-repellent finishes is a group of oil-stain-resistant and soil-release finishes. The soil resistance and soil release of wool is better than some other fibers, but it is desirable that modern textiles should have a high resistance to staining and soiling, and when soiled should be readily cleaned (i.e., their soil-release properties should be good). Application of silicones and other water-repellent finishes increases the stain repellency of wool fabrics, owing to their effect on the "pearling" of drops of aqueous solutions on the fabric. As would be expected from the data given in Table 8-1, it would be expected that "degradative" shrink-resist processes would improve overall soil resistance, but resin-based nonshrink finishes can adversely affect soiling because of the oily nature of the soil. Block copolymers of fluorocarbons and polyethylene oxides should improve soil repellency and soil release on washing.

## 8.10. Polymer Grafting

The grafting of polymers onto wool is important not only in respect to rendering the material nonfelting but also as a means of modifying both the bulk and surface properties of the fiber to give the fabrics other desirable properties. The grafting can be done in several ways. The most common chemical method involves free-radical polymerization and is specially applicable to acrylic polymers and related materials. The mechanism involved in these reactions is illustrated by reference to the iron-peroxide system, which proceeds as follows.

The fabric is impregnated with a dilute solution of ferrous ammonium sulfate, rinsed, squeezed, and then a solution or emulsion containing the monomer and hydrogen peroxide is added. When the latter diffuses into the fiber, hydroxyl radicals are produced and these initiate polymerization.

$$Fe^{2+} + H_2O_2 \longrightarrow Fe^{3+} + OH^- + \underset{\text{Hydroxyl radical}}{OH}$$

$$OH + CH_2{=}CHX + CH_2{=}CHX \longrightarrow HO{-}CH_2{-}CHX{-}CH_2{-}CHX$$

To obtain the best results, oxygen should be excluded, and this can be greatly facilitated by having THPC $[P^+(CH_2OH)_4Cl^-]$ present as an oxygen scavenger.

Persulfates can be applied to the wool to produce free radicals in the fiber by interaction with the —S—S— bonds, and effective polymerization can be brought about by the use of cerium salts. Of special interest is dye-sensitized polymerization. This is effective for polymerizing acrylamide, and anthraquinone or riboflavine can be used as sensitizers. Polymerization is brought about by irradiation in an ultraviolet light source for 2 h. The reaction involves radical abstraction of hydrogen from the wool by a radical dye intermediate, followed by initiation of photopolymerization from these sites on the wool. Polymers can also be formed by exposing the monomer-treated material to $\beta$ or $\alpha$ rays.

A more convenient way is, of course, to apply a solution or emulsion of a preformed polymer, but the effectiveness of this is greatly affected by the wetting behavior of the wool. The polymer should spread easily on the surface, and in this respect application from solutions in organic solvents is often preferred to treatment with aqueous dispersions. Spreading will occur if the critical surface tension ($\gamma_c$) of the polymer coating is less than that of wool. There is considerable evidence to support the presence of an adsorbed water layer on the fiber surface; and since water has a high value for $\gamma_c$, 22 dyn/cm, spreading of polymer is often difficult from aqueous media. The effect of chemical treatment on the critical surface tension of wool is therefore often important in determining the effectiveness of polymer finishes.

Some properties of polymer-coated fabrics are shown in Tables 8-10 to 8-13.

**Table 8-10. Reactions Used to Form Polymers on Wool by Interfacial Polymerization Techniques**

Polyamides

$$H_2N—R—NH_2 + Cl—CO—R'—CO—Cl \longrightarrow —CO—NH—R—NH—CO—R'—CO—NH—$$

Polyurethanes

$$H_2N—R—NH_2 + \underset{(bis\text{chloroformate})}{Cl—CO—O—R'—O—CO—Cl} \longrightarrow —NH—R—NH—CO—O—R'—O—CO—$$

Polyureas

$$H_2N—R—NH_2 + OCN—R'—NCO \longrightarrow —NH—R—NH—CO—NH—R'—NH—CO—$$

Polyesters

$$HO—R—OH + Cl—CO—R'—CO—Cl \longrightarrow —O—R—O—CO—R'—CO—$$

Polycarbonates

$$HO—R—OH + Cl—CO—O—R'—O—CO—Cl \longrightarrow —O—R—O—CO—O—R'—O—CO—$$

**Table 8-11. Some Reactions Used for PBLC of Polymers**

| Polymer | Crosslinking agent |
|---|---|
| Polyamide | Diisocyanate |
| Polyol | Diisocyanate |
| Polyolefine with reactive sulfonyl halide or carbonyl chloride groups | Diamine |
| Polyurethane with reactive isocyanate groups | Diamine |
| Polysiloxane with free primary amino groups | Diisocyanate |
| Polyfluoroacrylate with reactive carbonyl chloride groups | Diamine |

## 8.11. Carbonizing

The object of carbonizing wool piece goods is to remove cellulosic material, which may be present in the form of burrs or cellulosic fibers. The process consists in impregnating the fabric with sulfuric acid (4–8° Tw), drying at 70°C, baking at 110°C, and then running the fabric dry in a milling machine to convert the degraded cellulose into dust. The fabric is finally washed in water, and neutralized. Carbonizing is done either before or after dyeing. Clearly the more economical process, from a commercial point of view, is to carbonize fabric before it is dyed, because this facilitates rapid delivery of material dyed the color required by the customer. If the fabric is dyed before carbonizing, there may be considerable delay before delivery, because the piece has to be dyed, carbonized, and finished after receipt of the order. It is therefore desirable to stock carbonized pieces so that ranges of fashionable colors can be produced at short notice. It is, however, essential to ensure that minimum adverse modification takes place during storage. If dyed pieces are to be carbonized, the dyes must be fast to carbonizing.

The carbonizing process is difficult to control, but probably the most important aspect is to ensure that the pieces are clean before being carbonized. The presence of soap or oil residues is undesirable. Clearly the aim of good processing is to obtain maximum degradation of the cellulose with minimum adverse modification of the wool.

**Table 8-12. Effect of Polymers on the Felting Shrinkage of a Standard Worsted Fabric**

| Polymer (% on weight of wool) | dyn/cm | Method of application | Pretreatment | Area felting shrinkage after standard test |
|---|---|---|---|---|
| | | (Untreated fabric) | | (40) |
| | | (Pretreated with 0.4% bromate/salt) | | (33) |
| Very soft polyacrylate (Primal K3), 4.9 | 35 | Aqueous dispersion, 1% $H_2SO_4$ catalyst, dried at 100°C, cured 5 min at 130°C | — | 10<br>4 |
| Very soft polyacrylate (Primal K3), 1.2 | | | 0.4% bromate/salt | 4 |
| Polyacrylate (Primal E168), 7.1 | 47 | As for Primal K3 | — | 34 |
| Polyacrylate (Primal E168), 3.5 | | | 0.4% bromate/salt | 4 |
| Polyacrylate with low crosslinking capacity, 3.8 | 35 | As for Primal K3 | — | 32 |
| Polyacrylate with low crosslinking capacity, 3.0 | | | 0.4% bromate/salt | 19 |
| Polyethylene, 6.1 | 31 | Toluene solution dried at 100°C, cured 2 min at 160°C | — | 16 |
| Polyethylene, 4.2 | | | 0.4% bromate/salt | 5 |
| Polyvinyl acetate (Gelva Emulsion TS100), 5.2 | 38 | Aqueous dispersion dried at 100°C | — | 30 |
| Polyvinyl acetate (Gelva Emulsion TS100), 4.6 | | | 0.4% bromate/salt | 21 |
| Polyamide–epichlorhydrin (Kymene 557), 10.2 | | Aqueous solution dried at 100°C | — | 28 |
| Polyamide–epichlorhydrin (Kymene 557), 3.0 | | | 0.4% bromate/salt | 3 |
| Polyolefin (Surlyn T) | 28 | Toluene solution dried at 100°C | — | 5 |
| Polyolefin (Surlyn T), 1.6 | | | 0.4% bromate/salt | 2 |
| Polyethyleneimine, 9.4 | | Aqueous solution dried at 100°C | — | 31 |
| Polyethyleneimine, 2.8 | | | 0.4% bromate/salt | 5 |

**Table 8-13. Properties of Woollen Fabric Treated with Various Polymers**

| Polymer (% on weight of wool) | Pretreatment | Area shrinkage after 1 h washing test (%) | Warp tensile strength, 1-in grab test (kg wt) | Abrasion resistance, no. of rubs to hole on Martindale Tester | Tear strength (g wt) | Flexural rigidity[a] (mg cm) | Wrinkle recovery (%) |
|---|---|---|---|---|---|---|---|
| (Untreated fabric) | | (49) | (11.8) | (8,500) | (500) | (85) | (66) |
| (0.5% bromate/salt) | | (44) | (12.4) | (7,600) | (470) | (80) | (65) |
| Very soft polyacrylate (Primal K3), 3.5 | — | 5 | 12.9 | 20,200 | 486 | 104 | 64 |
| Very soft polyacrylate (Primal K3), 2.7 | 0.5% bromate/salt | 1 | 13.8 | 17,400 | 461 | 117 | 63 |
| Hard polyacrylate (Primal HA16), 2.8 | — | 6 | 15.9 | 14,700 | 395 | 391 | 51 |
| Hard polyacrylate (Primal HA16), 2.8 | 0.5% bromate/salt | 1 | 16.7 | 12,100 | 389 | 667 | 53 |
| Polyamide–epichlorhydrin (Kymene 557), 4.7 | — | 12 | 13.2 | 18,200 | 437 | 354 | 56 |
| Polyamide–epichlordydrin (Kymene 557), 2.4 | 0.5% bromate/salt | 4 | 15.0 | 15,500 | 448 | 272 | 55 |
| Polyolefin (Surlyn T), 2.6 | — | 1 | 13.7 | 14,000 | 441 | 216 | 59 |
| Polyethyleneimine, 2.8 | 0.5% bromate/salt | 2 | 12.9 | 11,000 | 464 | 124 | 63 |

[a] Measured on fabrics carefully wet-out by hand and air-dried after polymer application.

Under certain conditions (e.g., when the concentration of acid is high) the carbonized wool absorbs less dye than the uncarbonized material when the two are dyed together. This was formerly attributed to interaction between sulfuric acid and —$NH_2$ groups in the wool, but more recent work suggests that serine residues are sulfated, and there is reaction involving the tyrosine residues.[55] Both these increase the acid "environment" of the wool and make more difficult the reaction between acid dyes and the basic groups in the wool. The reaction between wool and sulfuric acid has been considered earlier, and in this connection the work of Elliott and his colleagues[56] is particularly important.

As to the change taking place in carbonizing, the work of Zahn et al.[57] throws considerable light on specific difficult problems. It was shown that carbonization of wool does not immediately increase its alkali solubility if determined immediately after baking. Provided that moisture is completely excluded, the carbonized material can be stored for several weeks without further adverse modification, but under normal humidity conditions, progressive increase in alkali solubility, as well as progressive increase in amino end groups, begins after 4 days of storage. There is also a decrease in the serine and threonine amino end groups. These changes are attributed to an N—O— peptidyl shift:

$$\begin{array}{ccc} \begin{array}{l} \quad | \\ \quad HC\text{———}CH(CH_3) \\ \quad | \qquad\qquad | \\ \quad NH \qquad\quad OH \\ \quad | \\ O{=}C \\ \quad | \end{array} & \underset{-H^+}{\overset{+H^+}{\rightleftharpoons}} & \begin{array}{l} | \\ HC\text{———}CH(CH_3) \\ | \qquad\qquad\quad | \\ {}^+NH_3 \qquad\quad O \\ \qquad\qquad\quad\;\; | \\ \qquad\qquad O{=}C \end{array} \\ N\text{-Peptidyl} & & O\text{-Peptidyl} \end{array}$$

Zahn was able to explain several practical observations in these terms; for example:

1. Preheating before baking results in low alkali solubility, because in the absence of moisture, no irreversible hydrolysis takes place. The sulfuric acid causes only a peptidyl shift, which is reversed on neutralization.
2. Storing carbonized pieces under humid conditions increases alkali solubility, because in the presence of moisture, the *O*-peptidyl or ester forms of the peptides in the wool are hydrolyzed by the acids.
3. Carbonized and carefully neutralized wool must be handled with care during subsequent processing because there is always some permanent breaking of peptide bonds.

To minimize degradation during carbonizing other workers advocate addition of surface-active agents. These also assist in "wetting out" the cellulosic impurities with the acid solution.

## 8.12. References

1. *British Wool Manual* (H. Spibey, ed.), pp. 409–486, Columbine Press, Buxton (1970).
2. N. K. Adam, *J. Soc. Dyers Colourists* **53**, 121 (1937); R. C. Palmer, *J. Soc. Chem. Ind.* (*Lond.*) **60**, 59 (1941).
3. K. Durham, *Surface Activity and Detergency*, Macmillan Ltd., London (1961).
4. J. C. Stewart and C. S. Whewell, *Textile Res. J.* **30**, 912 (1960).
5. J. B. Speakman and N. H. Chamberlain, *Trans. Faraday Soc.* **79**, 358 (1933).
6. C. S. Whewell, A. Charlesworth, and R. L. Kitchen, *J. Textile Inst.* **40**, T769 (1949).
7. H. Moxon, *J. Huddersfield Textile Soc.* 139, (1950–1951).
8. J. B. Speakman et al., *J. Soc. Dyers Colourists.* **52**, 335, 380 (1936); **57**, 73 (1941). R. S. Asquith and J. B. Speakman, *Nature* (*Lond.*) **180**, 502 (1957).
9. A. J. Farnworth, *Textile Res. J.* **28**, 453 (1957).
10. R. W. Burley, *Proc. Intern. Wool Textile Res. Conf., Australia* **D**, 88 (1955).
11. K. Ziegler, *Proc. 3rd Intern. Wool Textile Res. Conf., Paris* **2**, 403 (1965).
12. J. B. Speakman, *J. Soc. Dyers Colourists* **52**, 335 (1936).
13. C. S. Whewell, *J. Soc. Cosmetic Chem.* **12**, 207 (1961). S. Blackburn and H. Lindley, *J. Soc. Dyers Colourists* **64**, 305 (1948).
14. L. A. Holt, S. J. Leach, and B. Milligan, *Textile Res. J.* **39**, 705 (1969)
15. L. S. Bajpai, C. S. Whewell, and J. M. Woodhouse, *Nature* (*Lond.*) **187**, 603 (1960); *J. Soc. Dyers Colourists* **77**, 193 (1961); A. D. Jenkins and L. J. Wolfram, *J. Soc. Dyers Colourists* **79**, 35 (1961); L. J. Wolfram, *Proc. 3rd Intern. Wool Textile Res. Conf., Paris* **2**, 505 (1965).
16. J. B. Speakman, *J. Soc. Dyers Colourists* **52**, 335 (1936).
17. J. B. Speakman, J. L. Stoves, and H. Bradbury, *J. Soc. Dyers Colourists* **57**, 73 (1941).
18. L. S. Bajpai and C. S. Whewell, *J. Soc. Dyers Colourists* **77**, 350 (1961).
19. J. H. Elliott, C. B. Stevens, and C. S. Whewell, *Proc. 3rd. Intern. Wool Textile Res. Conf., Paris* **3**, 431 (1965).
20. A. J. P. Martin, *J. Soc. Dyers Colourists* **60**, 325 (1944).
21. J. B. Speakman and E. Whewell, *J. Textile Inst.* **36**, T48 (1945).
22. J. B. Speakman and H. M. S. Thompson, *Nature* (*Lond.*) **157**, 804 (1946).
23. J. B. Speakman, J. Menkart, and W. T. Liu, *J. Textile Inst.* **35**, T41 (1944).
24. J. Menkart and J. B. Speakman, *J. Soc. Dyers Colourists* **64**, 14 (1948).
25. J. B. Speakman, E. Stott, and H. Chang, *J. Textile Inst.* **24**, T273 (1933).
26. R. V. Peryman and J. B. Speakman, *J. Textile Inst.* **41**, T241 (1950).
27. C. S. Whewell, *J. Huddersfield Textile Soc.* 96 (1949–1950).
28. K. Dabanovitch and C. S. Whewell, unpublished observation.
29. A. Ogawa and C. S. Whewell, *J. Textile Inst.* **53**, 378 (1962).
30. R. C. Cheetham and C. S. Whewell, *Commun. 1st World Congr. Surface Active Agents, Paris* (1954).
31. *Wool Sci. Rev.* 18, 18 (1953); A. D. Sule, R. Y. Churi, V. S. Kenkare, and C. A. Shetty, *Proc. First All-India Wool Res. Conf.* **VI**, 1 (1970).
32. W.I.R.A., Brit. Pat. 417,719; 475, 742; 493,098.
33. C. S. Whewell and A. Selim, *J. Soc. Chem. Ind.* (*Lond.*) **63**, 121 (1944).
34. E. Clayton and W. A. Edwards, Brit. Pat. 537,671.
35. Brit. Pat. 569,730.
36. Ciba, Brit. Pat. 675,137; 704,896; 709,377; 711,861.
37. *Wool Sci. Rev.* 341 (July 1968).
38. R. E. Ashmore, *Wool Rec., Dyeing Finishing Surv.*, (March 15, 1968), 33.

39. K. Reincke, *Textil-Praxis* **21**, 684 (1966).
40. Brit. Pat. 583,396.
41. Council for Scientific and Industrial Research, Australia, *Pamphlet 94* (1940).
42. Brit. Pat. 614,966.
43. W. R. Middlebrook and H. Phillips, *J. Soc. Dyers Colourists* **57**, 137 (1941); Brit Pat. 513,919.
44. Stevenson's (Dyers) Ltd., Brit. Pat. 538,429; 692,258; 716,806.
45. J. R. McPhee, *Textile Res. J.* **30**, 249, 358, 538 (1960).
46. A. W. Baldwin, T. Barr, and J. B. Speakman, *J. Soc. Dyers Colourists* **62**, 4 (1946).
47. American Cyanamid, Brit. Pat. 623,361; 679,386.
48. Brit. Pat. 692,380; 692,381.
49. M. Lipson and J. B. Speakman, *Nature* (*Lond.*) **157**, 736 (1946).
50. B. K. Lohani, L. Valentine, and C. S. Whewell, *J. Textile Inst.* **49**, T265 (1958).
51. S. Howarth and J. R. Holker, *J. Soc. Dyers Colourists* **82**, 257 (1966).
52. *Wool Sci. Rev.* 36, 2 (May 1969).
53. *Wool Sci. Rev.* 37, 37 (Oct. 1969)
54. G. Laxer and C. S. Whewell, *Proc. Intern. Wool Textile Res. Conf., Australia*, **F**, 186 (1955).
55. R. L. Elliott, R. S. Asquith, and B. J. Jordan, *J. Soc. Dyers Colourists* **77**, 345 (1961).
56. R. L. Elliott, R. S. Asquith, and D. H. Rawson, *J. Soc. Dyers Colourists* **74**, 178 (1958); **73**, 424 (1957); **76**, 222 (1960).
57. H. Zahn et al., *J. Soc. Dyers Colourists* **76**, 226 (1960).

# 9

# Other Animal Fibers

**Fred Kidd**

## 9.1. Introduction

Many of the fibers discussed in this chapter are used or have been used as brush filling materials, and books about the brush industry are rare. Prior to 1957 only three are known to have been published and all are out of print.[1-3] Insofar as the chemistry of animal fibers is concerned, they are of very limited interest. Some brush filling materials are obtained from fur-bearing animals, and furs are described in books by Links[4] and Bachrach,[5] both of which provide much background information relating to the furriers' work; the latter also includes some information on degreasing, dyeing, and bleaching. Frohlich's[6] account covers all aspects of chemical processing in detail.

As would be expected, human hair has been extensively studied, but it is not necessarily typical of animal hairs, nor is the treatment it receives while on the human head by comparison with, say, wool on a sheep. Only limited chemical research work has been published on the specialty fibers used in textiles, except for mohair.

Since the funds available will largely determine the volume of research done, it is not surprising that Merino wool has tended to be regarded as typical of animal hairs and as a standard against which others are compared. It has no medulla and the cuticle is so thin that for many purposes it may be considered as almost pure cortex. Medullation is, however, common in most animal hairs and the cuticle forms a significant proportion of some fibers.

---

**Fred Kidd** • Scottish College of Textiles, Galashiels, Scotland

Rudall,[7] for example, showed that in kolinsky guard hairs, a cuticle 7 μm thick consisted of 18 scales; these could be separated by treating a cross section with sodium hypochlorite. Several authors[8–16] have drawn attention to the thick cuticle in pig's bristle. Sikorski and Simpson[17] stained cross-sections with sodium plumbite prior to electron microscopy and showed 15 scales in a cuticle 5 μm thick.

Such physical differences between fine wool and other fibers may well be chemically significant, and it is important that sources of information about physical structure should be summarized. Further than that, if one wishes to study a particular physical feature of animal hairs, it is helpful to know in which hair it is most readily seen. Wildman, Kraus, von Bergen and Kraus, Hardy and Hardy, Newman, and Kidd, in the publications already referred to,[8–16] have given photographs of the cross sections of various animal fibers and Klenk and Scholzel,[18] Slack,[19] Ryder,[20] Appleyard,[21] Bachrach,[5] and Stoves[22–25] have all made significant contributions. Stoves[24] drew particular attention to forensic aspects of the physical structure. The shape and physical structure of human hairs have been widely established by various authors[26–28] in addition to others mentioned earlier.

A summary of scale patterns is to be found in publications already mentioned and in papers by Hausman,[29] who also classified medullas; his scheme differs from Wildman's.[9] Burns[30] has suggested that rippled and crenate-waved patterns derive from smooth-margined scales by wear, basing her observations on wool, human, rat, and kolinsky hairs; weathering as well as abrasion must be a factor. Holmes[31] has carried out detailed studies of cuticular scales on human hair. Wolfram[32] has separated a tubular fiber sheath 200 nm thick from hair as a residue following treatment in ethylene glycol. If the hair is first treated with hypochlorite (to which epicuticle is highly resistant), no fiber sheath is left. From the sheath only 30% of the weight was recoverable after acid hydrolysis. Piper's[33] work with the electron microscope revealed numerous interlocking structures in vicuna and guanaco and human nostril hairs which serve to attach the cuticle to the cortex by protuberances from the latter. In the nostril hair, interlocking of the scales was also apparent. Stoves[34] undertook what is now regarded as a classical study of the chemical reactivity of the different structural parts of the fiber. The work involved using 5 μm cross sections of the primary hairs of the rabbit and hare, which consist mainly of medulla, and the primary hairs of mustelidae such as the kolinsky, which have well-defined cuticle, cortex, and medulla. These cross sections were treated with various reagents, and it was shown that the medulla was most resistant and the cortex least resistant to alkali and alkaline reducing agents. The medulla could be dissolved in 15 min in buffered trypsin solution at 35°C, whereas the cuticle and cortex were intact after 3 weeks of exposure.

While the Allwörden reagent on whole fibers produces the well-known sacs, treatment of cross sections caused the cuticle to swell markedly and to separate into platelets; the medulla collapsed and the cortex swelled. Treatment of cross sections with Millon's reagent (nitric and nitrous acid and their mercury salts) or with the Pauly diazo test gave an intense stain in the medulla but only a faint stain elsewhere, indicating that tyrosine was present in the medulla. When cross sections were boiled for 1 h in a solution of an acid dye (Kiton Red G), the cortex was stained preferentially, whereas a basic dye (Safranine B) stained the medulla preferentially, because it contains more acid groups than the cortex. However, at pH 3 the acid dye gave uniform staining, and the same was true of the basic dye at pH 8.9. With acid dye at pH 8.9 or basic dye at pH 3, little staining occurred. When cross sections of kolinsky, ermine, and horsehair were treated with a 5% solution of sodium sulfide or hydroxide, the cortex dissolved first and the medulla last.[35]

For many purposes, then, one should consider the chemistry of the different structural parts of the animal fiber and not just the whole fiber, although for commercial purposes the chemical behavior of the whole fiber or of its "weakest link" may be more important. In addition, Stoves[36,37] has drawn attention to the supercontraction and setting properties of primary mammalian hairs, an aspect that will be considered later in more detail. Their behavior was then considered to be "abnormal" in that they behaved differently from wool.

## 9.2. Hair Growth

Physical structure and hair growth are related, and while hair growth in the usual laboratory animals, such as rabbits,[38] rats, and mice, and wool growth in domesticated sheep has been widely studied, it is only recently that much has been done with other animals. Since this has been covered by Lyne and Short,[39] it will suffice to mention that information relating to mink, rabbit, nutria, otter, chinchilla, weasel, silver fox, stoat, seals, ferret, European wild horse, cattle, and rats and mice is to be found in that volume. Stoves[25] gives a summary of growth and structure.

Inner root sheath chemistry has been studied by Jarrett[40] and by Rogers,[41] who recorded only 1% cystine and a large proportion of citrulline.

Loss of hair, following injection of cyclophosphamide, in poodles, sheep, and guinea pigs is the subject of a recent paper by Holman et al.[42]

## 9.3. Fibers as Keratins

Wool is a convenient source of α-keratin for research purposes, and certain authors[43] suggest that parallel studies on hairs lead to the same conclusion with minor qualifications. Other workers have found it convenient to use fibers such as human hair or horsehair because of special features such as their greater diameter. Horsehair has been used by Elod and Zahn[44] to study elasticity and thermal expansion; by Lloyd and Garrod[45] to investigate the effect of temperature on the load-extension curve in formic acid; and by Bendit,[46] who found this hair to be a convenient fiber to study the ir absorption bands of the side chains in α-keratin by a technique involving deuteration. The same author used longitudinal sections of horsehair to study the uv absorption spectrum.[96]

Pautard and Speakman[47,48] prepared a continuous β-keratose fiber by oxidizing horsehair or human hair with peracetic acid and studied its reversible supercontraction in solutions of known pH. Bendit[49] also used horsehair to study the effect of concentrated formic acid and dilute monochloroacetic acid on keratin, neither of which destroyed the α-helix structure or the corresponding 5.15 Å main meridional reflection, although prolonged exposure to the formic acid weakened the reflections.

Human hair has been used by various authors[50,58] to represent α-keratin, of which it is one form; there is a considerable volume of evidence to indicate that it should not be considered to represent other forms, such as wool. Menkart et al.[51] noted very large differences in the physical and chemical properties of human hairs (Caucasian and Negro) compared with Lincoln wool, for example. Slower rates of diffusion of reagents into the hair and its greater chemical resistance are due to the thicker cuticle[52,53] and more extensive crosslinking in particular. Their investigations were particularly wide ranging. Holmes[54] has studied diffusion processes in human hair.

The differences in amino acid content between human hair (12.07% half-cystine) and Lincoln wool (7.18% half-cystine) are sufficient, according to Whitmore,[55] to account for significant differences in behavior on stretching. For example, human hair treated in urea and thioglycollic acid and extended by 80% gave a β pattern but no meridional 4.7-Å reflection. The conditions necessary to produce cross-β structure in Lincoln wool do not apply to human hair.

Barnard, Stam, et al.[56] used human hair to study the uptake of alkali metal chlorides and dyes by methods involving radioactive tracers. Barnard and White[57] observed the swelling from water to pH 1 in HCl to be only $3\frac{1}{2}$%, compared with 6% for Cotswold wool, as a result of restraining crosslinks.

Reese and Eyring[58] investigated the effect of acids, bases, and oxidizing and reducing agents to determine what contribution the various bonds made to mechanical strength of human hair; they included also some thermodynamic studies of the processes involved.

If one considers the effect of agents such as sunlight, soaps, detergents, and bath water on human hair, it is perhaps not surprising that results obtained by different researchers do not agree. Using one's own hair is no guarantee that the hair is unmodified or even approaching normal. While the bulk of a Merino wool staple is protected while on the sheep's back, human hair is exposed to all kinds of agents. Horsetail hair and cow tail hair in a chemically unmodified state are less common than supposed; the tip effect in such hairs can extend quite a long way along the hair, as has been shown by Kidd,[10,12] using a test developed specially to study damage and degradation in coarse fibers. If imported bristle is used for experimental work it is important to note that it may have been treated in steam or boiling water. Imported goat hair for brushmaking is disinfected. A knowledge of the previous history of the fiber is important, and for much human hair the previous history is complex. Mechanical strength, stress, and related phenomenon have often been studied by various authors[58–62] using horsehair or human hair.

Various types of horn have attracted the attention of research workers such as Ryder[63] (structure of Indian rhino horn), Rudall[64] (structure and keratinization of cow's horn), Mason[65] (thermal transition at 35°C, 55°C, and 84°C in African black rhino horn), and Zahn[66] (denaturation in superheated water).

### 9.3.1. X-Ray Diffraction Patterns and Crystallinity

The cortex, being $\alpha$-keratin, gives the expected $\alpha$ pattern as established by the early work of Astbury,[67,68,60] Woods,[68,60] and MacArthur.[69] High-angle photographs of mohair, alpaca, llama, cashmere, vicuna, and camel fibers show minor differences from wool but the same basic pattern.[70]

Much attention has been given to the cortex of porcupine quill tip, which gives a relatively clear $\alpha$ pattern because of its well-ordered crystallinity.[71] Astbury[71] has, however, pointed out that the sulfur content is much lower (2.5%) than for human hair (5.1%). It can be even lower (1.35%), which is abnormally low for a keratin.

Woods[72] has observed that mohair gives a pattern most nearly approaching that obtained from porcupine quill tip.

Sikorski[73] and Slack[19] have prepared or published X-ray photographs of bristle and horsehair, the photographs from the bristle being both clearer and sharper. This agrees with the observation that the fibrillar structure in bristle cortical cells is extremely well defined in low-resolution electron micrograph photographs reproduced by Woods[74] and by Jeffrey,[75] who confirmed by X-ray diffraction the well-orientated paracrystalline system in bristle. The occurrence of the $\alpha \rightarrow \beta$ transformation in human hair and horsehair was studied by Elliott[76] and in horsehair by Zahn et al.[77,78] who included thermal contraction, swelling anisotropy, and various chemical analyses and tests in their studies. Elöd and Zahn[79] also studied horsehair after supercontraction in formamide and other reagents. Zahn and others[80] studied horse and human hairs after treatment with nitric acid.

Rudall[81,64] has shown that lightly squashed medulla gives a partially orientated cross-$\beta$ pattern with traces of $\alpha$ pattern; Rogers[82] has confirmed this and has given a complete amino acid analysis of the trypsin digestible protein from African porcupine quill medulla.

Relative density measurements have been used by Fraser and MacRae[83] to estimate purely for purposes of comparison, the relative noncrystalline content as 64 (porcupine quill), 67 (human child hair), 81 (Merino wool), 86 (kid mohair), and 100 (cow's horn). Nicholls[84] carried out a Hailwood and Horrobin analysis on mohair and estimated the fraction accessible to water to be 0.60 compared with 0.63 for Merino wool and 0.70 for Lincoln wool. Adsorption and desorption regains of wide range of commercial brush filling materials have been published by Kidd,[10] and Sagar[85] has studied purified samples. She has shown the fraction accessible to water to be in the range 0.53–0.57 (bristle), 0.46–0.52 (horsehair), and 0.52–0.63 (soft hairs) but concluded that much further work would be needed since the fibers studied differed so widely in origin and morphology.

Von Bergen[204] recorded the moisture regains of a range of specialty hairs and fur fibers.

### 9.3.2. Infrared Spectroscopy

Infrared spectroscopy using polarized radiation has been applied to human hair by Fraser[86] and to many different forms of $\alpha$- and $\beta$-keratin by Keighley,[87] to determine the orientation in bristle, porcupine quill tip, elephant hair, and leopard and lion whiskers, which all show parallel dichroism. Wing feather calamus of the hen, peacock and turkey, all of which are poor flyers, show parallel dichroism whereas the sparrow hawk shows perpendicular dichroism. The study is of wider interest in that scales, hooves, horns, beaks, fingernails, and claws are included. Earland et al.[88] comment on the polyphase structure of feather calamus in relation to Keighley's findings, which must relate to the outer layer, which is birefringent in cross-section. Frohlich and Rau[89] compared rabbit fur before and after carrotting by different methods and confirmed the view that rupture of disulfide bonds is the critical reaction for both types of carrotting; no peptide-bond rupturing was detected.

Bendit[46] used longitudinal sections of horsetail hair prepared by a new technique; such sections are wide enough to permit the use of a standard ir spectrophotometer. $\alpha$-, $\beta$-, and supercontracted keratins were all studied and new features noted. Deuteration studies revealed new bands in the spectra of deuterated $\alpha$ and $\beta$ forms, and as a result of his work, the recorded bands due to side chains now include most of the important residues.

Wood and King[90] found longitudinal sections from softened cow tail hair (unmedullated) to be suitable to study ir absorption by stretched keratin. Mackenzie,[91] in his extensive study of keratin by ir spectroscopic methods, selected 16 keratinous materials to include at least two of each of the following: horns, hooves, claws, quills, whiskers, hairs, and wing feathers. For pig's bristle he recorded a particularly high value of the parallel dichroism, a finding in keeping with Jeffrey's[75] observations of very long, well-orientated fibrils in the cortical cells and X-ray observations of very well orientated "crystallites."

To explain his findings over the full range of samples, Mackenzie proposed that in the predominantly noncrystalline regions the molecules exist mainly in the enol-tautomeric form. In all cases the keratins were heterogeneous, being deposited in a layerwise fashion.

### 9.3.3. Electron Spin Resonance

Electron spin resonance spectroscopy of fibers following the method of Gordy et al.[92] has included bristle and feather quill, which gave spectra similar to that of cystine. Keighley[93] included bristle, human hair, horse mane and tail, wolf hair, cow hair, and feather calamus as well as wool in his electron spin resonance studies.

### 9.3.4. Differential Thermal Analysis

Crighton and Happey[94] extended their thermal analyses of wool in air to include cashmere, kid mohair, and human hair and their analyses *in vacuo* to kid mohair and rhino (unicornis) horn. Felix et al.[95] have reported results for horn, mohair, and wool, the latter two giving similar results.

### 9.3.5. Ultraviolet Absorption Spectroscopy

The uv absorption spectrum of solid keratin in the form of longitudinal sections of horsehair is given in a paper by Bendit.[96] The effects of stretching, of uv damage, and various chemical treatments are discussed.

### 9.3.6. Chemical Analyses

While earlier results are ultimate analyses of a few hairs[97,98] and whalebone[99] and of limited value for materials that are heterogeneous, much effort has gone into determining sulfur contents, some of which are listed in Table 9-1. Lindley[146] fractionated the sulfur in a range of fibers. Such results are average results for all the component parts of the fiber throughout its length. Barritt and King[103] found higher sulfur contents at the tip of wool than elsewhere, in spite of losses by weathering. Results of cystine determinations on some of the fibers listed in Table 9-2 indicate a lower sulfur content at the tip than in the main shaft.

Frohlich[109] recorded corresponding alkali solubilities of 14.3% (tip) and only 10.2% (shaft) for the rabbit hair, as well as more marked swelling behavior of the tip. Carrotting with $HCl/H_2O_2$ enhanced these differences, as expected.

Partial or complete amino acid analyses of human hair are fairly common,[114,116–119] such hair being important both from a personal point of view and a commercial one. While Clay et al.[120] found more cystine in dark hair than light and more cysteine in male hair than in female, no consistent relationship between age and chemical composition was apparent for the samples of hair analyzed for these acids and for nitrogen and sulfur contents. Scott[114] found the methionine content of human hair to be 0.6%, which is similar to that of Merino wool. The ratio of polar to nonpolar groups is similar to that in wool, although numbers present in the human hair are lower. Cysteine contents are greater in camel hair and alpaca than in mohair and wool, according to Satlow et al.[111] Satlow[121] also quotes values for cysteic acid and tryptophane for goat hair, camel hair, alpaca, and mohair. Springell and Leach[115] determined mercurial binding in body hair from cattle of various breeds in Australia which serves as an index of cystine content, which is 10.8–12.6% for British breeds. Brahman cattle hair gave lower values.

With the advent of autoanalyzers, complete amino acid analyses are now relatively common for such materials, as, for example: rabbit hair (before and after carrotting), Frohlich;[109] porcupine quill tip, Fraser et al.[122] Gillespie,[101] and Bendit and Feughelman;[123] whale baleen, Fincham;[124] mohair, Swart et al.[125] and Gillespie;[101] human hair, Ward and Lundgren,[126] Simmonds,[117] and Menkart et al.[51] guinea pig hair, Gillespie;[101] and cattle and rhino horn, Gillespie.[101]

Simmonds[117] has recorded the variation between Merino wools even though they are relatively protected during growth. Greater variations might be expected for other fibers, which are not so well protected.

Fincham[124] showed whale baleen, after demineralizing with ethylene

**Table 9-1. Sulfur Content of Various Fibers**

| Fiber | Sulfur content (%) | Reference |
|---|---|---|
| Bristle (Gortorg) | 3.7 | 100 |
| Whalebone or baleen | 3.6 | 99 |
| | 3.0 | 101 |
| Cow's horn | 3.4 | 99 |
| | 4.1 | 101 |
| Rhino horn | 1.9 | 101 |
| Porcupine quill | 1.35–2.5 | 71 |
| descaled tip | 2.7 | 102 |
| Rabbit hair | | |
| — | 4.2 | 101 |
| — | 5.2 | 102 |
| Russian | 3.84 | 98 |
| Agouti | 4.30 | 103 |
| Black | 4.14 | 103 |
| Racoon fur | 5.78 | 98 |
| | 5.6 | 101 |
| Dog hair | 5.1 | 101 |
| Dog wool, white | 5.07 | 103 |
| Muskrat fiber | 4.68 | 98 |
| Goat hair, Tunisian | 3.5 | 102 |
| Guinea pig hair | 4.3 | 101 |
| Alpaca | | |
| — | 4.17 | 98 |
| white | 3.93 | 103 |
| brown | 4.35 | 103 |
| black | 3.90 | 103 |
| — | 3.85 | 104 |
| Mohair | | |
| — | 3.22 | 104 |
| Turkish | | |
| fine | 3.36 | 97 |
| coarse | 3.03 | 97 |
| Texas, kid | 2.92 | 98 |
| Turkish fleece | 3.58 | 98 |
| Vicuña | 4.10 | 98 |
| Camel hair | 3.41 | 98 |
| Cashmere | 3.39 | 98 |
| Human hair | | |
| — | 5.0 | 99 |
| — | 5.1 | 105 |
| — | 4.9 | 106 |
| — | 4.7 | 107 |
| — | 5.0 | 108 |
| — | 3.9 | 102 |
| black | 4.9 | 103 |
| Horsetail hair | 3.1–3.9 | 99 |
| | 3.6 | 108 |

**Table 9-2. Cystine Content of Various Fibers**

| Fiber | Cystine content (%) | Reference |
|---|---|---|
| Rabbit hair | | |
| shaft | 11.8 | 109 |
| tip | 8.8 | 109 |
| Alpaca | | |
| Suri, white | | |
| shaft | 15.1 | 110 |
| tip | 13.8 | 110 |
| Huacayo, white | | |
| shaft | 14.4 | 110 |
| tip | 11.1 | 110 |
| — | 12.0–13.0 | 111 |
| — | 12.2–12.9 | 112 |
| Cashmere | | |
| white | 11.8 | 110 |
| — | 11.5–12.0 | 111 |
| — | 11.6–12.0 | 112 |
| Camel hair | 10.8 | 111 |
| | 10.3–11.1 | 112 |
| Mohair | | |
| kid | 10.4 | 110 |
| — | 10.4–10.9 | 111 |
| — | 10.2–10.9 | 112 |
| Vicuña, dark | 13.6 | 110 |
| Rabbit hair | | |
| Angora | 13.1–13.8 | 112 |
| blown | 11.2–11.7 | 112 |
| Hare hair, blown | 13.9–14.2 | 112 |
| Goat hair | | |
| clipped | 9.8–11.2 | 112 |
| clipped | 9.4–12.0 | 113 |
| lime-soaked | 2.2–9.6 | 113 |
| Human hair | | |
| — | $17.6 \pm 0.9$ | 107 |
| Italian female | 14.5 | 114 |
| African Negro male | 15.5 | 114 |
| Cattle body hair (British breeds) | 10.8–12.6 | 115 |

diamine tetraacetic acid (EDTA) or other acid, to be a classical keratin. Cystine residues per thousand ranged from 68 to 90 compared with 110 for 64's Merino wool and 113 for nonmedullated horsetail hair.

Chemical analyses of individual structural components of animal fibers are still rare. They include the work of Lustig et al.[127] (human hair); Bradbury, Chapman, and King[128] (wool, mohair, human hair, alpaca,

Australian possum fur and rabbit fur); Rogers and Simmonds[129] (rat's whiskers); Rogers[130] (porcupine quill medulla and inner root sheath, which are chemically similar).

Rogers'[130] analyses of porcupine quill medulla were compared with those calculated by Swart et al.[131] for mohair kemp medulla, and apart from the much higher arginine content and the complete absence of citrulline, the results were generally similar.

The C-terminal residues in human hair are glycine, alanine, valine, serine, threonine, aspartic acid, and glutamic acid, according to Blackburn and Lee.[132] For human hair, horsehair, wool, and feathers the C-terminal residues of the keratin showed no significant differences, according to Kerr and Godin;[133] the same is also true of the N-terminal residues.

The grease contents of commercial brush filling materials have been reported by Kidd.[10] The composition of the grease extracted from human hair has been investigated by several authors.[134,135] A comprehensive study of the grease contents of a wide range of specialty fibers was undertaken by Frohlich[112] using methylene chloride. Other values are to be found widely scattered throughout the references at the end of this chapter.

Bolliger[136] studied the water-extractable constituents of hair and found much less pentose in human hair than in the faster-growing rabbit hair.

### 9.3.7. Kerateines

It is now well established that α-keratin contains a cystine-rich high-sulfur matrix in which are embedded low-sulfur α-helixes. Fractionation of *S*-carboxymethyl kerateines is a convenient method for investigation which has been studied by Gillespie and Lennox,[137] Crewther et al.[43,138] and by Robson,[139] following on the work of Goddard and Michaelis.[140]

Crewther et al.[43] give a detailed account of methods used to study the chemistry of keratins, particularly wool and, to a lesser extent, feathers. Gillespie[101,141] has made a detailed study of the high-sulfur proteins, as well as low-sulfur proteins, from raccoon, guinea pig, rabbit and cattle hairs, whale baleen, and several horns. The contribution to the total sulfur from the low-sulfur proteins is remarkably constant; the contribution of the high-sulfur proteins varies by a factor of $\times 11$.

Gillespie recorded the electrophoretic patterns of the high-sulfur proteins SCMKB from cashmere and mohair as being similar to those from wool. Human hair, porcupine quills, and whale baleen gave much simpler patterns.

Results for mohair have been published by Swart et al.[125] and Joubert[142] has continued the work to confirm the heterogeneous nature of the SCMKB of mohair, which consist of five major molecular-weight species which could be separated by gel filtration following a preliminary chromatographic separation. Each group contained a number of closely related proteins separable by further chromatography on DEAE cellulose.

### 9.3.8. Keratoses

The method used is based on the work of Alexander and Earland[143] and has been developed by Corfield et al.[144] Results for fibers other than wool and mohair are rare and are shown in Table 9-3.

Fincham and Graham[124,145] showed by amino acid analysis that the keratose in the baleens had a composition similar to that of the α-keratose from wool but the β- and γ-keratoses were not the same as those from wool. This lends support to the theory of a common low-sulfur microfibril yielding the α-keratose. Since medullary cell, if present, will appear among the β-keratose fraction, along with insoluble membranes and nuclear remnants, care is necessary in interpreting results from medullated fibers.

The high-sulfur proteins of mohair were fractionated by column electrophoresis and subsequent gel filtration on Sephadex S-100 and compared with similar subfractions of the γ-keratose from wool. Differences in amino acid composition were observed by Swart et al.,[125] who also showed that oxidized mohair contains more α- and less β-keratose than wool; the γ-keratose contents are almost the same. The amino acid compositions of the keratoses from wool and mohair showed only minor differences. Comparison of adult mohair, kid mohair, and mohair kemp also revealed differences in amino acid composition of the fiber and of the γ-keratose subfractions.[131]

### 9.3.9. Abnormal Human Hairs

As would be expected, certain abnormal states of health will affect the growth and structure of hair. Gillespie[147] found that hair from a child

**Table 9-3. Percentage of Keratose in Various Fibers**

| Fiber | α | β | γ | Reference |
|---|---|---|---|---|
| Merino wool | 56 | 10 | 25 | 144 |
| Caucasian human hair | 43 | 15 | 33 | 51 |
| Five species of whale baleen | 39–46 | 25–33 | 15–18 | 145 |

suffering from the protein-deficiency disease kwashiorkor had a lower content of cystine and sulfur in the high sulfur protein fraction than is normal, but that this state of affairs was rectified by a correct diet. Kwashiorkor produces changes in pigmentation and crimp in human hair reminiscent of those apparent in steely wool, according to Underwood.[148] Cystine deficiency is never found in mammals having an adequate diet.

Pollit et al.[149] studied *Tichorrhesis nodosa* in human hair when parts of the hair shaft split longitudinally and breaks occur; this condition can arise from various causes. The normal scale pattern is absent, the hair is twisted along the main axis, and the section is flattened. Amino acid analyses show an abnormally low cystine content, a deficiency of high-sulfur proteins, and a deficiency of copper. The protein resembles low-sulfur proteins from wool. Significant differences in size and arrangement of the microfibrils were noted by Robson and Sikorski[150] in the hair of children with *Homocysturina*. They observed in the cortex deep transverse fissures which crossed cell boundaries; the surface was rough as if most of the scales had been torn off.

## 9.4. Inorganic Constituents

The presence of traces of several metals in animal fibers is not unexpected. There are significant amounts of Ca, Mg, Zn, Cu, Pb, Fe, and traces of other metals as well as silicon and phosphorus in black human hair;[151] the following have been recorded: Pb, 5–40 $\mu$g/g (Av17); Arsenic, 0.8 $\mu$g/g; and Zn, 200 $\mu$g/g, by three other groups of authors, respectively.[152–154]

Dutcher and Rothman[155] give iron, copper, and ash contents for human hairs of different colors. Salaman and Luttrell[156] established that the disease kwashiorkor does not necessarily mean a reduction in copper content from its normal value in the range 8–20$\mu$g/g; the copper content and color are not related.

Traces of phosphorus were recorded by Blackburn[157] in 1948 [e.g., 390 $\mu$g/g (black horsetail hair); 110 $\mu$g/g in porcupine quill cortex but only 20 $\mu$g/g in medulla]. Stoves[158] showed phosphates to be present in horsehair medullary cells.

Pautard[159–161] opened up a new field with his investigations into calcification of keratins such as whale baleen. In the X-ray reflections, the "salt" (calcium orthophosphate) reflections suggested the presence of hydroxyapatite in crystalline or densely organized form; a high ash content is not in itself evidence of calcification or mineralization. The location of the hydroxyapatite is such as to suggest a functional significance. Milling of baleen, horsetail hair, porcupine quill, and bear claw yielded a powder which was fractionated by relative density flotation, the most dense being highly calcified.[162,163] Calcification serves a functional purpose by making the fibers, quills, and claws more resistant to abrasion in use. For example, nonmedullated horsetail hairs had a particularly high mineral content at the tips, and horsetails are used as switches. The highest recorded hydroxyapatite content was 14.5% in Sei whale finner. The calcium salts were laid down in concentric layers which were most common near the central, C-type medulla.

Fincham[124,164] continued the investigations and concluded that the mineralized deposits in baleen are indistinguishable from those found in bone in dimension, appearance, ash composition, and X-ray diffraction behavior. In baleen and hair, however, the deposits are intracellular and not all of them are crystalline. The location of the deposits can be revealed by staining with von Kossa silver nitrate stain.

Blakey et al.[165] examined a large number of X-ray diffraction photographs of a wide range of keratins. Strong apatite reflections were observed for Indian rhino horn, platypus body (but not tail) hair, lion's whiskers butt end (but not tip), and in the ash from kemp fibers (but not from the kemp itself). In the lion's whiskers the apatite is situated either in the medulla or in the adjacent region, which shows Maltese cross birefringence in cross sections.[88]

Blakey and Lockwood[166] have located and identified the hydroxyapatite and other calcium salts using dark-field electron microscopy and electron diffraction. At the same time they have revealed important new information about the structure of the cortex. For example, the compressed elongated cortical cells, which form a membranelike structure nearest the medulla, stain with pyronin in the methyl green/pyronin stain and also give a positive Feulgen reaction, suggesting that DNA is present; electron diffraction confirmed this view. In the adjacent region of flattened cortical cells, macrofibrils tangential to the medulla were seen as well as many intermediate orientations, in addition to the more usual axial orientation. This region contains the crystalline inorganic material and does not give the X-ray diffraction picture characteristic of keratin. The earlier observations[88] on Maltese cross birefringence in cat, tiger, and lion whiskers, whale baleen, rhino hair, and very coarse horsehairs, all containing a C-type medulla, were thus completely explained. In cattle hair the outer layers, amounting to one third of the thickness, are birefringent, owing to tangential orientation. All hairs other than the very coarse horse hair studied had the expected orientation parallel to the main axis.

No apatite was found in the head hairs of geriatric patients, although it was present in the fingernails of about one in six of the patients. It was noted, however, that damaged follicles can produce calcified hairs.[167]

## 9.5. Morphology and Chemistry

This topic is conveniently discussed under three subheadings, which relate to the medulla, the cortex, and to pigmentation.

### 9.5.1. Medullation

Medullas have not received the attention they deserve; this may be because medullation in wool is regarded as detrimental, in that the finest nonmedullated wools are most desirable, whereas strongly medullated wool fibers, and kemps in particular, are to be avoided. Bliss et al.[168] have studied wool and mohair kemps. Auber[169] included medullated fibers in his classical study of wool follicles. Wildman,[9] Appleyard,[21] Rudall,[81] Roth and Clark,[170] and Ballin and Happey[171] have also contributed to our limited knowledge. Medullation in animal fibers other than wool is not unusual. In pigs' (or hogs') bristle, for example, the medulla increases in area as the cortex decreases in area toward the tip, so that inevitably the tip is split, giving a soft "flag." The extent of flagging, and the type of flagging, varies with origin.[172,10] The presence of a large medulla can have a significant effect on the structure of the cortex, and therefore on hair properties.[172]

Osman Hill et al.[173] record an unusual alveolar texture in the medulla of certain monkey hairs from a specialized area in *Aotes humboldt*. It extends close to the cuticle at some points and facilitates splitting (i.e., flagging) of the hair. With monkey hairs most alveoli were almost empty and absorb very little stain, in contrast to the behavior of the wool and other animal fibers, where the medullary cell remnants stain readily.

Medullary cells can be separated out by treating whole fibers in, for example, caustic potash or caustic soda solution (Stoves[158,35,37] and Ross[174]) overnight at room temperature, or in dilute peracetic acid followed by ammonia, since they are essentially devoid of sulfur.[157,174,175] The cells may then cohere or not, according to Wildman.[9] Muskrat hair gives a coherent medulla; rabbit hair does not. These cells may become more spherical, even the flattened ones, when liberated. The behavior of cells so separated from horsehair when treated with tannic acid, sodium oleate, and saponin according to the method of Rideal and Schulman[176] suggested the presence of a lipoprotein–cholesterol complex in the cell wall. The presence of phosphate was established, as was the presence of sterols by the Liebermann Borchard reaction.[35]

In cross section the medulla can be differentiated by staining with, for example, ammoniated picrocarmine (picric acid and indigo-carmine), which stains the acidophilic cortex yellow and the basophilic medulla pink,[177,25,37,178] or with Auber's SACPIC stain,[169] giving yellow cortex and dark green medulla; this staining technique includes staining in Krause's picro-indigo-carmine. Eosin, safranine, gentian violet, and methyl red all differentiate medulla; so does Janus Green B.[179] Such differential staining is of particular value when the medulla is indistinct; Kidd and Sagar [178] used ammoniacal picro-carmine to establish the presence of basophilic medullary cells in the lightly pigmented central irregularly shaped area in pigmented horsehair. Kidd[172] has recorded this type of medulla, which he refers to as B type, in a wide range of hairs, staining tests using methylene blue having confirmed the results of microscopic observation, including treatment of cross sections with *o*-chlorophenol.

The medullas of whalebone from baleen whales show multiple layers when stained with haemotoxylin and eosin, so that Yablokov and Andreyeva,[180] who observed this, suggested that it might be used to determine the age of the animal.

In goat hair, Lloyd and Marriott[181] found 13.5% of the weight of the hair to be medullary cells; these had a sulfur content of 0.23%. Bekker and King[182] dissected out the medulla from porcupine quill for sulfur determinations and also showed that, in cross sections, it did not stain with sodium plumbite. Thorsen[183] observed no staining of medullary cells with lead nitrate solution.

Rogers[82] and Rudall[184] have given complete amino acid analyses of medullary cells as well as inner root sheath. While only a trace of cystine is present, the percentages of glutamic acid and citrulline are high, the latter being very rare in proteins. Both medullary cells and inner root sheath are digested by trypsin and are similar in many ways.

### 9.5.2. Types of Cortical Cell

The staining, chemical and electron microscope methods used to differentiate ortho- from para-cells are common to all fibers and need not be discussed further. Perkin and Appleyard's[185] use of fluorescent stains is worthy of particular mention, since it included mohair, llama hair, cow tail, and horsetail hair cross sections stained for periods up to 4 h at room temperature. In all cases some differentiation was observed. In the tail hairs, the paracortex surrounded the ortho, which occupied the central zone, which confirmed Auber and Ryder's[186] findings using Janus Green. Some ortho-cells were found among the para and Piper's[187] work encourages the belief that ortho- and para-cells are randomly distributed in many fibers which lack crimp. Evidence from many sources suggests that heterotype cells are also present, which stain to a degree intermediate between ortho and para with EM stains as well as with dyes.

Earlier, Kidd[12,172] found Snyman's[188] method using methylene blue at room temperature at pH 7.4 particularly useful for studying brush filling materials, which being coarse fibers were, as expected, all para-cell. Whalebone plate and finner, however, stained throughout the section.

Various opinions have been put forward regarding the structure of mohair. Mohair from adult goats consists of a center of ortho cortex surrounded by para (i.e., a radial distribution, according to Scheepers.[189]). On the basis of staining with methylene blue at pH 7.4, Onions[190] found the center to be a mixture of ortho- and para-cells. Fraser and Macrae[191] showed some para-type staining in both kid and adult mohair, which other tests indicated was pure ortho. The ortho/para ratio is higher than in Merino wool, so mohair fiber dyes more rapidly; the ratio is particularly high in fine young goat mohair (Veldsman[192]). A density fractionation of mohair cortical cells was carried out by Lundgren.[193] Kid mohair is low in sulfur content and predominantly ortho, according to Lundgren and Ward[194] and to Dusenbury and Menkart[107]; human hair, with its high sulfur content, is predominantly para according to them and according to Menkart and Coe,[195] resembling the para cortex of wool in chemical composition and properties.

Caucasian human hair which is not crimped is, as expected, all para-cell.[142] Scott's[114] more recent work is also of interest. If there is a bilateral structure in crimped Negro hair as suggested by Mercer,[196] its magnitude is much smaller than that found in wool.[51] Spearman and Barnicot[197] investigated hairs from different races. The cortex of highly crimped Negro hairs behaved like ortho cortex in wool in its staining behavior, but in its resistance to supercontraction and in trypsin digestion after autoclaving in steam, its behavior lay between that of ortho-cells and para-cells. Bilateral pigmentation does not mean bilateral keratin.[12,198,10]

Fine cashmere and camel hairs show ortho/para differentiation; alpaca, which is coarser, does not.

Kassenbeck et al.[199,200] used thallium carbonate staining for optical microscopy of mohair and silver for electron micrograph studies. While differential staining methods appear to be satisfactory in some instances, different techniques and different substrates do not agree. The use of transmission electron microscopy is desirable in certain cases.

Kassenbeck[201] has suggested a relationship between cell type and rate of growth. Slow growth of fur in winter gives para cortical cells in the lower part of the shaft, whereas rapid growth yields ortho-cells at the tip of the hair, with associated thicker cuticle. Kassenbeck[202] has also related diameter of fiber to scale pattern independently of the animal.

Even with transmission electron micrograph techniques, the whole problem cannot be considered to be resolved and further research is indicated; it may be that electron diffraction will provide the answer (Dobb[203]).

Dusenbury[110] sought to characterize the cortical cell structure of a wide range of fibers by their solubility in urea-bisulfite solution (see Table 9-4). Low values would indicate all para-cells with a cystine content about twice that of ortho-cells.

### 9.5.3. Pigmentation

Whereas most wool is white and pigmented wools are less desirable, many of the fibers considered in this chapter are pigmented. The way in which pigment is produced in the follicle and introduced into the fiber has been described by various authors: Auber,[169] Birbeck and Barnicott,[205] Mason,[206] Moyer,[207] Fitzpatrick et al.[208] Swift,[209] Mercer,[210] and Piper.[211]

Two types of high-molecular-weight polymer are distinguished: brown-black melanin and yellow-red pheomelanin. Only one of these is produced at one time and there are no intermediate forms. The yellow-red granules are slightly smaller and rounder than the brown-black and have irregular ill-defined boundaries. A pigment, trichosiderin, occurs, however, in red human hair (Flesch and Rothman[212] and Barnicot[213]). The granules can be isolated by boiling the fibers in 5 *N* or 6 *N* HCl but are degraded.[215–217] Human hair yields significantly more residue after such treatment than Lincoln wool (Caucasian brown hair, 2.5%; Negro black hair, 2.7%; Lincoln wool, 0.06%; Caucasian white hair, 1.0% consisting mainly of cell membrane).[219] Booth[214] showed that wool protein would dissolve in a solution of thioglycollic acid, and Ross[174] found that the pigment granules could be isolated by refluxing the fibers in this reagent and Laxer[215] developed the method further; the granules were isolated without destroying their shape. Swift[218] improved the method to make it quantitative by collecting the granules on a glass-fiber filter cloth. In all, over 100 mammalian, avian, and reptilian keratins were studied. Avian keratins yield rodlike granules, but those from mammalian keratins are more spherical or ellipsoidal ant-egg-shaped.

The fine structure of the pigment granules has been studied by electron microscopy, as have the dimensions.[215,218–222] The density is high for the black/brown granules: 1.70–1.74 according to Swift,[218] and 1.70 according to Ballin.[220] A relatively transparent membrane surrounds electron dense matrix containing even denser rods.[218,220] Laxer et al.[219] found that no internal structure was visible when isolated granules were studied. Mercer showed a laminated structure in cross sections.[221] The yellow-red is less electron dense than the brown-black (Hagége).[222] X-ray diffraction shows them to contain disorientated "crystallites" (Swift[218]).

Alkali degrades those from red-brown hairs, but the black-brown melanins are only slightly affected. Hydrogen peroxide disintegrates the granules into small particles.[218] The iron content is inversely related to their size, and the color is directly proportional to their size, which Swift[218] suggested is the effect of a colorless premelanin core surrounded by colored melanin. Copper is present but is only loosely bound.

Several authors have concerned themselves with the chemistry of pigments (Mason,[206] Laxer and Whewell,[217] and others which follow). Green and Happey[223] have compared the ir spectra of melanin pigment from brown Welsh mountain wool with the spectra of synthetic melanins to elucidate the chemical structure. The spectra resemble those of amorphous carbons, particularly low-rank coals. They suggest that the dark color is due to an extensively conjugated heteroaromatic structure and not to quinonoid carbonyl groups. They showed that melanin contains carboxylic acid groups; it is slowly leached out of pigmented fibers by alkaline solutions, as shown for karakul wool by Kriel et al.[224] Bristles and horsehair used in alkaline solution show a similar reduction in color intensity; alkali-damaged fibers are significantly lighter in color, especially when wet, where the effect is enhanced by increased swelling of the fiber consequent upon disulfide-bond breakdown.[10]

Pigmented fibers absorb iron from appropriate solutions, and this has been used to develop processes for bleaching pigmented fibers.[215] It was not found to be satisfactory for bristles or horsehair for brushmaking since the flags or tips are damaged during the bleaching process.

The relationship between pigmentation and chemical stability merits further work. Richter and Stary[225] found the alkali solubility of human hair to be lower for pigmented hair than unpigmented hair. Reducing agents affect pigmented fibers less, according to Goddard and Michaelis.[226] Pigmented wools are more resistant than unpigmented wools to solution in copper-catalyzed hydrogen peroxide solution. Laxer et al.[198] studied bilateral pigmentation in vicuña and wools. Laxer and Whewell[217] found pigmented fibers to be more resistant to supercontraction in metabisulfite solutions than unpigmented fibers. Laxer et al.[198] found that while the less stable ortho cortex in white wools was dissolved by Mercer's supercontraction/alkaline trypsin treatment,[227] no such differential behavior occurred with vicuña fibers, even lightly pigmented ones, nor with the Welsh pigmented mountain wools showing bilateral pigmentation. Bilateral pigmentation occurs in many types of bristle studied by Kidd,[12,172] but he found no evidence of cortical differentiation in staining tests; only para cells were found.

Most pigment granules are to be found in the cortex and are readily seen in cross section; bilateral pigmentation is not uncommon and certain bristles show this feature particularly well (Kidd[10,12]). Walsh and Chapman[27] found bilateral pigmentation in 24 of 56 black or brown human hair samples but none in 9 red hair samples. Some are to be found in the medulla (Burgess[228] and Auber[169]).

In certain mice hairs the pigment in the medulla is dense and the hairs have a banded appearance (McGrath and Quevedo[229]). In the hairs of a certain species of monkey (*Saimiri sciuirea*) Osman Hill et al.[173] noted pigment granules in the medulla which differed in color from those in the cortex.

They may occur in the cuticle (Swift[218] and Ballin[220]); Kassenbeck[230] observed them in mink, cashmere, and karakul fibers; Piper[211] observed them quite frequently in guanaco, vicuña, and swamp wallaby fibers. Ballin[220] found pigment granules in the outer but not the inner root sheath.

Boyd[231] and Billingham[232] have shown unpigmented dendritic cells to be present in unpigmented areas of skin. Dark skin grafts on to white areas of cattle skin or guinea pig skin will spread pigment to adjacent areas, but the hairs on the latter remain white. The corresponding wool on sheep was found to become pigmented.[233]

Copper deficiency leads to loss of pigmentation and crimp in black highly crimped wool fibers, according to Crewther.[234]

## 9.6. Effect of Chemical Reagents

### 9.6.1. Behavior during Processing

Von Bergen[235] comments that specialty hair fibers are generally similar to wool in their behavior during chemical processing. Cashmere, because of its fineness, is more sensitive to alkali than wool, and so too is mohair. Damage to tips of mohair by uv light is mentioned. Onions[190] stresses their sensitivity to alkali and refers to the Burgess test for camel hair or cashmere in blends with wool, based on their more rapid rate of wetting out when placed on a gelatine slide.

The behavior of human hair in alkali, reducing agents, (sulfite and metabisulfite) and oxidizing agents (hydrogen peroxide, per salts, and nitric acid) is the subject of a series of three papers by Stoves.[236] A mention in the third paper that only the root ends were used since the tips of human hair have been exposed to light and air during growth is significant, since it is not taken into account in so many other papers concerned with human hair. The effect of metallic catalysts such as cobalt, manganese, nickel, and copper ions in accelerating the rate of oxidation by hydrogen peroxide was studied, and copper was shown to increase the rate fourfold.

Lower acid binding of human hair compared with wool is due to lower basic amino acid content. It binds 7% less HCl than wool at pH 1.16, according to Speakman and Elliott.[237] Its maximum binding capacity is about 90 milliequivalents per 100 g of dry hair, according to Speakman and Hirst.[238]

The adsorption isotherms for five phenolic compounds on human hair have been determined by Breuer;[239] he suggests a hydrogen-bonding mechanism.

Swift[240] has studied diffusion of $Ag(NH_3)_2^+$ and PTA into human hair, and Kassenbeck and Hagége[241] have investigated the fixation of silver by alpaca, mohair, human hair, and wools.

Mohair dyes more rapidly than wool of similar diameter, and for the same uptake of dye the mohair appears deeper in color.[192] Unpublished work by Hill and Bell[242] showed that acid dyes were invariably faster to washing when dyed on mohair than on wool and that loss of dye from the wool component of wool/mohair fabric could produce a speckly appearance, thus giving rise to a mistaken belief that the dye was not fast on mohair. Industrial experience with solvent dyeing shows that mohair is more readily dyed than Lincoln wool. Ahmad[243] has investigated the adsorption of alcohols by the two fibers. The results support industrial experience. Equilibrium absorption may be attained, however, only after 8 months of immersion in isopropanol, for example, compared with 1 month for the straight-chain *n*-propanol.

Using alkali solubility as a measure of the degree of modification, human hair (all para cortex) was shown to be much less affected by treatment by boiling in sulfuric acid solution than either Rambouillet wool or kid mohair (all ortho cortex), according to Dusenbury and Menkart.[107]

Fur carrotting with mercuric and nonmercuric agents has been the subject of intensive investigation by Watt.[244,245] A relationship between alkali solubility (in 0.1 *N* NaOH at 65°C) of mercury carrotted wild fur and its felting rate was established and formed the basis of the HAFRA test and standards.[244] For nonmercuric (i.e., nitric acid and hydrogen peroxide) carrotting of the type described in British Patent 532,370, the HAFRA alkali solubility test was shown to be equally valuable as a control, and appropriate standards were defined.[245] Quantitative relationships between degree of carrotting, its depth and intensity, and the conditions of treatment have been derived by Watt,[246] and factors influencing the accuracy of the test have been further studied by Barr et al.[247]

Aston et al.[248] extended the investigation and derived an equation relating percentage weight loss in acid to the temperature and normality of the acid during a given time of acid treatment; the equation holds for each HAFRA number and for both types of carrotting. This is important with reference to felting and dyeing processes. Haigh[249] had earlier begun a study of the effect of the dyeing process on the fur felt.

Frohlich[109] has confirmed that only those agents which break cystine linkages are effective; either oxidation or reducing agents may be used. The felting power of rabbit hair is not increased if peptide, salt, or hydrogen linkages are broken unless cystine linkages are also broken. Cystine, amino acid, amino-end-group, and alkali-solubility determinations are recorded. The research has been continued and extended by Frohlich and Hille[250] to elucidate the precise reactions which occur.

Whereas Satlow et al.[111] report that mohair is more sensitive to damage by alkali (or acid) than BA wool, Kriel[251] finds BSFH Mohair (38 $\mu$m in diameter) to be significantly more resistant to alkali than Merino wool of average 21 $\mu$m. The grease on greasy mohair exerts a protective action during scouring. Alkali treatment converts carboxyl groups to the sodium salt, and dyeing with reactive dyes is accelerated; the effect is reversible, as was shown by rinsing in dilute acetic acid and then in water.

### 9.6.2. Tests for Damage

Solubility tests in alkali, urea–bisulfite, and in hydrochloric acid have been applied to fibers other than wool and a selection of these are summarized in Table 9-4. Frohlich[112] and Satlow[113,121] have also given results for various fibers produced by fellmongers. Further tables of results for solubilities in alkali, hydrochloric acid, and urea–bisulfite have been given by Frohlich[264] in relation to felting properties. While figures for the solubility of these fibers in standard reagents are now becoming available, it is apparent that previous history can be a critical factor, as with many other properties of commercial fibers.

Staining tests are of limited value when testing pigmented fibers, and many of the fibers discussed in this chapter are pigmented. Other methods of testing for damage to brush filling materials in particular have been described by Kidd. These involve microscopic examination; a modified form of the Kornreich[252] procedure using Krais reagent, and a new test based on the change in stiffness on wetting giving a % IDW (increase in deflection of a loaded cantilever of the fiber when wetted) which has been fully described in publications by Kidd,[10,12,253] together with its application to a wide range of processing, as a control on quality.

Table 9-4. Solubilities of Various Fibers

| Fiber | Alkali solubility[a] (%) | Urea–bisulfite solubility[b] (%) | 4.5 N HCl solubility[c] (%) | Reference |
|---|---|---|---|---|
| Cashmere | | | | |
| — | 13–17 | 38–45 | — | 111 |
| white | 12–18 | 43–47 | 12–23 | 112 |
| — | | 31 | | 110 |
| Alpaca | 9–16 | 47–60 | — | 111 |
| Suri | 7–13 | 47–58 | 7–10 | 112 |
| shaft | — | 39 | — | 110 |
| tips | | 25 | | 110 |
| Huacayo | | | | |
| shaft | — | 73 | — | 110 |
| tips | | 54 | | 110 |
| Camel hair | 14–15 | 29–52 | — | 111 |
| | 10–17 | 37–52 | 11–20 | 112 |
| Vicuña, dark | | 46 | | 110 |
| Mohair | | | | |
| — | 11–15 | 55–65 | — | 111 |
| — | 9–27 | 44–69 | 4–13 | 112 |
| — | — | 70 | — | 251 |
| kid | | 75 | | 110 |
| Rabbit hair | | | | |
| Angora | 7–12 | 60–69 | 5–11 | 112 |
| blown | 9–15 | 52–60 | 10–14 | 112 |
| Hare hair, blown | 7–11 | 60–65 | 7–10 | 112 |
| Goat hair | | | | |
| shorn | 13–17 | 43–44 | 9–11 | 112 |
| shorn | 10–20 | 29–56 | 8–15 | 113 |
| Cattle hair, clipped | 6–7 | 0.7–1.2 | 12 | 112 |
| Calf hair, clipped | 6–8 | 2–10 | 9–14 | 112 |
| Human hair | | | | |
| Caucasian | 5.0 | 27 | | 51 |
| Negro | 4.1 | 37 | | 51 |
| — | | 12 | | 110 |
| Lincoln wool for comparison | 10 | 53 | | 51 |

[a] Method of Harris and Smith.(254)
[b] Method of Lees and Elsworth.(255)
[c] Method of Zahn and Würz.(256)

### 9.6.3. Supercontraction and Setting Behavior

As would be expected, much of the early work on fibers other than wool involved human hair. Reference to the well-known work on supercontraction in various reagents which was carried out by Speakman, Whewell, and Woods[257–259] will show that it was soon known that human hair does not behave in the same way as wool. Woods,[259] for example, recorded 50% supercontraction of Cotswold wool, whereas the corresponding figure for human hair was only 30%. Stoves[37,236] has studied a wide range of fibers, including human hair. He also observed the axial contraction after boiling for 1 h in solutions of sodium metabisulfite, thioglycollic acid, formic acid, saturated urea, *m*-cresol, chlorine water, and cuprammonium hydroxide of a range of hairs (human, horse, kolinsky, and skunk), pig's bristle, and Cotswold wool.[23,260,261] As expected, there were marked differences in behavior. White skunk hairs, for example, supercontract more in sodium metabisulfite than do human hairs. However, after setting in 2% borax solution they not only retain a greater permanent set than do human hairs, but the set is not diminished by boiling in 5% metabisulfite solution.[36]

Stoves[37] observed the contraction in chlorine water of bristle, horsehair, human, skunk, and kolinsky hairs. Elod and Zahn[79] studied horsehair supercontraction in formamide solution and Zahn[262] its behavior in caustic soda, in bisulfite, and in phenol solutions after prior treatment in dilute caustic soda or baryta water, as an extension of his work with Brauckoff[263] on setting in steam. In addition, Zahn[108] has made an extended study of the supercontraction of wool; horse, human, and dog hairs; and China bristle in 50% phenol solution. Wool contracts 20% at 84°C, whereas bristle requires a temperature of 97.5°C, the other fibers requiring intermediate temperatures to cause softening and contraction. A range of phenolic compounds was included by Zahn[265] in an earlier study of horsehair supercontraction. Laxer and Whewell[217] found that pigmentation increased the resistance to supercontraction in boiling metabisulfite solution. Menkart et al.[51] observed the effect of prior stretching on the supercontraction of human hair as part of a detailed and wide-ranging investigation.

Speakman[266] observed in 1936 that human hair set to a lesser extent than Lincoln wool and Menkart et al.[51] report similar findings. Whewell[267] found that different types of wool gave setting values in the range 6–21% after setting in boiling water, so it was not unexpected that he also found human hair to behave differently from certain wools. His value for human hair was 12% compared with 21% for Merino wool and only 8% for Scottish Blackface wool.

Whewell[268] and Bell and Whewell[119] have reported detailed studies on the setting behavior of human hair in the context of hair chemistry and have given a concise summary of the chemical changes brought about by the action of various chemical reagents on keratins; this is an extended version of Lennox's[269] table. Scott[114] included setting in his studies of human hair.

Whewell[270] has investigated the setting of human hair under many different conditions, including prior esterification, oxidation, reduction, deamination, and iodination. The setting behavior of oxidized fibers is very dependent upon the oxidizing agent used; peroxide-bleached hair will not take a set, but some treatments with peracetic acid, periodic acid, or permonosulfuric acid facilitate subsequent setting. Farnworth[(271)] has shown that a mixture of reducing agents and hydrogen-bond breakers, such as urea, are more effective setting agents than either reagent alone.

Cold setting treatments for human hair are now well established. Recently, Bones[(272)] studied the setting of human hair in thioglycollic acid solutions followed by sodium bromate solution and observed that the cuticle puckered but did not shed its scales, although their distal edges curled, when the hair was set without tension; when set under very slight tension, there was no puckering, but shedding of scales occurred. Subsequent swelling in water was 36% (set without tension) and 14.3% (set with tension) compared with 5.1% for the original hair.

Mitchell and Feughelman[(273)] showed that set obtained in boiling water or in borax solution decreased with increasing diameter of a *limited* range of fibers from Merino wool through Lincoln wool to human hair. Bajpai's[(274)] work, covering a wider range of fibers, did not confirm their conclusion.

Stoves[(275)] found marked differences in setting behavior in borax solution between one group, consisting of skunk, Russian hare, kolinsky, and mink guard hairs, and another group, consisting of reindeer hair, whitecoat hair, and pig's bristle. He concluded that the ratios of cuticle to cortex to medulla is not the determining factor. After being set at zero extension, fibers from the first group do not contract in boiling 5% metabisulfite solution, whereas those in the other group do. Linkages other than hydrogen bonds account for the behavior of only two out of the four fibers from the first group, namely skunk and mink, judging from their behavior in lithium bromide solution.

The research work of Bajpai[(274)] and Gandhi[(276)] was concerned with both supercontraction and set, and they both concluded that supercontraction

**Table 9-5. Contraction or Set of Fibers Treated in Boiling Water at 40% Extension**

| | Treatment period | | | |
|---|---|---|---|---|
| Fiber | 2 min | 30 min | 60 min | 120 min |
| Lincoln wool | −27.9 | −6.3 | 8.4 | 14.9 |
| Mohair | −26.9 | 12.9 | 17.5 | 26.6 |
| Alpaca | −5.9 | −0.2 | 8.6 | 16.2 |
| Human hair | −7.9 | −2.2 | 5.6 | 14.5 |
| Cattle tail hair | −18.9 | 2.2 | 10.7 | 20.4 |

and the ability to take a set are not directly related. Some results obtained at 40% extension by Bajpai are given in Table 9-5, and his results for various pig bristles and animal body and tail hairs treated at 30% extension are similarly varied.

Descaling the human hair had no effect on the results obtained. Bajpai et al.[277] subsequently published details of the effect of tetrakis(hydroxymethyl)phosphonium chloride on the setting of chemically modified Lincoln wool and Chinese human hair. Oxidized, iodinated, and deaminated fibers can all be set in boiling THPC solutions, but not in boiling water.

Bajpai and Whewell[278] found solutions of thiourea dioxide to be very effective setting agents for wool and human hair; it can be used to set deaminated fibers but not fibers previously treated with potassium cyanide.

Gandhi[276] treated human hair, mohair, alpaca, and various wools at 40% extension in boiling water for times up to 6 h. Once again, marked differences in behavior were recorded in the early stages, but after 6 h the degree of set obtained was in a range from about 23% to 29%. The degree of medullation did not appear to be related to the ability to take set, nor was the latter related to the water of imbibition. If the transition time (i.e. the time of treatment) for which there is no supercontraction and no set is noted, then apart from the human hair, the transition time is least for fibers with the highest urea–bisulfite solubility. Mohair set more readily than wool, human hair less readily. Onions[190] comments that the ease with which mohair sets by comparison with wool accounts for its use in curled pile rugs and simulated astrakhan fabrics.

Setting of fibers other than wool is commercially very important. For example, the setting of pigs' bristle in a straightened form, or with a desired slight curvature, is essential for certain types of brush; other hairs, such as horsehair, cattle tail hair, and certain soft hair, may also be straightened. On the other hand, bristles and hair for upholstery may be treated to increase the curvature. The setting is carried out with the fibers only slightly stressed, and relaxation is conveniently carried out at 60°C rather than at 100°C (Kidd[10,12]). Some set has been observed in acid solution. The investigations by Kidd covered a wide range of conditions and fibers. The IDW test developed by Kidd[10,12,253] has been used to establish optimum times of treatment and to study the effect of, for example, residual acid from prior sulfur stoving of white bristle on its quality after the straightening process.

### 9.6.4. Resin Impregnation and Sterilization

The properties of animal fibers can be changed by impregnation with the precondensates of thermosetting resins (Calva,[279] Teller,[280] Kidd[10]), treatment with formaldehyde (Calva,[279] Stonehill[281]), or treatment with vinyl monomers followed by polymerization (Wolfram[282]). Increased stiffness, increased resiliency, reduction in moisture absorption, improved set-holding, and possibly some degree of thermoplasticity[282] are some of the properties looked for.

The methods used to sterilize certain brush filling materials have been fully described in publications by Slack[19,283] and Kidd.[10]

### 9.6.5. Egg Capsules

Because some of the animals yielding fibers are only partially domesticated, it is not unexpected that infestation with lice may occur and the egg capsules may be found attached by a sheath to fibers and, although quite harmless, can give rise to processing problems. Because the capsule of the pig louse *Haematopinus suis* L. is relatively large, it is most suitable for scientific investigation. It has been found that freezing at −72°C did not make the capsules brittle. Direct chemical treatments with acid, alkali, oxidizing agents, or lithium bromide designed to dissolve the sheath which attaches the egg capsule to the fiber have not been successul, the sheath being more resistant than the fiber. Tanning with a vegetable tanning agent, Gambier extract, followed by heating in an attempt to harden the sheath and make it brittle was only partially successful.[10]

Selective treatment with a mixture of enzymes (papain and ficin) in the presence of sodium thioglycollate has been shown to be effective, particularly at 37°C.[284] Greenwood[285] continued this work using lice egg capsules plus sheath from bristles and carried out a complete amino acid analysis using thin layer chromatography. Of the 29% of the sample which he showed to be protein, alanine constituted 14.3%, valine 4.5%, histidine 3.9%, and serine 2.8%; the remainder was carbohydrate. He confirmed that ficin was the most effective single enzyme to degrade the mucoprotein: this enzyme needed 7 days at room temperature in a 0.001 *M* solution. Enzymes such as ficin also attack the incompletely keratinized root end of bristle.

Works by Rudall, Millard, Kenchington, and Flower[286–290] have elaborated the production and fine details of the structure of the ootheca of the praying mantis, and it is possible that similar research into pig lice would yield results of great interest and value.

## 9.7. Effect of Bacteria and Enzymes

In common with wool, the various fibers described in this chapter are susceptible to attack by bacteria[291,10] and by appropriate enzymes such as trypsin[292,228] or papain, in the presence of reducing agents[293,294]; such bacteria and enzymes digest the intercellular membrane and liberate cortical cells. The fibers do vary widely in their resistance to attack. Burgess[292,228] found that human, horse, pig, calf, and dog hairs were more resistant than wool, whereas others, such as mohair, camel hair, vicuna, and cashmere were less resistant; all wools did not, however, have the same resistance, nor did all rabbit hairs.

An extended study of the behavior of brush filling materials in the laboratory and in use has been carried out by Kidd.[10,12] Microbiological attack normally liberates cortical cells but not invariably so; Noval, Nickerson, et al.[295] report that *Streptomyces fradiae* will digest keratin. Usually, however, cortical cells are liberated, and the rate of liberation can vary widely within a fiber, between fibers, and between species of animal. Using trypsin, for example, the main shaft of bristle was very resistant to attack, although the root, where present, yielded shorter and fatter cells quite readily.[12]

The most resistant of the fibers tested[12] to attack by trypsin were the shafts of wild boar bristle, wildebeest, and coatimundi hairs, which were unaffected after 18 days of exposure. Variations in resistance are to be found in fibers from the same species of animal. Previous history is important as well as histology.

The tips of hairs such as horse or cattle tail are invariably more readily attacked than the butt ends. Subsequent chemical modification, caused, for example, by overexposure to oxidizing agents during bleaching, can open the way for rapid attack by enzymes. Overbleached tips are therefore particularly susceptible to attack, Mechanical damage to the cuticle, allowing access of the agent to the cortex, is an important factor, and such damage may occur before or during use in a brush. It cannot be emphasized too strongly that in all studies with enzymes a knowledge of the previous history of the fiber is essential or the results obtained can be misleading. Stary[296] and Laxer and Whewell[297] showed pigmented fibers to be more resistant to attack than unpigmented fibers; most animal brush filling materials are pigmented and are not readily attacked.

Not all cortical cells liberated by the action of enzymes or bacteria are "spindle-shaped." It has long been known that flattened cells are obtained from such fibrous materials as cow's horn, sheep's horn, whalebone, and the surface of fingernail (Burgess).[228] Auber[169] recorded flattened cortical cells in heavily medullated wool fibers, and Kidd[298,172,12] found them in several other heavily medullated types of animal hair (squirrel, reindeer, goat); partially flattened cortical cells were seen in other hairs (skunk, rabbit, hare, ringcat, horse body, coatimundi). For cells that were not flattened, Kidd[12] recorded average breadths in the range 7–11 $\mu$m. Satlow et al.[111] quote breadths of 9 μm (cashmere), 7–8 μm (camel hair), 8–9 μm (alpaca and mohair), and $8\frac{1}{2}$ μm (Merino wool), whereas most values quoted in the literature are not greater than 5 μm for wool. Appleyard[(299)] has continued his studies of the effect of *o*-chlorophenol on cross sections to include bristle and mohair, alpaca, cashmere, human, goat, llama, and camel hairs and has observed various types of flattened cortical cells.

The intercellular membrane is well defined in cross sections of cow tail and wildebeest tail hair; Manogue, Moss, and Elliott[(300)] used enzyme treatment of the former after oxidation to show the continuous nature of the membrane. Holmes[(301)] described the chemical properties of an ethanol-soluble component of human hair which restricted the rate of attack by papain/sodium bisulfite.[(31)] Earlier, Zahn[(302)] postulated the presence of a submicroscopic membrane resistant to 1 *N* NaOH and to the enzyme. The loss in weight on trypsin digestion is the basis of a test developed by Elöd and Zahn[(303)] and used by Satlow[(121)] in a modified form. Trypsin is one of the enzymes present in pancreatin, the effect of which has been quantitatively studied by Elöd and Zahn et al.[(304)] in relation to horse, human, dog, rabbit, calf, and pony hair.

## 9.8. References


1. K. Koope, *Der Bürsten und Pinsel fabrikant*, Voigt, Weimar (1834).
2. W. Kiddier, *The Brushmaker*, Sir Isaac Pitman & Sons Ltd., London (1912).
3. W. Kiddier, *National Society of Brushmakers*, Sir Isaac Pitman & Sons Ltd., London, (1931).
4. J. G. Links, *The Book of Fur*, James Barrie, London, (1956).
5. M. Bachrach, *Fur*, Prentice-Hall, Inc., Englewood Cliffs, N. J. (1953).
6. H. G. Fröhlich, in: *Ullman's Encyklopedie der Technischen Chemie*, 3rd ed., Vol. 14, p. 574, Verlag Urban & Schwarzenberg, Munich (1963).
7. K. M. Rudall, *Proc. Leeds Phil. Soc.*, **IV**, Pt. 1, 13 (1941).
8. A. B. Wildman, *Brushes* **25**, 14 (Oct. 1939).
9. A. B. Wildman, *The Microscopy of Animal Textile Fibres*, Wool Industries Research Association, Leeds (1954).
10. F. Kidd, *Brushmaking Materials*, British Brush Manufacturers Research Association, London (1957).
11. F. Kidd, *Product Finishing* **17**, No. 4, 68 (1954).
12. F. Kidd, Ph.D. thesis, Pts. I and II, University of Leeds (1964).
13. T. M. P. Hardy and J. L. Hardy, *Animal Fibres Used in Brushes*, U. S. Dept. Agr. Circ. 802 (1949).
14. W. von Bergen and W. Kraus, *Textile Fibre Atlas*, Textile Book Publishers, New York (1945).
15. W. Kraus, in: *Matthew's Textile Fibres* (H. R. Mauersberger, ed.), 5th ed., p. 1002, John Wiley & Sons, Inc., New York (1947).
16. S. B. Newman, RP2315, *J. Res. Natl. Bur. Std.* **48**, 387 (1952).

17. J. Sikorski and W. S. Simpson, *J. Roy. Microscopical Soc.* **78**, Pts. 1 and 2, 35 (1959); *Nature* (*Lond.*) **182**, 1235 (1958).
18. H. Klenk and G. Scholzel, *Melliand Textilber.* **31**, 451 (1950).
19. E. B. Slack, *Coarse Fibres*, Wheatland Journals, London (1957).
20. M. L. Ryder, in: *Fibre Structure* (J. W. S. Hearle and R. H. Peters eds.), p. 534, Butterworth for the Textile Institute, Manchester (1963).
21. H. M. Appleyard, *Guide to the Identification of Animal Fibres*, Wool Industries Research Association, Leeds (1960).
22. J. L. Stoves, *Proc. Roy. Soc. Edinburgh* **B62**, 99 (1944).
23. J. L. Stoves, *J. Soc. Dyers Colourists* **63**, 65 (1947).
24. J. L. Stoves, *Medicolegal Criminological Rev.* **2**, Pt. 4, 185 (Oct.–Dec. 1943); *Fibers* **11**, No. 4, 105 (1950).
25. J. L. Stoves, *Fibre Microscopy*, National Trade Press, London (1957).
26. M. Trotter, *Amer. J. Phys. Anthropol.* **14**, 433 (July/Sept. 1930).
27. R. J. Walsh and R. E. Chapman, *Man* **1**, 226 (June 1966).
28. S. Chowduri and B. Battacharyya, *Curr. Sci.* **33**, 784 (1964).
29. L. A. Hausman, *Amer. Naturalist* **54**, 496 (1920); **58**, 544 (1924).
30. M. Burns, *Nature* (*Lond.*) **195**, 512 (1962).
31. A. W. Holmes, *J. Textile Inst.* **50**, T422 (1959).
32. L. J. Wolfram, *Textile Res. J.* **38**, 1144 (1968).
33. L. P. S. Piper, *J. Textile Inst.* **57**, T185 (1966).
34. J. L. Stoves, *Proc. Roy. Soc. Edinburgh*, **63**, Sect. B, 132 (1945).
35. J. L. Stoves, *J. Soc. Leather Trades Chem.* **32**, 254 (1948).
36. J. L. Stoves, *Nature* (*Lond.*) **151**, 304 (1943).
37. J. L. Stoves, *Fibrous Proteins Symposium, Leeds*, p. 58, Society of Dyers and Colourists, Bradford, England (1946).
38. A. Durward and K. M. Rudall, *Proc. Intern. Wool Text. Res. Conf., Australia* **F**, 112 (1955).
39. A. G. Lyne and B. F. Short (eds.), *Biology of the Skin and Hair Growth*, Angus & Robertson Ltd., Sydney (1965).
40. A. Jarrett, *Brit. J. Dermatol.* **70**, 271 (1958).
41. G. E. Rogers, *Biochem. Biophy. Act* **29**, 33 (1958); *Ann. N.Y. Acad. Sci.* **83**, 408 (1959).
42. E. R. Holman, R. P. Zenozian, W. M. Busey, and D. P. Rall, *Nature* (*Lond.*) **221**, 1058 (1969)
43. W. G. Crewther, R. D. B. Fraser, F. G. Lennox, and H. Lindley, *Advan. Protein Chem.* **20**, 191 (1965).
44. E. Elod and H. Zahn, *Kolloid-Z.* **113**, 10 (1949).
45. D. J. Lloyd and M. Garrod, *Fibrous Proteins Sympsium, Leeds*, p. 24, Society of Dyers and Colourists, Bradford, England (1946).
46. E. G. Bendit, *Biopolymers* **4**, 539, and 561 (1966); and in: *Symposium on Fibrous Proteins* (W. G. Crewther, ed.), p. 386, Butterworth & Co. (Australia) Ltd., Sydney (1968).
47. F. G. E. Pautard and P. T. Speakman, *Nature* (*Lond.*) **185**, 176 (1960).
48. P. T. Speakman, *J. Textile Inst.* **51**, T792 (1960).
49. E. G. Bendit, *Nature* (*Lond.*) **211**, 1257 (1966).
50. R. D. B. Fraser, T. P. MacRae, and G. E. Rogers, *J. Textile Inst.* **51**, T497 (1969).
51. J. Menkart, L. J. Wolfram, and I. Mao, *J. Soc. Cosmetic Chem.* **17**, 769 (1966).
52. J. B. Speakman and S. G. Smith, *J. Soc. Dyers Colourists* **52**, 121 (1936).
53. J. Menkart and J. B. Speakman, *J. Soc. Dyers Colourists* **63**, 322 (1947).

54. A. W. Holmes, *J. Soc. Cosmetic Chem.* **15**, 595 (1964); *Proc. 3rd Intern. Wool Textile Conf., Paris* **3**, 79 (1965).
55. P. G. Whitmore, Ph.D. thesis, University of Leeds (1969).
56. W. S. Barnard, A. Palm, P. B. Stam, D. L. Underwood, and H. J. White, *Textile Res. J.* **24**, 863 (1954); P. B. Stam and H. J. White, *Textile Res. J.* **24**, 785 (1954).
57. W. S. Barnard and H. J. White, *Textile Res. J.* **24**, 695 (1954).
58. C. E. Reese and H. Eyring, *Textile Res. J.* **20**, 743 (1950).
59. J. E. Algie, *J. Polymer Sci.* **35**, 535 (1959).
60. W. T. Astbury and H. J. Woods, *Phil. Trans. Roy. Soc.* **A230**, 75 (1931); **A232**, 333 (1933).
61. H. M. Burte, *Textile Res. J.* **23**, 405 (1953); **24**, 414 (1954); **24**, 423 (1954).
62. G. C. Wood *J. Textile Inst.* **45**, T462 (1954).
63. M. L. Ryder, *Nature (Lond.)* **193**, 1199 (1962).
64. K. M. Rudall, *Proc. Intern. Wool Textile Res. Conf., Australia* **F**, 176 (1955).
65. P. Mason, *Textile Res. J.* **34**, 913 (1964).
66. H. Zahn, *Melliand Textilber.* **31**, 481 (1950).
67. W. T. Astbury, *Trans. Faraday Soc.* **29**, 193 (1933).
68. W. T. Astbury and H. J. Woods, *Nature (Lond.)* **126**, 913 (1930).
69. I. MacArthur, *Nature (Lond.)* **152**, 38 (1943).
70. A. N. J. Heyn, in: *Matthew's Textile Fibres* (H. R. Mauersberger, ed.), p. 1184, 6th ed., John Wiley & Sons, Inc., New York (1954).
71. W. T. Astbury, *Proc. Intern. Wool Textile Res. Conf., Australia* **B**, 206 (1955).
72. H. J. Woods, *Proc. 3rd. Intern. Wool Textile Res. Conf., Paris* **1** 505 (1965).
73. J. Sikorski, private communication.
74. H. J. Woods, *Physics of Fibers*, Fig. 14, Chapman & Hall Ltd. for the Institute of Physics, London (1955).
75. G. M. Jeffrey, Ph.D. thesis, Plate 16, University of Leeds, (1954).
76. A. Elliott, *Textile Res. J.* **22**, 783 (1952).
77. H. Zahn, *Naturwissenschaften* **31**, 137 (1943).
78. E. Elod, H. Nowoty, and H. Zahn, *Melliand Textilber* **25**, 73 (1944).
79. E. Elod and H. Zahn, *Kolloid-Z.* **108**, 94 (1944); *Melliand Textilber.* **28**, 2 (1947).
80. O. Kratky, A. Sekora, H. Zahn, and E. R. Fritze, *Z. Naturforsch.* **10b**, 68 (1955).
81. K. M. Rudall, *Biochem. Biophys. Acta.* **1**, 549 (1947).
82. G. E. Rogers, in: *The Epidermis* (W. Montagna and W. C. Lobitz, eds.), p. 179, Academic Press, New York (1964).
83. R. D. B. Fraser and T. P. MacRae, *Textile Res. J.* **27**, 379, 384 (1957).
84. C. H. Nicholls, Ph.D. thesis, University of Leeds, (1953).
85. C. E. Sagar, private communication.
86. R. D. B. Fraser, *Proc. Intern. Wool Textile Res. Conf., Australia* **B**, 120, 130 (1955).
87. J. H. Keighley, M.Sc. thesis, University of Liverpool (1958).
88. C. Earland, P. F. Blakey, and J. G. P. Stell, *Biochem. Biophys. Acta* **56**, 268 (1962); *Nature (Lond.)* **196**, 1287 (1962).
89. H. G. Fröhlich and J. Rau, *Textile Res. J.* **33**, 865 (1963); H. G. Fröhlich, *SVF Fachorgan* **17**, 917 (1962).
90. F. Wood, and G. King, *Proc. 3rd. Intern. Wool Textile Res. Conf., Paris* **1**, 529 (1965).
91. J. P. MacKenzie, M.Phil. thesis, University of Leeds (1968).
92. W. Gordy et al., *Rad. Res.* **9**, 611 (1958).
93. J. H. Keighley, Ph.D. thesis, University of Leeds (1966); *J. Text. Inst.* **59**, 470 (1968).

94. J. S. Crighton and F. Happey, in: *Symposium of Fibrous Proteins Australia, 1967*, (W. G. Crewther, ed.), p. 409, Butterworth & Co. (Australia) Ltd., Sydney (1968).
95. W. D. Felix, M. A. McDowall, and H. Eyring, *Textile Res. J.* **33**, 465 (1963).
96. E. G. Bendit, *J. Textile Inst.* **51**, T544 (1960).
97. J. Barritt and A. T. King, *J. Textile Inst.* **17**, T386 (1926); see also M. Bachrach, in: *Matthew's Textile Fibres* (H. R. Mauersberger, ed.), 6th ed., p. 732, John Wiley & Sons, Inc., New York (1954).
98. W. von Bergen, in: *Matthew's Textile Fibres* (H. R. Mauersberger, ed.), 6th ed., pp. 628 and 694, John Wiley & Sons, Inc., New York (1954), quoting M. Harris and A. L. Smith, *J. Res. Natl. Bur. Std.* **18**, 623 (May 1937); Reprint RP998 and M. Harris, private communication.
99. R. H. Bogue, *The Chemistry and Technology of Gelatine and Glue*, p. 64, McGraw-Hill Book Company, New York (1922).
100. R. C. Hoather and P. G. T. Hand, *J. Soc. Chem. Ind.* (*Lond.*) **57**, T93 (1938).
101. J. M. Gillespie, in: *Biology of the Skin and Hair Growth* (A. G. Lyne and B. F. Short, eds.), p. 377, Angus & Robertson Ltd., Sydney (1965).
102. H. P. Lundgren and W. H. Ward, in: *Ultrastructure of Protein Fibres* (R. Borasky, ed.), p. 52, Academic Press, New York (1963).
103. J. Barritt and A. T. King, *J. Textile Inst.* **20**, T151 (1929).
104. P. Kassenbeck, *Proc. 3rd. Intern Wool Textile Res. Conf., Paris* **1** 367 (1965).
105. N. H. Chamberlain, *J. Textile Inst.* **23**, T13 (1932).
106. T. Tadokoro and H. Ugami, *J. Biochem.* (*Japan*), **15**, 257 (1932).
107. J. H. Dusenbury and J. Menkart, *Proc. Intern. Wool Textile Res. Conf., Australia* **F**, 142 (1955).
108. H. Zahn, *Kolloid-Z.* **113**, 157 (1949).
109. H. G. Fröhlich, *J. Textile Inst.* **51**, T1237 (1960); *SVF Fachorgan* **17**, 917 (1962).
110. J. H. Dusenbury, *J. Textile Inst.* **51**, T756 (1960).
111. G. Satlow, S. Cieplik, and G. Fichtner, *Faserforsch. Textiltech.* **16**, 143 (1965).
112. H. G. Fröhlich, *Z. Ges. Textil-Ind.* **71**, 39 (1969); **71**, 318 (1969); **71**, 588 (1969); **71**, 837 (1969); **72**, 349 (1970).
113. G. Satlow, *Z. Ges. Textil-Ind.* **67**, 943 (1965).
114. P. B. Scott, M.Phil. thesis, University of Leeds (1968).
115. P. H. S. Springell and S. J. Leach, *Nature* (*Lond.*) **208**, 1326 (1965).
116. W. T. Astbury, *J. Chem. Soc.* **337** (1942).
117. D. H. Simmonds, *Textile Res. J.* **28**, 314 (1958).
118. J. M. Lang and C. C. Lucas, *Biochem. J.* **52**, 84 (1952).
119. J. W. Bell and C. S. Whewell, in: *Handbook of Cosmetic Science* (H. W. Hibbott, ed.), p. 23, Pergamon Press, Oxford (1963).
120. R. C. Clay, K. Cook, and J. I. Routh, *J. Amer. Chem. Soc.* **62**, 2709 (1940).
121. G. Satlow, *Forschungber. Landes Nordrhein-Westfalen* **1890** (1967).
122. R. D. B. Fraser, T. P. MacRae, and D. H. Simmonds, *Biochem. Biophys. Acta* **25**, 654 (1951).
123. E. G. Bendit and M. Feughelman, in: *Encyclopedia of Polymer Science and Technology* (H. F. Mark, N. G. Gaylord, and N. M. Bikales, eds.), Vol. 8, pp. 1–44, John Wiley & Sons, Inc., New York (1968).
124. A. G. Fincham, Ph.D. thesis, University of Leeds, (1966).
125. L. S. Swart, F. J. Joubert, et al., *Proc. 3rd Intern. Wool Textile Res. Conf., Paris* **1**, 493 (1965).
126. W. H. Ward and H. P. Lundgren, *Advan. Protein Chem.* **9**, 243 (1954).
127. B. Lustig, A. A. Kondritzer, and D. H. Moore, *Arch. Biochem.* **8**, 57 (1945).

128. J. H. Bradbury and G. V. Chapman, *Aust. J. Biol. Sci.* **17**, 960 (1964); J. H. Bradbury, G. V. Chapman, and N. L. R. King, in: *Symposium on Fibrous Proteins, Australia, 1967*, (W. G. Crewther, ed.), Butterworth & Co. (Australia) Ltd., Sydney (1968); J. H. Bradbury, *Nature (Lond.)* **210**, 1333 (1966); J. H. Bradbury, *Proc. 3rd Intern. Textile Res. Conf., Paris* **1**, 359 (1965).
129. G. E. Rogers and D. H. Simmonds, *Nature (Lond.)* **182**, 186 (1958).
130. G. E. Rogers, *Ann. N.Y. Acad. Sci.* **83**, 408 (1959); in: *The Epidermis* (W. Montagna and W. C. Lobitz, eds.), pp. 202, 214, Academic Press, New York (1964); *Exptl. Cell. Res.* **33**, 264 (1964).
131. L. S. Swart, F. J. Joubert, and T. Haylett, *J. South African Chem. Inst.*, **20**, 132 (1967).
132. S. Blackburn and G. R. Lee, *J. Textile Inst.* **45**, J487 (1954).
133. M. F. Kerr and C. Godin, *Can. J. Chem.* **37**, 11 (1959).
134. E. Lederer, *Ind. Parfum. Cosmet.* **5**, 552 (1950).
135. A. W. Weitkamp, A. M. Smiljanik, and S. Rothman, *J. Amer. Chem. Soc.* **69**, 1936 (1947).
136. A. Bolliger, *J. Invest. Dermatol.* **17**, 79 (1951).
137. J. M. Gillespie and F. G. Lennox, *Biochim. Biophy. Acta* **12**, 481 (1953); *Aust. J. Biol. Sci.* **8**, 378 (1955).
138. W. G. Crewther, J. M. Gillespie, B. S. Harrap, and A. S. Inglis, *Biopolymers* **4**, 905 (1966).
139. A. Robson, *Textile Inst. Ind.* **4**, 37 (1966).
140. D. R. Goddard and L. Michaelis, *J. Biol. Chem.* **112**, 361 (1935–1936).
141. J. M. Gillespie, *Aust. J. Biol. Sci.* **17**, 282 (1964); J. M. Gillespie and A. S. Inglis, *J. Comp. Biochem. Physiology*, **15**, 175 (1965); *Nature (Lond.)* **207**, 1293 (1965).
142. F. J. Joubert, *J. S. African Chem. Inst.* **24**, 61 (1971).
143. P. Alexander and C. F. Earland, *Textile Res. J.* **20**, 298 (1950); *Nature (Lond.)* **166**, 396 (1950).
144. M. C. Corfield, A. Robson, and B. Skinner, *Biochem. J.* **68**, 348 (1958).
145. A. G. Fincham and G. N. Graham, *Biochem. J.* **95**, 6P (1965).
146. H. Lindley, *Biochem. J.* **42**, 481 (1948).
147. J. M. Gillespie, in (W. G. Crewther, ed.), *Symposium on Fibrous Proteins, Australia, 1967*, p. 362, Butterworth & Co. (Australia) Ltd., Sydney (1968).
148. E. J. Underwood, *Trace Elements in Human and Animal Nutrition*, 2nd ed., Academic Press, New York (1962).
149. R. J. Pollit, F. A. Jenner, and M. Davies, *Arch. Dis. Child.* **43**, 211 (1968).
150. R. M. Robson and J. Sikorski, *Ann. Ital. Dematol. Clin.* **22**, 340 (1968).
151. R. W. Goldblum, S. Derby, and A. B. Lerner, *J. Invest. Dermatol* **20**, 13 (1953).
152. H. Kraut and M. Weber, *Biochem. Z.* **317**, 133 (1944).
153. S. Forshufrud, H. Smith, and A. Wassen, *Nature (Lond.)* **192**, 103 (1961).
154. W. H. Addink and L. J. P. Frank, *Nature (Lond.)* **193**, 1190 (1962).
155. T. F. Dutcher and S. Rothman, *J. Invest. Derm.* **17**, 65 (1951).
156. C. M. Salaman and V. A. S. Luttrell, *Nature (Lond.)* **206**, 413 (1965).
157. S. Blackburn, *Biochem. J.* **43**, 114 (1948).
158. J. L. Stoves, *Nature (Lond.)* **157**, 230 (1946).
159. F. G. E. Pautard, *J. Dent. Res.* **40**, 1285 (1961).
160. F. G. E. Pautard, *Clin. Orthopaedics* **24**, 230 (1962).
161. F. G. E. Pautard, *Nature (Lond.)* **199**, 531 (1963).
162. F. G. E. Pautard, in *Progress in the Biological Sciences in Relation to Dermatology*

(A. Rook and R. H. Champion, eds.), Vol. 2, p. 227, Cambridge University Press, Cambridge (1964).
163. F. G. E. Pautard, *Proc. 2nd European Symp. Calcified Tissues*, p. 347, University of Liège, Liège (1964).
164. A. G. Fincham, *Proc. 3rd Intern. Wool Textile Res. Conf., Paris*, **1** p. 519 (1965).
165. P. F. Blakey, C. Earland, and J. G. P. Stell, *Nature (Lond.)* **198**, 481 (1963).
166. P. R. Blakey and P. Lockwood, *Calc. Tissue Res.* **2**, 361 (1968).
167. P. R. Blakey, C. Earland, J. G. P. Stell, and D. Swift, *Nature (Lond.)* **207**, 190 (1965).
168. H. J. W. Bliss, J. E. Duerden, J. A. F. Roberts, J. S. S. Blyth, H. R. Hirst, and A. T. King, *J. Textile Inst.* **17**, T264–T304 (1926).
169. L. Auber, *Trans. Roy. Soc. Edinburgh* **62**, Pt. I, No. 7, 191 (1950–1951).
170. S. T. Roth and W. H. Clark, J. R., in: *The Epidermis* (W. Montagna and W. C. Lobitz, eds), Academic Press, New York (1964).
171. R. H. M. Ballin and F. Happey, *Proc. 3rd Intern. Wool Textile Res. Conf., Paris* **1**, 181 (1965).
172. F. Kidd, *Proc. 3rd Intern. Wool Textile Res. Conf., Paris* **1**, 221 (1965).
173. W. C. Osman Hill, H. M. Appleyard, and L. Auber, *Trans. Roy. Soc. Edinburgh*, **63**, Pt. III, 53 (1958–1959).
174. D. A. Ross, Ph.D. thesis, University of Leeds (1955).
175. A. T. King, *J. Textile Inst.* **18**, T365 (1927).
176. E. K. Rideal and J. H. Schulman, *Nature (Lond.)* **144**, 100 (1939).
177. L. A. Hausman, *Amer. J. Anat.* **27**, 463 (1920).
178. F. Kidd and C. E. Sagar, *J. Textile Inst.* **44**, T245 (1953).
179. L. Auber and M. Ryder, *Proc. Intern. Wool Textile Res. Conf., Australia* **F**, 61 (1955).
180. A. V. Yablokov and T. V. Andreyeva, *Nature (Lond.)* **205**, 413 (1965).
181. D. J. Lloyd and R. H. Marriott, *Biochem. J.* **27**, 911 (1933).
182. J. G. Bekker and A. T. King, *Biochem. J.* **25**, 1077 (1931).
183. W. J. Thorsen, *Textile Res. J.* **28**, 185 (1958).
184. K. M. Rudall, in: *Extracellular and Supporting Structures in Comprehensive Biochemistry*, (M. Florkin and E. H. Stotz, eds.), Vol. 26B, p. 584, Elsevier Publishing Company, Amsterdam (1968).
185. M. E. A. Perkin and H. M. Appleyard, *J. Textile Inst.* **59**, 117 (1968).
186. L. Auber and M. L. Ryder, *Proc. Intern. Wool Textile Res. Conf., Australia* **F**, 36 (1955).
187. L. P. S. Piper, private communication (1970).
188. J. G. Snyman, *Textile Res. J.* **33**, 219 (1963).
189. G. E. Scheepers, *Southern Africa Textiles* **16**, 53 (Mar. 1967).
190. W. J. Onions, *Wool*, p. 216, Ernest Benn Ltd., London (1962).
191. R. D. B. Fraser and T. P. MacRae, *Textile Res. J.* **26**, 618 (1956).
192. D. P. Veldsman, *From Mohair Fleece to Fabric*, Sawtri, Port Elizabeth (1969).
193. H. P. Lundgren, *Proc. Intern. Wool Textile Res. Conf., Australia* **F**, 200 (1955).
194. H. P. Lundgren and W. H. Ward, in: *Ultrastructure of Protein Fibres* (R. Borasky, ed.), p. 65, Academic Press, New York (1963).
195. J. Menkart and A. B. Coe, *Textile Res. J.* **28**, 218 (1958).
196. E. H. Mercer, *Textile Res. J.* **24**, 39 (1954).
197. R. I. Spearman and N. A. Barnicot, *Amer. J. Phys. Anthropol.* **18**, 91 (1960).
198. G. Laxer, C. S. Whewell, and H. J. Woods, *J. Textile Inst.* **45**, T483 (1954).
199. P. Kassenbeck, J. Jacquemart, and R. Monrocq, *Proc. 3rd Intern Wool Textile Res. Conf., Paris* **1**, 209 (1965).

200. P. Kassenbeck, *Proc. 3rd Intern. Wool Textile Res. Conf., Paris* **1**, 115 and 135 (1965).
201. P. Kassenbeck, *Ciba Rev.* (1962–1964), 23, 25.
202. P. Kassenbeck, *Textes et discussions du colloque*, p. 51, Institut Textile de France, Paris (1961).
203. M. G. Dobb, *J. Text. Inst.* **61**, 232 (1970); *7th Congr. Intern. Microscopie Électronique, Grenoble*, p. 645, Société Française de Microscopie Électronique, Paris (1970).
204. W. von Bergen, *Textile Res. J.* **29**, 586 (1959).
205. M. S. C. Birbeck and N. A. Barnicot, in: *Pigment Cell Biology* (M. Gordon, ed.), p. 549, Academic Press, New York (1959).
206. H. S. Mason, ibid. p. 563. H. S. Mason, in: *Pigment Cell Growth* (M. Gordon, ed.), p. 277, Academic Press, New York (1953).
207. F. Moyer, in: *Structure of the Eye* (G. K. Smelser, ed.), p. 325, Academic Press, New York (1961).
208. T. B. Fitzpatrick, P. Brunet, and A. Kukita, in: *Biology of Hair Growth* (W. Montagna and R. A. Ellis, eds.), p. 255, Academic Press, New York (1958).
209. J. A. Swift, *Nature* (*Lond.*) **203**, 976 (1964).
210. E. H. Mercer, *Textile Res. J.* **23**, 388 (1953).
211. L. P. S. Piper, *Nature* (*Lond.*) **213**, 596 (1967).
212. P. Flesch and S. Rothman, *J. Invest. Dermatol.* **6**, 257 (1945).
213. N. A. Barnicot, *Nature* (*Lond.*) **177**, 528 (1956).
214. B. D. Booth, Ph.D. thesis, University of Leeds (1952).
215. G. Laxer, Ph.D. thesis, University of Leeds (1955).
216. G. Laxer and D. A. Ross, *Textile Res. J.* **24**, 672 (1954).
217. G. Laxer and C. S. Whewell, *Proc. Intern. Wool Textile Res. Conf., Australia* **F**, 186 (1955).
218. J. A. Swift, Ph.D. thesis, University of Leeds (1963); *Proc. 3rd Intern. Wool Textile Res. Conf., Paris* **1**, 300 (1965).
219. G. Laxer, J. Sikorski, C. S. Whewell, and H. J. Woods, *Biochim. Biophys. Acta* **15**, 174 (1954).
220. R. H. M. Ballin, private communication.
221. E. H. Mercer, *Proc. Intern. Wool Textile Res. Conf., Australia* **F**, 222 (1955).
222. R. Hagége, *Proc. 3rd Intern. Wool Textile Res. Conf., Paris* **1**, 301 (1965).
223. D. B. Green and F. Happey, *Proc. 3rd Intern. Wool Textile Res. Conf., Paris* **1**, 283 (1965).
224. W. J. Kriel, D. Albertyn, and O. A. Swanepoel, *Sawtri Bulletin* **3**, No. 1 (Mar. 1969).
225. R. Richter and Z. Stary, *Z. Physiol. Chem.* **253**, 159 (1938).
226. D. R. Goddard and L. Michaelis, *J. Biol. Chem.* **106**, 605 (1934).
227. E. H. Mercer, *Textile Res. J.* **23**, 388 (1953).
228. R. Burgess, *J. Textile Inst.* **25**, T391 (1934).
229. E. P. McGrath and W. C. Quevedo, Jr., in: *Biology of the Skin and Hair Growth* (A. G. Lyne and B. F. Short, eds.), p. 727, Angus & Robertson Ltd., Sydney (1965).
230. P. Kassenbeck, private communication.
231. E. Boyd, *Proc. Roy. Soc. Edinburgh* **52**, 218 (1932).
232. R. E. Billingham, *J. Anat.* **82**, 93 (1948).
233. M. H. Hardy, A. S. Fraser, and B. F. Short, *Nature* (*Lond.*) **170**, 849 (1952).
234. W. G. Crewther, *Proc. Intern. Wool Textile Res. Conf., Australia* **F**, 223 (1955).
235. W. von Bergen, in: *Matthew's Textile Fibres* (H. R. Mauersberger, ed.), 6th ed., pp. 688, 694, 700, 714, John Wiley & Sons, Inc., New York (1954).

236. J. L. Stoves, *Trans. Faraday Soc.* **38**, 254 (1942); **38**, 261 (1942); **38**, 501 (1942).
237. J. B. Speakman and G. H. Elliott, *Fibrous Proteins, Symposium Leeds*, p. 116, Society of Dyers and Colourists, Bradford, England (1946).
238. J. B. Speakman and M. C. Hirst, *Trans. Faraday Soc.* **29**, 148 (1933).
239. M. M. Breuer, *J. Phy. Chem.* **68**, iii, 2067, 2074, 2081 (1964).
240. J. A. Swift, *Proc. 3rd Intern. Wool Textile Res. Conf., Paris* **1**, 265 (1965).
241. A. Kassenbeck and R. Hagége, *Proc. 3rd Intern. Wool Textile Res. Conf., Paris* **1**, 245 (1965).
242. S. E. Hill and J. W. Bell, Project Report, University of Leeds, Textile Department (June 1967).
243. N. Ahmad, private communication, University of Leeds, Textile Department (1970).
244. J. A. C. Watt, *Carrotting of Fur: A Method of Standardization*, Pt. I, BHAFRA Publication TR3, Leeds (Sept. 1949).
245. J. A. C. Watt, *Carrotting of Fur: A Method of Standardization*, Pt. II, BHAFRA Publication TR5, Manchester (May 1952).
246. J. A. C. Watt, *The HAFRA Test*, BHAFRA Publication TR10, Manchester (July 1954).
247. T. A. Barr, J. A. C. Watt, and R. Withington, *Carrotting of Fur: A Method of Standardization*, Pt. III, BHAFRA Publication TR15, Manchester (Feb. 1956).
248. C. E. J. Aston, T. Barr, and D. Haigh, *Hatters' Fur: The Effect of Treatment with Sulphuric Acid*, BHAFRA Publication TR17, Manchester (Dec. 1956).
249. D. Haigh, *J. Soc. Dyers Colourists* **70**, 539, 547 (1954).
250. H. G. Fröhlich and E. Hille, *Z. Ges. Textil-Ind.* **64**, 48 (1962).
251. W. Kriel, *SAWTRI Tech. Rept.* **60** (1965).
252. E. Kornreich, *J. Textile Inst.* **41**, T321 (1950); **47**, T486 (1956).
253. F. Kidd and C. S. Whewell, *J. Soc. Dyers Colourists* **68**, 396 (1952).
254. M. Harris and A. L. Smith, *Amer. Dyestuff Reporter* **25**, 542 (1936).
255. K. Lees and F. F. Elsworth, *J. Soc. Dyers Colourists* **68**, 207 (1952).
256. H. Zahn and H. Würz, *Textil-Praxis* **8**, 971 (1953).
257. J. B. Speakman and C. S. Whewell, *J. Soc. Dyers Colourists* **52**, 380 (1936).
258. C. S. Whewell and H. J. Woods, *Fibrous Proteins Symp. Univ. Leeds*, p. 50, Society of Dyers and Colourists, Bradford, England (1946)
259. H. J. Woods, *Proc. Roy. Soc.* (*Lond.*) **116A**, 76 (1938).
260. J. L. Stoves, *Proc. Leeds Phil. Soc. Sci. Sec.* **4**, 83 (1943).
261. J. L. Stoves, *J. Soc. Leather Trades Chem.* **32**, 254 (1948).
262. H. Zahn, *Kolloid-Z.* **117**, 102 (1950).
263. H. Zahn and H. Brauckoff, *Biochem. Z.* **318**, 401 (1948).
264. H. G. Fröhlich, *Das Filzvermögen von Wolle und Tierhaaren in Deutscher Färberkalender*, Franz Eder Verlag, Munich (1967).
265. H. Zahn, *Z. Naturforschung* **26**, 286 (1947); *Melliand Textilber.* **30**, 517 (1949).
266. J. B. Speakman, *J. Soc. Dyers Colourists* **52**, 423 (1936).
267. C. S. Whewell, *J. Soc. Dyers Colourists* **62**, 423 (1946).
268. C. S. Whewell, *J. Soc. Cosmetic Chem.* **15**, 423 (1964).
269. F. G. Lennox, *Textile Res. J.* **25**, 677 (1955).
270. C. S. Whewell, *J. Soc. Cosmetic Chem.* **12**, 207 (1961).
271. A. Farnworth, *Textile Res. J.* **27**, 632 (1957).
272. R. M. Bones, Ph.D. thesis, University of Leeds (1966).
273. T. Mitchell and M. Feughelman, *Textile Res. J.* 28, 453 (1958).
274. L. S. Bajpai, Ph.D. thesis, University of Leeds (1960).

275. J. L. Stoves, *J. Textile Inst.* **51**, T603 (1960).
276. R. S. Gandhi, Ph.D. thesis, University of Leeds (1965).
277. L. S. Bajpai, C. S. Whewell, and J. M. Woodhouse, *J. Soc. Dyers Colourists* **77**, 93 (1961).
278. L. S. Bajpai and C. S. Whewell, *J. Soc. Dyers Colourists* **77**, 350 (1961).
279. J. B. Calva, U.S. Pat. 2,240,388 (1941).
280. W. K. Teller, U.S. Pat. 2,055,322.
281. A. H. Stonehill, U.S. Pat. 2,309,021.
282. L. J. Wolfram, *J. Soc. Cosmetic Chem.* **20**, 539 (1969).
283. E. B. Slack, *Brushes* **34**, 58 (1948).
284. Private communication.
285. B. D. Greenwood, Internal Research Report, Scottish College of Textiles (1968).
286. K. M. Rudall, in: *Lectures on the Scientific Basis of Medicine*, Vol. 5, p. 217, University of London, The Athlone Press, London, (1957).
287. K. M. Rudall, in: *Comparative Biochemistry* (M. Florkin and H. S. Mason, eds.), Vol. 4, p. 397, Academic Press, New York (1962).
288. A. Millard and K. M. Rudall, *Proc. 5th Intern. Cong. Electron Microscopy, Philadelphia* (1962).
289. W. Kenchington and N. E. Flower, *J. Microscopy* **89**, Pt. 2, 263 (1969).
290. W. Kenchington, Ph.D. thesis, University of Leeds (1965).
291. E. Race, *Fibrous Proteins Symposium, Leeds*, p. 80, Society of Dyers and Colourists, Bradford, England (1946); *Wool Sci. Rev.* **6**, 31 (1950).
292. R. Burgess, *J. Textile Inst.* **25**, T289 (1934).
293. W. R. Middlebrook and H. Phillips, *J. Soc. Dyers Colourists* **57**, 137 (1941).
294. S. Blackburn, *Biochem. J.* **47**, 443 (1950).
295. J. J. Noval and W. J. Nickerson, *J. Bacteriol.* **77**, 251 (1959); W. J. Nickerson, J. J. Noval, and R. S. Robinson, *Biochem. Biophys. Acta* **77**, 73 (1963); W. J. Nickerson and S. C. Durand, *Biochim. Biophys. Acta* **77**, 87 (1963).
296. Z. Stary, *Bull. Fac. Med. Univ. Istanbul* **12**, 329 (1946).
297. G. Laxer and C. S. Whewell, *Proc. Intern. Wool Textile Res. Conf., Australia* **F**, 194 (1955).
298. F. Kidd, *Nature (Lond.)* **202**, 416 (1964).
299. H. M. Appleyard, *J. Roy. Microscopical Soc.* **87**, Pt. I, 1 (1967).
300. B. Manogue, M. S. Moss, and R. L. Elliott, *J. Soc. Dyers Colourists* **70**, 502 (1954).
301. A. W. Holmes, *Textiles Res. J.* **34**, 777 (1964).
302. H. Zahn, *Melliand Textilber.* **30**, 275, 294 (1949).
303. E. Elöd and H. Zahn, *Melliand Textilber.* **25**, 361 (1944).
304. E. Elöd and H. Zahn, *Melliand Textilber.* **25**, 73, 361 (1944); *Z. Naturforsch.* **26**, 286 (1947).

# Index